Lecture Notes in Computer Science   16567

Founding Editors

Gerhard Goos
Juris Hartmanis

The series Lecture Notes in Computer Science (LNCS), including its subseries Lecture Notes in Artificial Intelligence (LNAI) and Lecture Notes in Bioinformatics (LNBI), has established itself as a medium for the publication of new developments in computer science and information technology research, teaching, and education.

LNCS enjoys close cooperation with the computer science R & D community, the series counts many renowned academics among its volume editors and paper authors, and collaborates with prestigious societies. Its mission is to serve this international community by providing an invaluable service, mainly focused on the publication of conference and workshop proceedings and postproceedings. LNCS commenced publication in 1973.

Jörg Desel · Anna Kalenkova

Editors

Application and Theory of Petri Nets and Concurrency

47th International Conference, PETRI NETS 2026
Hamburg, Germany, June 22–26, 2026
Proceedings

 Springer

Editors
Jörg Desel
FernUniversität
Hagen, Germany

Anna Kalenkova
Adelaide University
Adelaide, SA, Australia

ISSN 0302-9743   ISSN 1611-3349 (electronic)
Lecture Notes in Computer Science
ISBN 978-3-032-27878-4   ISBN 978-3-032-27879-1 (eBook)
https://doi.org/10.1007/978-3-032-27879-1

This Springer imprint is published by the registered company Springer Nature Switzerland AG
The registered company address is: Gewerbestrasse 11, 6330 Cham, Switzerland

If disposing of this product, please recycle the paper.

Preface

This volume contains the proceedings of the 47th International Conference on Application and Theory of Petri Nets and Concurrency (Petri Nets 2026), which took place in Hamburg, Germany, June 22–26, 2026. The aim of this series of conferences is to create an annual opportunity to discuss and disseminate the latest results in the fields of Petri nets and related models of concurrency, including tools, applications, and theoretical progress. The 47th conference and affiliated events were organized jointly by the Hamburg University of Applied Science and the University of Hamburg, where Carl Adam Petri held an honorary professorship. The 100th anniversary of Carl Adam Petri's birth was chosen as an occasion to return to this city where he shared and discussed his groundbreaking concepts and insights with colleagues and students. Michael Köhler-Bußmeier from the Hamburg University of Applied Sciences and Daniel Moldt from the University of Hamburg were responsible for the organization and did an excellent job.

This year, 42 papers were submitted to the conference and single-blind reviewed by three or four reviewers. The discussion phase and final selection process by the Program Committee were supported by the EasyChair conference system, which ensured that Program Committee members were not involved in the evaluation or discussion of their own work.

From 34 regular papers and 8 tool papers, the program committee selected 14 regular papers and 6 tool papers. After the conference, some of the authors were invited to submit an extended version of their contribution for consideration in a special issue of a journal. We thank the members of the Program Committee and further reviewers for their careful and timely evaluation of the submissions and the fruitful and constructive discussions that resulted in the final selection of the papers. The Springer LNCS team provided excellent support in the preparation of this volume.

Keynote presentations were given by

- Véronique Cortier, Université de Lorraine, CNRS, Inria, LORIA, Nancy, France, on *Electronic Voting: Design, Attacks and Formal Verification* and
- Jan Mendling, Institut für Informatik, Humboldt-Universität zu Berlin, Germany, on *Empirical Research on Petri Nets.*

This conference proceedings volume contains summaries of these two keynote speeches, which were not peer-reviewed.

Another highlight of the conference was a scientific symposium in honor of Carl Adam Petri, chaired by Laure Petrucci, Sorbonne Paris Nord University, France. Speakers included Rüdiger Valk, University of Hamburg; Wolfgang Reisig, Humboldt University of Berlin; Jörg Desel, FernUniversität in Hagen; and Wil van der Aalst, RWTH Aachen University (all in Germany).

Alongside Petri Nets 2026, the following further events and workshops took place:

- The Petri Net Course and Advanced Tutorials, organized by Lars Kristensen, Western Norway University of Applied Sciences, Norway, and Karsten Wolf, University of Rostock, Germany
- The International Workshop on Petri Nets and Software Engineering (PNSE 2026), organized by Michael Köhler-Bußmeier, Hamburg University of Applied Sciences, Germany, and Heiko Röhlke, Graubünden University of Applied Sciences, Switzerland
- The International Workshop on Algorithms & Theories for the Analysis of Event Data (ATAED 2026), organized by Robin Bergenthum, FernUniversität in Hagen, Germany, and Sander Leemans, RWTH Aachen University, Germany
- The International Workshop on Petri Nets, Higher-Dimensional Automata, Partial Orders, and Concurrency (PHOCON 2026), organized by Uli Fahrenberg, Université Paris-Saclay, France, Loïc Hélouët, INRIA Rennes, France, Philipp Schlehuber-Caissier, RST, Télécom SudParis, France, and Krzysztof Ziemiański, University of Warsaw, Poland
- The International Workshop on Petri Nets for Adaptive Systems (PNAS 2026), organized by Lorenzo Capra, University of Milan, Italy and Michael Köhler-Bußmeier, Hamburg University of Applied Sciences, Germany

June 2026   Jörg Desel
                                                                       Anna Kalenkova

Organization

Program Committee Chairs

Jörg Desel	FernUniversität in Hagen, Germany
Anna Kalenkova	Adelaide University, Australia

Steering Committee

Wil van der Aalst	RWTH Aachen University, Germany
Gianfranco Ciardo	Iowa State University, USA
Jörg Desel	FernUniversität in Hagen, Germany
Susanna Donatelli	University of Turin, Italy
Giuliana Franceschinis	University of Eastern Piedmont, Italy
Serge Haddad	Université Paris-Saclay, France
Fabrice Kordon (Co-chair)	Sorbonne University, France
Maciej Koutny	Newcastle University, UK
Lars Kristensen	Western Norway University of Applied Sciences, Norway
Łukasz Mikulski	Nicolaus Copernicus University, Poland
Wojciech Penczek	Polish Academy of Sciences, Poland
Laure Petrucci (Co-chair)	Sorbonne Paris Nord University, France
Artem Polyvyanyy	University of Melbourne, Australia
Jan Martijn van der Werf	Utrecht University, The Netherlands
Karsten Wolf	University of Rostock, Germany
Alex Yakovlev	Newcastle University, UK

Program Committee

Abel Armas-Cervantes	University of Melbourne, Australia
João Paulo Barros	Polytechnic Institute of Beja, Portugal
Luca Bernardinello	Università degli Studi di Milano-Bicocca, Italy
José-Manuel Colom	University of Zaragoza, Spain
Silvano Dal Zilio	LAAS-CNRS, France
Jörg Desel	FernUniversität in Hagen, Germany
Dirk Fahland	Eindhoven University of Technology, The Netherlands

Peter Fettke	DFKI and Saarland University, Germany
João Miguel Fernandes	University of Minho, Portugal
Guiliana Franceschinis	University of Eastern Piedmont, Italy
Luís Gomes	NOVA University of Lisbon, Portugal
Stefan Haar	Inria and Université Paris-Saclay, France
Xudong He	Florida International University, USA
Gabriel Juhás	Pan-European University, Slovakia
Anna Kalenkova	Adelaide University, Australia
Kaïs Klai	Sorbonne Paris Nord University, France
Lars Kristensen	Western Norway University of Applied Sciences, Norway
Didier Lime	Centrale Nantes, France
Robert Lorenz	University of Augsburg, Germany
Lisa Mannel	RWTH Aachen University, Germany
Andrew Miner	Iowa State University, USA
Marco Montali	Free University of Bozen-Bolzano, Italy
Wojciech Penczek	Polish Academy of Sciences, Poland
Guillermo-Alberto Pérez	University of Antwerp, Belgium
Marta Pietkiewicz-Koutny	Newcastle University, UK
Artem Polyvyanyy	University of Melbourne, Australia
Andrey Rivkin	Technical University of Denmark
Jeremy Sproston	University of Turin, Italy
Jiří Srba	Aalborg University, Denmark
Nathalie Sznajder	Sorbonne University, France
Jan Martijn van der Werf	Utrecht University, The Netherlands
Remigiusz Wiśniewski	University of Zielona Góra, Poland

Additional Reviewers

Nicolas Amat	Rémi Parrot
Samik Basu	Lucia Pomello
Enrico Bini	Nikolaj Rossander Kristensen
Oliver Bøving	Olivier H. Roux
Guillaume Dupont	Arnaud Sangnier
Goran Faisal	Patrizia Schalk
Carlo Ferigato	Teofil Sidoruk
Vanessa Flügel	Victoria Sonnemans
Lucie Guillou	Gaëtan Staquet
Piotr Hofman	Andrei Tour
Akram Idani	Antti Valmari
Andrei Karatkevich	Marcin Wojnakowski
Marc Kimmel	Agnieszka Zbrzezny
Didier Le Botlan	Mengchu Zhou

Electronic Voting: Design, Attacks and Formal Verification (Keynote)

Véronique Cortier

Université de Lorraine, CNRS, Inria, LORIA, France
`veronique.cortier@loria.fr`

Abstract of the Talk

Electronic voting aims at guaranteeing apparently conflicting properties: no one should know how I voted and yet, I should be able to check that my vote has been properly counted. Many more properties may be considered such as everlasting privacy, coercion-resistance, or accountability. In this talk, we will first survey how voting protocols work through the example of the French Legislative elections in 2022 [4, 5].

Electronic voting belongs to the large family of security protocols, that aim at securing communications against powerful adversaries that may read, block, and modify messages. Many techniques and tools [1, 6–8] have been developed to formally *prove* the security of protocols. Yet, voting protocols push such techniques at their limits. We will see how to model and analyze the security of voting protocols using formal methods and in particular with the tool ProVerif [2, 3], in order to (automatically) detect attacks at an early stage, or to prove security, yielding a better understanding of the security guarantees and the threat model.

Acknowledgment. This work benefited from funding managed by the French National Research Agency under the France 2030 programme with the reference ANR-22-PECY-0006.

References

1. Blanchet, B.: Automatic verification of security protocols in the symbolic model: the verifier proverif. In: Aldini, A., Lopez, J., Martinelli, F. (eds.) Foundations of Security Analysis and Design VII. LNCS, FOSAD FOSAD 2013 2012, Vol. 8604, pp. 54–87. Springer, Cham (2014). https://doi.org/10.1007/978-3-319-10082-1_3
2. Blanchet, B., Cheval, V., Cortier, V.: Proverif with lemmas, induction, fast subsumption, and much more. In: Proceedings of the 42nd IEEE Symposium on Security and Privacy (S&P'22). IEEE Computer Society Press (2022)
3. Cortier, V., Debant, A., Cheval, V.: Election verifiability with proverif. In: 36th IEEE Computer Security Foundations Symposium (CSF'23). Dubrovnik, Croatia (2023)

4. Cortier, V., Gaudry, P., Glondu, S., Ruhault, S.: French 2022 legislatives elections: a verifiability experiment. In: The International Conference for Electronic Voting (E-Vote-ID'23). Luxembourg City, Luxembourg (2023)
5. Debant, A., Hirschi, L.: Reversing, breaking, and fixing the french legislative election e-voting protocol. In: 32nd USENIX Security Symposium, USENIX Security 2023, pp. 6737–6752. USENIX Association (2023)
6. Escobar, S., et al.: Protocol analysis in maude-NPA using unification modulo homomorphic encryption. In: Proceedings of the 13th International ACM SIGPLAN Conference on Principles and Practice of Declarative Programming - PPDP'11, pp. 65–76. ACM (2011)
7. Schmidt, B., Meier, S., Cremers, C., Basin, D.: Automated analysis of Diffie-Hellman protocols and advanced security properties. In: 25th IEEE Computer Security Foundations Symposium, CSF 2012, pp. 78–94. IEEE (2012)
8. Swamy, N., Chen, J., Fournet, C., Strub, P., Bhargavan, K., Yang, J.: Secure distributed programming with value-dependent types. In: Proceeding of the 16th ACM SIGPLAN International Conference on Functional Programming, pp. 266–278 (2011)

Contents

Tool Papers

Keynote

Empirical Research on Petri Nets

Jan Mendling[1,2,3(✉)] ⓘ, Benoit Depaire[4] ⓘ, and Henrik Leopold[5] ⓘ

[1] Department of Computer Science, Humboldt-Universität zu Berlin, Berlin,
Germany
[2] Department of Information Systems and Operations Management, Vienna
University of Economics and Business, Vienna, Austria
[3] Weizenbaum Institute, Berlin, Germany
jan.mendling@hu-berlin.de
[4] Hasselt University, Hasselt, Belgium
[5] Kühne Logistics University, Hamburg, Germany

Abstract. Petri net research often builds on the definition of formal
properties and corresponding analysis methods. A larger share of Petri
net papers provides such formal contributions, but there are also works
that offer empirical insights. Such empirical insights can be highly effective in stimulating major breakthroughs even in more formal areas of
computer science, such as graph algorithms or automatic image recognition. In this paper, we discuss the benefits and challenges of complementing formal research on Petri nets with an empirical research agenda.
To that end, we build on a methodological framework for algorithm engineering and present selected examples of empirical works on Petri nets.
These examples illustrate the spectrum of potential contributions and
emphasize the salience of validity concerns. These works can serve as
pillars for advancing empirical research on Petri nets.

Keywords: Petri nets · Methodology · Algorithm engineering ·
Process Mining · Process Models · Validity

1 Introduction

Petri net research looks back at a proud history of developing formal properties
and corresponding analysis methods for analyzing behavior in various domains
of application [80]. These include properties such as reachability, boundedness,
or liveness, and methods like the coverability tree, incidence matrix and state
equation, or reduction rules [56]. Various survey papers [56,58], lecture notes
[68], and textbooks [67] present these concepts, properties and techniques in a
comprehensive and accessible way.

J. Mendling—The research of the author was supported by the Einstein Foundation
Berlin under grant EPP-2019-524, by the German Federal Ministry of Research, Technology and Space under grant 16DII143, and by Deutsche Forschungsgemeinschaft
under grants 496119880 (VisualMine), 531115272 (ProImpact), SFB 1404/2 (FONDA).

J. Desel and A. Kalenkova (Eds.): PETRI NETS 2026, LNCS 16567, pp. 3–18, 2026.
https://doi.org/10.1007/978-3-032-27879-1_1

Though a larger share of Petri net research papers focuses on formal contributions, there are also works that provide empirical insights. The general benefits of empirical research for computer science are often difficult to trace. They can for instance be illustrated by reference to graph algorithms and the traveling salesperson problem. It is well known that problems in these areas often belongs to the class of NP-hard problems. Recent advancements for these problems are connected with empirical works inspired by the availability of realistic problem instances and benchmark datasets such as the TSP World Tour with almost two million cities [71]. Also other areas of computer science such as image recognition have benefited from empirical research on benchmark datasets [70]. So far, there has been limited exchange between different areas of computer science on how experiences with empirical research designs can be transferred.

In this paper, we discuss the benefits and challenges of complementing Petri net research on mathematical concepts with an empirical research agenda. To that end, we build on a methodological framework for algorithms engineering and present select examples of empirical contributions to Petri net research. These examples illustrate the spectrum of potential contributions and emphasize the salience of validity concerns. We discuss how to address these concerns in turn.

This paper is partially based on [19] and structured as follows. Section 2 presents the overarching framework of algorithm engineering. Section 3 describes its ontological perspective. Section 4 discusses the epistemological perspective and the different types of knowledge related to phenomenological areas in which Petri nets are applied. Section 5 presents several examples of empirical research related to Petri nets. Section 6 investigates methodological approaches with a focus on implementation validity, external validity, and internal validity. Section 7 closes the paper with an outlook on future research.

2   Framework for Algorithm Engineering

Even though much of Petri nets research is mathematical, restricting research on Petri nets to exclusively mathematical contributions would miss important epistemological angles. Such a restriction would not only lead to an incomplete stock of knowledge–it would also fail to reap benefits from the combination of different types of knowledge. To develop this argument, we follow a framework for the methodology of algorithm engineering [54]. In this context, an *algorithm* is a finite sequence of computational steps that transforms some input into some output [5,14]. The notion of algorithm in this framework extends to what for instance Murata [56] or Reisig [66] call an *analysis method* in a Petri net context.

The framework distinguishes three perspectives: an ontological, an epistemological, and a methodological perspective [19]. First, the ontological perspective asks what the key entities are when we speak of an algorithm, namely, how real-world problems are turned into algorithmic tasks, designs, and implementations. Second, the epistemological perspective considers what we can know about an algorithm. To this end, we distinguish the notion of knowledge *of* and *about* tasks as well as knowledge *of* and *about* designs. Third, the methodological perspective reflects on how we can proceed when designing, testing, and refining algorithms with a particular focus on threats to validity. Together, these perspectives describe the basis for understanding and studying algorithms in general, and *analysis methods* more specifically in a Petri net context.

Figure 1 provides an overview of the ontological, epistemological, and methodological perspective, and their relationships to the different forms of knowledge and their associated validity concerns. The figure summarizes how real-world problems are abstracted into algorithmic tasks, how designs and implementations instantiate these tasks, and how various validity concerns arise along these relationships. We discuss each perspective in turn.

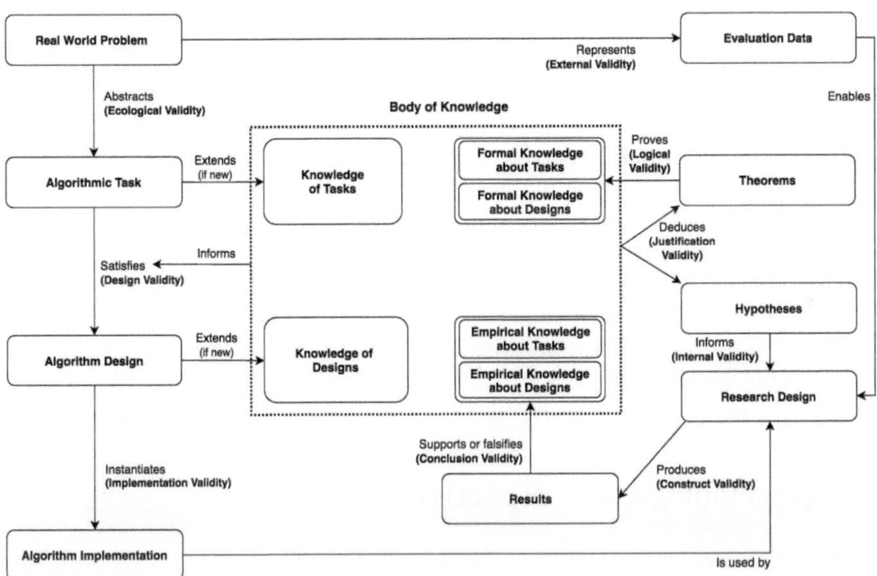

Fig. 1. Overview of the algorithm engineering framework, showing the ontological, epistemological, and methodological perspectives and their associated validity concerns (adapted from [54]). Analysis methods in a Petri net context are specific types of algorithms.

3   Ontological Perspective

Building on the ontological model of [74], the framework characterizes algorithm engineering through four ontological entities and their relations: 1) a real-world problem situated in a concrete context, 2) an algorithmic task that abstracts this problem by making its assumptions explicit and setting its goal, 3) an algorithm design that satisfies the task through principled design choices, and 4) an algorithm implementation that instantiates the design and produces concrete results when executed on data. This view highlights that algorithms are engineered artifacts whose identity and value depend on their relationship to the real world. We illustrate this view for Petri net analysis methods.

The starting point according to the framework is a *real-world problem*. For example, an organization may want to understand how one of its business processes is working. To tackle such a problem, one first defines an *algorithmic task*: a carefully delimited abstraction that captures the essence of the practical need. A suitable abstraction is to assume that the business process can be modeled as a workflow net, and that such a specific Petri net should be consistent with a notion of soundness [1,2,17]. The next step is developing an *algorithm design* that will address or solve that task. For the soundness of workflow nets, van der Aalst defines transformation rules that can help to check soundness for workflow nets [1]. Finally, the *algorithm implementation* realizes the design as a concrete piece of software. There are several implemented soundness checkers for workflow nets, including Woflan [77] or WoPeD [31]. Only at this stage with this implementation being available, the algorithm can be executed on actual Petri nets, producing a tangible output that allows researchers to evaluate how well the algorithm meets the original need. Here, we can distinguish two dimensions of performance: effectiveness, meaning the discovered algorithm truly captures that the business process design works well, and efficiency, meaning the program runs with acceptable time and resource use.

This example illustrates that the ontological perspective clarifies how real-world needs are transformed into algorithmic artifacts. It provides a vocabulary for reasoning about where scientific claims belong, whether about the problem, the algorithmic task, the algorithmic design choices, or the final implementation.

4   Epistemological Perspective

While the ontological perspective clarifies what entities exist in algorithm engineering, the epistemological perspective asks what we know about these entities. We distinguish between *knowledge of* a task or design and *knowledge about* them, resulting in four types of knowledge:

Knowledge of tasks captures what tasks exist and how they are structured. This notion of knowledge is important since algorithmic tasks are not connected with a real-world problem via a simple isomorphic mapping. Real-world problems are often "wicked problems" [69], meaning that they are complex and can be addressed in various ways. To illustrate this, reconsider the problem of an organization that would like to check their business process design. One task that might be derived from this problem is indeed the analysis task whether the workflow net of the business process is sound. However, there exist numerous other tasks that could be derived, for instance whether the graph of the business process is structured according to notions of the refined process structure tree [60]. While many tasks are already known, researchers might also come up with new ones for a given real-world problem. What the example above highlights is that algorithmic tasks explicate assumptions regarding the input, the processing of the input, and the goal the algorithm is meant to achieve in terms of output. The definition of the soundness notion in the paper by van der Aalst [1] provides this knowledge of the task.

Knowledge about tasks relates to insights into the properties of specific algorithmic tasks. Such analysis can yield both formal and empirical knowledge. Formal knowledge has been established for a wide range of tasks, often described as computational problems. A classical example is the classification of problems by complexity. Many well-known problems have been shown to belong to the class NP, the set of problems that can be solved in polynomial time on a non-deterministic Turing machine. An important subclass is the set of NP-complete problems, whose members are, loosely speaking, at least as difficult to solve as any other problem in NP. For soundness, van der Aalst provides theorems and corresponding proofs that free-choice workflow nets can be checked in polynomial time, referencing [21]. Empirical knowledge, in contrast, is obtained through experimentation. It emerges from observing patterns in typical input data that can be exploited when designing algorithms. Improvements for many algorithmic tasks have been inspired by the availability of benchmark data sets. For instance, Mendling reports that soundness-based analysis methods, though being polynomial in complexity, are practically feasible for real-world process models [53].

Knowledge of designs refers to understanding what an algorithm design is, how it operates, and how it addresses a given algorithmic task. This kind of knowledge is prescriptive: it enables humans to implement concrete algorithms by following the design's blueprint. Established designs range from concrete designs that address specific tasks to more general procedures that are applicable for a wide range of tasks. For instance, van der Aalst describes transformation rules by which sound workflow nets can be constructed [1]. At the same time, this design draws on broader principles. It relates to other works on reduction rules for free-choice Petri nets [21] and operations preserving liveness and boundedness [56], which can be seen as specific instances of the divide-and-conquer principle that guide the construction of solutions for a wide variety of algorithmic problems. Whether specific or general, knowledge of designs captures the *how* for realizing an algorithm as program code to solve the intended task. It is important to note that knowledge of a concrete algorithm design can be expressed at varying levels of abstraction, including pseudocode, diagrams, or formal mathematical specifications. Each offers different trade-offs in terms of clarity, detail, and suitability for re-implementation [54].

Knowledge about designs concerns the properties and characteristics of an algorithmic design. It can be substantiated in both formal and empirical ways. Building on work by Santner at al. [72] on computer experiments, we differentiate four broad types of such knowledge: performance, sensitivity, uncertainty, and explanatory knowledge. *Performance knowledge* addresses how well the algorithm design meets its task requirements. It can be framed as a question of satisfaction: Does the design fulfill the requirements? or to what extent does the design satisfy the requirements better than alternative designs [34,74,78]?

For the soundness of workflow nets, van der Aalst identifies liveness and bound-edness as necessary and sufficient conditions [1]. In this way, existing formal analysis methods for analyzing liveness and boundedness can be reused. *Sensitivity knowledge* examines how robust an algorithm's performance is to changes in internal design decisions, such as parameter settings. It also assesses whether the algorithm remains effective when these parameters are not optimally chosen or assumptions are not fully met. For example, reduction rules offer an efficient way for checking properties such as soundness, and the gain in performance appears to be increasing with the size of the Petri net [53]. *Uncertainty knowledge* considers the performance of the algorithm relative to the assumptions of the task environment and seeks to determine how expected performance varies across different problem instances, such as soundness checking for Petri net that are not fully free-choice [53]. Finally, *explanatory knowledge* provides insight into the mechanisms by which task assumptions and design decisions interact to influence performance, often by comparing alternative configurations. Together these four types of knowledge make explicit what can be claimed and justified about an algorithm design in general, and for a Petri net analysis method in particular.

5   Empirical Research Related to Petri Nets

While formal contributions are typical for Petri net papers, there are also various empirical contributions. This section presents examples that relate to analyzing Petri net properties, generating Petri nets using process mining algorithms, and human comprehension of Petri nets.

5.1   Analyzing Petri Net Properties

We know that certain properties such as wellformedness can be decided in poly-nomial time for free-choice Petri nets [20] while others that relate to the state explosion problem are exponential [75]. From advancements on the traveling salesperson problem we also know that results on such unfavorable complex-ity classes still offer various opportunities for developing efficient algorithms for classes of practically relevant problem instances [71]. Some empirical results for Petri net analysis relate to these observations.

Some early work on the verification of business process models from practice use Petri net based analysis methods. Mendling et al. use an implementation of these methods for analyzing soundness of business process models of the SAP reference model and other model collections. They find that 215 of these 2003 models are not sound and that unsound models tend to be more complex [52,53]. They also find that reduction rules fully reduce almost 85% of all models and that half of the models can be reduced in less than one second [52]. There are also further analysis results on how often which reduction rules and error patterns are observed [52].

These empirical works point to opportunities for further improvements on efficiency of checking soundness. One stream builds on the connection between structure and behavior [23,45]. Work by Vanhatalo et al. [76] and by Polyvyanyy et al. [62] present algorithms inspired by Hopcroft and Tarjan [36] for parsing a process model into a tree of its components. If these components are structurally well-formed, then also the bahavior is sound. This refined process structure tree can be computed in linear time. Research by Fahland et al. [26,27] integrates these parsing techniques with verification techniques. Their results demonstrate that business process models from practice can be checked in a few milliseconds.

5.2   Generating Petri Nets Using Process Mining Algorithms

Process mining defines various algorithmic tasks including automatic process discovery. Algorithmic designs addressing this task generate a Petri net from an event log. Various algorithmic designs have been proposed, ranging from the alpha-miner [3], the inductive miner [46], to the Split miner [8], and beyond. Some of them offer guarantees. The alpha-algorithm was developed under several simplifying assumptions, such as completeness in terms of directly-follows relations and the absence of short loops or duplicate tasks. If these assumptions hold, the algorithm rediscovers a sound and structured WF-nets from event logs. The inductive miner recursively partitions logs to obtain a sound, block-structured workflow model with perfect fitness.

Several studies have been conducted to compare the performance of process mining algorithms that automatically discover process models. These algorithms are often assessed by help of fitness and precision [61]. De Weerdt et al. [16] were among the first to conduct a systematic comparative study in this area. They use various event log sets from industry and many competing algorithms. They find that the heuristics miner performs well, but also suffers like other algorithms from degrading performance for complex event logs. Augusto et al. extend this study with more recent algorithms and a larger set of event logs [7]. They find that some algorithms can cope better with certain input data characteristics. Such an interaction between algorithm properties and input data characteristics and its joint effect on output properties is more systematically investigated in [9].

5.3   Human Comprehension of Petri Nets

The research on soundness checking for business process models from practice hinted towards an empirical connection between complexity and issues with soundness [52]. The presumable explanation is that humans manually creating complex business process models loose track of split and join semantics and introduced errors. Reijers et al. explicitly investigated this connection using a correlational design [65]. Their study offered the basis for experimental studies with systematic control, e.g. [55], which eventually laid the foundation for a empirical understanding of process model comprehension [28,50].

Next to studies that focus on process model characteristics, there are also contributions that compare business process modeling languages. Among the first is Sarshar et al. who compare Petri nets with EPCs from a user perspective [73]. Studies in this area also investigate the effect of different symbols for splits and joins on comprehension [29,30]. In essence, different sets of symbols are variations of so-called secondary notation. The experiment by Freytag et al. studies how the systematic coloring of splits and joins affects the comprehension of complex Petri net models [64]. That study demonstrates how secondary notation of Petri nets can be modified for the benefit of the user.

6   Methodological Perspective and Validity Concerns

There are some lessons from these empirical works. An empirical research agenda for Petri nets requires the consideration of a large spectrum of methodological guidelines and corresponding validity concerns. In this context, research contributes in four broad ways: 1) by creating new or improved knowledge of tasks, 2) by developing new or improved knowledge of designs, 3) by establishing new or stronger formal knowledge about tasks and designs, and 4) by generating new or stronger empirical knowledge about tasks and designs. As illustrated by Fig. 1, these contributions are typically part of iterative processes in which new tasks or designs expand their respective knowledge categories: new theorems and proofs establish formal insights about algorithms and new experimental results enrich the empirical understanding of algorithmic behavior. Each type of knowledge is associated with characteristic research methods, such as inductive methods for identifying and formulating tasks, design methods for crafting new algorithms, formal methods for deriving theorems and proofs, and empirical methods for experimentation and evaluation. Across all of these methods, careful attention to validity concerns is essential to ensure that the generated knowledge established with methodologically rigor. Figure 1 gives an overview. In the following, we will focus on those three of the concerns in detail that are directly connected with the instrumentation of an empirical research design: implementation validity, external validity, internal validity, and construct validity.

6.1   Implementation Validity

Many evaluations in process mining build on the comparison of different implementations of competing algorithmic designs. These implementations are run with evaluation data as input to produce performance measurements of the computational process and of the generated output. Implementation validity is the extent to which an algorithm implementation faithfully instantiates the intended design and behaves as expected [47,48,54].

Designs do not simply map to an implementation. Any implementation uses a specific programming language on a specific operating system running on a specific microprocessor family, which all affect the performance of the algorithm [54]. We also know that there are bugs in machine learning code with a negative effect on statistical analysis properties [11]. Kriegel et al. demonstrate that implementations of the same algorithm design can drastically vary in performance [42].

Table 1. Validity concerns in algorithm engineering, excerpt from [54].

Validity Concern	Explanation
Ecological validity	The extent to which an algorithmic task or setup reflects real-world conditions and problem contexts (based on [6,35]).
Design validity	The degree to which the internal structure and logic of an algorithm design is coherent, justified, and explainable [44].
Implementation validity	The extent to which an algorithm implementation faithfully instantiates the intended design and behaves as expected (based on [47,48]).
External validity	The degree to which results generalize across data sets of interest (based on [13]).
Justification validity	The degree how convincingly a hypothesis or theorem is supported by a deductive argument (based on [43]).
Logical validity	The degree to which the syllogisms used in a proof preserve truth (based on Aristotle and reflection in [24]).
Internal validity	The extent to which observed effects can be attributed to the treatment rather than to confounding factors [79].
Construct validity	The degree to which the measure of a construct accurately measures the intended property [57].
Conclusion validity	The degree to which the results can reasonably be regarded as revealing the hypothesized connection [12,32].

They find run time differences of DBScan implementations of four orders of magnitude. Even different versions of the same frameworks like ELKI and WEKA yielded substantial performance differences.

While only a few works have paid attention to implementation validity in process mining so far, a good example that stands out is the ProM framework[1] that was initiated at TU Eindhoven by van der Aalst. In order to manage the various plug-ins, there is the option to include unit tests. Every ProM package comes with a default test file. Another good example is research on trace alignment [37]. The conceptual foundations of alignment are known to be computationally expensive. Advances in this area focus not only on algorithmic designs, but also on efficient implementation strategies. González-Montesino and Grass-Boada use various optimization strategies with the effect of improving the speed of alignment by three orders of magnitude [33]. This example highlights the need to reflect on implementation concerns whenever empirical evaluation is the primary source of knowledge about algorithmic performance.

Today, comparative evaluation has become a prominent strategy to provide evidence that a new algorithm improves performance. Both the BPM and the ICPM conference call for making implementations available in order to foster reproducibility. Publishing code is generally recommended to establish imple-

---

[1] https://promtools.org.

mentation validity [42]. This is even more important for research that increasingly uses AI-based techniques, for example, for process prediction where formal guarantees cannot be devised. As Kriegel et al. observe that implementations strongly differ in quality, authors should find the fastest implementation of an algorithm, debug and optimize it before conducting comparative evaluations [42].

6.2 External Validity

Already the article introducing the alpha algorithm discusses, though only very briefly, its application for mining practically relevant processes in hospitals in Tilburg and Maastricht and at a Dutch authority in Leeuwarden [4]. The article by van der Aalst et al. on the industrial application of process mining demonstrates the benefits and challenges of working with realistic data [15]. External validity refers to the extent to which empirical research findings can be generalized beyond a specific study context to other datasets, problem instances, populations, and conditions. This concern relates to characteristics of the data being used for evaluating algorithms.

The introduction of the BPI challenge logs significantly facilitated a shift towards the use of real-life data [22]. On the one hand, this shift enhanced external validity by ensuring the data more accurately reflected actual processes. On the other hand, these logs often represent a convenience sample and are likely subject to selection bias, which limits their representativeness. Recent research showed that the available real-life event logs indeed only cover a limited set of possible process variants [49]. To address these limitations, Jouck et al. [40] introduced a methodology for sampling synthetic event logs from a predefined population of processes. By defining such a population, researchers can randomly generate an unlimited number of event logs, thus improving representativeness and external validity with respect to that population. Recent research has further extended this approach to ensure that synthetic data covers a broad range of possible process populations [49].

A key takeaway from neighboring research communities is the strategic focus on the creation of benchmark data sets. Entire research projects in areas such as computer vision or time-series modeling have evolved around the systematic construction of large and publicly available data collections. Well-known examples include ImageNet [18] and TSM-Bench [41]. In these fields, it is common practice to publish dedicated papers on data sets, whose primary contribution is the careful design and open release of a data resource that others can use for evaluation and comparison. Such an option exists at the BPM conference and its call for demonstrations and resources. Although there is still a way to go to match other fields, there have been several submissions of resource papers over the past years, among others by the team from RWTH Aachen [25,59].

Today, many process mining papers use BPI Challenge datasets for their evaluation. It has also become good practice to conduct evaluations using both synthetic data and available BPI data. Generators like GEDI support the construction of synthetic data [49]. Furthermore, event logs are always preprocessed and involve a potentially large number of selection choices or pragmatic design decision [63]. It is therefore worth investigating how datasets can be constructed from real databases, like the MIMIC-III clinical database [39], and how pre-analysis event construction like surveyed in [63] affects algorithmic performance.

6.3 Internal Validity

Process mining research increasingly builds on empirical evaluations. Internal validity is the extent to which observed effects can be attributed to the treatment rather than to confounding factors [79]. In process mining, this often concerns knowledge about algorithm designs. Here, we have to distinguish studies that focus on performance and sensitivity analysis for a comparative set of algorithms from studies that investigate uncertainty and explanatory claims.

For studies dealing with knowledge claims of performance and sensitivity, different algorithms or their variants are typically applied to the same set of event logs [7]. Here, observed performance differences can be directly attributed to the algorithms under comparison, since they are the only element changing in the experimental setup. In essence, such a setup is a within-subject design. While challenges with human subjects such as carryover, learning effects, or fatigue [79] do not apply for computer experiments, we have to carefully consider implementation validity.

For studies dealing with knowledge claims about uncertainty or explanation, more caution is necessary. These studies often focus on effects obtained from combinations of algorithm designs and corresponding input data. They require a systematic variation of both algorithms and data features. When operating on real-life event logs, a systematic controlled variation of input data is often not directly possible. Such a study is therefore observational, suffering from challenges associated with measured and unmeasured confounding factors. One method proposed to address these challenges is the use of synthetic event logs. Systematic variation of synthetic input data transforms the study into a controlled experiment, thereby allowing researchers to manage confounding factors more effectively [10, 38].

Today, various studies in process mining conduct comparative evaluations using a large set of BPI Challenge datasets. It is important to bear in mind that some algorithms perform better on input data with specific characteristics [9, 38]. Therefore, it is a good idea to perform focused evaluations with synthetic data for internal validity and additionally with real-world data for external validity, as for example done the article by de Medeiros et al. on genetic process mining [51].

7   Conclusion

This paper has discussed the benefits and challenges of complementing formal research on Petri nets with an empirical research agenda. We developed our argument based on a methodological framework for algorithm engineering and selected examples of empirical contributions to Petri net research. These examples illustrated the spectrum of potential empirical contributions to Petri net research and challenges of addressing validity concerns. We discussed how future empirical research on Petri nets can address these validity concerns.

Such an empirical research agenda entails various opportunities for Petri net research. Recent advancements in neighboring areas of computer science are closely connected with empirical works inspired by the availability of realistic problem instances and benchmark datasets [70,71]. Partially, Petri net research is already benefiting from the feedback of its applied sibling communities that research business process modeling and process mining. Strengthening the ties between these communities is of mutual benefit and bears the potential that empirical observations continue to inspire new formal properties and analysis methods for Petri nets.

Acknowledgments. This research was supported by the Einstein Foundation Berlin under grant EPP-2019-524, by the Federal Ministry of Research, Technology and Space under the grant 16DII143, and by Deutsche Forschungsgemeinschaft under grants 496119880 (VisualMine), 531115272 (ProImpact), SFB 1404/2 (FONDA).

References

1. Aalst, W.M.P.: Verification of workflow nets. In: Azéma, P., Balbo, G. (eds.) ICATPN 1997. LNCS, vol. 1248, pp. 407–426. Springer, Heidelberg (1997). https://doi.org/10.1007/3-540-63139-9_48
2. van der Aalst, W.M.P., et al.: Soundness of workflow nets: classification, decidability, and analysis. Formal Aspects Comput. **23**(3), 333–363 (2011)
3. van der Aalst, W.M.P., Weijters, T., Maruster, L.: Workflow mining: discovering process models from event logs. IEEE Trans. Knowl. Data Eng. **16**(9), 1128–1142 (2004)
4. van der Aalst, W.M.P., Weijters, T., Maruster, L.: Workflow mining: discovering process models from event logs. IEEE Trans. Knowl. Data Eng. **16**(9), 1128–1142 (2004)
5. Aho, A.V., Hopcroft, J.E., Ullman, J.D.: The design and analysis of computer algorithms. Addison-Wesley Series in Computer Science and Information Processing, Reading, MA: Addison-Wesley, 1974 (1974)
6. Ashcraft, M.H., Radvansky, G.A.: Cognition. Pearson Education India (2010)
7. Augusto, A., et al.: Automated discovery of process models from event logs: review and benchmark. IEEE Trans. Knowl. Data Eng. **31**(4), 686–705 (2018)
8. Augusto, A., Conforti, R., Dumas, M., La Rosa, M., Polyvyanyy, A.: Split miner: automated discovery of accurate and simple business process models from event logs. Knowl. Inf. Syst. **59**(2), 251–284 (2019)
9. Augusto, A., Mendling, J., Vidgof, M., Wurm, B.: The connection between process complexity of event sequences and models discovered by process mining. Inf. Sci. **598**, 196–215 (2022)
10. vanden Broucke, S.K., Delvaux, C., Freitas, J., Rogova, T., Vanthienen, J., Baesens, B.: Uncovering the relationship between event log characteristics and process discovery techniques. In: International Conference on Business Process Management, pp. 41–53. Springer, Cham (2013). https://doi.org/10.1007/978-3-319-06257-0_4

11. Cheng, D., Cao, C., Xu, C., Ma, X.: Manifesting bugs in machine learning code: An explorative study with mutation testing. In: 2018 IEEE International Conference on Software Quality, Reliability and Security (QRS), pp. 313–324. IEEE (2018)
12. Cook, T.D., Campbell, D.T.: The design and conduct of true experiments and quasi-experiments in field settings. In: Reproduced in part in Research in Organizations: Issues and Controversies. Goodyear Publishing Company (1979)
13. Cook, T.D., Campbell, D.T., Shadish, W.: Experimental and Quasi-experimental Designs for Generalized Causal Inference, vol. 1195. Houghton Mifflin Boston, MA (2002)
14. Cormen, T.H., Leiserson, C.E., Rivest, R.L., Stein, C.: Introduction to Algorithms. MIT Press (2009)
15. De Medeiros, A.A.: Business process mining: an industrial application. Inf. Syst. **32**(5), 713–732 (2007)
16. De Weerdt, J., De Backer, M., Vanthienen, J., Baesens, B.: A multi-dimensional quality assessment of state-of-the-art process discovery algorithms using real-life event logs. Inf. Syst. **37**(7), 654–676 (2012)
17. Dehnert, J., van der Aalst, W.M.P.: Bridging the gap between business models and workflow specifications. Int. J. Cooperative Inf. Syst. **13**(03), 289–332 (2004)
18. Deng, J., Dong, W., Socher, R., Li, L.J., Li, K., Fei-Fei, L.: Imagenet: a large-scale hierarchical image database. In: 2009 IEEE Conference on Computer Vision and Pattern Recognition, pp. 248–255 (2009)
19. Depaire, B., Leopold, H., Mendling, J.: Threats to validity of process mining research. In: Mining a Scientist's Process: Essays Dedicated to Wil van der Aalst on the Occasion of His 60th Birthday, pp. 551–568. Springer (2026)
20. Desel, J.: A proof of the rank theorem for extended free choice nets. In: Jensen, K. (ed.) Application and Theory of Petri Nets 1992, pp. 134–153. Springer, Berlin Heidelberg, Berlin, Heidelberg (1992)
21. Desel, J., Esparza, J.: Free choice Petri nets. No. 40, Cambridge university press (1995)
22. van Dongen, B.F., Verbeek, E.: Challenges and contests in process mining. In: Mining a Scientist's Process: Essays Dedicated to Wil van der Aalst on the Occasion of His 60th Birthday, pp. 333–348. Springer, Cham (2026)
23. Dumas, M., La Rosa, M., Mendling, J., Mäesalu, R., Reijers, H.A., Semenenko, N.: Understanding business process models: the costs and benefits of structuredness. In: International Conference on Advanced Information Systems Engineering, pp. 31–46. Springer, Cham (2012)
24. Durand-Guerrier, V.: Truth versus validity in mathematical proof. ZDM **40**(3), 373–384 (2008)
25. Engelberg, G., Hadad, M., Pegoraro, M., Soffer, P., Hadar, E., van der Aalst, W.M.: An uncertainty-aware event log of network traffic. In: BPM (Demos/Resources Forum), pp. 67–71 (2023)
26. Fahland, D., Favre, C., Jobstmann, B., Koehler, J., Lohmann, N., Völzer, H., Wolf, K.: Instantaneous Soundness Checking of Industrial Business Process Models. In: Dayal, U., Eder, J., Koehler, J., Reijers, H.A. (eds.) BPM 2009. LNCS, vol. 5701, pp. 278–293. Springer, Heidelberg (2009). https://doi.org/10.1007/978-3-642-03848-8_19
27. Fahland, D., Favre, C., Koehler, J., Lohmann, N., Völzer, H., Wolf, K.: Analysis on demand: instantaneous soundness checking of industrial business process models. Data Knowl. Eng. **70**(5), 448–466 (2011)
28. Figl, K.: Comprehension of procedural visual business process models: a literature review. Bus. Inf. Syst. Eng. **59**(1), 41–67 (2017)

29. Figl, K., Mendling, J., Strembeck, M.: The influence of notational deficiencies on process model comprehension. J. Assoc. Inf. Syst. **14**(6), 1 (2013)
30. Figl, K., Recker, J., Mendling, J.: A study on the effects of routing symbol design on process model comprehension. Decis. Support Syst. **54**(2), 1104–1118 (2013)
31. Freytag, T., Allgaier, P., Burattin, A., Danek-Bulius, A.: Woped - A "proof-of-concept" platform for experimental BPM research projects. In: Clarisó, R., et al. (eds.) Proceedings of the BPM Demo Track and BPM Dissertation Award co-located with 15th International Conference on Business Process Modeling (BPM 2017), Barcelona, Spain, September 13, 2017. CEUR Workshop Proceedings, CEUR-WS.org (2017). https://ceur-ws.org/Vol-1920/BPM_2017_paper_190.pdf
32. García-Pérez, M.A.: Statistical conclusion validity: Some common threats and simple remedies. Front. Psychol. **3**, 325 (2012)
33. González-Montesino, L., Grass-Boada, D.H.: Trace alignment algorithm optimization. Data Min. Knowl. Disc. **39**(5), 59 (2025)
34. Hall, J.G., Rapanotti, L.: A design theory for software engineering. Inf. Softw. Technol. **87**, 46–61 (2017)
35. Holleman, G.A., Hooge, I.T.C., Kemner, C., Hessels, R.S.: The 'Real-World Approach' and Its Problems: A Critique of the Term Ecological Validity. Front. Psychol. **11** (2020)
36. Hopcroft, J.E., Tarjan, R.E.: Dividing a graph into triconnected components. SIAM J. Comput. **2**(3), 135–158 (1973)
37. Jagadeesh Chandra Bose, R.P., van der Aalst, W.: Trace Alignment in Process Mining: Opportunities for Process Diagnostics. In: Hull, R., Mendling, J., Tai, S. (eds.) BPM 2010. LNCS, vol. 6336, pp. 227–242. Springer, Heidelberg (2010). https://doi.org/10.1007/978-3-642-15618-2_17
38. Janssenswillen, G., Donders, N., Jouck, T., Depaire, B.: A comparative study of existing quality measures for process discovery. Inf. Syst. **71**, 1–15 (2017)
39. Johnson, A.E., et al.: Mimic-iii, a freely accessible critical care database. Sci. Data **3**(1), 1–9 (2016)
40. Jouck, T., Depaire, B.: Generating artificial data for empirical analysis of control-flow discovery algorithms: a process tree and log generator. Bus. Inf. Syst. Eng. **61**(6), 695–712 (2019)
41. Khelifati, A., Khayati, M., Dignös, A., Difallah, D., Cudré-Mauroux, P.: Tsmbench: Benchmarking time series database systems for monitoring applications. Proc. VLDB Endow. **16**(11), 3363–3376 (2023)
42. Kriegel, H.P., Schubert, E., Zimek, A.: The (black) art of runtime evaluation: Are we comparing algorithms or implementations? Knowl. Inf. Syst. **52**(2), 341–378 (2017)
43. Lannin, J.K.: Generalization and justification: the challenge of introducing algebraic reasoning through patterning activities. Math. Think. Learn. **7**(3), 231–258 (2005)
44. Larsen, K.R., Lukyanenko, R., Mueller, R.M., Storey, V.C., VanderMeer, D., Parsons, J., Hovorka, D.S.: Validity in Design Science Research. In: Hofmann, S., Müller, O., Rossi, M. (eds.) DESRIST 2020. LNCS, vol. 12388, pp. 272–282. Springer, Cham (2020). https://doi.org/10.1007/978-3-030-64823-7_25
45. Laue, R., Mendling, J.: Structuredness and its significance for correctness of process models. IseB **8**(3), 287–307 (2010)

46. Leemans, S.J.J., Fahland, D., van der Aalst, W.M.P.: Discovering block-structured process models from incomplete event logs. In: Ciardo, G., Kindler, E. (eds.) Application and Theory of Petri Nets and Concurrency, pp. 91–110. Springer International Publishing, Cham (2014)
47. Lukyanenko, R., Evermann, J., Parsons, J.: Guidelines for Establishing Instantiation Validity in IT Artifacts: A Survey of IS Research. In: Donnellan, B., Helfert, M., Kenneally, J., VanderMeer, D., Rothenberger, M., Winter, R. (eds.) DESRIST 2015. LNCS, vol. 9073, pp. 430–438. Springer, Cham (2015). https://doi.org/10.1007/978-3-319-18714-3_35
48. Lukyanenko, R., Parsons, J.: Design theory indeterminacy: What is it, how can it be reduced, and why did the polar bear drown? J. Assoc. Inf. Syst. (2020)
49. Maldonado, A., Frey, C.M., Tavares, G.M., Rehwald, N., Seidl, T.: Gedi: generating event data with intentional features for benchmarking process mining. In: International Conference on Business Process Management, pp. 221–237. Springer, Cham (2024)
50. Malinova Mandelburger, M., Mendling, J.: Cognitive diagram understanding and task performance in systems analysis and design. MIS Q. **45**(4), 2101–2157 (2021)
51. de Medeiros, A.K.A., Weijters, A.J., van der Aalst, W.M.P.: Genetic process mining: an experimental evaluation. Data Min. Knowl. Disc. **14**(2), 245–304 (2007)
52. Metrics for Process Models. LNBIP, vol. 6. Springer, Heidelberg (2008). https://doi.org/10.1007/978-3-540-89224-3
53. Mendling, J.: Empirical Studies in Process Model Verification. In: Jensen, K., van der Aalst, W.M.P. (eds.) Transactions on Petri Nets and Other Models of Concurrency II. LNCS, vol. 5460, pp. 208–224. Springer, Heidelberg (2009). https://doi.org/10.1007/978-3-642-00899-3_12
54. Mendling, J., Leopold, H., Meyerhenke, H., Depaire, B.: Methodology of algorithm engineering. ACM Comput. Surv. (2025)
55. Mendling, J., Strembeck, M., Recker, J.: Factors of process model comprehension—findings from a series of experiments. Decis. Support Syst. **53**(1), 195–206 (2012)
56. Murata, T.: Petri nets: properties, analysis and appl kat ions. Proc. IEEE **77**(4), 541 (1989)
57. O'Leary-Kelly, S.W., J. Vokurka, R.: The empirical assessment of construct validity. J. Oper. Manag. **16**(4), 387–405 (1998)
58. Peterson, J.L.: Petri nets. ACM Comput. Surv. (CSUR) **9**(3), 223–252 (1977)
59. Pohl, T., Berti, A., Qafari, M.S., van der Aalst, W.M.P.: A collection of simulated event logs for fairness assessment in process mining. arXiv preprint arXiv:2306.11453 (2023)
60. Polyvyanyy, A., García-Bañuelos, L., Dumas, M.: Structuring acyclic process models. Inf. Syst. **37**(6), 518–538 (2012)
61. Polyvyanyy, A., Solti, A., Weidlich, M., Ciccio, C.D., Mendling, J.: Monotone precision and recall measures for comparing executions and specifications of dynamic systems. ACM Trans. Softw. Eng. Methodol. (TOSEM) **29**(3), 1–41 (2020)
62. Polyvyanyy, A., Vanhatalo, J., Völzer, H.: Simplified Computation and Generalization of the Refined Process Structure Tree. In: Bravetti, M., Bultan, T. (eds.) WS-FM 2010. LNCS, vol. 6551, pp. 25–41. Springer, Heidelberg (2011). https://doi.org/10.1007/978-3-642-19589-1_2
63. Pradhan, S.K., Jans, M., Martin, N.: Getting the data in shape for your process mining analysis: An in-depth analysis of the pre-analysis stage. ACM Comput. Surv. **57**(6), 1–37 (2025)
64. Reijers, H.A., Freytag, T., Mendling, J., Eckleder, A.: Syntax highlighting in business process models. Decis. Support Syst. **51**(3), 339–349 (2011)

65. Reijers, H.A., Mendling, J.: A study into the factors that influence the understandability of business process models. IEEE Trans. Syst. Man Cybern.-Part A Syst. Hum. **41**(3), 449–462 (2010)
66. Reisig, W.: Understanding petri nets. Springer
67. Reisig, W.: Understanding Petri Nets: Modeling Techniques. Case Studies. Springer Science & Business Media, Analysis Methods (2013)
68. Reisig, W., Rozenberg, G.: Lectures on petri nets i: basic models: advances in petri nets. Springer Science & Business Media (1998)
69. Rittel, H.W., Webber, M.M.: Dilemmas in a general theory of planning. Policy Sci. **4**(2), 155–169 (1973)
70. Russakovsky, O., et al.: Imagenet large scale visual recognition challenge. Int. J. Comput. Vision **115**, 211–252 (2015)
71. Sanders, P.: Algorithm engineering–an attempt at a definition. In: Efficient Algorithms, pp. 321–340. Springer, Cham (2009)
72. Santner, T.J., Williams, B.J., Notz, W., Williams, B.J.: The design and analysis of computer experiments, vol. 1. Springer, Cham (2003)
73. Sarshar, K., Loos, P.: Comparing the Control-Flow of EPC and Petri Net from the End-User Perspective. In: van der Aalst, W.M.P., Benatallah, B., Casati, F., Curbera, F. (eds.) BPM 2005. LNCS, vol. 3649, pp. 434–439. Springer, Heidelberg (2005). https://doi.org/10.1007/11538394_36
74. Staples, M.: Critical rationalism and engineering: ontology. Synthese **191**(10), 2255–2279 (2014). https://doi.org/10.1007/s11229-014-0396-3
75. Valmari, A.: The state explosion problem. In: Reisig, W., Rozenberg, G. (eds.) ACPN 1996. LNCS, vol. 1491, pp. 429–528. Springer, Heidelberg (1998). https://doi.org/10.1007/3-540-65306-6_21
76. Vanhatalo, J., Völzer, H., Koehler, J.: The refined process structure tree. In: International Conference on Business Process Management, pp. 100–115. Springer, Cham (2008)
77. Verbeek, H.M., Basten, T., van der Aalst, W.M.: Diagnosing workflow processes using woflan. Comput. J. **44**(4), 246–279 (2001)
78. Wieringa, R.J.: Design science methodology for information systems and software engineering. Springer, Cham (2014)
79. Wohlin, C., Runeson, P., Höst, M., Ohlsson, M.C., Regnell, B., Wesslén, A.: Experimentation in software engineering. Springer Science & Business Media (2012)
80. Zurawski, R., Zhou, M.: Petri nets and industrial applications: A tutorial. IEEE Trans. Industr. Electron. **41**(6), 567–583 (1994)

Regular Papers

Asymptotic Analysis of Expected Complexity in VASS MDPs

Michal Ajdarów$^{(\boxtimes)}$ (iD)

University of Liverpool, Liverpool, UK
`ajdarow@liverpool.ac.uk`

Abstract. We study the asymptotics of various types of expected complexity in Vector Addition Systems with States over Markov Decision Processes (VASS MDP). We provide a full classification of the asymptotics of expected termination, counter, and transition complexities in one-counter VASS MDPs, and we show this class can be fully classified in **PTIME**. We also show **PSPACE**-hardness for multiple problems related to the asymptotics of expected termination and counter complexity in VASS MDPs. Namely, that any non-trivial instance of deciding whether such complexity is in $\mathcal{O}\big(f(n)\big)$ or $\Omega\big(f(n)\big)$ is **PSPACE**-hard for general VASS MDPs. For the class of strongly connected VASS MDPs we show **PSPACE**-hardness for deciding whether such expected complexity is in $2^{\mathcal{O}\big(f(n)\big)}$ for given $f \in \Omega(n)$, or $2^{\Omega\big(f(n)\big)}$ for given $f \in \omega(n)$. Finally, we also show that deciding whether such expected complexity for a given VASS MDP is finite is **PSPACE**-hard already for the class of strongly connected VASS MDPs.

Keywords: VASS · MDP · expected complexity · asymptotic complexity

1 Introduction

Vector addition systems with states (VASS) [11] are a generic formalism expressively equivalent to Petri nets [20]. Intuitively, a VASS with $d \geq 1$ counters is a finite directed graph where the transitions are labeled by d-dimensional vectors of integers representing *counter updates*. A computation starts in some state for some initial vector of non-negative counter values and proceeds by selecting transitions non-deterministically and performing the associated counter updates. The computation terminates whenever any of the counters becomes negative.

In program analysis, VASS are used as abstractions for programs operating over unbounded integer variables. Input parameters are represented by initial counter values, and more complicated arithmetical functions, such as multiplication, are modeled by VASS gadgets computing these functions in a weak sense (see, e.g., [15]). Branching constructs, such as **if-then-else**, are usually replaced with non-deterministic choice. Various questions about the original systems can thus be reduced to the corresponding problems for VASS. Unfortunately, the

J. Desel and A. Kalenkova (Eds.): PETRI NETS 2026, LNCS 16567, pp. 21–42, 2026.
https://doi.org/10.1007/978-3-032-27879-1_2

scalability of this approach is limited by the high computational complexity of
the relevant VASS problems.

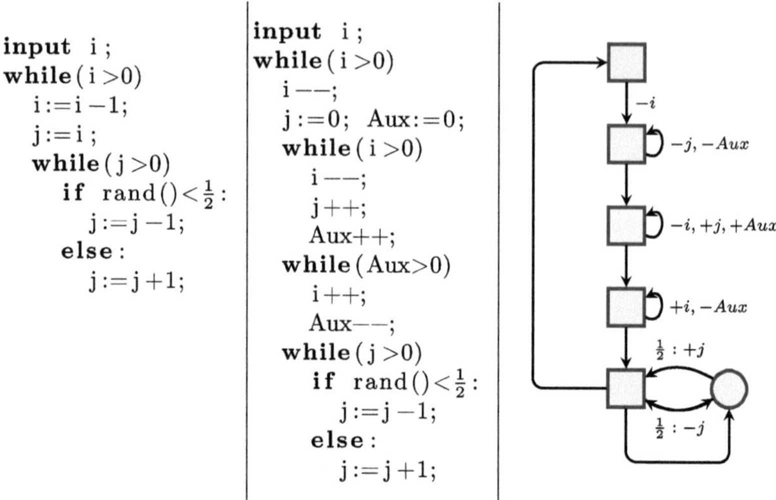

Fig. 1. A skeleton of a simple program (left) and its overapproximating VASS MDP
model (right). Note that upper bounds on the complexity of this VASS MDP carry
over to the original program. For example, if this VASS MDP were to terminate with
probability 1 (equivalent to finite upper asymptotic estimate [2]) or had linear/finite
expectation then so does the original program.

Traditional VASS decision problems such as reachability, liveness, or boundedness are computationally hard [10, 17, 18], and other verification problems such
as equivalence-checking [12] or model-checking [19] are even undecidable. In contrast to this, decision problems related to the asymptotic growth of VASS complexity measures are solvable with low complexity and for many important subclasses even in *polynomial time* [3, 7, 14, 16, 23]; see [13] for a recent overview.
Due to this VASS are particularly useful for evaluating the *asymptotic complexity* of infinite-state programs, i.e., the dependency of the running time (and other
complexity measures) on the size of the program input [21, 22] (see e.g. Fig. 1).
The complexity measures of VASS for which the asymptotic growth is usually
considered are *termination complexity*, which can be seen as an analogy of time
complexity, and *c-counter complexity*, which is an analogy of space complexity
of a single variable c.

The existing results about VASS asymptotic analysis are applicable to programs with non-determinism (in *demonic* or *angelic* form, see [9]), but cannot be used to analyze the complexity of *probabilistic programs*. This motivates the study of VASS Markov decision process (VASS MDP) with both non-
deterministic and probabilistic states, where transitions in probabilistic states
are selected according to fixed probability distributions. Here, the problems of

asymptotic complexity analysis become even more challenging because VASS MDPs subsume infinite-state stochastic models that are notoriously hard to analyze. So far, there are only three existing papers about asymptotic VASS MDP analysis. First is [8] where it is shown that for VASS MDPs with DAG-like maximal end-component (MEC) decomposition the expected termination complexity is either linear or at least quadratic, and it is decidable in polynomial time which case holds. The second is [4], which discusses limitations of expected values for asymptotic analysis and to counter them suggests a new notion of *asymptotic estimates* for analyzing the asymptotic behavior of probabilistic systems which consists of a bound on all but ϵ-ratio of runs for all $\epsilon > 0$. Then it shows how we can classify one-dimensional VASS MDPs in terms of asymptotic estimates in polynomial time. Furthermore [4] also shows that for VASS MDPs with DAG-like MEC decomposition the termination/c-counter complexity has either a linear tight asymptotic estimate or a quadratic lower asymptotic estimate, and it is decidable in polynomial time which case holds. The paper [2] then further studies properties of asymptotic estimates and their relation to the expectation measure, while also showing that the linear vs at least quadratic dichotomy does not hold for expected c-counter complexity in VASS MDPs by describing a VASS MDP with expected c-counter complexity in $\Theta(n \cdot \log n)$. This shows that the expected c-counter complexity in VASS MDPs can also take values that are not possible in VASS as this dichotomy does hold for VASS [3,23].

Our Contribution: In this paper we study the asymptotics of the expected complexity of VASS MDPs. We consider exclusively VASS MDPs with demonic non-determinism.

Our first result (Sect. 3) is a full classification of expected asymptotic complexity for one-dimensional VASS MDPs, where we show that the expected termination/c-counter complexity is either infinite or linear. Furthermore, it is decidable in *polynomial time* which case holds.

For the second result (Sect. 4) we show **PSPACE**-hardness for several problems about the asymptotics of the expected complexity of VASS MDPs. Concretely, we show the following problems are **PSPACE**-hard:

- deciding whether the expected termination/c-counter complexity of a given strongly connected VASS MDP is finite;
- deciding whether the expected termination/c-counter complexity is in $\mathcal{O}\big(f(n)\big)$ for a given VASS MDP and a given $f \in \Omega(n)$;
- deciding whether the expected termination/c-counter complexity is in $\Omega\big(f(n)\big)$ for a given VASS MDP and a given $f \in \omega(n)$;
- deciding whether the expected termination/c-counter complexity is in $2^{\mathcal{O}\big(f(n)\big)}$ for a given *strongly connected* VASS MDP and a given $f \in \Omega(n)$;
- deciding whether the expected termination/c-counter complexity is in $2^{\Omega\big(f(n)\big)}$ for a given *strongly connected* VASS MDP and a given $f \in \omega(n)$.

Note that as the expected termination/c-counter complexity is trivially in $\Omega(n)$ this shows that any question regarding asymptotic lower/upper bounds for general VASS MDPs is either trivial or **PSPACE**-hard.

Additionally, let $g(n)$ be a function such that there exists a VASS with termination complexity $g(n)$. We also show **PSPACE**-hardness for:

- deciding whether the expected termination/c-counter complexity is in $\Theta\big(g(n)\big)$ for a given VASS MDP;
- deciding whether the expected termination/c-counter complexity is in $2^{\Theta\big(g(n)\big)}$ for a given *strongly connected* VASS MDP.

2   Preliminaries

We use $\mathbb{N}$, $\mathbb{Z}$, $\mathbb{Q}$, and $\mathbb{R}$ to denote the sets of positive integers, integers, rational numbers, and real numbers, respectively.

Let A be a finite index set. The vectors of $\mathbb{R}^A$ are denoted by bold letters such as $\mathbf{u}, \mathbf{v}, \mathbf{z}, \ldots$. The component of $\mathbf{v}$ of index $i \in A$ is denoted by $\mathbf{v}(i)$. For every $n \in \mathbb{R}$, we use $\vec{n}$ to denote the constant vector where all components are equal to n. The other standard operations and relations on $\mathbb{R}$ such as $+$, $\leq$, or $<$ are extended to $\mathbb{R}^A$ in the component-wise way. In particular, $\mathbf{v} \leq \mathbf{u}$ if $\mathbf{v}(i) \leq \mathbf{u}(i)$ for every index i.

Given functions $f(n), g(n) : \mathbb{R} \to \mathbb{R}$ we use $f \in \mathcal{O}(g(n))$, $f \in \Omega(g(n))$, and $f \in \omega\big(g(n)\big)$ to denote that $\lim_{n \to \infty} \frac{f(n)}{g(n)} \in \mathbb{R}$, $\lim_{n \to \infty} \frac{f(n)}{g(n)} > 0$, and $\lim_{n \to \infty} \frac{f(n)}{g(n)} = \infty$, respectively. We use $f \in \Theta\big(g(n)\big)$ to denote that both $f \in \mathcal{O}\big(g(n)\big)$ and $f \in \Omega\big(g(n)\big)$.

2.1   VASS Markov Decision Processes

Definition 1. *A* Vector Addition System with States over Markov Decision Process *(VASS MDP) with counters* $Count(\mathcal{A})$ *is a tuple*

$$\mathcal{A} = \Big(St(\mathcal{A}), \big(St_n(\mathcal{A}), St_p(\mathcal{A})\big), Trns(\mathcal{A}), P \Big)$$

where

- *$Count(\mathcal{A})$ is a finite set of counters,*
- *$St(\mathcal{A}) \neq \emptyset$ is a finite set of* states *split into two disjoint subsets* $St_n(\mathcal{A})$ *and* $St_p(\mathcal{A})$ *of* non-deterministic *and* probabilistic *states, respectively,*
- *$Trns(\mathcal{A}) \subseteq St(\mathcal{A}) \times \mathbb{Z}^{Count(\mathcal{A})} \times St(\mathcal{A})$ is a finite set of* transitions,
- *P is a function assigning to each $t \in Out(p)$, where $p \in St_p(\mathcal{A})$ and $Out(p) \subseteq Trns(\mathcal{A})$ is the set of all transitions of the form $(p, \mathbf{u}, q)$, a positive rational probability so that $\sum_{t \in Out(p)} P(t) = 1$.*

A VASS MDP $\mathcal{A}$ is a *VASS Markov chain* if $St_n(\mathcal{A}) = \emptyset$, and a *VASS* if $St_p(\mathcal{A}) = \emptyset$. We say $\mathcal{A}$ is d-*dimensional* if $|Count(\mathcal{A})| = d$. For every $p \in St(\mathcal{A})$, we use $In(p) \subseteq Trns(\mathcal{A})$ to denote the set of all transitions of the form $(q, \mathbf{u}, p)$. The update vector $\mathbf{u}$ of a transition $t = (p, \mathbf{u}, q)$ is also denoted by $\mathbf{u}_t$.

The encoding size of $\mathcal{A}$ is denoted by $\|\mathcal{A}\|$, where the integers representing counter updates are written in binary and probability values are written as fractions of binary numbers.

A *configuration* of $\mathcal{A}$ is a pair $p\mathbf{v}$, where $p \in St(\mathcal{A})$ and $\mathbf{v} \in \mathbb{Z}^{Count(\mathcal{A})}$. If some component of $\mathbf{v}$ is negative then $p\mathbf{v}$ is *terminal*.

A *computation* in $\mathcal{A}$ is a finite sequence of the form $\pi = p_0\mathbf{v}_0, p_1\mathbf{v}_1, \ldots, p_k\mathbf{v}_k$ where $(p_i, \mathbf{v}_{i+1} - \mathbf{v}_i, p_{i+1}) \in Trns(\mathcal{A})$ for all $i < k$. The *length* of α is defined as $len(\alpha) = k$. An *infinite computation* in $\mathcal{A}$ is an infinite sequence $\pi = p_0\mathbf{v}_0, p_1\mathbf{v}_1, p_2\mathbf{v}_2, \ldots$ such that every finite prefix of π is a computation. Let $Term(\pi)$ be the least j such that $p_j\mathbf{v}_j$ is terminal. If there is no such j, we put $Term(\pi) = len(\pi)$.

A *strategy* of $\mathcal{A}$ is a function σ assigning to every computation $p_0\mathbf{v}_0, p_1\mathbf{v}_1, \ldots, p_k\mathbf{v}_k$ such that $p_k \in St_n(\mathcal{A})$ a probability distribution over $Out(p_k)$. A strategy is *counterless-memoryless (cM)* if it depends only on the last state p_k, and *deterministic (D)* if it always returns a Dirac distribution. We denote by $cMD(\mathcal{A})$ the set of all cMD strategies of $\mathcal{A}$.

Every initial configuration $p\mathbf{v}$ and every strategy σ determine the probability space over computations initiated in $p\mathbf{v}$ in the standard way.[1] We use $\mathbb{P}_{p\mathbf{v}}^{\sigma}$ to denote the associated probability measure. For a measurable function X over computations, we use $\mathbb{E}_{p\mathbf{v}}^{\sigma}[X]$ to denote the expected value of X.

We say that q is *reachable* from p in $\mathcal{A}$ if there exists a computation $p\mathbf{v}, \ldots, q\mathbf{v}'$. We say $\mathcal{A}$ is *strongly connected* if q is reachable from p for all $p, q \in St(\mathcal{A})$.

An *end component (EC)* of $\mathcal{A}$ is a pair (C, L) where $C \subseteq St(\mathcal{A})$ and $L \subseteq Trns(\mathcal{A})$ such that the following conditions are satisfied:

- $C \neq \emptyset$;
- if $p \in C \cap St_n(\mathcal{A})$, then $Out(p) \cap L \neq \emptyset$;
- if $p \in C \cap St_p(\mathcal{A})$, then $Out(p) \subseteq L$;
- if $(p, \mathbf{u}, q) \in L$, then $p, q \in C$;
- for all $p, q \in C$ we have that q is reachable from p and vice versa using only transitions from L.

Note that if (C, L) and (C', L') are ECs such that $C \cap C' \neq \emptyset$, then $(C \cup C', L \cup L')$ is also an EC. Hence, every $p \in St(\mathcal{A})$ either belongs to a unique *maximal end component* (MEC), or does not belong to any EC. Also observe that each MEC can be seen as a strongly connected VASS MDP. (intuitively, a MEC is a "largest" strongly connected sub-VASS MDP where you can resolve nondeterminism so as never to leave).

---

[1] See e.g. [6] for details on the "standard way".

2.2   Complexity of VASS Computations

Let $\mathcal{A}$ be a VASS MDP, $c \in Count(\mathcal{A})$, and $t \in Trns(\mathcal{A})$. For every computation $\pi = p_0\mathbf{v}_0, p_1\mathbf{v}_1, p_2\mathbf{v}_2, \ldots$, we put

$$\mathcal{L}(\pi) = Term(\pi)$$
$$\mathcal{C}[c](\pi) = \sup\{\mathbf{v}_i(c) \mid 0 \leq i < Term(\pi)\}$$
$$\mathcal{T}[t](\pi) = \big|\{i \in \mathbb{N}_0 \mid 0 \leq i < Term(\pi) \text{ and } (p_i, \mathbf{v}_{i+1}-\mathbf{v}_i, p_{i+1}) = t\}\big|$$

We refer to the functions $\mathcal{L}$, $\mathcal{C}[c]$, and $\mathcal{T}[t]$ as *termination, c-counter,* and *t-transition complexity* respectively.

Note that $\mathcal{L}$, $\mathcal{C}[c]$, and $\mathcal{T}[t]$ are the complexity measures for VASS runs used in previous works [3, 4, 7, 14, 16, 23]. These functions can be seen as variants of the standard time/space complexities for Turing machines.

Let $\mathcal{F}$ be one of the complexity functions defined above. In VASS abstractions of computer programs, the input is represented by initial counter values, and the input size corresponds to the maximal initial counter value. The existing works on *non-probabilistic* VASS concentrate on analyzing the asymptotic growth of the functions $\mathcal{F}^{\max} : \mathbb{N} \to \mathbb{N}_\infty$ where

$$\mathcal{F}^{\max}(n) = \sup\{\mathcal{F}(\pi) \mid \pi \text{ is a computation initiated in } p\overrightarrow{n} \text{ where } p \in St(\mathcal{A})\}$$

For VASS MDPs, we can generalize $\mathcal{F}^{\max}$ into the *expected complexity* $\mathcal{F}^{\exp} : \mathbb{N} \to \mathbb{R}_\infty$ as follows:

$$\mathcal{F}^{\exp}(n) = \sup\{\mathbb{E}^{\sigma}_{p\overrightarrow{n}}[\mathcal{F}] \mid \sigma \text{ is a strategy of } \mathcal{A}, p \in St(\mathcal{A})\}$$

When the underlying VASS MDP is not clear, we use $\mathcal{F}_{\mathcal{A}}$ instead of just $\mathcal{F}$ (e.g., $\mathcal{C}^{\exp}_{\mathcal{A}}[c]$, etc.).

3   One-Dimensional VASS MDP

In this section, we give a full and effective classification of the asymptotics for $\mathcal{L}^{\exp}$, $\mathcal{C}^{\exp}[c]$, and $\mathcal{T}^{\exp}[t]$ for 1-dimensional VASS MDPs. More precisely, we prove the following theorem:

Theorem 1. *Let $\mathcal{A}$ be a 1-dimensional VASS MDP. The following holds:*

- *One of the following possibilities holds:*
  - $\mathcal{L}^{\exp}(n) = \infty$ *for all but finitely many n.*
  - $\mathcal{L}^{\exp}(n) \in \Theta(n)$.
- *Let c be the only counter of $\mathcal{A}$. Then one of the following possibilities holds:*
  - $\mathcal{C}^{\exp}[c](n) = \infty$ *for all but finitely many n.*
  - $\mathcal{C}^{\exp}[c](n) \in \Theta(n)$.
- *Let t be a transition of $\mathcal{A}$. Then one of the following possibilities holds:*
  - $\mathcal{T}^{\exp}[t](n) = \infty$ *for all but finitely many n.*
  - $\mathcal{T}^{\exp}[t](n) \in \Theta(n)$.

- $\mathcal{T}^{\exp}[t](n) \in \Theta(1)$.

It is decidable in polynomial time which of the above cases hold.

We use the technique of pointing MDP combined with simple components developed in [4] to provide full classification of 1-dimensional VASS MDPs with respect to asymptotic estimates. This shows that this technique can be used to analyze the asymptotics of expected complexities of VASS MDPs as well. The technique consists of defining simple components, which are a "probabilistic analogy of simple cycles", and then showing that any computation on a VASS MDP can be (up to some asymptotically irrelevant part) seen as an interweaving of a finite number of computations iterating simple components (see Lemma 31 of [5] for details). Hence towards proving Theorem 1 it suffices to analyze the simple components of $\mathcal{A}$.

We proceed with defining simple components. For the rest of this section, we fix a 1-dimensional VASS MDP $\mathcal{A}$ and some linear ordering $\sqsubseteq$ on $St(\mathcal{A})$. Recall that each $\sigma \in cMD(\mathcal{A})$ selects the same outgoing transition in every $p \in St_n(\mathcal{A})$ whenever p is visited, and hence we can "apply" σ to $\mathcal{A}$ by removing the other outgoing transitions. The resulting VASS Markov chain is denoted by $\mathcal{A}_\sigma$. Note that every SCC $\mathcal{B}$ of $\mathcal{A}_\sigma$ can also be seen as a VASS Markov chain, as a MEC of $\mathcal{A}_\sigma$, and as an EC of $\mathcal{A}$.

For every SCC $\mathcal{B}$ of $\mathcal{A}_\sigma$, let $p_\mathcal{B}$ be the least state of $\mathcal{B}$ with respect to $\sqsubseteq$, and let $\mathbf{y}_\mathcal{B}$ be a solution to

$$
\begin{aligned}
&\mathbf{y}_\mathcal{B} \in \mathbb{Q}^{Trns(\mathcal{A})} \text{ such that}\\[2mm]
&\qquad\qquad \mathbf{y}_\mathcal{B} \geq \vec{0}\\[2mm]
&\quad \mathbf{y}_\mathcal{B}\big(Trns(\mathcal{A}) \setminus Trns(\mathcal{B})\big) = \vec{0}\\[2mm]
&\qquad\qquad \sum_{t \in Out(p_\mathcal{B})} \mathbf{y}_\mathcal{B}(t) = 1\\[2mm]
&\text{for all } p \in St(\mathcal{B})\\[2mm]
&\qquad \sum_{t \in Out(p)} \mathbf{y}_\mathcal{B}(t) = \sum_{t \in In(p)} \mathbf{y}_\mathcal{B}(t)\\[2mm]
&\text{and for all } p \in St_p(\mathcal{B}),\ t \in Out(p)\\[2mm]
&\qquad \mathbf{y}_\mathcal{B}(t) = P(t) \cdot \sum_{t' \in Out(p)} \mathbf{y}_\mathcal{B}(t')
\end{aligned}
$$

Note that $\mathbf{y}_\mathcal{B}$ is defined uniquely. We call $\mathbf{y}_\mathcal{B}$ a *simple component* of $\mathcal{A}$.

For a simple component $\mathbf{y}$ we use $\sigma_\mathbf{y}$ to denote the corresponding strategy σ (i.e., such that $\mathbf{y} = \mathbf{y}_\mathcal{B}$ for an SCC $\mathcal{B}$ of $\mathcal{A}_\sigma$). We use $\mathcal{A}_\mathbf{y}$ to denote the corresponding VASS Markov chain $\mathcal{A}_{\sigma_\mathbf{y}}$, and $p_\mathbf{y}$ to denote the state $p_\mathcal{B}$. We define the *effect* of $\mathbf{y}$ as $\Delta(\mathbf{y}) = \sum_{t \in Trns(\mathcal{A})} \mathbf{u}_t \cdot \mathbf{y}(t)$.

We use $\mathrm{It}_\mathbf{y}$ to denote a random variable representing the effect of a computation on $\mathcal{A}_\mathbf{y}$ started from $p_\mathbf{y}$ until the first time $p_\mathbf{y}$ is revisited. We call such computation a single *iteration* of $\mathbf{y}$.

Definition 2. *Let c be a counter of $\mathcal{A}$. We say a simple component $\mathbf{y}$ of $\mathcal{A}$ is*

- *increasing on c if $\Delta(\mathbf{y})(c) > 0$,*
- *decreasing on c if $\Delta(\mathbf{y})(c) < 0$,*
- *zero-bounded on c if $\Delta(\mathbf{y})(c) = 0$ and $\mathbb{P}^{\sigma_{\mathbf{y}}}_{p_{\mathbf{y}}\,\vec{n}}[\mathrm{It}_{\mathbf{y}}(c) = 0] = 1$,*
- *zero-unbounded on c if $\Delta(\mathbf{y})(c) = 0$ and $\mathbb{P}^{\sigma_{\mathbf{y}}}_{p_{\mathbf{y}}\,\vec{n}}[\mathrm{It}_{\mathbf{y}}(c) = 0] < 1$.*

Note that the above definition does not depend on the concrete choice of $\sqsubseteq$ as per the following lemma.

Lemma 1. *Let $\mathbf{y}$ be a simple component and c a counter. Let $\mathbf{w}$ be the long-run average effect of $\mathcal{A}_{\mathbf{y}}$. Then there exist $a, b > 0$ such that $\mathbf{w}(c) = a \cdot \mathbb{E}^{\sigma_{\mathbf{y}}}_{p_{\mathbf{y}}\,\vec{n}}[\mathrm{It}_{\mathbf{y}}(c)] = b \cdot \Delta(\mathbf{y})(c)$.*

The poof of Lemma 1 is purely technical and can be found in the full version of this paper [1].

We prove the following results relating the asymptotics of $\mathcal{L}^{\exp}$, $\mathcal{C}^{\exp}[c]$, and $\mathcal{T}^{\exp}[t]$ to the existence of simple components with certain properties. Note that, as per Observation 5 of [4], if $\mathcal{F}$ is unbounded in the asymptotic estimates metric then it holds $\mathcal{F}^{\exp}(n) = \infty$ for almost all n. Hence these cases follow directly from the corresponding lemmas of [5] (which is the full version of [4]) and we only have to further consider the cases in which the asymptotic estimate is not unbounded. More concretely:

- For $\mathcal{L}^{\exp}$, we show that
    - $\mathcal{L}^{\exp}[c](n) = \infty$ for all but finitely many n if $\mathcal{A}$ contains a simple component that is either increasing on c (Lemma 23 of [5]), zero-unbounded on c (Lemma 2), or zero-bounded on c (Lemma 25 of [5]);
    - otherwise $\mathcal{L}^{\exp}[c](n) \in \Theta(n)$ (as per Observation 5 of [4] and Lemma 1 this follows from Lemma 3 of [8]).
- For $\mathcal{C}^{\exp}[c]$, we show that
    - $\mathcal{C}^{\exp}[c](n) = \infty$ for all but finitely many n if $\mathcal{A}$ contains a simple component that is either increasing on c (Lemma 23 of [5]) or zero-unbounded on c (Lemma 2);
    - otherwise $\mathcal{C}^{\exp}[c](n) \in \Theta(n)$ (Lemma 3).
- For $\mathcal{T}^{\exp}[t]$, we distinguish two cases:
    - If t is not contained in any MEC of $\mathcal{A}$ then $\mathcal{T}^{\exp}[t](n) \in \Theta(1)$ (Lemma 6).
    - If t is contained in a MEC $\mathcal{B}$ of $\mathcal{A}$, then
        * $\mathcal{T}^{\exp}[t](n) = \infty$ for all but finitely many n if one of the following holds:
            - there exists a simple component $\mathbf{y}$ of $\mathcal{A}$ such that $p_{\mathbf{y}}$ is in $\mathcal{B}$ and $\mathbf{y}$ is increasing on c (Lemma 23 of [5]);
            - there exists a simple component $\mathbf{y}$ with $\mathbf{y}(t) > 0$ that is either zero-unbounded (Lemma 2) or zero-bounded (Lemma 25 of [5]) on c.
            - there exists a simple component $\mathbf{y}$ of $\mathcal{A}$ that is increasing on c and $\mathcal{B}$ is reachable from $p_{\mathbf{y}}$ (Lemma 23 of [5]);
        * otherwise, $\mathcal{T}^{\exp}[t](n) \in \Theta(n)$ (Lemma 5).

Remark 1. We comment on the difference between the three distinct cases for $\mathcal{T}^{\mathrm{exp}}[t](n) = \infty$ (see Fig. 2). The VASS MDP on the left contains a single MEC containing a simple component $\mathbf{y}$ increasing on c, but $\mathbf{y}$ does not contain t. The VASS MDP in the middle contains a single MEC containing a single simple component which is zero-bounded on c. Whereas the VASS MDP on the right contains three MECs, one with a simple component increasing on c which allows to pump c arbitrarily high before moving to the MEC which contains t. Note that in the first two examples t can be iterated infinitely many times with probability 1 whereas in the last example t can only ever be iterated finitely many times.

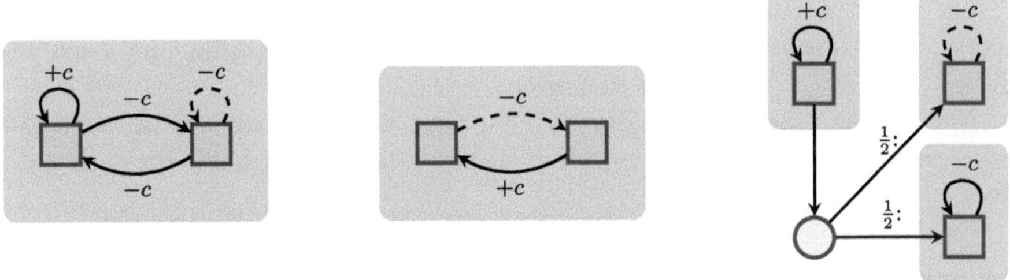

Fig. 2. The three possibilities for $\mathcal{T}^{\mathrm{exp}}[t](n) = \infty$. The dashed transition represents t. The grey areas are the MECs.

The polynomial time bound of Theorem 1 is then obtained the same way as the polynomial bound for the analysis of asymptotic estimates for 1-dim VASS MDPs from [4]. For $\mathcal{L}^{\mathrm{exp}}$ and $\mathcal{C}^{\mathrm{exp}}[c]$ it suffices to decide the existence of a simple component that is increasing, zero-bounded, or zero-unbounded on c. This can be done in polynomial time using linear programming [4]. For $\mathcal{T}^{\mathrm{exp}}[t]$ we first have to decide existence of a simple component increasing on c which can similarly be done in polynomial time using linear programming [4]. Then if no increasing simple component exists we need to decide existence of simple component containing t that is zero-bounded, or zero-unbounded on c, which can once again be done using linear programming in polynomial time [4].

Remark 2. Compared to the analysis of the asymptotic estimates in 1-dim VASS MDPs from [4], the main difference is that zero-unboundedness gives infinite expectation while having n^2 as a tight asymptotic estimate for each of $\mathcal{L}, \mathcal{C}[c]$, and $\mathcal{T}[t]$. In all other cases the expectation measure gives same bounds as the asymptotic estimates measure.

In the rest of this section we proceed with the individual lemmas referenced above.

Lemma 2. *If $\mathcal{A}$ contains a simple component $\mathbf{y}$ that is zero-unbounded on c, then $\mathcal{C}^{\exp}[c](n) = \infty$ and $\mathcal{L}^{\exp}(n) = \infty$ for all n. Additionally, for any t with $\mathbf{y}(t) > 0$ also $\mathcal{T}^{\exp}[t](n) = \infty$ for all n.*

Proof. Since $\mathbf{y}$ is zero-unbounded on c and is the only simple component of $\mathcal{A}_{\mathbf{y}}$, from Lemma 7 of [4] there exists $\mathbf{a} \geq 1$ and $\mathbf{b} \in \mathbb{Z}^{St(\mathcal{A}_{\mathbf{y}})}$ such that the ranking function $rank(\mathbf{pv}) = \mathbf{b}(p) + \mathbf{a}(c) \cdot \mathbf{v}(c)$ changes by 0 in expectation at every single computational step. I.e., formally, let $P_0 V_0, P_1 V_1, \ldots$ be such that P_i denotes the state and V_i the counter vector at the i-th computational step in $\mathcal{A}_{\mathbf{y}}$. Let $R_i = rank(P_i V_i)$. Then $R_0, R_1, \ldots$ is a supermartingale. Furthermore, we can wlog. assume that $R_0 = n$ and that $R_i > 0$ implies $V_i(c) > 0$ (see [4] for details).

Let τ be the stopping time such that $R_\tau \leq 0$ for the first time. Let $R_{max} = \max_{i=1}^{\tau} R_i$. Additionally, for each $x \in \mathbb{N}$ let $\tau[x]$ be a stopping time such that $R_{\tau[x]} \leq 0$ or $R_{\tau[x]} \geq x$ for the first time.

Notice also that there exists $a \in \mathbb{N}$, that depends only on $\mathcal{A}$, such that $|R_i - R_{i+1}| \leq a$. Hence from Optional stopping theorem it holds for each $x \in \mathbb{N}$ that

$$n = R_0 = R_{\tau[x]} = \sum_{i=0}^{a} -i \cdot \mathbb{P}^{\sigma_{\mathbf{y}}}_{p\overrightarrow{n}}[R_{\tau[x]} = -i] + \sum_{i=0}^{a} (x+i) \cdot \mathbb{P}^{\sigma_{\mathbf{y}}}_{p\overrightarrow{n}}[R_{\tau[x]} = x+i]$$

Thus it holds

$$n \geq -a \cdot \mathbb{P}^{\sigma_{\mathbf{y}}}_{p\overrightarrow{n}}[R_{\tau[x]} \leq 0] + x \cdot \mathbb{P}^{\sigma_{\mathbf{y}}}_{p\overrightarrow{n}}[R_{\tau[x]} \geq x]$$

$$n \geq -a + x \cdot \mathbb{P}^{\sigma_{\mathbf{y}}}_{p\overrightarrow{n}}[R_{\tau[x]} \geq x]$$

$$\frac{n+a}{x} \geq \mathbb{P}^{\sigma_{\mathbf{y}}}_{p\overrightarrow{n}}[R_{\tau[x]} \geq x]$$

and in the other direction it holds

$$n \leq 0 + (x+a) \cdot \mathbb{P}^{\sigma_{\mathbf{y}}}_{p\overrightarrow{n}}[R_{\tau[x]} \geq x]$$

$$\frac{n}{x+a} \leq \mathbb{P}^{\sigma_{\mathbf{y}}}_{p\overrightarrow{n}}[R_{\tau[x]} \geq x]$$

Hence we can write

$$\mathbb{E}^{\sigma_{\mathbf{y}}}_{p\overrightarrow{n}}[R_{max}] = \sum_{i=n}^{\infty} i \cdot \mathbb{P}^{\sigma_{\mathbf{y}}}_{p\overrightarrow{n}}[R_{\tau[i]} = i]$$

$$\geq \sum_{i=n}^{\infty} i \cdot \mathbb{P}^{\sigma_{\mathbf{y}}}_{p\overrightarrow{n}}[R_{\tau[i]} \geq i] \cdot (1 - \mathbb{P}^{\sigma_{\mathbf{y}}}_{p\overrightarrow{n}}[R_{\tau[i+1]} \geq i+1])$$

$$\geq \sum_{i=n}^{\infty} i \cdot \frac{n}{i+a} \cdot \left(1 - \frac{n+a}{i+1}\right) = \infty$$

As $\mathcal{C}^{\exp}_{\mathcal{A}_{\mathbf{y}}}[c] = R_{max}$ this immediately gives us $\mathcal{C}^{\exp}_{\mathcal{A}}[c](n) = \infty$ for all $n \geq 1$. Since $\mathcal{L} \geq \frac{\mathcal{C}[c]}{u}$ for some constant u that depends only on $\mathcal{A}$ this also implies $\mathcal{L}^{\exp}_{\mathcal{A}}(n) = \infty$. Similarly we obtain $\mathcal{T}^{\exp}_{\mathcal{A}}[t](n) = \infty$, for any transition t with $\mathbf{y}(t) > 0$. $\qquad\square$

Lemma 3. *If every simple component of $\mathcal{A}$ is either decreasing or zero-bounded on c then $\mathcal{C}^{\exp}[c](n) \in \mathcal{O}(n)$.*

Proof. From Lemma 31 of [5] it can be shown that any computation on $\mathcal{A}$ can be seen as an interweaving of at most $2^{|St(\mathcal{A})| \cdot |Trns(\mathcal{A})|}$ computations on $\mathcal{A}_\sigma$ for various $\sigma \in cMD(\mathcal{A})$. Each such $\mathcal{A}_\sigma$ reaches a MEC in constant expected time, and each MEC of $\mathcal{A}_\sigma$ can be seen as a simple component. As per Lemma 4 the computations which fall into MECs corresponding to zero-bounded simple components cannot ever change the counter by more than a constant in total. From Lemma 1 we then obtain that any strategy on $\mathcal{A}$ which does not iterate exclusively zero-bounded simple components has negative long-term average effect on the counter that depends only on $\mathcal{A}$. Hence $\mathcal{C}^{\exp}[c](n) \in \mathcal{O}(n)$. $\qquad\square$

We say a computation $\pi = p_0 \mathbf{v}_0, \ldots, p_k \mathbf{v}_k$ is a *cycle* if $p_0 = p_k$, and we define the *effect* of π as $\Delta(\pi) = \mathbf{v}_k - \mathbf{v}_0$.

Lemma 4. *A simple component $\mathbf{y}$ is zero-bounded on c iff every cycle π in $\mathcal{A}_{\mathbf{y}}$ satisfies $\Delta(\pi)(c) = 0$.*

The proof of Lemma 4 is purely technical and can be found in the full version of this paper [1].

Lemma 5. *Let $t = (p, \mathbf{u}, q) \in Trns(\mathcal{A})$ be such that every simple component $\mathbf{y}$ with $\mathbf{y}(t) > 0$ is decreasing on c. If $\mathcal{A}$ contains no simple component increasing on c then $\mathcal{T}^{\exp}[t](n) \in \mathcal{O}(n)$.*

Proof. Since $\mathcal{A}$ contains no simple component increasing on c, from Lemma 7 of [4] we get that there exists $0 < \mathbf{a} \in \mathbb{Q}$ and $\vec{0} \leq \mathbf{b} \in \mathbb{Q}^{St(\mathcal{A})}$ such that the ranking function $rank(q\mathbf{v}) = \mathbf{b}(q) + \mathbf{v}(c) \cdot \mathbf{a}$ is strictly non-increasing on average in $\mathcal{A}$, and furthermore, since every simple component $\mathbf{y}$ with $\mathbf{y}(t) > 0$ is decreasing on c it also holds that $rank$ is strictly decreasing on average from the state p by some constant that depends only on $\mathcal{A}$.

Let $R_0, R_1, \ldots$ be such that R_i denotes the value of the ranking function $rank$ after i computational steps (see proof of Lemma 2). As per the above $R_0, R_1, \ldots$ is a sub-martingale.

Let τ be a stopping time such that $R_\tau < 0$ for the first time, and let κ denote the number of times the state p was visited within the first τ steps. Notice that $R_\tau < 0$ implies the counter c is also negative. Hence $\mathcal{T}[t] \leq \kappa$.

From optional stopping theorem, for all σ, q we obtain

$$\mathbb{E}^{\sigma}_{q\vec{n}}[R_0] = \mathbb{E}^{\sigma}_{q\vec{n}}[R_\tau] + \sum_{i=1}^{\tau} \mathbb{E}^{\sigma}_{q\vec{n}}[R_{i-1} - R_i]$$

which gives us

$$\mathbf{b}(q) + n \cdot \mathbf{a} \geq \sum_{i=1}^{\tau} \mathbb{E}_{q\,\vec{n}}^{\sigma}[R_{i-1} - R_i]$$

But since $\sum_{i=1}^{\tau} \mathbb{E}_{q\,\vec{n}}^{\sigma}[R_{i-1} - R_i] \geq m \cdot \mathbb{E}_{q\,\vec{n}}^{\sigma}[\kappa]$, where m is a constant that depends only on $\mathcal{A}$, it also holds

$$\mathbf{b}(q) + n \cdot \mathbf{a} \geq m \cdot \mathbb{E}_{q\,\vec{n}}^{\sigma}[\kappa]$$

But this gives us that $\mathcal{T}^{\exp}[t] \leq \sup_{q,\sigma} \mathbb{E}_{q\,\vec{n}}^{\sigma}[\kappa] \in \mathcal{O}(n)$. $\qquad\square$

Lemma 6. *If* $t = (p, \mathbf{u}, q) \in Trns(\mathcal{A})$ *is not contained in any MEC of* $\mathcal{A}$ *then* $\mathcal{T}^{\exp}[t](n) \in \Theta(1)$.

Proof. If the current state is not in a MEC, then there exists some $a > 0$, that depends only on $\mathcal{A}$, such that the probability of the computation reaching some MEC within the next $|St(\mathcal{A})|$ steps is at least a. Thus $\mathbb{P}_{p\,\vec{n}}^{\sigma}\big[\mathcal{T}[t] > i\big] \leq (1 - a)^{\frac{i}{|St(\mathcal{A})|}}$ for any strategy σ and $p \in St(\mathcal{A})$. Hence $\mathcal{T}^{\exp}[t](n) \leq \sum_{i=1}^{\infty} i \cdot (1 - a)^{\frac{i-1}{|St(\mathcal{A})|}} \in \mathcal{O}(1)$. $\qquad\square$

4   PSPACE-hardness

In this section we prove the following theorem.

Theorem 2. *Let* $f, g : \mathbb{R} \to \mathbb{R}$ *be such that* $f \in \Omega(n)$ *and* $g \in \omega(n)$. *Let* $\mathcal{F}$ *be either of* $\mathcal{L}, \mathcal{C}[c]$, *or* $\mathcal{T}^{\exp}[t]$. *Then, for a given VASS MDP* $\mathcal{A}$, *the following problems are* **PSPACE**-*hard:*

- *deciding whether there exists* $n \in \mathbb{N}$ *such that* $\mathcal{F}_{\mathcal{A}}^{\exp}(n) = \infty$;
- *deciding whether* $\mathcal{F}_{\mathcal{A}}^{\exp} \in \mathcal{O}\big(f(n)\big)$;
- *deciding whether* $\mathcal{F}_{\mathcal{A}}^{\exp} \in \Omega\big(g(n)\big)$.

Additionally, for a given strongly connected VASS MDP $\mathcal{A}'$, *the following problems are* **PSPACE**-*hard:*

- *deciding whether there exists* $n \in \mathbb{N}$ *such that* $\mathcal{F}_{\mathcal{A}'}^{\exp}(n) = \infty$;
- *deciding whether* $\mathcal{F}_{\mathcal{A}'}^{\exp} \in 2^{\mathcal{O}\big(f(n)\big)}$;
- *deciding whether* $\mathcal{F}_{\mathcal{A}'}^{\exp} \in 2^{\Omega\big(g(n)\big)}$.

Furthermore, let $\mathcal{B}$ *be a (strongly connected) VASS. Then deciding whether a given (strongly connected) VASS MDP* $\mathcal{A}$ *has* $\mathcal{F}_{\mathcal{A}}^{\exp} \in \Theta\big(\mathcal{L}_{\mathcal{B}}^{\max}(n)\big)$ *(*$\mathcal{F}_{\mathcal{A}}^{\exp} \in 2^{\Theta\big(\mathcal{L}_{\mathcal{B}}^{\max}(n)\big)}$*) is* **PSPACE**-*hard.*

We divide the proof of Theorem 2 into three parts. We begin by proving the first part for general VASS MDPs in Sect. 4.1. Then we extend this proof also to strongly connected VASS MDPs in Sect. 4.2. Finally, in Sect. 4.3 we show how we can effectively initialize each counter in a given VASS MDP to $\mathcal{L}_{\mathcal{B}}^{\max}(n)$ instead of n, for a given VASS $\mathcal{B}$. This then gives us the last part of Theorem 2 as per Lemmas 7 and 11.

One important implication of Theorem 2 is that deciding whether $\mathcal{L}^{\exp}$ is finite for a given VASS MDP is **PSPACE**-hard already in the case of s.c. VASS MDPs. This suggests that analyzing the expected complexity of VASS MDPs is indeed computationally hard. Note that to the best of our knowledge this problem is not yet even known to be decidable.

Another implication is that the results of [8], which show that given a DAG-like VASS MDP it is decidable in **P** whether $\mathcal{L}^{\exp} \in \Theta(n)$ or $\mathcal{L}^{\exp} \in \Omega(n^2)$ cannot be extended onto general VASS MDPs without a blowup in complexity (unless **P** = **PSPACE**). Note that it still may be decidable in **P** whether $\mathcal{L}^{\exp}$ for a given DAG-like VASS MDP is polynomial.

Finally, since the expected termination and c-counter complexities are both trivially in $\Omega(n)$ this shows that any non-trivial question regarding asymptotic lower/upper bounds for these for general VASS MDPs is **PSPACE**-hard.

Remark 3. We comment on the possible values for $\mathcal{L}_{\mathcal{B}}^{max}(n)$. From [3] we have that either $\mathcal{L}_{\mathcal{B}}^{max} \in \Theta(n^k)$ where $k \in \mathbb{N}$ or $\mathcal{L}_{\mathcal{B}}^{max} \in 2^{\Omega(n)}$. Furthermore, for each $k \in \mathbb{N}$ there exists a s.c. VASS with $\mathcal{L}_{\mathcal{B}}^{max} \in \Theta(n^k)$ [23]. In the case that $\mathcal{L}_{\mathcal{B}}^{max} \in 2^{\Omega(n)}$, it is shown in [14] that for each $k \in \mathbb{N}$ either $\mathcal{L}_{\mathcal{B}}^{max}$ lies in the k-th level of the Grzegorczyk hierarchy, or it grows at least as fast as the generator for the $(k+1)$-st level of the Grzegorczyk hierarchy.

Also note that while [2] shows there exists a VASS MDP with $\mathcal{C}^{\exp}[c] \in \Theta(n \cdot \log n)$ it is not possible to obtain this asymptotic bound in VASS for either of $\mathcal{L}^{max}, \mathcal{C}^{max}[c], \mathcal{T}^{max}[t]$ [3,23], nor in VASS MDP for $\mathcal{L}^{\exp}$ [8].

4.1 General VASS MDP

We obtain the **PSPACE**-hardness by reduction from the standard **PSPACE**-complete problem of deciding whether a given QBF formula is valid or not. The first part of Theorem 2 follows directly form the following lemma.

Lemma 7. *For every quantified Boolean formula ϕ in prenex 3-CNF there exists a VASS MDP $\mathcal{A}$, with counter c, constructible in time polynomial in $\|\phi\|$ such that*

- *if ϕ is valid, then $\mathcal{L}^{\exp}(n) = \infty$ and $\mathcal{C}^{\exp}[c](n) = \infty$ for all $n \in \mathbb{N}$;*
- *if ϕ is not valid, then $\mathcal{L}^{\exp} \in \Theta(n)$ and $\mathcal{C}^{\exp}[c] \in \Theta(n)$.*

Remember that a QBF formula in prenex 3-CNF form can be written as

$$\phi \quad \equiv \quad \forall x_1 \exists x_2 \forall x_3 \ldots \exists x_l \; C_1 \wedge \ldots \wedge C_m$$

where each clause $C_i = \ell_i^1 \vee \ell_i^2 \vee \ell_i^3$ contains precisely three literals.

To describe the VASS MDP we utilize VASS programs (introduced in [3]) utilizing the gadgets of Fig. 3.

One can easily verify that all gadgets implement the operations associated to their labels up to some "irrelevant side effects". More precisely,

- the $\{x_1, \ldots, x_i\} \leftarrow k \cdot y$ gadget ensures that the Demon can increase the value of every counter from $\{x_1, \ldots, x_i\}$ by $k \cdot val(y)$ (but not more) if he plays optimally, where $val(y)$ is the value of the counter y when initiating the gadget (while also decreasing the value of y).
- the **Return** c gadget is simply a loop which decreases c and eventually leads to termination.
- the **P-choose:** ins_1; **or** ins_2 represents a probabilistic choice with equal probabilities of choosing either ins_1 or ins_2.
- the **Repeat** ins; **WITH probability** $\frac{1}{k}$ represents a probabilistic loop where every time ins is exited it is repeated with probability $\frac{1}{k}$.
- the $z \leftarrow \min(x, y)$ gadget allows the Demon to increase z by the minimum of $val(x)$ and $val(y)$, respectively (but not more).
- the $ins_1; ins_2$ is a concise representation of the corresponding sequence of instructions. We use **foreach** $i = 1, \ldots, a$ **do** ins_i **end** where a is a constant as a shorthand for $ins_1; \ldots; ins_a$.
- the **foreach** $i = 1 \ldots, n$ **do** ins **end** gadget represents a loop repeating ins for at most $val(\alpha) + 1$ times. Note that this is a different gadget from the similar shorthand notation from the previous bullet point (as this one is parameterized by n instead of a constant). α is a special counter that is used Sect. 4.2.
- the **D-choose:** ins_1; **or** $\cdots$ **or** ins_j represents a non-deterministic choice of one of $ins_1, \ldots, ins_j$.

We show the VASS MDP $\mathcal{A}$ defined by the VASS Program 1 satisfies Lemma 7. $\mathcal{A}$ is parameterized by a constant k that we define as $k = 2^{(l+5)\cdot(m+5)}$.[2] Note that the value of k depends only on ϕ and is computable in time polynomial in $\|\phi\|$. Also note that on the lines 10 and 12 the counters ℓ_i^j are actually the counters x_r^i (or $\bar{x}_r^i$) such that $\ell_i^j \equiv x_r$ (or $\ell_i^j \equiv \neg x_r$).

Intuitively, $\mathcal{A}$ consists of a probabilistic loop, which is repeated with probability $\frac{1}{k}$. Each loop encodes some valuation of the variable $x_1, \ldots, x_l$ where the value $true$ is encoded by adding $k \cdot c_i$ to the corresponding counters. Then the VASS MDP evaluates $C_1 \wedge \cdots \wedge C_m$ for this valuation, encoding the result in s_m before adding this value to the counters $c_1, \ldots, c_l$. Hence when, and only when, $C_1 \wedge \cdots \wedge C_m$ is true in a given loop the c counters are effectively multiplied by k. Since ϕ being valid ensures $C_1 \wedge \cdots \wedge C_m$ can be true in every iteration this leads to infinite expectation, whereas if ϕ is not valid then there is a positive probability of a given loop evaluating as false and thus the value of counters c_i grows slower enough to make the expectation linear in n.

---

[2] Note that the exact value of k does not matter much. We could have chosen arbitrary value which is sufficiently large and computable in polynomial time.

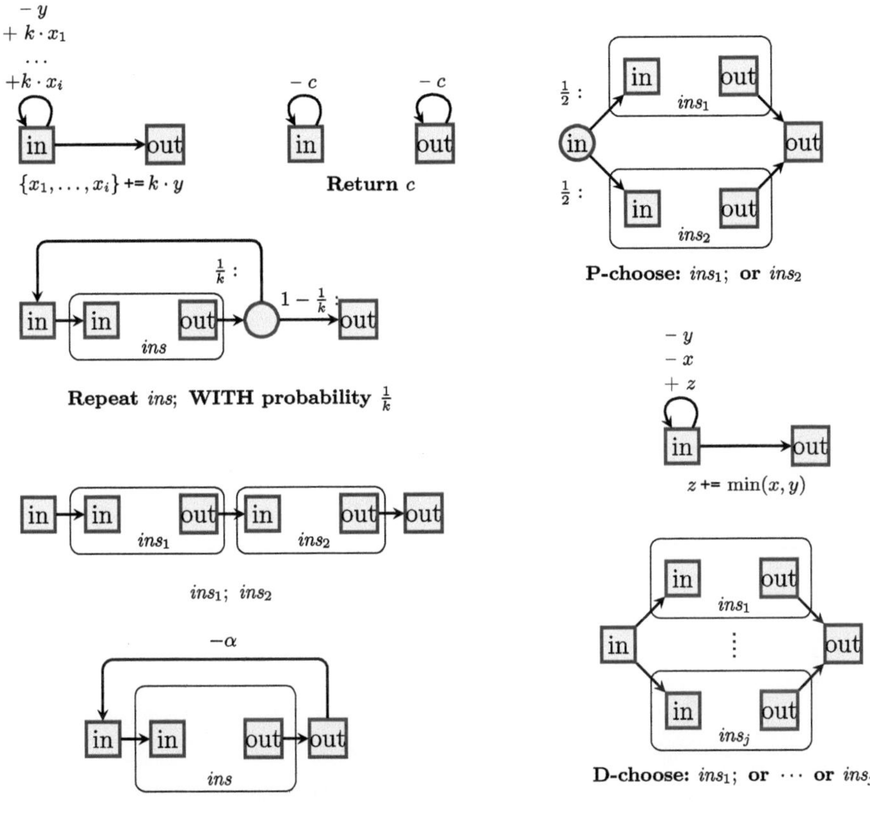

Fig. 3. The gadgets used in VASS programs.

For each counter χ let $\chi(i)$ denote the value of χ upon the i-th visit of the probabilistic state of the *REPEAT* gadget (i.e., line 15 of the program). We start with the case when ϕ is valid.

Lemma 8. *If ϕ is valid then $\mathcal{L}_{\mathcal{A}}^{\exp}(n) = \infty$ and $\mathcal{C}_{\mathcal{A}}^{\exp}[c_1](n) = \infty$ for all $n \in \mathbb{N}$.*

Proof. Notice that if the computation starts on line 2 and ϕ is satisfiable, then Demon can resolve the non-deterministic choices such that $c_i(j) \geq k \cdot c_i(j-1)$ with probability 1. This is because the demon can resolve the choices on lines 2–9 such that the variables x_i and $\bar{x}_i$ that have $k \cdot c_i$ added to them encode a true evaluation (where these are set to true), and thus the Demon can reach line 14 with each of $s_0, s_1, \ldots, s_m$ having $k \cdot \min_{i=1}^{l} c_i(j-1)$ added to them as well. Since each c_i has the value of s_m added to it on line 14 it holds $c_i(j) \geq k \cdot c_i(j-1)$.

VASS Program 1: the VASS MDP $\mathcal{A}$

```
 1  Repeat
 2  │  foreach i = 1, ..., l do
 3  │  │  if x_i is universal then
 4  │  │  │  P-choose:  {x_i^1, ..., x_i^m} ← k · c_i or {x̄_i^1, ..., x̄_i^m} ← k · c_i
 5  │  │  end
 6  │  │  else
 7  │  │  │  D-choose:  {x_i^1, ..., x_i^m} ← k · c_i or {x̄_i^1, ..., x̄_i^m} ← k · c_i
 8  │  │  end
 9  │  end
10  │  D-choose:  {s_1} ← 1 · ℓ_1^1 or {s_1} ← 1 · ℓ_1^2 or {s_1} ← 1 · ℓ_1^3
11  │  foreach i = 2, ..., m do
12  │  │  D-choose:  s_i ← min(ℓ_i^1, s_{i-1}) or s_i ← min(ℓ_i^2, s_{i-1}) or s_i ← min(ℓ_i^3, s_{i-1})
13  │  end
14  │  {c_1, ..., c_l} ← 1 · s_m
15  WITH probability 1/k
16  return c_1
```

And since $c_i(0) = n$ this gives us $c_i(j) \geq k^j \cdot n$. But this gives us

$$\mathcal{C}_{\mathcal{A}}^{\exp}[c_1](n) \geq \sum_{i=1}^{\infty} \mathbb{P}_{p\overrightarrow{n}}^{\sigma}[k^{i+1} \cdot n > \mathcal{C}[c_1] \geq k^i \cdot n] \cdot k^i \cdot n$$

$$\geq \sum_{i=1}^{\infty} \left(\frac{1}{k^i} - \frac{1}{k^{i+1}}\right) \cdot k^i \cdot n = n \cdot \sum_{i=1}^{\infty}\left(1 - \frac{1}{k}\right) = \infty$$

Since $\mathcal{L}_{\mathcal{A}} \geq \frac{\mathcal{C}_{\mathcal{A}}[c_1]}{k}$ this also implies $\mathcal{L}_{\mathcal{A}}^{\exp}(n) = \infty$. □

Now we consider the case when ϕ is not valid.

Lemma 9. *If ϕ is not valid then $\mathcal{L}_{\mathcal{A}}^{\exp} \in \Theta(n)$ and $\mathcal{C}_{\mathcal{A}}^{\exp}[c_1] \in \Theta(n)$.*

$\mathcal{L}_{\mathcal{A}}^{\exp} \in \Omega(n)$ and $\mathcal{C}_{\mathcal{A}}^{\exp}[c_1] \in \Omega(n)$ holds trivially. Furthermore, since $\mathcal{C}_{\mathcal{A}}[c] \leq k \cdot \mathcal{L}_{\mathcal{A}}$ it suffices to show that $\mathcal{L}_{\mathcal{A}}^{\exp} \in \mathcal{O}(n)$.

We say the τ-th loop of the *REPEAT* gadget assigns the value *true* to the variable x_i if on lines 4 or 7 the computation visited the gadget increasing $x_i^1, ..., x_i^m$. Notice that if x_i is not assigned *true* then the computation visits the gadget increasing $\bar{x}_i^1, ..., \bar{x}_i^m$, thus in such case we can say $\neg x_i$ is assigned to be true. We say the τ-th loop resolves ϕ as true if the assignment of truth values to $x_1, ..., x_l$ during this loop satisfies $C_1 \wedge \cdots \wedge C_m$. Otherwise the τ-th loop resolves ϕ as false.

We assign each loop to be either type I or II as follows:

- 1-st loop is type I;
- if $(\tau - 1)$-st loop is type I and τ-th loop resolves ϕ as true then τ-th loop is type I;

- if $(\tau - 1)$-st loop is type I and τ-th loop resolves ϕ as false then τ-th loop is type II;
- if $(\tau - 1)$-st loop is type II then τ-th loop is type I.

Let $I(\tau)$ and $II(\tau)$ denote the number of loops of type I and II within the first τ loops, respectively. We have the following regarding the behavior of $\mathcal{A}$.

Lemma 10. *Let $1 \leq \tau$. If the τ-th loop is type I then the following holds:*

- $\kappa(\tau) \leq (k + 3)^{I(\tau)} \cdot (m + 4)^{II(\tau)} \cdot n$ *for all* $\kappa \in Count(\mathcal{A})$.

and if the τ-th loop is type II then the following holds:

- $\kappa(\tau) \leq (k + 3)^{I(\tau)+1} \cdot (m + 4)^{II(\tau)-1} \cdot n$ *for all* $\kappa \in Count(\mathcal{A}) \setminus \{c_1, \ldots, c_l\}$;
- $c_i(\tau) \leq (k + 3)^{I(\tau)} \cdot (m + 4)^{II(\tau)-1} \cdot (m + 2) \cdot n$ *for all* $1 \leq i \leq l$.

The proof of Lemma 10 is obtained using an induction over τ and a careful analysis of the behavior of $\mathcal{A}$. A full proof can be found in the full version of this paper [1].

Let p be the maximal probability with which Demon can enforce that a given loop evaluates as true. Since ϕ is not valid it holds $p \leq 1 - 2^{-l}$.

Notice also that no transition can be fired in the τ-th loop more than the size of the largest counter during this loop. Thus using Lemma 10 we can upper bound $\mathcal{L}_{\mathcal{A}}^{\exp}$ as (here i represents the number of type I loops that do not happen right after a type II loop, j is the number of type II loops, and r is the number of type I loops right after a type II loop):

$$\mathcal{L}_{\mathcal{A}}^{\exp}(n) \leq |Trns(\mathcal{A})| \cdot \max_{c \in Count(\mathcal{A})} \mathcal{C}_{\mathcal{A}}^{\exp}[c](n)$$

$$\leq |Trns(\mathcal{A})| \cdot \sum_{i=1}^{\infty} \sum_{j=0}^{\infty} \sum_{r=j-1}^{j} \frac{1 - \frac{1}{k}}{k^{i+j+r-1}} \cdot p^{i-1} \cdot (1 - p)^j \cdot (k + 3)^{i+r+1} \cdot (m + 4)^j \cdot$$

$$\cdot n \cdot \binom{ni + j - 1}{j} \in \mathcal{O}(n)$$

where the last inclusion can be obtained by evaluating the sum. It's evaluation can be found in the full version of this paper [1]. $\qquad\square$

This finishes the proof of Lemma 9.

4.2   Strongly Connected VASS MDP

We now extend the proof from the previous section to strongly connected VASS MDPs to prove also the part of Theorem 2 about s.c. VASS MDPs. It follows directly from the following lemma.

Lemma 11. *For every quantified Boolean formula ϕ in prenex 3-CNF there exists a strongly connected VASS MDP $\mathcal{A}'$, with counter c, constructible in time polynomial in $\|\phi\|$ such that*

VASS Program 2: the strongly connected VASS MDP $\mathcal{A}'$

---

1 **foreach** $i = 1, \ldots, n$ **do**
2 | **Repeat**
3 | | **foreach** $i = 1, \ldots, l$ **do**
4 | | | **if** x_i *is universal* **then**
5 | | | | P-**choose:** $\{x_i^1, \ldots, x_i^m\} \leftarrow k \cdot c_i$ **or** $\{\bar{x}_i^1, \ldots, \bar{x}_i^m\} \leftarrow k \cdot c_i$
6 | | | **end**
7 | | | **else**
8 | | | | D-**choose:** $\{x_i^1, \ldots, x_i^m\} \leftarrow k \cdot c_i$ **or** $\{\bar{x}_i^1, \ldots, \bar{x}_i^m\} \leftarrow k \cdot c_i$
9 | | | **end**
10 | | **end**
11 | | D-**choose:** $\{s_1\} \leftarrow 1 \cdot \ell_1^1$ **or** $\{s_1\} \leftarrow 1 \cdot \ell_1^2$ **or** $\{s_1\} \leftarrow 1 \cdot \ell_1^3$
12 | | **foreach** $i = 2, \ldots, m$ **do**
13 | | | D-**choose:** $s_i \leftarrow \min(\ell_i^1, s_{i-1})$ **or** $s_i \leftarrow \min(\ell_i^2, s_{i-1})$ **or** $s_i \leftarrow \min(\ell_i^3, s_{i-1})$
14 | | **end**
15 | | $\{c_1, \ldots, c_l\} \leftarrow 1 \cdot s_m$
16 | **WITH** probability $\frac{1}{k}$
17 **end**

- if ϕ is valid, then $\mathcal{L}_{\mathcal{A}'}^{\exp}(n) = \infty$ and $\mathcal{C}_{\mathcal{A}'}^{\exp}[c](n) = \infty$ for all $n \in \mathbb{N}$;
- if ϕ is not valid, then $\mathcal{L}_{\mathcal{A}'}^{\exp} \in 2^{\Theta(n)}$ and $\mathcal{C}_{\mathcal{A}'}^{\exp}[c] \in 2^{\Theta(n)}$.

We modify the VASS MDP from the VASS program 1 from the previous section to be strongly connected as depicted in the VASS program 2. Note that the only difference from the VASS program 1 is that once the REPEAT gadget is exited it is repeated again after decreasing a new counter α by 1. Hence the computation can exit the REPEAT gadget exactly $n+1$ times before terminating (as opposed to just once in the VASS program 1).

If ϕ is valid then the expectation remains infinite from Lemma 8 as then ∞ is achieved already before exiting the REPEAT gadget for the first time. Hence we only have to consider the case when ϕ is not valid.

We prove the upper bounds of Lemma 11 in the following lemma, and then afterwards in Lemma 13 we prove the matching lower bounds.

Lemma 12. *If ϕ is not valid, then $\mathcal{L}_{\mathcal{A}'}^{\exp} \in 2^{\mathcal{O}(n)}$ and $\mathcal{C}_{\mathcal{A}'}^{\exp}[c_1] \in 2^{\mathcal{O}(n)}$.*

Proof. Notice that the upper bounds from Lemma 10 are still valid in $\mathcal{A}'$. Since at least $n + 1$ loops are guaranteed it hence holds

$$\mathcal{L}_{\mathcal{A}'}^{\exp}(n) \leq |\mathit{Trns}(\mathcal{A}')| \cdot \max_{c \in \mathit{Count}(\mathcal{A}') \setminus \{\alpha\}} \mathcal{C}_{\mathcal{A}'}^{\exp}[c]$$

$$\leq |\mathit{Trns}(\mathcal{A}')| \cdot \sum_{i=1}^{\infty} \sum_{j=\max(0, \lceil \frac{n+1-i}{2} \rceil)}^{\infty} \sum_{r=j-1}^{j} \frac{(1 - \frac{1}{k})^{n+1}}{k^{i+j+r-n-1}} \cdot p^{i-1} \cdot (1-p)^j \cdot (k+3)^{i+r+1} \cdot$$

$$\cdot (m+4)^j \cdot n \cdot \binom{i+j-1}{j} \in 2^{\mathcal{O}(n)}$$

where the last inclusion can be obtained by evaluating the sum. It's evaluation can be found in the full version of this paper [1]. $\square$

In the next lemma we show the matching lower bound. Note this lower bound does not depend on whether ϕ is valid or not.

Lemma 13. *It holds* $\mathcal{L}_{\mathcal{A}'}^{\exp} \in 2^{\Omega(n)}$ *and* $\mathcal{C}_{\mathcal{A}'}^{\exp}[c_1] \in 2^{\Omega(n)}$.

Proof. We describe a strategy σ which over multiple consecutive loops tries to set all $x_i^j, \bar{x}_i^j$ to true simultaneously, and then uses these to multiply $c_1, \ldots, c_l$ by $\frac{k}{2}$. This is done as follows.

Let τ be the last loop in which $c_1, \ldots, c_l$ have been multiplied by $\frac{k}{2}$ (or $\tau = 0$ if this has not happened yet). Starting from the $(\tau + 1)$-st loop σ tries to visit every single gadget pumping counters x_i^j as well as every gadget pumping counters $\bar{x}_i^j$. When such gadget is visited the first time since the τ-th loop σ uses it to increase the corresponding counters by $\frac{k}{2} \cdot c_i(\tau)$. Hence after visiting every such gadget, all the $x_i^j, \bar{x}_i^j$ counters have value at least $\frac{k}{2} \cdot c_i(\tau)$, and thus on the loop the last gadget gets visited σ can pump $s_1, \ldots, s_m$ to at least $\frac{k}{2} \cdot c_i(\tau)$ which then allows σ to pump $c_1, \ldots, c_l$ to at least $\frac{k}{2} \cdot c_i(\tau)$ as well. This iteration hence becomes the new τ and this procedure is repeated until termination.

We proceed to show that $\mathbb{E}_{p\vec{n}}^{\sigma}[\mathcal{C}[c_1]] \in 2^{\Omega(n)}$ where p is the *in* state of the REPEAT gadget. It is easy to see that under σ it holds $c_1(i) \geq (\frac{k}{2})^{f(i)} \cdot n$ where $f(i)$ is the number of times $c_1, \ldots, c_l$ have been multiplied by $\frac{k}{2}$ within the first i loops.

For simplicity, let us assume that σ only checks whether all the gadgets were visited on loops whose number is divisible by $l+1$. Then there are at least $l+1$ loops between each check. Both gadgets for existential variables are guaranteed to be visited in the first two loops, whereas the gadgets corresponding to universal variables have one of the two gadgets visited on the first loop, and then probability of $\frac{1}{2}$ of the other gadget being visited every loop. Hence probability of at least one of the gadgets not being visited within $l+1$ loops is upper bounded by $l \cdot 2^{-l}$.

Since the number of loops is guaranteed to be at least $n+1$ it holds (here $i = f(n+1)$):

$$\mathbb{E}_{p\vec{n}}^{\sigma}[\mathcal{C}[c_1]] \geq \sum_{i=0}^{\lfloor \frac{n+1}{l+1} \rfloor} (\frac{k}{2})^i \cdot n \cdot (1 - \frac{l}{2^l})^i \cdot (\frac{l}{2^l})^{\lfloor \frac{n+1}{l+1} \rfloor - i} \cdot \binom{\lfloor \frac{n+1}{l+1} \rfloor}{i}$$

$$\geq \sum_{i=\lfloor \frac{n+1}{l+1} \rfloor}^{\lfloor \frac{n+1}{l+1} \rfloor} (\frac{k}{2})^i \cdot n \cdot (1 - \frac{l}{2^l})^i \cdot (\frac{l}{2^l})^{\lfloor \frac{n+1}{l+1} \rfloor - i} \cdot \binom{\lfloor \frac{n+1}{l+1} \rfloor}{i}$$

$$= n \cdot (\frac{k}{2} \cdot (1 - \frac{l}{2^l}))^{\lfloor \frac{n+1}{l+1} \rfloor} \geq n \cdot (\frac{k}{4})^{\lfloor \frac{n+1}{l+1} \rfloor} \in 2^{\Omega(n)}$$

Where the second to last inequality follows from $\frac{l}{2^l} \leq \frac{1}{2}$. $\square$

4.3   Initializing Counters to $\mathcal{L}^{\max}(n)$

In this section we show how to modify a given VASS MDP $\mathcal{A}$ so as to effectively have each counter initialized to $\mathcal{L}_{\mathcal{B}}^{\max}(n)$ instead of n, for a given VASS $\mathcal{B}$. We do this in the following lemma.

Lemma 14. *Let $\mathcal{B}$ be a (strongly connected) VASS and $\mathcal{A}$ a (strongly connected) VASS MDP. Let $\mathcal{F}$ be one of $\mathcal{L}, \mathcal{C}[c]$, or $\mathcal{T}[t]$. Then there exists a (strongly connected) VASS MDP $\mathcal{C}$ with $\mathcal{F}_{\mathcal{C}}^{\exp}(n) \in \Theta\left(\mathcal{F}_{\mathcal{A}}^{\exp}\left(\mathcal{L}_{\mathcal{B}}^{\max}(n)\right)\right)$.*

Proof. We can obtain $\mathcal{C}$ by constructing a VASS MDP which effectively simulates computing on $\mathcal{A}$ but where each counter is initialized to $\mathcal{F}_{\mathcal{B}}^{\max}(n) + n$ instead of n. This is achieved by adding a new gadget to $\mathcal{A}$ which first adds $\mathcal{L}_{\mathcal{B}}^{max}(n)$ to each counter before starting computing in $\mathcal{A}$.

We can wlog. assume that $\mathcal{A}$ and $\mathcal{B}$ share no counters, states, nor transitions. Let $\mathcal{B}'$ be created from $\mathcal{B}$ by setting $Count(\mathcal{B}') = Count(\mathcal{B}) \cup Count(\mathcal{A})$ and modifying every transition to increase every counter of $\mathcal{A}$ by 1. Then the VASS MDP $\mathcal{C}$ is created as a union of $\mathcal{A}$ and $\mathcal{B}'$ while also adding new transitions allowing to move between the states of $\mathcal{B}'$ and $\mathcal{A}$ (and modifying the transitions of $\mathcal{A}$ to have effect 0 on the counters from $\mathcal{B}$).

If either of $\mathcal{A}$ or $\mathcal{B}$ is not strongly connected then, for each $p \in St(\mathcal{B}')$ and $q \in St(\mathcal{A})$, we add a transition $(p, \overrightarrow{0}, q)$ to $\mathcal{C}$. It is easy to see such resulting VASS MDP $\mathcal{C}$ results in all the counters being effectively initialized to $n + \mathcal{L}_{\mathcal{B}}^{max}(n)$ instead of n in $\mathcal{A}$.

If both $\mathcal{A}$ and $\mathcal{B}$ are strongly connected, we can make $\mathcal{C}$ strongly connected as well, but we have to be careful so that the newly added transitions do not create a useful way for the computation to move from one state to another. To do this, notice that since $\mathcal{A}$ has a finite number of states, there exists $q \in St(\mathcal{A})$ such that

$$\sup\{\mathbb{E}_{q\overrightarrow{m}}^{\sigma}[\mathcal{F}] \mid \sigma \text{ is a strategy of } \mathcal{A}\} = \sup\{\mathbb{E}_{p\overrightarrow{m}}^{\sigma}[\mathcal{F}] \mid \sigma \text{ is a strategy of } \mathcal{A}, p \in St(\mathcal{A})\}$$

for infinitely many n, where $m = \mathcal{L}_{\mathcal{B}}^{\max}(n)$.

Hence it suffices to assume that q is the initial state of $\mathcal{A}$. In this case we add to $\mathcal{C}$, for each $p \in St(\mathcal{B})$, the transitions $(p, \overrightarrow{0}, q)$ and $(q, -\overrightarrow{a}, p)$, where a is a sufficiently large constant such that for every computation π in $\mathcal{B}$, with $len(\pi) \leq |St(\mathcal{B})|$ it holds $\Delta(\pi) \geq -\overrightarrow{a}$. It is easy to see that such VASS MDP $\mathcal{C}$ is strongly connected, and it is always sub-optimal for Demon to ever use any of the transitions $(q, -\overrightarrow{a}, p)$.

Hence in both cases it holds $\mathcal{F}_{\mathcal{C}}^{\exp}(n) \in \Theta\left(\mathcal{F}_{\mathcal{A}}^{\exp}\left(\mathcal{L}_{\mathcal{B}}^{\max}(n)\right)\right)$. $\qquad\square$

5   Conclusions

We have shown **PSPACE**-hardness for multiple important problems regarding the asymptotics for expected termination, c-counter, and t-transition complexities in both general and strongly connected VASS MDPs. We have also provided full classification of these expected complexities in 1-dimensional VASS MDPs.

A natural continuation of our work would be to try to classify expected complexity also in VASS MDPs of higher dimension. Alternatively, answering whether deciding the polynomiality of expected complexity in DAG-like VASS MDPs is decidable in $\mathbf{P}$.

Acknowledgments. This work is supported by the EPSRC project EP/X042596/1.

Disclosure of Interests. The authors have no competing interests to declare that are relevant to the content of this article.

References

1. Ajdarów, M.: Asymptotic analysis of expected complexity in VASS MDPs (2026). arXiv
2. Ajdarów, M., Kučera, A., Novotný, P.: Asymptotic analysis of probabilistic programs: when expectations do not meet our expectations, pp. 85–97. Springer Nature Switzerland, Cham (2025). https://doi.org/10.1007/978-3-031-75783-9_4
3. Ajdarów, M., Kučera, A.: Deciding polynomial termination complexity for VASS programs. In: Haddad, S., Varacca, D. (eds.) 32nd International Conference on Concurrency Theory (CONCUR 2021). Leibniz International Proceedings in Informatics (LIPIcs), vol. 203, pp. 30:1–30:15. Schloss Dagstuhl - Leibniz-Zentrum für Informatik, Dagstuhl, Germany (2021). https://doi.org/10.4230/LIPIcs.CONCUR.2021.30
4. Ajdarów, M., Kučera, A.: Asymptotic complexity estimates for probabilistic programs and their VASS abstractions. In: Pérez, G.A., Raskin, J.F. (eds.) 34th International Conference on Concurrency Theory (CONCUR 2023). Leibniz International Proceedings in Informatics (LIPIcs), vol. 279, pp. 12:1–12:16. Schloss Dagstuhl - Leibniz-Zentrum für Informatik, Dagstuhl, Germany (2023). https://doi.org/10.4230/LIPIcs.CONCUR.2023.12
5. Ajdarów, M., Kučera, A.: Asymptotic complexity estimates for probabilistic programs and their VASS abstractions (2023). https://arxiv.org/abs/2307.04707
6. Baier, C., Katoen, J.: Principles of Model Checking. MIT Press, United States (2008)
7. Brázdil, T., et al.: Efficient algorithms for asymptotic bounds on termination time in VASS. In: Proceedings of the 33rd Annual ACM/IEEE Symposium on Logic in Computer Science (2018). https://api.semanticscholar.org/CorpusID:13742070
8. Brázdil, T., Chatterjee, K., Kučera, A., Novotný, P., Velan, D.: Deciding fast termination for probabilistic VASS with nondeterminism. In: Chen, Y.F., Cheng, C.H., Esparza, J. (eds.) Automated Technology for Verification and Analysis, pp. 462–478. Springer International Publishing, Cham (2019)
9. Broy, M., Wirsing, M.: On the algebraic specification of nondeterministic programming languages. In: Astesiano, E., Böhm, C. (eds.) CAAP 1981. LNCS, vol. 112, pp. 162–179. Springer, Heidelberg (1981). https://doi.org/10.1007/3-540-10828-9_61
10. Czerwiński, W., Lasota, S., Lazić, R., Leroux, J., Mazowiecki, F.: The reachability problem for petri nets is not elementary. In: Proceedings of the 51st Annual ACM SIGACT Symposium on Theory of Computing, pp. 24–33. STOC 2019, Association for Computing Machinery, New York, NY, USA (2019). https://doi.org/10.1145/3313276.3316369

11. Hopcroft, J., Pansiot, J.J.: On the reachability problem for 5-dimensional vector addition systems. Theoret. Comput. Sci. **8**(2), 135–159 (1979). https://doi.org/10.1016/0304-3975(79)90041-0
12. Jančar, P.: Undecidability of bisimilarity for petri nets and some related problems. Theoret. Comput. Sci. **148**(2), 281–301 (1995). https://doi.org/10.1016/0304-3975(95)00037-W
13. Kučera, A.: Algorithmic analysis of termination and counter complexity in vector addition systems with states: a survey of recent results. ACM SIGLOG News **8**(4), 4–21 (2022). https://doi.org/10.1145/3527372.3527374
14. Kučera, A., Leroux, J., Velan, D.: Efficient analysis of VASS termination complexity. In: Proceedings of the 35th Annual ACM/IEEE Symposium on Logic in Computer Science, pp. 676–688. LICS '20, Association for Computing Machinery, New York, NY, USA (2020). https://doi.org/10.1145/3373718.3394751
15. Leroux, J., Schnoebelen, P.: On functions weakly computable by petri nets and vector addition systems. In: Ouaknine, J., Potapov, I., Worrell, J. (eds.) Reachability Problems, pp. 190–202. Springer International Publishing, Cham (2014)
16. Leroux, J.: Polynomial vector addition systems with states. In: Chatzigiannakis, I., Kaklamanis, C., Marx, D., Sannella, D. (eds.) 45th International Colloquium on Automata, Languages, and Programming (ICALP 2018). Leibniz International Proceedings in Informatics (LIPIcs), vol. 107, pp. 134:1–134:13. Schloss Dagstuhl - Leibniz-Zentrum für Informatik, Dagstuhl, Germany (2018). https://doi.org/10.4230/LIPIcs.ICALP.2018.134
17. Lipton, R.: The reachability problem requires exponential space, p. 62. (1976). Technical report
18. Mayr, E.W., Meyer, A.R.: The complexity of the finite containment problem for petri nets. J. ACM **28**(3), 561–576 (1981). https://doi.org/10.1145/322261.322271
19. Mayr, R.: Decidability and complexity of model checking problems for infinite state systems (1998). https://api.semanticscholar.org/CorpusID:30301617
20. Petri, C.: Kommunikation mit automaten. Schriften des Institutes für Instrumentelle Mathematik **3** (1962)
21. Sinn, M., Zuleger, F., Veith, H.: A simple and scalable static analysis for bound analysis and amortized complexity analysis. In: Biere, A., Bloem, R. (eds.) Computer Aided Verification, pp. 745–761. Springer International Publishing, Cham (2014)
22. Sinn, M., Zuleger, F., Veith, H.: Complexity and resource bound analysis of imperative programs using difference constraints. J. Autom. Reason. **59**(1), 3–45 (2017). https://doi.org/10.1007/s10817-016-9402-4
23. Zuleger, F.: The polynomial complexity of vector addition systems with states. In: Goubault-Larrecq, J., König, B. (eds.) Foundations of Software Science and Computation Structures, pp. 622–641. Springer International Publishing, Cham (2020)

Old and New Perspectives on Petri Nets Flows

Elvio G. Amparore[1]([✉])[ID], Gianfranco Ciardo[2][ID], Susanna Donatelli[1][ID], and Lea Terracini[1][ID]

[1] Università degli Studi di Torino, Turin, Italy
{elviogilberto.amparore,susanna.donatelli,lea.terracini}@unito.it
[2] Iowa State University, Ames, IA, USA
ciardo@iastate.edu

Abstract. Structural analysis of Petri nets has played an important role in the success of Petri nets for the analysis of discrete-event dynamic systems. Most structural Petri net theory has focused on P- (and T-) semiflows, the left (and right) non-negative integer annullers of the Petri net incidence matrix, which capture invariant laws for markings and cyclic behaviors, while little attention has been devoted to the general case of integer annullers, leading us to our first question: *Are there useful invariant laws based on arbitrary integer flows?* Moreover, the set of semiflows is typically characterized (in theory and tools) by the *minimal* semiflows, whose entries are relatively prime and whose support does not contain that of other semiflows. Minimal semiflows form a $\mathbb{Q}_{\geq 0}$-generator: any semiflow can be written as a linear combination with positive rational coefficients of minimal semiflows. Then, our second question is: *Can we characterize the set of (semi)flows through an $\mathbb{N}$-generating set, and do we gain better or additional insights by doing so?* To answer these questions, we consider the lattice of integer flows and shed light on the role of the Hilbert and Graver bases of a lattice vs. the notion of minimal flows. Moreover, we use various examples to discuss invariant laws based on flows in the Graver basis, and the role of arbitrary integer flows.

Keywords: Semiflows · Flows · Hilbert bases · Graver bases

1 Introduction

Structural analysis of Petri nets uses its incidence matrix, and possibly its initial marking, to derive behavioral properties of the net without actually exploring its reachability set, which can be extremely large or even infinite. The (left or right) non-negative integer annullers of the incidence matrix, i.e., the P-semiflows and the T-semiflows, are at the basis of much of these analysis techniques (for the rest of this section, we mostly refer to left annullers and P-semiflows or P-flows, but the same observations hold for right annullers and T-semiflows or T-flows).

As any multiple or sum of P-semiflows is a P-semiflow, a net has either zero or an infinite number of P-semiflows. In the latter case, it is important to finitely

J. Desel and A. Kalenkova (Eds.): PETRI NETS 2026, LNCS 16567, pp. 43–63, 2026.
https://doi.org/10.1007/978-3-032-27879-1_3

represent this infinite set, and this is traditionally done by defining a *generating set* of *minimal* P-semiflows such that any other P-semiflows can be obtained as a linear non-negative *rational* combination of the generating set elements.

From an algebraic point of view, the set of (left) non-negative integer annullers of the incidence matrix, thus the P-semiflows, forms an additive monoid over $\mathbb{N}^m$, and such structure can be finitely and uniquely characterized through its Hilbert basis, widely studied in linear programming. Any non-negative integer annuller can then be expressed as a linear combination with non-negative *integer* (as opposed to rational, as for minimal semiflows) coefficients of the Hilbert basis vectors. A Hilbert basis provides an alternative to the set of minimal semiflows when seeking a finite characterization of this monoid, thus we explore the relations between these two and their relative advantages and disadvantages.

The incidence matrix of a Petri net may also admit arbitrary left integer annullers (i.e., whose entries are not all non-negative; indeed, if $\mathbf{y}$ is annuller, so is $-\mathbf{y}$), these are the P-flows (or T-flows in the case of right annullers), and have received much less attention in the literature. As the set of P-flows is a vector space over the rationals, it is immediate to conclude that it can be finitely characterized as the rational linear span of a basis of flows of size equal to the dimension of this space. However, we are not aware of a proper definition of a unique generator set of P-flows using *non-negative* rational coefficients, analogous to the unique generator set of P-semiflows. We provide such a definition.

Also in this case, though, an alternative finite representation is available from linear programming. The set of arbitrary left integer annullers forms an integral lattice, which can be finitely and uniquely characterized by its Graver basis: any such annuller can be obtained as a linear combination with non-negative integer coefficients of the Graver basis vectors. Again, we explore the relations between these two finite representations and their relative advantages and disadvantages.

As for the rest of the paper, Sect. 2 presents the required background on Petri nets and their semiflows and flows, as well as linear algebra concepts related to the annullers of an integer matrix and the corresponding monoids and lattices they form. Section 3 defines the bases for these algebraic structures, Hilbert bases for monoids and Graver bases for lattices. Section 4 discusses semiflows and flows in the context of the algebraic structures of the preceding section, and provides illustrative examples. Section 5 examines how to perform structural analysis of Petri nets using minimal semiflows vs. Hilbert bases vs. minimal flows vs. Graver bases. Sections 6 and 7 illustrate such analysis using small examples and the well-known dining philosopher model, respectively. Finally, Sect. 8 concludes.

2   Preliminaries

This section presents a short summary of standard linear algebra definitions, which we use to study the algebraic structure of the set of annullers in the next section. Then, it briefly recalls the important role of the left and right annullers of a Petri net's incidence matrix in the structural analysis of nets, and the classical characterization of the set of these annullers using the minimal semiflows.

2.1 Linear Algebra Background

For completeness, we first recall standard results of linear algebra [13]. Let $\mathbf{A} \in \mathbb{Z}^{n \times m}$ be a matrix over integers (e.g., the incidence matrix of a Petri net with n places and m transitions), and consider the (right) null space of $\mathbf{A}$:

$$\mathcal{N} = \{\mathbf{x} \in \mathbb{R}^m \mid \mathbf{A} \cdot \mathbf{x} = \mathbf{0}\}. \tag{1}$$

The set of all right integral annullers of $\mathbf{A}$ forms an *integral lattice*[1] $\mathcal{L}$, which is a subgroup of $\mathbb{Z}^m$ built from the generating set $\mathbf{A}$:

$$\mathcal{L} = \mathcal{N} \cap \mathbb{Z}^m = \{\mathbf{z} \in \mathbb{Z}^m \mid \mathbf{A} \cdot \mathbf{z} = \mathbf{0}\}. \tag{2}$$

Let $t \leq m$ be the rank of $\mathcal{L}$, i.e., the dimension of its $\mathbb{R}$-span, defined as $\mathrm{span}_{\mathbb{R}}(\mathcal{L}) = \{\sum_{i=1}^{k} \alpha_i \cdot \mathbf{z}_i \mid k \in \mathbb{N}, \alpha_i \in \mathbb{R}, \mathbf{z}_i \in \mathcal{L}\}$.

If $\mathbb{D}$ is $\mathbb{N}$, $\mathbb{Z}$, $\mathbb{Q}$, or $\mathbb{Q}_{\geq 0}$, a set $G = \{\mathbf{g}_1, ..., \mathbf{g}_k\} \subseteq \mathcal{L}$, with $k \geq t$, is a $\mathbb{D}$-*generating set* of $\mathcal{L}$ if every $\mathbf{z} \in \mathcal{L}$ can be expressed as a $\mathbb{D}$-combination of elements of G:

$$\forall \mathbf{z} \in \mathcal{L}, \ \exists \alpha_1, ..., \alpha_k \in \mathbb{D} \quad \text{s.t.} \quad \mathbf{z} = \sum_{i=1}^{k} \alpha_i \cdot \mathbf{g}_i. \tag{3}$$

A $\mathbb{Q}$-basis B of $\mathcal{L}$ is a $\mathbb{Q}$-generating set of $\mathcal{L}$ of size t, implying that its elements are linearly independent. Every lattice admits a $\mathbb{Q}$-basis B and, if B is also a $\mathbb{Z}$-generating set of $\mathcal{L}$, then B is also a $\mathbb{Z}$-basis. Not all $\mathbb{Q}$-basis are $\mathbb{Z}$-basis, but every rational lattice $\mathcal{L}$ admits (infinitely many) $\mathbb{Z}$-bases. An $\mathbb{N}$-generating set $\mathcal{F}$ of $\mathcal{L}$ can be trivially obtained by *symmetrizing* a $\mathbb{Z}$-generating set G of $\mathcal{L}$:

$$\mathcal{F} = G \cup -G = \{\mathbf{g}, -\mathbf{g} \mid \mathbf{g} \in G\}.$$

Among all possible annullers of the (incidence) matrix $\mathbf{A}$ we might be interested only those whose elements have given signs. Let $\boldsymbol{\sigma} \in \{+1, -1\}^m$ be a vector of signs, and $\mathbb{R}^{\sigma} \subseteq \mathbb{R}^m$ the corresponding orthant:

$$\mathbb{R}^{\sigma} = \{\mathbf{x} \in \mathbb{R}^m \mid \forall h \in \{1, ..., m\}, \boldsymbol{\sigma}[h] \cdot \mathbf{x}[h] \geq 0\}.$$

Then, let $\mathcal{M}^{\sigma}$ be the set of all integral annullers of matrix $\mathbf{A}$ in orthant $\mathbb{R}^{\sigma}$:

$$\mathcal{M}^{\sigma} = \mathbb{R}^{\sigma} \cap \mathcal{L}.$$

$\mathcal{M}^{\sigma}$ is no longer a lattice, but only a monoid (indeed, it is an additive submonoid of $\mathbb{Z}^m$, since $\mathbf{0} \in \mathcal{M}^{\sigma}$ and, if $\mathbf{z}, \mathbf{w} \in \mathcal{M}^{\sigma}$, then $\mathbf{z} + \mathbf{w} \in \mathcal{M}^{\sigma}$). We can define an $\mathbb{N}$-, $\mathbb{Z}$-, $\mathbb{Q}$-, or $\mathbb{Q}_{\geq 0}$-generating set also for $\mathcal{M}^{\sigma}$. A fundamental property of an $\mathbb{N}$-generating set for $\mathcal{M}^{\sigma}$ is that its elements are vectors in orthant $\mathbb{R}^{\sigma}$ such that we can generate any annuller of $\mathbf{A}$ in $\mathbb{R}^{\sigma}$ by repeatedly adding (some of) them.

To investigate our second research question, about the definition of an $\mathbb{N}$-generating set for the annullers of the incidence matrix, the following section studies the algebraic structure of the set of annullers of the incidence matrix.

---

[1] Not to be confused with a lattice in order theory.

2.2 Petri Net Background

We assume standard definitions of Petri net and reachability set as given below.

Definition 1. *A* Petri net *is a directed bipartite multigraph over sets P of n places and T of m transitions. Matrices $\mathbf{C}^-, \mathbf{C}^+ \in \mathbb{N}^{n \times m}$ capture the multiplicity of (input) edges from P to T and (output) edges from T to P, respectively, e.g., the edge from $p \in P$ to $t \in T$ has multiplicity $\mathbf{C}^-[p, t]$ (if $\mathbf{C}^-[p, t] = 0$, the edge is absent). As a special case, the net is* ordinary *if $\mathbf{C}^-, \mathbf{C}^+ \in \{0, 1\}^{n \times m}$. A marking is a vector $\mathbf{m} \in \mathbb{N}^n$ assigning a number of tokens to each place. A marked* Petri net *(which we assume) specifies an* initial *marking $\mathbf{m}_0$. A transition $t \in T$ is enabled in marking $\mathbf{m}$ iff $\mathbf{m}[p] \geq \mathbf{C}^-[p, t]$ for each $p \in P$ and, if so, it may* fire *and change $\mathbf{m}$ into $\mathbf{m}'$ s.t. $\mathbf{m}'[p] = \mathbf{m}[p] + \mathbf{C}[p, t]$ for each $p \in P$, where $\mathbf{C} = \mathbf{C}^+ - \mathbf{C}^- \in \mathbb{Z}^{n \times m}$ is the* incidence matrix*; we write this as $\mathbf{m} \xrightarrow{t} \mathbf{m}'$, and extend this notation to (fireable) firing sequences $\sigma \in T^*$. The* reachability set *of the Petri net is then $RS(\mathbf{m}_0) = \{\mathbf{m} \in \mathbb{N}^P : \exists \sigma \in T^*, \mathbf{m}_0 \xrightarrow{\sigma} \mathbf{m}\}$.*

We briefly recall the invariant laws at the core of Petri net structural analysis, which motivate our interest in the left and right annullers of the incidence matrix. Most structural analysis stems from the observation that, if a marking $\mathbf{m}$ is reachable from marking $\mathbf{m}'$ through a firing sequence $\sigma \in T^*$ with firing (Parikh) count $\boldsymbol{\sigma} \in \mathbb{N}^m$, then

$$\mathbf{m} = \mathbf{m}' + \mathbf{C} \cdot \boldsymbol{\sigma}.$$

This is known as the *state equation* of a Petri net. It is important to remark that the opposite does not hold: if, for a given pair $\mathbf{m}$ and $\mathbf{m}'$, there exists a vector of non-negative integers $\boldsymbol{\sigma}$ that satisfies the state equation, we cannot conclude that $\mathbf{m}$ is reachable from $\mathbf{m}'$.

The *left annullers* of the incidence matrix $\mathbf{C}$ are at the foundations of the *marking invariant law*, derived from the state equation: if $\mathbf{y}$ is a left annuller of $\mathbf{C}$, then any marking $\mathbf{m}$ reachable from the initial marking $\mathbf{m}_0$ satisfies

$$\forall \mathbf{y}, \; \mathbf{y} \cdot \mathbf{C} = 0 \;\Rightarrow\; \mathbf{y} \odot \mathbf{m} = \mathbf{y} \odot \mathbf{m}_0, \tag{4}$$

i.e., the weighted sum of tokens in the support places of $\mathbf{y}$ is a constant determined by the initial marking (we will denote with $[\![\mathbf{x}]\!] \in \{0, 1\}^m$ the support of vector $\mathbf{x}$, where $[\![\mathbf{x}]\!][i] = 0$ iff $\mathbf{x}[i] = 0$).

When considering instead the *right annullers* of $\mathbf{C}$, we observe that, if $\boldsymbol{\sigma}$ is a T-semiflow (a non-negative right annuller), the state equation simplifies to $\mathbf{m} = \mathbf{m}'$ and we can define a *cyclic law*: any firing sequence σ whose Parikh vector $\boldsymbol{\sigma}$ is a right annuller of the incidence matrix takes the net back to the the same marking. In other words, consider a firing sequence σ with Parikh vector $\boldsymbol{\sigma}$ from marking $\mathbf{m}$ such that $\mathbf{m} \xrightarrow{\sigma} \mathbf{m}'$; then, if $\boldsymbol{\sigma}$ is a right annuller of $\mathbf{C}$, we know that $\mathbf{m}' = \mathbf{m}$:

$$\forall \boldsymbol{\sigma} : \mathbf{C} \cdot \boldsymbol{\sigma} = 0 \;\Rightarrow\; \mathbf{m} = \mathbf{m} + \mathbf{C} \cdot \boldsymbol{\sigma}. \tag{5}$$

If we take $\mathbf{m} = \mathbf{m}_0$, the state equation implies that the firing vector of any firing sequence that takes the net back to the initial marking must be a right annuller of the matrix. However, the reverse is again not true: given a right annuller $\boldsymbol{\sigma}$ of $\mathbf{C}$, there may be no firing sequence σ with firing count $\boldsymbol{\sigma}$ that takes the net back to the initial marking: the existence of a non-trivial non-negative right annuller is just a necessary condition for the initial marking to be a home state. Of course, if we can modify the initial marking so that it is sufficiently large, then a firing sequence with firing count $\boldsymbol{\sigma}$ will exist that takes the net back to its initial marking (in fact, for a large enough initial marking, *any* sequence of transitions with firing count $\boldsymbol{\sigma}$ will be a firing sequence that takes the net back to its initial marking).

For Petri nets analysis, non-negative integer annullers, the so-called P- and T- semiflows play a fundamental role. Given that they are infinite in nature, the notion of minimal P- and T- semiflows has been defined, see for example [3,14]:

Definition 2. *A P-semiflow (or T-semiflow) is* minimal *if its elements are relatively prime (the g.c.d. of its non null elements is 1) and it has minimal support (i.e., its support does not contain that of any other semiflow).*

Definition 3. *A P-semiflow (or T-semiflow) generator set is a set* $\Psi = \{\mathbf{y}_1, ..., \mathbf{y}_s\}$ *containing the least number of P-semiflows (or T-semiflows) such that any P-semiflow (or T-semiflow) $\mathbf{y}$ can be expressed as*

$$\mathbf{y} = \sum_{1 \leq j \leq s} \alpha_j \cdot \mathbf{y}_j, \quad \alpha_j \in \mathbb{Q}_{\geq 0}$$

It is well-known [3, corollary 2.3] that the set of minimal P-semiflows (or T-semiflows) is finite and unique, and it is a P-semiflow (resp. T-semiflow) generator set, according to Definition 3.

3   Algebraic Structure of the Null Space of the Incidence Matrix

We now examine the algebraic structure of the (left or right) null space of the incidence matrix, which is the essential notion for P- and T-(semi)flows and of the invariant laws derived from them.

3.1   Hilbert Basis of a Monoid and Graver Basis of a Lattice

Definition 4. *Let $\mathcal{M}$ be an additive monoid. A subset $\mathcal{H} \subseteq \mathcal{M}$ is a* Hilbert basis *[13, Sec. 16.4] of $\mathcal{M}$ iff it is an $\mathbb{N}$-generating set of $\mathcal{M}$ and it is minimal w.r.t. set inclusion, i.e., no proper subset of $\mathcal{H}$ is a generating set of $\mathcal{M}$.*

An alternative characterization of the Hilbert basis for $\mathcal{M}^\sigma$ can be given based on a notion of irreducibility [7]. Given $\mathbf{x}, \mathbf{y} \in \mathbb{R}^m$, let

$$\mathbf{x} \sqsubseteq \mathbf{y} \quad \text{iff} \quad \forall i \in \{1, ..., m\} : |\mathbf{x}[i]| \leq |\mathbf{y}[i]| \ \wedge \ \mathbf{x}[i] \cdot \mathbf{y}[i] \geq 0, \tag{6}$$

i.e., each entry of $\mathbf{x}$ is zero or has the same sign as the corresponding entry of $\mathbf{y}$, and it is not greater in absolute value than the corresponding entry of $\mathbf{y}$. Clearly $\mathbf{x} \sqsubseteq \mathbf{y}$ can hold only if there exists (at least) an orthant $\mathbb{R}^\sigma$ s.t. $\mathbf{x}, \mathbf{y} \in \mathbb{R}^\sigma$. Let $\mathbf{x} \sqsubsetneq \mathbf{y}$ mean $\mathbf{x} \sqsubseteq \mathbf{y} \wedge \mathbf{x} \neq \mathbf{y}$.

Definition 5. *An element $\mathbf{z} \in \mathcal{L}$ is said to be $\sqsubseteq$-irreducible in $\mathcal{L}$ if it cannot be written as the sum of two elements $\mathbf{x}, \mathbf{y} \in \mathcal{L} \setminus \{\mathbf{0}\}$ satisfying $\mathbf{x}, \mathbf{y} \sqsubseteq \mathbf{z}$.*

Given the definition of $\sqsubseteq$, we can also say that an element of a monoid is irreducible if it cannot be written as the sum of two non-zero elements of the monoid [13]. Algorithms [6,10] for Hilbert basis computation are then based on the following equivalent characterization, based on reducibility.

Proposition 1. *The Hilbert basis $\mathcal{H}^\sigma$ of $\mathcal{M}^\sigma$ can be identified as:*

$$\mathcal{H}^\sigma = \left\{ \mathbf{z} \in \mathcal{M}^\sigma \setminus \{\mathbf{0}\} \ \middle| \ \mathbf{z} \, is \ \sqsubseteq\text{-}irreducible \ in \ \mathcal{M}^\sigma \right\}. \tag{7}$$

[2, prop 2.9] provides a detailed explanation of the equivalence of the two definitions.

The Hilbert bases of $\mathcal{M}^\sigma$, for all orthants $\mathbb{R}^\sigma$, can then be used to define the *Graver basis* of $\mathcal{L}$.

Definition 6. *Let $\mathcal{M}^\sigma$, for $\sigma \in \{\pm 1\}^m$, be an orthant decomposition of $\mathcal{L}$ and let $\mathcal{H}^\sigma$ be the Hilbert basis of $\mathcal{M}^\sigma$. The* Graver basis *[5] $\mathcal{G}$ of $\mathcal{L}$ is defined as*

$$\mathcal{G} = \bigcup_{\sigma \in \{\pm 1\}^m} \mathcal{H}^\sigma. \tag{8}$$

Due to Eq. 7 we can also equivalently define the Graver basis as

$$\mathcal{G} = \left\{ \mathbf{z} \in \mathcal{L} \setminus \{\mathbf{0}\} \ \middle| \ \mathbf{z} \text{ is } \sqsubseteq\text{-irreducible in } \mathcal{L} \right\}. \tag{9}$$

Due to this notion of irreducibility, the Graver basis $\mathcal{G}$ is an N-generating set for $\mathcal{L}$ and can differ from an N-generating set $\mathcal{F}$ for $\mathcal{L}$ obtained by symmetrizing a $\mathbb{Z}$-basis of $\mathcal{L}$ as in Eq. 2.1 (in fact, the projection $\mathcal{F}^\sigma$ of $\mathcal{F}$ over an orthant $\mathbb{R}^\sigma$ may not even be an N-generating set for $\mathcal{M}^\sigma$). This is a very interesting characterization of the flows in a given orthant $\mathbb{R}^\sigma$: every integer annuller of a Petri net incidence matrix in $\mathbb{R}^\sigma$ can be obtained by summing *only* elements of $\mathcal{H}^\sigma$. This does not hold for $\mathcal{F}$, as it is possible that, to obtain an integer annuller in $\mathcal{M}^\sigma$, we must sum elements of $\mathcal{F}$ belonging to different orthants.

A related property then follows: $\mathcal{G}$ can be directly used to enumerate the integral annullers of matrix $\mathbf{A}$ inside an orthant $\mathbb{R}^\sigma$ in a particular order, for instance according to their $|| \cdot ||_1$ norm.

But what is the relation between Hilbert and Graver bases and the set of minimal P- and T-semiflows? To investigate their relationship we consider what are minimal P-semiflows in the theory of convex geometry. The work of Colom and Silva in [3], in Table 1, points out that minimal semiflows corresponds to what in convex geometry and linear programming is known as the directions (extremal rays) of a positive cone.

3.2 Cones and Extremal Vectors

Let $\mathcal{N}$ be the vector space of real annullers of a matrix, which is also the vector space of the span of the lattice of the integer annullers of that matrix. When interested in elements with given signs σ, we can consider the intersection of the vector space $\mathcal{N}$ with $\mathbb{R}^\sigma$ to obtain

$$\mathcal{C}^\sigma = \mathcal{N} \cap \mathbb{R}^\sigma = \mathrm{span}_\mathbb{R}(\mathcal{L}) \cap \mathbb{R}^\sigma$$

which is a cone (more precisely, an orthant cone). We briefly recall the definition of cones and associated extremal rays and extremal vectors to show that minimal semiflows are primitive extremal vectors of $\mathcal{C}^\sigma$ when $\mathbb{R}^\sigma$ is the positive orthant.

A cone is a subset C of a vector space that is closed under positive scalar multiplication, i.e., for any $\mathbf{x} \in C$ and $\lambda \geq 0$, $\lambda \cdot \mathbf{x} \in C$. A convex cone is a cone that is also closed under addition, or, equivalently, a subset of a vector space that is closed under linear combinations with positive scalars.

Definition 7 (Faces and Extremal Rays of a Cone). *A* face *of a convex cone C is a subset F of C s.t. F is also a convex cone and, for any $\mathbf{x}, \mathbf{y} \in C$, if $\mathbf{x}+\mathbf{y} \in F$, then $\mathbf{x}$ and $\mathbf{y}$ must also be in F [11]. A* ray *of C is the set of all non-negative multiples of a non-zero vector of C. A ray is* extremal *if it is a face of C.*

A convex cone is polyhedral if it has a finite number of faces. The intersection of two polyhedral convex cones is a polyhedral convex cone. By Farkas theorem, the notions of polyhedral cone and finitely generated cone are equivalent [13, p. 87]. An orthant $\mathbb{R}^\sigma$ is obviously a pointed[2] polyhedral convex cone. A null space $\mathcal{N}$ is a (non-pointed) polyhedral convex cone. The intersection of $\mathbb{R}^\sigma$ and $\mathcal{N}$ (i.e., two polyhedral convex cones) is a polyhedral convex cone, and since the former is pointed, the intersection must also be pointed. Therefore $\mathcal{C}^\sigma$ is a pointed polyhedral convex cone. When we consider the monoid $\mathcal{M}^\sigma = \mathcal{C}^\sigma \cap \mathcal{L}$, since $\mathcal{C}^\sigma$ is pointed, we can say that its minimal $\mathbb{N}$-generating set (i.e., its Hilbert basis $\mathcal{H}^\sigma$) is finite and unique [13, Theorem 16.4].

Definition 8. *An* extremal vector *of a convex cone $\mathcal{C}^\sigma$ is any non-null vector lying on an extremal ray of $\mathcal{C}^\sigma$.*

A convex cone is said to be rational if each extremal ray contains a vector with rational coordinates. Since we are concerned with integral structures embedded in Euclidean spaces, we restrict ourselves to rational polyhedral convex cones ("convex cones" for short).

An equivalent formulation of extremal vector is then the following.

---

[2] A cone C is pointed if $C \cap -C = \{\mathbf{0}\}$.

Proposition 2. *A vector* $\mathbf{x} \in C^\sigma$ *is an extremal vector of* C^σ *iff it has a* minimal support *w.r.t. set inclusion [8, Prop. 2.2.13].*

An extremal vector $\mathbf{x} \in C^\sigma$ is said to be *primitive* if it is integral and its entries are relatively prime. Since we assume rational cones, the *primitive* extremal vector of an extremal ray always exists. Note that a primitive extremal vector $\mathbf{x}$ may not belong to $\mathcal{M}^\sigma$, but some integral multiple of $\mathbf{x}$ surely does. The smallest of such multiples is called the *smallest representative* in $\mathcal{M}^\sigma$ of that extremal ray.

In general, the smallest representative in $\mathcal{M}^\sigma$ of an extremal ray may not be a primitive extremal vector, if $\mathcal{L}$ has cotorsion[3]. As an example of a cotorsion lattice with non primitive extremal rays, consider the lattice $\mathcal{L} \subset \mathbb{R}^3$ generated by the integral vectors $(1, 2, 3)$ and $(1, 4, 5)$. $\mathcal{L}$ is contained in the linear subspace defined by the equation $x + y = z$ and has cotorsion: the second coordinate of each element in $\mathcal{L}$ is even, thus $\mathcal{L}$ does not contain, for example, the vector $(0, 1, 1)$. For $\boldsymbol{\sigma} = (1, 1, 1)$, vectors $(1, 0, 1)$ and $(0, 2, 2)$ are extremal vectors: the first one is primitive, while the second one is the smallest representative of the ray generated by $(0, 2, 2)$, since the vector $(0, 1, 1)$ does not belong to $\mathcal{M}^\sigma$. Note that, when $\mathcal{M}^\sigma$ is the set of annullers of a matrix, it is cotorsion-free, thus smallest representative and primitive extremal vector of an extremal ray always coincide (this is the case for our application, where we focus on the annullers of the incidence matrix of a Petri net).

We can then define the *extremal set* as the set of smallest representatives of the extremal rays of C^σ:

Definition 9. *The extremal set* $\mathcal{X}^\sigma$ *of a monoid* $\mathcal{M}^\sigma$ *is defined as:*

$$\mathcal{X}^\sigma = \{\mathbf{z} \in \mathcal{M}^\sigma \mid \mathbf{z} \text{ is the smallest representative of an extremal ray of } C^\sigma\},$$

In the following, we use the superscript "+" instead of "σ" if $\boldsymbol{\sigma}$ refers to the important case of the positive orthant, i.e., when all the entries of $\boldsymbol{\sigma}$ are "+"; also, since we discuss sets containing either left or right annullers of the incidence matrix $\mathbf{C}$ of a Petri net, thus either P-(semi)flows or T-(semi)flows, respectively, we add the subscript "P" or "T" as needed, for clarity. For example, $\mathcal{X}_P^+$ refers to the extremal set of monoid $\mathcal{M}_P^+$ (the non-negative left annullers of $\mathbf{C}$).

Proposition 3. $\mathcal{X}_P^+$ *(resp.* $\mathcal{X}_T^+$*) is exactly the set of minimal P-semiflows (resp. T-semiflows) of Definition 2.*

Proof. By definition, the set of minimal P-semiflows includes all P-semiflows with minimal support and gcd equal to 1. According to Proposition 2, the elements of $\mathcal{X}_P^+$ also have minimal support; moreover, since they are the smallest representatives of the extremal rays, and the lattice we consider does not have co-torsion, the notions of smallest representative of an extremal vector and of primitive extremal vector coincide, therefore the elements of the extremal vector are integral and relatively prime. The proof is analogous for T-semiflows. $\square$

---

[3] By definition, $\mathcal{L}$ does not have cotorsion in $\mathbb{Z}^m$ if, for every $\mathbf{z} \in \mathbb{Z}^m$ and $k \in \mathbb{Z} \setminus \{0\}$, it holds that $k \cdot \mathbf{z} \in \mathcal{L} \Rightarrow \mathbf{z} \in \mathcal{L}$.

Proposition 4. *The following definition is equivalent to Definition 9*

$$\mathcal{X}^{\sigma} = \{\mathbf{z} \in \mathcal{H}^{\sigma} \mid \text{the support } [\![\mathbf{z}]\!] \text{ is minimal }\}.$$

Proof. Combining Proposition 1 with Proposition 2, we see that extremal rays are characterized by minimal support, and that their smallest lattice representatives are irreducible, thus they belong to the Hilbert basis.   □

Note that, since $\mathcal{H}^{\sigma}$ is finite and unique, then so is $\mathcal{X}^{\sigma}$.

The extremal set $\mathcal{X}^{\sigma}$ of each monoid $\mathcal{M}^{\sigma}$ can then be used to define the extremal set $\mathcal{X}$ of lattice $\mathcal{L}$, according to an orthant decomposition, as done when defining the Graver basis from the set of Hilbert bases.

Definition 10. *The extremal set $\mathcal{X}$ of lattice $\mathcal{L} = \bigcup_{\sigma \in \{0,1\}^m} \mathcal{M}^{\sigma}$ is*

$$\mathcal{X} = \bigcup_{\sigma \in \{\pm 1\}^m} \mathcal{X}^{\sigma}. \tag{10}$$

Definitions 9 and 10 imply the relations

Proposition 5.
$$\mathcal{X}^{\sigma} \subseteq \mathcal{H}^{\sigma} \subseteq \mathcal{G} \qquad and \qquad \mathcal{X} \subseteq \mathcal{G}.$$

We can now state the first important result of this paper.

Proposition 6. *P-flows (resp. T-flows) have a minimal generating set that is finite and unique, which is equivalent to $\mathcal{X}_P$ (resp. $\mathcal{X}_T$).*

Proof. Uniqueness and finiteness follows from (10).   □

Unlike the Hilbert basis $\mathcal{H}^{\sigma}$, the set of extremal vectors $\mathcal{X}^{\sigma}$ is not in general an $\mathbb{N}$-generating set for $\mathcal{M}^{\sigma}$. It is, however, a $\mathbb{Q}_{\geq 0}$-generating set for $\mathcal{M}^{\sigma}$, i.e., every $\mathbf{z} \in \mathcal{M}^{\sigma}$ can be expressed as a conical combination of $\mathcal{X}^{\sigma} = \{\mathbf{r}_1, ..., \mathbf{r}_v\}$ over $\mathbb{Q}_{\geq 0}$:

$$\forall \mathbf{z} \in \mathcal{M}^{\sigma}, \ \exists q_1, ..., q_v \in \mathbb{Q}_{\geq 0} \quad \text{s.t.} \quad \mathbf{z} = \sum_{i=1}^{v} q_i \mathbf{r}_i.$$

The Hilbert basis $\mathcal{H}^{\sigma}$ for $\mathcal{M}^{\sigma}$ conveys more information than $\mathcal{X}^{\sigma}$, if one is interested in considering only integer combinations. This is exactly our case, since integer combinations are particularly relevant in Petri nets, where the conservation laws for the number of tokens in places and the circuits and branches for transition firings are naturally expressed using integral values.

Observe that the size of $\mathcal{X}^{\sigma}$ is at most that of the largest set of non-comparable supports, thus of non-comparable subsets of the n places (in the case of left annullers), while the size of $\mathcal{H}^{\sigma}$ is at most that of the largest set of noncomparable vectors of size n where entries corresponding to place p_i range from 0 to $\mu_i = \max\{\mathbf{x}[i] : \mathbf{x} \in \mathcal{X}^{\sigma}\}$ (this is because each vector $\mathbf{x}'$ in $\mathcal{H}^{\sigma} \setminus \mathcal{X}^{\sigma}$ is obtainable as a rational convex combination of vectors in $\mathcal{X}^{\sigma}$, so that its entry $\mathbf{x}'[i]$ corresponding to place p_i is an integer between 0 and μ_i). The size of $\mathcal{X}^\sigma$ is then bounded by Sperner's formula [15], $\binom{n}{\lfloor n/2 \rfloor}$, which is asymptotically $\mathcal{O}(2^n/\sqrt{n})$ [4]; the size of $\mathcal{H}^\sigma$, assuming for simplicity that $\mu_i = \mu$ for all places p_i, is instead bounded by Sander's formula [12], $\sum_{k=0}^{\lfloor(\lfloor n\mu/2\rfloor+1)/(\mu+1)\rfloor}(-1)^k\binom{n}{k}\binom{n+\lfloor n\mu/2\rfloor-(\mu+1)k}{n-1}$, which for a fixed μ is asymptotically $\mathcal{O}(\mu^n/\sqrt{n})$. Of course, $\mathcal{X}^\sigma = \mathcal{H}^\sigma$ when $\mu = 1$, i.e., when all vectors in $\mathcal{X}^\sigma$ have only entries with value in $\{-1, 0, 1\}$.

Let us summarize the differences between $\mathcal{X}$ and $\mathcal{G}$ for the lattice of the left (or right) annullers of a Petri net's incidence matrix, as well as for $\mathcal{X}^+$ and $\mathcal{H}^+$ of the monoid $\mathcal{M}^+$ of the non-negative left (or right) annullers of a Petri net's incidence matrix:

- The notion of minimality for $\mathcal{X}$ and $\mathcal{X}^+$ is the set inclusion of the support, while for $\mathcal{G}$ and $\mathcal{H}^+$ it is the $\sqsubseteq$ ordering.
- $\mathcal{X}$ (resp. $\mathcal{X}^+$) is a $\mathbb{Q}_{\geq 0}$-generator of flows (resp. semiflows), while $\mathcal{G}$ (resp. $\mathcal{H}^+$) is an $\mathbb{N}$-generator of flows (resp. semiflows).
- $\mathcal{G}$ (resp $\mathcal{H}^+$) provides a straightforward way of enumerating the integer (resp. non-negative) annullers of the incidence matrix in a norm-increasing order (e.g., $\|\cdot\|_1$), while this is not the case for $\mathcal{X}$ (resp $\mathcal{X}^+$).

4   A Taxonomy of Flows

Since the annullers are infinite in number, it is essential to have a finite characterization for them. In the previous literature, the characterization of the set of non-negative integer annullers has been based on the notion of minimal semiflow.

As we will see in the following, there can be value in characterizing the entire set of annullers, not just the non-negative ones, thus we will discuss how to extend the notion of minimal semiflows to that of minimal flows in a straightforward manner. Moreover, there exist alternative characterizations of the set of (non-negative) integer annullers, and we propose one based on Hilbert and Graver bases. To highlight similarities and differences, this section proposes a taxonomy of integer annullers that "gives a name" to different types of annullers of the incidence matrix. This taxonomy will be the foundation for the definition of a taxonomy of invariant laws. Let us first define a generic notion of flow.

Definition 11. *A P-flow of a Petri net with incidence matrix* $\mathbf{C} \in \mathbb{Z}^{n \times m}$ *is a non-zero left integer annuller of* $\mathbf{C}$, *i.e., a vector*

$$\mathbf{y} \in \mathbb{Z}^n \setminus \{\mathbf{0}\} : \mathbf{y} \cdot \mathbf{C} = 0.$$

Definition 12. *A T-flow of a Petri net with incidence matrix* $\mathbf{C} \in \mathbb{Z}^{n \times m}$ *is a non-zero right integer annuller of* $\mathbf{C}$, *i.e., a vector*

$$\mathbf{x} \in \mathbb{Z}^m \setminus \{\mathbf{0}\} : \mathbf{C} \cdot \mathbf{x} = 0.$$

Table 1. Taxonomy of the unique bases for the set of flows.

	$\mathbb{Q}_{\geq 0}$-**generators**		$\mathbb{N}$-**generators**
annullers in $\mathbb{N}^m$	$\mathcal{X}^+$: minimal semiflows	$\subseteq$	$\mathcal{H}^+$: Hilbert semiflows
	$\text{I}\cap$		$\text{I}\cap$
annullers in $\mathbb{Z}^m$	$\mathcal{X}$: minimal flows	$\subseteq$	$\mathcal{G}$: Graver flows

We will use the term flow to refer to either P- or T- flows, depending on context, and the prefix "semi" when the elements of the flow are natural numbers.

Table 1 summarizes various finite characterization of Petri net flows. A first main distinction is whether we consider flows over non-negative integer values (first row, semiflows) or over integers (second row, general flows). A second one is whether the identified set is a $\mathbb{Q}_{\geq 0}$-generator or an $\mathbb{N}$-generator of the entire set of flows (or semiflows). Let $\mathcal{X}^+$ be the set of extremal vectors of the positive orthant, i.e., the set of minimal (P- or T-) semiflows (which are known to be a $\mathbb{Q}_{\geq 0}$-generator), and $\mathcal{H}^+$ be the set of semiflows in the Hilbert basis of $\mathcal{M}^+$, the monoid for the positive orthant, an $\mathbb{N}$-generator of the (P- or T-) semiflows. If we relax the requirement of being in the positive orthant, we can define $\mathcal{X}$, the union of the extremal vectors over all orthants, again a $\mathbb{Q}_{\geq 0}$-generator of all flows, and $\mathcal{G}$, the set of the flows in the Graver basis of the lattice of the (P- or T-) flows, which is an $\mathbb{N}$-generator of the (P- or T-) flows. As described in the previous sections, sets $\mathcal{X}^+, \mathcal{H}^+, \mathcal{X}$, and $\mathcal{G}$ are all finite and uniquely identified.

Minimal Semiflows and Their Relation with Extremal Rays. We now ask: What are minimal semiflows in linear algebra terms? [3] provides an interesting comparison table assimilating minimal semiflows with the extremal directions (actually the extremal vectors) of a cone, as we discuss in the following section. Minimal semiflows provide a balance between looking for more concise information (the number of minimal semiflows is in general much smaller than the size of the Hilbert basis) and having "only" a $\mathbb{Q}_{\geq 0}$-generator for the set of semiflows, instead than an $\mathbb{N}$-generator. This is achieved by moving from a discrete structure (a lattice in $\mathbb{Z}^m$) to a structure over $\mathbb{R}$ (the cone) as illustrated in Sect. 3.2.

We now illustrate the difference between the Hilbert basis $\mathcal{H}^+$ of the non-negative orthant and the classical minimal P-semiflows (or T-semiflows) $\mathcal{X}^+$.

4.1   Two Small Illustrative Examples of the Difference Between Minimal and Hilbert Semiflows

We use a small example to highlight the difference between using, as representative of the T-semiflows, either the set of minimal T-semiflows $\mathcal{X}_T^+$ or the set of Hilbert T-semiflows $\mathcal{H}_T^+$; later we do the same for P-semiflows.

Consider the small Petri net in Fig. 1, which shows $\mathcal{X}_T^+$ and $\mathcal{H}_T^+$ on the right. A T-semiflow (either from $\mathcal{X}_T^+$ or from $\mathcal{H}_T^+$) tells us that any firing sequence with that Parikh count, if fireable from $\mathbf{m}$, takes the net back to $\mathbf{m}$. However, the T-semiflows from $\mathcal{X}_T^+$ only represent the "extreme" behaviors, where, upon a possible choice of transitions, we only take a minimum subset of transitions, but may have to fire the transitions in that subset more times. Therefore, those T-semiflows miss "intermediate" behaviors, where we can choose combinations of more transitions but fire them fewer times. These are captured by the T-semiflows in $\mathcal{H}_T^+$, which represent all minimal cycles (i.e., cycles that do not contain a proper subsequence that is also a cycle), including all choices. As a consequence, any Parikh vector of a firing sequence that is a cycle is $\mathbb{N}$-generated by $\mathcal{H}_T^+$ (but not necessarily by $\mathcal{X}_T^+$).

T-semiflow $[1, 3, 0]$ characterizes a cyclic behavior where T_2 does not fire and T_1 fires three times more than T_0; T-semiflow $[1, 0, 3]$ is similar, but with T_1 and T_2 swapped. A different cyclic behavior is identified by the T-semiflows in $\mathcal{H}_T^+$: T-semiflow $[1, 1, 2]$ in $\mathcal{H}_T^+$ represents cycles where T_1 fires once and T_2 fires twice for each single firing of T_0; T-semiflow $[1, 2, 1]$ in $\mathcal{H}_T^+$ represents the corresponding cycles with T_1 and T_2 swapped.

The elements of $\mathcal{X}_T^+$ are minimal w.r.t. their support, which means that there is no cyclic firing sequence that includes a proper subset of the transitions in the support of a minimal T-semiflow. Instead, the elements of the Hilbert basis $\mathcal{H}_T^+$ are the *smallest circuits in the $\sqsubseteq$sense*: there is no cyclic firing sequence σ whose Parikh vector $\boldsymbol{\sigma}$ is $\sqsubseteq \mathbf{x}$, for any $\mathbf{x} \in \mathcal{H}_T^+$. Thus, for example, we can conclude right away that $[1, 2, 0]$ does not correspond to a cyclic behavior (of course, one can also reach this conclusion by checking that $[1, 2, 0]$ is not a right annuller of the incidence matrix).

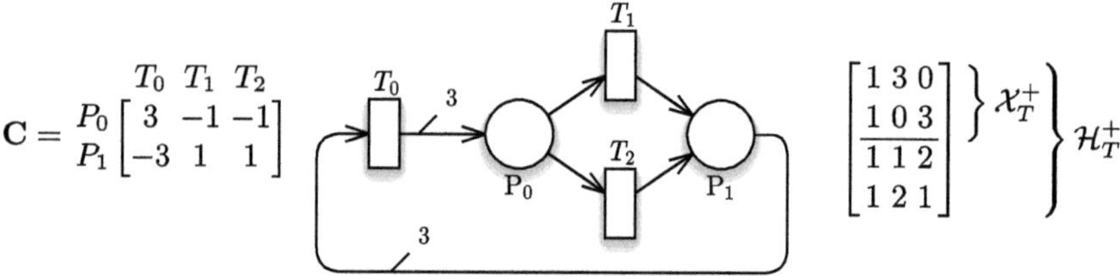

Fig. 1. A simple Petri net and its minimal and Hilbert T-semiflows.

Let us now consider the difference between $\mathcal{X}_P^+$ and $\mathcal{H}_P^+$ as generators of P-semiflows. Figure 2 shows a second simple Petri net and its incidence matrix $\mathbf{C}$. As this net is the dual of the one in Fig. 1, its sets $\mathcal{X}_P^+$ and $\mathcal{H}_P^+$ for the P-semiflows are the same as sets $\mathcal{X}_T^+$ and $\mathcal{H}_T^+$ for the T-semiflows of Fig. 1, again reported on the right.

The fact that $\mathcal{X}_P^+$ is only a $\mathbb{Q}_{\geq 0}$-generator of the P-semiflows can be easily exemplified by observing that the annuller $[1, 1, 2]$ cannot be obtained as a linear combination with natural coefficient of the vectors in $\mathcal{X}_P^+$: rational coefficients are needed, $[1, 1, 2] = 1/3 * [1, 3, 0] + 2/3 * [1, 0, 3]$. As already mentioned, the elements of $\mathcal{H}_P^+$ can be used to to enumerate all others P-semiflows in a structured manner, for example in $||\cdot||_1$ norm order. To achieve this we can simply sum each

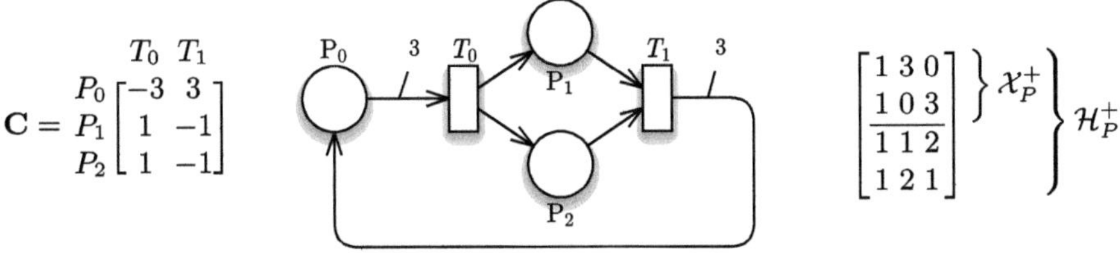

Fig. 2. The dual of the Petri net of Fig. 1 and its minimal and Hilbert P-semiflows.

vector in $\mathcal{H}_P^+$ in norm order, to obtain $[2, 6, 0]$, $[2, 0, 6]$, $[2, 3, 3]$, $[2, 2, 4]$, $[2, 4, 2]$, $[2, 5, 1]$, and $[2, 1, 5]$ (of norm 8), $[3, 9, 0]$, $[3, 6, 3]$, etc. (of norm 12), and so on.

We now use this small example to highlight the difference between using $\mathcal{X}_P^+$ or $\mathcal{H}_P^+$ P-semiflows to derive marking invariant laws. If the initial marking $\mathbf{m}_0$ is $[3, 0, 0]$, the marking invariant law obtained considering $\mathcal{X}^+$ states that, in any reachable marking $\mathbf{m}$,

$$\mathbf{m}[P_0] + 3 * \mathbf{m}[P_1] = 3 \qquad \text{and} \qquad \mathbf{m}[P_0] + 3 * \mathbf{m}[P_2] = 3,$$

where the value 3 on the right-hand side of these two equations depends on the initial marking. Considering instead the marking invariant law obtained from P-semiflow $[1, 1, 2] \in \mathcal{H}_P^+ \setminus \mathcal{X}_P^+$, we can state that, in any reachable marking $\mathbf{m}$,

$$\mathbf{m}[P_0] + \mathbf{m}[P_1] + 2 * \mathbf{m}[P_2] = 3,$$

where 3 is again uniquely determined by the initial marking $\mathbf{m}_0$ and the flow $\mathbf{y}$ being considered. The invariant based on P-semiflow $[1, 1, 2]$ involves more places than the ones based on $\mathcal{X}_P^+$, since minimal semiflows have minimal support. In other words, invariants based on Hilbert establish a relation about the marking of more places, thus a finer information than minimal semiflows, but sometimes this information may even be too fine, as the cardinality of $\mathcal{H}_P^+$ may be much larger than the cardinality of $\mathcal{X}_P^+$. In our small example, if we change the weight of the non ordinary arcs from 3 to N, $\mathcal{H}_P^+$ will include all vectors of form $[1, i, N - i]$, for $i \in \{0..N\}$, while the cardinality of $\mathcal{X}_P^+$ does not change: it remains equal to 2. We will revisit the relation between $\mathcal{X}_P^+$ and $\mathcal{H}_P^+$ in Sect. 5, when considering the use of the marking invariant law to compute place bounds.

5   Flows and Invariant-Based Analysis

We now consider how analysis based on the annullers of $\mathbf{C}$ may depend on whether one considers those in $\mathcal{X}^+$, $\mathcal{H}^+$, $\mathcal{X}$, or $\mathcal{G}$. Given a Petri net with initial marking $\mathbf{m}_0$, we indicate with $b(p)$ the bound of place p, defined as

$$b(p) = \max\{\mathbf{m}(p) : \mathbf{m} \in RS(\mathbf{m}_0)\}.$$

P-semiflows can be used to derive place bounds, and we can then define a "bound invariant law" as follows.

Proposition 7 (Bound Invariant Law). *Let* $\mathbf{y}$ *be a P-semiflow, and* $K_\mathbf{y} = \mathbf{y}\cdot\mathbf{m}_0$ *be the token count of the marking invariant law derived using* $\mathbf{y}$*. Then, for any place* p *such that* $\mathbf{y}[p] > 0$*, we have* $b(p) \leq K_\mathbf{y}/\mathbf{y}[p]$*, thus we can conclude that*

$$b(p) \leq \min\{K_\mathbf{y}/\mathbf{y}[p] : \mathbf{y} \cdot \mathbf{C} = 0, \mathbf{y} \in \mathbb{N}^m \wedge \mathbf{y}[p] > 0\}.$$

The above bound on place p is in principle computed be considering *every* P-semiflow, which is of course impossible as the set of P-semiflows is infinite (if not empty), but the following proposition shows that it is sufficient to consider the semiflows $\mathcal{X}_P^+$ to obtain this bound.

Proposition 8. *The place bound computed considering only the P-semiflows in* $\mathcal{X}_P^+$ *cannot be improved by considering additional P-semiflows, i.e.,*

$$\min\{K_\mathbf{y}/\mathbf{y}[p] : \mathbf{y} \cdot \mathbf{C} = 0, \mathbf{y} \in \mathbb{N}^m \wedge \mathbf{y}[p] > 0\} = \min\{K_\mathbf{y}/\mathbf{y}[p] : \mathbf{y} \in \mathcal{X}_P^+ \wedge \mathbf{y}[p] > 0\}.$$

Proof. Let $\mathbf{x}$ be a P-semiflow in $\mathcal{X}_P^+$ for which the bound on place p is the most tight, i.e., for any other P-semiflow $\mathbf{y}_j \in \mathcal{X}_P^+$ with $\mathbf{y}_j[p] > 0$, we have

$$\frac{K_{\mathbf{y}_j}}{\mathbf{y}_j[p]} = \frac{\sum_{q\in P} \mathbf{m}_0[q] \cdot \mathbf{y}_j[q]}{\mathbf{y}_j[p]} \geq \frac{K_\mathbf{x}}{\mathbf{x}[p]} = \frac{\sum_{q\in P} \mathbf{m}_0[q] \cdot \mathbf{x}[q]}{\mathbf{x}[p]},$$

which implies

$$\sum_{q\in P} \mathbf{m}_0[q] \cdot \mathbf{y}_j[q] \geq \mathbf{y}_j[p] \cdot K_\mathbf{x}/\mathbf{x}[p].$$

Consider now an arbitrary P-semiflow $\mathbf{y} \in \mathbb{N}^m$ satisfying $\mathbf{y} \cdot \mathbf{C} = 0$ and $\mathbf{y}[p] > 0$. Since $\mathcal{X}_P^+ = \{\mathbf{y}_1, ..., \mathbf{y}_s\}$ is a $\mathbb{Q}_{\geq 0}$-generating set, there exist $\alpha_1, .., \alpha_s \in \mathbb{Q}_{\geq 0}$ such that $\mathbf{y} = \sum_{1\leq j\leq s} \alpha_j \cdot \mathbf{y}_j$. Then, the bound computed using $\mathbf{y}$ satisfies:

$$\frac{K_\mathbf{y}}{\mathbf{y}[p]} = \frac{\sum_{q\in P} \mathbf{m}_0[q] \cdot \mathbf{y}[q]}{\mathbf{y}[p]} = \frac{\sum_{q\in P} \mathbf{m}_0[q] \cdot \sum_{1\leq j\leq s} \alpha_j \cdot \mathbf{y}_j[q]}{\sum_{1\leq j\leq s} \alpha_j \cdot \mathbf{y}_j[p]} =$$

$$\frac{\sum_{1\leq j\leq s} \alpha_j \cdot \sum_{q\in P} \mathbf{m}_0[q] \cdot \mathbf{y}_j[q]}{\sum_{1\leq j\leq s} \alpha_j \cdot \mathbf{y}_j[p]} \geq \frac{\sum_{1\leq j\leq s} \alpha_j \cdot \mathbf{y}_j[p] \cdot K_\mathbf{x}/\mathbf{x}[p]}{\sum_{1\leq j\leq s} \alpha_j \cdot \mathbf{y}_j[p]} = \frac{K_\mathbf{x}}{\mathbf{x}[p]}. \qquad \square$$

We stress that, while we just showed that considering the P-semiflows in $\mathcal{X}_P^+$ suffices to determine structural place bounds, this does not mean that other P-semiflows cannot provide useful insight. In our small example, using the minimal P-semiflow $\mathbf{y} = [1, 3, 0]$ and letting $\mathbf{m}_0 = [k, 0, 0]$, we obtain $K = \mathbf{y} \cdot \mathbf{m}_0 = k$, thus $b(P_0) \leq k$ and $b(P_1) \leq \lfloor k/3 \rfloor$. Using instead Hilbert P-semiflow $\mathbf{y} = [1, 2, 1]$ and the same $\mathbf{m}_0$, we obtain again $K = k$, thus $b(P_0) \leq k$, $b(P_1) \leq \lfloor k/2 \rfloor$, and $b(P_2) \leq k$. The bound $b(P_1) \leq \lfloor k/2 \rfloor$ is less tight the one computed from the minimal P-semiflow; however, the marking invariant law derived from the Hilbert P-semiflow $[1, 2, 1]$ indicates that there is a *direct* dependency among the markings of the three places.

Proposition 9 (Shortest Cyclic Law). *Let* $\mathbf{m} \xrightarrow{\sigma} \mathbf{m}$*, and let* $\boldsymbol{\sigma}$ *be the Parikh vector of* σ*. Clearly* $\boldsymbol{\sigma}$ *is a T-semiflow. If* $\boldsymbol{\sigma} \in \mathcal{X}_T^+$*, then we know that no cyclic behavior with fewer distinct transitions can exists for the net. This is a direct consequence of* $\mathcal{X}_T^+$ *containing only minimal support T-semiflows.*

Proposition 10 (Smallest Cyclic Law).*Let* $\mathbf{m} \xrightarrow{\sigma} \mathbf{m}$*, and let* $\boldsymbol{\sigma}$ *be the Parikh vector of* σ*. Clearly* $\boldsymbol{\sigma}$ *is a T-semiflow. If* $\boldsymbol{\sigma} \in \mathcal{H}_T^+$ *then we know that no other cyclic behavior with a number of firings smaller or equal than in the corresponding entry of* $\boldsymbol{\sigma}$ *can exists for the net. This is a direct consequence of* $\mathcal{H}_T^+$ *containing T-semiflows that are* $\sqsubseteq$ *than any other T-semiflows.*

We now consider invariant laws for markings and for transition firings based on flows. Here the main difference is that we consider flows in any orthant (the second line of the taxonomy in Table 1).

Proposition 11 (Equal Behavior Law). *Given T-flow* $\mathbf{x}$*, write it as* $\mathbf{x} = \mathbf{x}^+ - \mathbf{x}^-$ *with* $\mathbf{x}^+, \mathbf{x}^- \in \mathbb{N}^m$*. Let* $\sigma_{\mathbf{x}}^+$ *be a firing sequence fireable from* $\mathbf{m}$ *with Parikh vector* $\mathbf{x}^+$*, and* $\sigma_{\mathbf{x}}^-$ *be a firing sequence fireable from* $\mathbf{m}$ *with Parikh vector* $\mathbf{x}^-$*. Since* $\mathbf{x}$ *is a T-flow (thus, a right annuller), the state equation* $\mathbf{m} = \mathbf{m} + \mathbf{C} \cdot \mathbf{x}$ *can be rewritten by substitution as* $\mathbf{m} = \mathbf{m} + \mathbf{C} \cdot (\mathbf{x}^+ - \mathbf{x}^-)$*, from which we derive:*

$$\mathbf{m} + \mathbf{C} \cdot (\mathbf{x}^+) = \mathbf{m} + \mathbf{C} \cdot (\mathbf{x}^-).$$

This equal behavior law tell us that the negative and positive components of a T-flow, if fireable, produce the same change in the marking. This is relevant to identify bags of transitions that correspond to different combinations of transition firings but produce the same overall effect. The current version of the GreatSPN tool [1] can compute $\mathcal{X}_T$ and $\mathcal{G}_T$ and visualize the derived equal behavior law. An example is in Fig. 3: the box on the left lists all T-flows in $\mathcal{X}_T$ and $\mathcal{G}_T$ (those in $\mathcal{G}_T$ have at least one negative entry; also, since technically $\mathcal{G}_T$ contains both $\mathbf{x}$ and $-\mathbf{x}$, GreatSPN shows only one of the two, the one whose first nonzero component is positive). By clicking on a T-flow, the graphical interface shows the corresponding equal behavior law on the Petri net: the change of marking produced by the orange transitions (T_0 and T_3) equals that produced by the pink ones (T_1 and T_4).

Note that the reverse also holds: if two firing sequences produce the same marking change in a net, then the difference of their Parikh vectors must be a right annuller of the incidence matrix.

While the Petri net community has focused on P-semiflows, P-flows can also be of interest, as they provide an invariant law that balances the weighted sums of tokens in disjoint subsets of places. The only reference we could find that considers flows over integers is the work in [14], which uses P-flows to define a marking balance invariant law.

Proposition 12 (Marking Balance Invariant Law). *Given P-flow* $\mathbf{y}$*, write it as* $\mathbf{y} = \mathbf{y}^+ - \mathbf{y}^-$ *with* $\mathbf{y}^+, \mathbf{y}^- \in \mathbb{N}^n$*, from which we derive the marking invariant law*

$$\mathbf{y}^+ \odot \mathbf{m} = \mathbf{y}^- \odot \mathbf{m} + \mathbf{y} \odot \mathbf{m}_0.$$

This means that, when the weighted number of tokens in the places corresponding to the support of $\mathbf{y}^+$ increases, the weighted number of tokens in the places corresponding to the support of $\mathbf{y}^-$ must increase by the same amount.

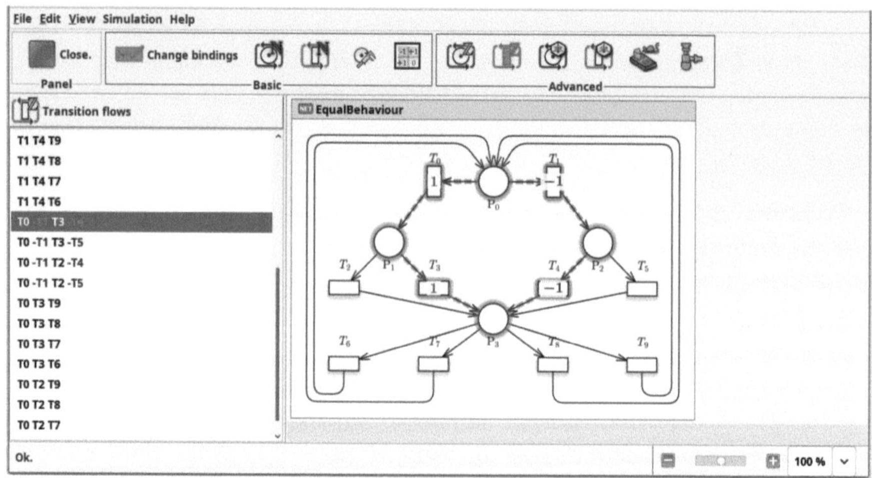

Fig. 3. Visualization of the equal behavior law in the GreatSPN tool.

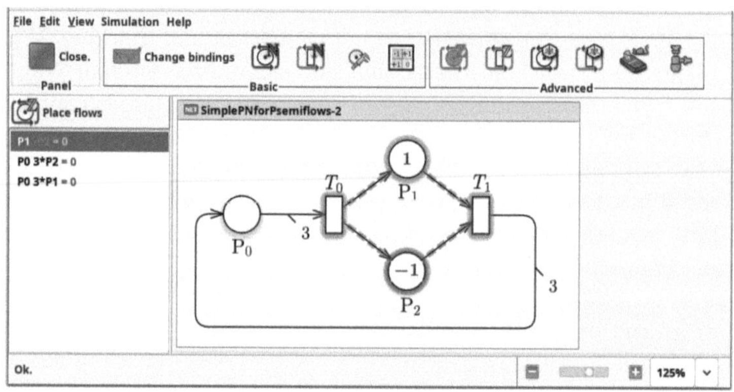

Fig. 4. Visualization of the marking balance law in the GreatSPN tool.

The marking invariant law of Eq. 4 has been often used in the literature to test non reachability or to compute bounds (as in Proposition 7). Indeed, if marking $\mathbf{m}$ does not satisfy the invariant of Eq. 4, then $\mathbf{m}$ is not reachable from $\mathbf{m}_0$. We stress that there is no need to limit ourselves to P-semiflows. On the contrary, the larger the number of flows considered, the larger the number of invariant laws and the easier may be to determine that a marking is non reachable. In the survey of [14] it is stated that minimal semiflows are the set of all flows useful to prove properties as any other flow can be generated from minimal semiflows as linear combination with non-negative rational coefficient. However, considering as starting point $\mathcal{X}_P$ instead of $\mathcal{X}_P^+$ can be convenient. For example consider the subnet of the net in Fig. 2 consisting only of transition T_0 and places P_1 and P_2. For this small net $\mathcal{X}_P^+$ is empty, while $\mathcal{X}_P$ include the vector $[1, -1]$, leading to a

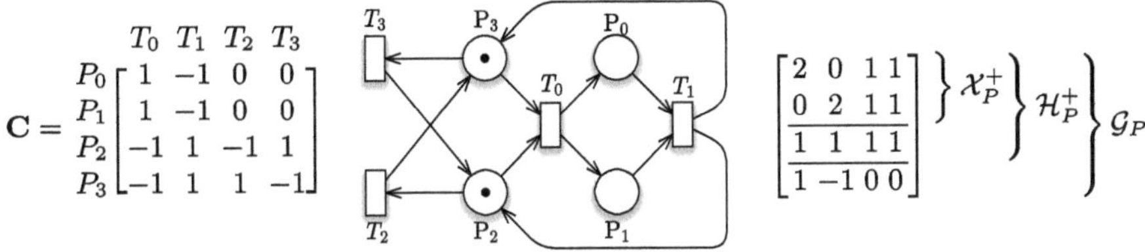

$$\mathbf{C} = \begin{array}{c} \\ P_0 \\ P_1 \\ P_2 \\ P_3 \end{array} \begin{bmatrix} -1 & 0 \\ -1 & 0 \\ 1 & -1 \\ 1 & 1 \end{bmatrix}$$

$$\left. \left. \left[\begin{array}{cccc} 2 & 0 & 1 & 1 \\ 0 & 2 & 1 & 1 \\ \hline 1 & 1 & 1 & 1 \\ \hline 1 & -1 & 0 & 0 \end{array} \right] \begin{array}{l} \Big\} \mathcal{X}_P^+ \\ \\ \end{array} \right\} \mathcal{H}_P^+ \right\} \mathcal{G}_P$$

Fig. 5. An ordinary but dead net where the P-semiflows in $\mathcal{X}_P^+$ and $\mathcal{H}_P^+$ differ.

$$\mathbf{C} = \begin{array}{c} \\ P_0 \\ P_1 \\ P_2 \\ P_3 \end{array} \begin{bmatrix} 1 & -1 & 0 & 0 \\ 1 & -1 & 0 & 0 \\ -1 & 1 & -1 & 1 \\ -1 & 1 & 1 & -1 \end{bmatrix}$$

$$\left. \left. \left[\begin{array}{cccc} 2 & 0 & 1 & 1 \\ 0 & 2 & 1 & 1 \\ \hline 1 & 1 & 1 & 1 \\ \hline 1 & -1 & 0 & 0 \end{array} \right] \begin{array}{l} \Big\} \mathcal{X}_P^+ \\ \\ \end{array} \right\} \mathcal{H}_P^+ \right\} \mathcal{G}_P$$

Fig. 6. An ordinary and live net where the P-semiflows in $\mathcal{X}_P^+$ and $\mathcal{H}_P^+$ differ.

balance marking invariant law that states that the number of tokens in P_1 and in P_2 is the same in any reachable marking. Thus, flows may lead to marking invariant laws that can make the test for non-reachability more powerful.

6 P and T-Flows Based Invariants on Small Examples

We now reconsider the small examples of Fig. 1 and 2 to show the conclusions that can be drawn from $\mathcal{X}_P$ and $\mathcal{G}_P$ flows, i.e., when we also consider flows with negative entries. The set of Graver P-flows for the net in Fig. 2 is as follows,

$$\left. \left. \left[\begin{array}{ccc} 1 & 3 & 0 \\ 1 & 0 & 3 \\ \hline 1 & 1 & 2 \\ 1 & 2 & 1 \\ \hline 0 & 1 & -1 \end{array} \right] \begin{array}{l} \Big\} \mathcal{X}_P^+ \\ \\ \end{array} \right\} \mathcal{H}_P^+ \right\} \mathcal{G}_P.$$

Flow $[0, 1, -1] \in \mathcal{G}_P$ is also an element of $\mathcal{X}_P \setminus \mathcal{X}_P^+$, i.e., it is a minimal P-flow, as its support is minimal, but not a P-semiflow, as it has negative entries. Consider the marking balance invariant law of Proposition 12 instantiated using the flow $\mathbf{y} = [0, 1, -1]$, so that $\mathbf{y}^+ = [0, 1, 0]$, $\mathbf{y}^- = [0, 0, 1]$; letting $\mathbf{m}_0 = [K, 0, 0]$ results in $\mathbf{m}[P_1] = \mathbf{m}[P_2]$, telling us that any reachable marking has the same number of tokens in P_1 and P_2. If we consider instead a different initial marking $\mathbf{m}_0 = [K, L, 0]$, we obtain the invariant $\mathbf{m}[P_1] = \mathbf{m}[P_2] + L$, telling us that any reachable marking has L more tokens in P_1 than in P_2. Figure 4 shows the visualization of this marking balance law inside the GreatSPN tool.

To reason about the role of T-flows let's consider again right annullers of the net in Fig. 1. The minimal T-flow $[0, 1, -1] \in \mathcal{X}_T \setminus \mathcal{X}_T^+$ clearly shows that T_0 and T_1 have the same effect. This is obvious in such a small example, but could be much harder to recognize in a large net where two different weighted subsets of transitions have the same effect.

Before moving on to examples based on the dining philosophers net, we make an additional observation. The examples we used to illustrate the difference between $\mathcal{X}_P$ and $\mathcal{G}_P$ are non-ordinary nets (some arcs have weight greater than one), but this is not necessary: there are ordinary nets whose Hilbert basis $\mathcal{H}_P^+$ strictly contain $\mathcal{X}_P^+$. Figure 5 shows such an example using a not "well-behaved" net (it is structurally dead), while Fig. 6 shows a well-behaved net (its reachability graph is a single, non trivial, strongly connected component) with the same $\mathcal{H}_P^+$, $\mathcal{X}_P^+$, and $\mathcal{G}_P$ as the one of Fig. 5, so, again $\mathcal{H}_P^+$ strictly contain $\mathcal{X}_P^+$.

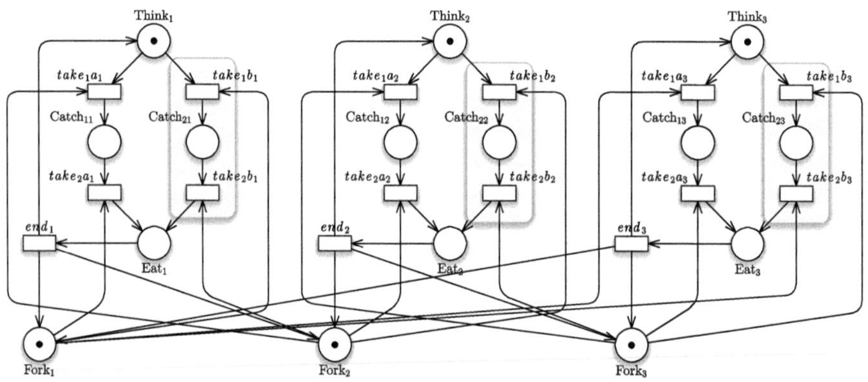

Fig. 7. The dining philosophers net (for the case three philosophers).

7   Studying the Dining Philosopher Model

Consider the model of the Dining Philosophers, shown in Fig. 7 for the case of $N = 3$ philosophers (ignore for now the presence of the yellow boxes).

This net has six minimal P-semiflows, determining the following place invariants:

$Think_i + Catch_{1i} + Catch_{2i} + Eat_i = 1$, for $i \in \{1, 2, 3\}$, expressing the fact that each philosopher i must be in one of four states, i.e., these four places must contain a total of one token (marking invariant law).

$Fork_i + Catch_{1\,(i \bmod 3)+1} + Catch_{2\,i} + Eat_i + Eat_{(i \bmod 3)+1} = 1$, for $i \in \{1, 2, 3\}$, expressing the fact that each fork i is either available, or taken by the philosopher to its left or the philosopher its right (marking invariant law).

The number of minimal P-flows is much larger, 67. We discuss a couple of them.

$\sum_{i=1}^{3}(Fork_i + Eat_i) = \sum_{i=1}^{3} Think_i$, expressing the fact that the number of thinking philosophers equals the number of available forks plus the number of eating philosophers (marking balance invariant law).

$2*Eat_1 + Catch_{11} + Catch_{21} + Fork_1 + Fork_2 + (Catch_{22} + Catch_{13}) = 1 + Fork_3 +$
$Catch_{12} + Catch_{23}$, expressing the fork availability from the point of view of philosopher 1: on the left-hand side, fork 1 is available, or taken by philosopher 1, or taken by philosopher 3; fork 2 is available, or taken by philosopher 1, or taken by philosopher 2; however, if both forks are taken by philosopher 1, then he is eating (thus place Eat_1 has weight 2); on the right-hand side, fork 3 is available, taken by philosopher 2, or taken by philosopher 3, and the additional constant 1 is needed because the right-hand side describes the position of one fork, while the left-hand side describes the position of two forks (marking balance invariant law).

Considering now the right annullers of this net, we find six minimal T-semiflows, all corresponding to the same type of cyclic behavior:

$[take_1 x_i, take_2 x_i, end_i]$, for $i \in \{1, 2, 3\}$ and $x \in \{a, b\}$, expressing the fact that the net returns to the same marking if each philosopher i takes the "a" or "b" branch (taking the fork to his left first, then the fork to his right, or vice versa), then finishes eating and returns the two forks (smallest cyclic law).

The list of minimal T-flows contains, in addition to the six minimal T-semiflows above, also three T-flows of the form

$[take_1 a_i, take_2 a_i, -take_1 b_i, -take2 b_i]$, for $i \in \{1, 2, 3\}$, expressing the fact each philosopher can take the forks to his left and his right in either order, following the "a" or "b" branch, with the same effect (equal behavior law).

Consider now a modified net without the "b" branches in the yellow boxes, so that each philosopher always takes the two forks in the same order.
This modified net still has six minimal P-semiflows, determining the following place invariants:

$Think_i + Catch_{1i} + Eat_i = 1$, for $i \in \{1, 2, 3\}$, expressing the fact that each philosopher i must be in one of three states, i.e., the three places must contain a total of one token (marking invariant law).
$Fork_i + Catch_{1\,(i-1)\bmod 3} + Eat_i + Eat_{(i-1)\bmod 3} = 1$, for $i \in \{1, 2, 3\}$, expressing the fact that each fork i is either available, or taken by the philosopher to its left or the philosopher to its right (marking invariant law).

Thus, removing the "b" branches simply removes the (now non-existent) places $Catch_{2i}$ from each corresponding minimal P-semiflow of the unmodified net.
The number of minimal P-flows is again much larger, as there are 30 additional minimal P-flows in addition to the six minimal P-semiflows above. Again, we only discuss a couple of them.

$Think_1 = Eat_2 + Fork_2$, expressing the fact that, if philosopher 1 thinks, fork 2 is either available or in use by the eating philosopher 2 while, if philosopher 1 is not thinking, then he has fork 2, which cannot then be available nor in use by philosopher 2 (marking balance invariant law).

$2 * Eat_1 + Fork_1 + Fork_2 + Catch_{11} + Catch_{13} = 1 + Fork_3 + Catch_{12}$, expressing the same constraint about the fork usage from the point of view of philosopher 1 as for the unmodified net, except that now, places corresponding to the "b" branches are not present (marking balance invariant law).

Considering now the right annullers of this modified net, we find only three minimal T-semiflows, all corresponding to the same type of cyclic behavior:

$[take_1a_i, take_2a_i, end_i]$, for $i \in \{1,2,3\}$, these are simply the same as the three T-semiflows following the "a" branch in the unmodified net, while those following the "b" branch in the unmodified net have been eliminated.

Finally, the list of minimal T-flows now contains only the three minimal T-semiflows above, as there are no more "b" branches equivalent to the "a" branches.

We conclude this section by observing that all P-semiflows of the unmodified or the modified net have entries in $\{0,1\}$, thus $\mathcal{X}_P^+$ and $\mathcal{H}_P^+$ coincide. Instead, $\mathcal{X}_P$ contains vectors with entries equal to 2, but $\mathcal{X}_P$ and $\mathcal{G}_P$ still coincide.

8   Conclusion

Non-negative integer annullers of the incidence matrix of a Petri net (the so-called P- and T-semiflows) are at the basis of structural theory and analysis of nets. Traditionally, the infinite sets of (P- or T-)semiflows have been characterized through the minimal (P- or T-)semiflows, which have minimal support and gcd equal to 1. The set of minimal semiflows is a $\mathbb{Q}_{\geq 0}$ generator of the entire set of semiflows. In this paper, we propose an alternative characterization based on the notion of Hilbert basis, which provides an $\mathbb{N}$-generator of semiflows.

Considering instead arbitrary integers annullers of the incidence matrix, in any orthant (P- and T-flows), we give a characterization of the infinite set of P- or T-flows through the Graver basis of the lattice of these annullers, which is finite and unique, and constitutes an $\mathbb{N}$-generator of the lattice. Then, we provide a characterization in terms of flows with minimal support and gcd 1, and prove it to be finite and unique, as it had been previously done for semiflows.

We also discuss how Hilbert and Graver basis and minimal flows can be used to derive new invariant laws: the Equal behavior law (Proposition 11), the Marking balance invariant law (Proposition 12), and the Smallest cyclic law (Proposition 10).

Computational Costs. It is well-known that there could be exponentially many minimal semiflows (in the number n of places, or transitions). For example, Silva [14] gives $\binom{n}{\lceil n/2 \rceil}$ as an upper bound, which is asymptotically $\mathcal{O}(2^n/\sqrt{n})$ [4]. $\mathcal{H}^+$ coincides with $\mathcal{X}^+$ when all semiflows have entries 0 or 1, but can be much larger if the semiflows contain entries up to $\mu \in \mathbb{N}$, an upper bound for its size is $\mathcal{O}(\mu^n/\sqrt{n})$. Finally, we observe that, while Fourier-Motzkin [13] is the standard method used to compute semiflows, more recent and effective methods have appeared in the linear algebra literature [6,9] for the computation of Hilbert and Graver basis, as well as for the set extremal vectors.

References

1. Gilberto Amparore, E., et al.: "years of -GreatSPN". In: Principles of Performance and Reliability Modeling and Evaluation, pp. 227–254. Springer (2016)
2. Bruns, W., Gubeladze, J.: Polytopes, rings, and K-theory. vol. 27. Springer (2009)
3. Manuel Colom, J., Silva, M.: Convex geometry and semiflows in P/T nets. A comparative study of algorithms for computation of minimal P-semiflows. In: - 10th International Conference on Application and Theory of Petri Nets, pp. 79–112. Springer (1989)
4. Graham, R.L., Knuth, D.E., Patashnik, O.: Concrete Mathematics: A Foundation for Computer Science. 2nd, pp. 440–475. Addison-Wesley (1994)
5. Graver, J.E.: On the foundations of linear and integer linear programming I. Math. Program. $9(1)$, 207–226 (1975)
6. Hemmecke, R.: On the computation of Hilbert bases of cones. In: Mathematical software. World Scientific, pp. 307–317 (2002)
7. Henk, M., Weismantel, R.: On Hilbert bases of polyhedral cones. eng. Tech. rep. SC-96-12. Takustr. 7, 14195 Berlin: ZIB (1996)
8. Malkin, P.: Computing Markov bases, Gröbner bases, and extreme rays. PhD thesis. Catholic University of Louvain, Louvain-la-Neuve, Belgium (2007)
9. Pottier, L.: Minimal solutions of linear Diophantine systems: bounds and algorithms. In: Rewriting Techniques and Applications (Como, 1991). vol. 488. Lecture Notes in Comput. Sci. Springer, Berlin, pp. 162–173 (1991)
10. Pottier, L.: The Euclidean algorithm in dimension n. In: Proceedings of the 1996 International Symposium on Symbolic and Algebraic Computation, pp. 40–42 (1996)
11. Rockafellar, R.T.: Convex Analysis. Princeton, Princeton U (1970)
12. Sander, J.W.: On maximal antihierarchic sets of integers. Discrete Math. $113(1\text{-}3)$, 179–189 (1993)
13. Schrijver, A.: Theory of linear and integer programming. Wiley-Interscience Series in Discrete Mathematics (1986)
14. Silva, M., Teruel, E., Manuel Colom, J.: Linear algebraic and linear programming techniques for the analysis of place/transition net systems. In: Advanced Course on Petri Nets, pp. 309–373. Springer (1996)
15. Sperner, E.: Ein satz über untermengen einer endlichen menge. Mathematische Zeitschrift $27(1)$, 544–548 (1928)

Aligning Observed Timed Traces with Timed Stochastic Models

Sofia Bellotti[1,2]($\boxtimes$), Thomas Chatain[1]($\boxtimes$) (ID), and Paolo Ballarini[3] (ID)

[1] ENS Paris Saclay, Gif-sur-Yvette, France
`thomas.chatain@ens-paris-saclay.fr`
[2] Politecnico di Torino, Turin, Italy
`s328866@studenti.polito.it`
[3] Université Paris Saclay, CentraleSupélec, MICS, Gif-sur-Yvette, France
`paolo.ballarini@centralesupelec.fr`

Abstract. Aligning observed and modeled behavior is a central task in conformance checking. We propose a novel approach to alignments in the setting of stochastic timed process models, where transitions fire after an exponentially distributed random delay. In the spirit of alignments based on combinations of log moves and model moves, we propose a setting where the alignment may suggest adjustments of the observed timestamps (considering that they might have been recorded with errors) in order to make it closer to the most likely trace that the model, according to its stochastic parameters, can produce.

Keywords: Conformance checking · Alignments · Timestamps · Timed Stochastic Petri nets · Continuous-time Markov chains

1 Introduction

Most attention has been paid in recent years on stochastic process models. The main idea is that process models should reflect the frequency distribution of the observed traces, taking into account that some of them may have been observed very frequently while other rarely. Stochastic workflow nets allow one to capture this reality by means of the *weight* parameter associated with each transition of the net, given that transition weights affect the probability with which transitions occur in the model. This induces a probability distribution on the model runs (over the language of the model) which can then be compared with the frequency of the observed traces through stochastic conformance measures [19, 26]. If considerable advancements have been achieved in the field of *untimed* stochastic process modelling [6–8, 11, 17] the extensions to the timed dimension of stochastic process mining are still relatively little explored.

In this paper, we indeed focus on the *time* dimension of the observed traces and therefore we consider timed stochastic Petri nets as the class of models

---

This work was partially funded by the Farman institute of ENS Paris-Saclay.

J. Desel and A. Kalenkova (Eds.): PETRI NETS 2026, LNCS 16567, pp. 64–86, 2026.
https://doi.org/10.1007/978-3-032-27879-1_4

for capturing not only the frequency but also the timing of the observed executions. Specifically, we consider, when the parameters associated to a net's transitions are understood as the rates of exponential distributions laws, the language induces a continuous probability distribution over an infinite domain of *timed* traces, which represents not only the probability of observing a sequence of actions, but also the likelihood of observing these actions at given dates.

Hence, these timed stochastic process models are relevant for a study of the *timestamps* recorded in logs. We propose original definitions for aligning recorded timed traces with timed stochastic process models. Aligning observed behavior with process models is a central task in conformance checking [9]. It is a necessary step for estimating fitness of models but also has many other applications, e.g. for model repair.

Contribution. We present a novel framework for aligning timed traces of a log to runs of a timed stochastic workflow net. Initially under the assumption of perfect fitness (i.e. the model can replay the considered trace) and considering a Markovian semantics for the timed transitions of the net (i.e., transitions delay follow an exponential distribution), we define the timed stochastic alignment as the optimization of an objective function in which the likelihood with which the model reproduces timed runs compatible with the observed trace is traded off with the distance between the occurrence time of the transitions in the net runs and the timestamps of the corresponding events in the observed trace. We then extend this formulation to handle models that do not perfectly fit the observed traces. We start with the simple case of alignments in which the order of the events in the trace is respected in the runs, we then extend the objective function to align partial orders (i.e. allowing for different ordering of concurrent events) in absence of conflicts and end up with the generalization of alignments to extended free choice nets.

Organization of the Paper. In Sect. 2 we recall the preliminary definitions of labelled timed stochastic Petri nets and their connection with continuous-time Markov chains. We recall results about the likelihood of a trace in these models. In Sect. 3, we formalize the alignment problem considering fully observable and perfectly fitting Timed Stochastic Petri Nets, analyze the structure of its solutions, and present an efficient algorithm for computing. In Sect. 4, we extend our approach to handle more realistic cases, where the model and the observation are not always perfectly aligned. We first show how to handle silent transitions, then we deal with not perfectly fitting models. In Sect. 5 we extend further this framework to partial-order traces, adapting our definition of alignment accordingly. We then derive a likelihood formula that is invariant over all traces equivalent under the partial-order semantics and, based on this formulation, introduce the corresponding optimization problem. In Sect. 6 we detail the computational resolution of the optimization problem introduced in Sects. 3 and 4.

Related Work. There is relatively little research focused on alignment analysis specifically for timed stochastic Petri nets.

Stochastic Conformance Checking. Leemans et al. [19] applied the *Earth Movers' Distance* to compare event logs with stochastic process models by measuring the cost of transforming one trace distribution to another. Cost-based fitness analysis was used by Adriansyah et al. [1] for conformance checking, although time was not considered. Similarly, Li et al. [22] performed stochastic alignments without incorporating temporal information. More recently, Incerto et al. [15] proposed stochastic conformance checking leveraging variable-length Markov chains, but also without accounting for timing aspects.

In recent work, Li, Polyvyanyy, and Leemans [23] studied the problem of matching an observed trace to a stochastic process model by formulating it as an optimization problem that jointly considers the likelihood of the model trace and its edit distance to the observed trace. Their approach is similar to the one we will present in this paper, but it is limited to an untimed setting. In [2] Alkhammash, Polyvyanyy, Moffat and García-Bañuelos introduce *entropic relevance*, a conformance measure that evaluates how well a stochastic process model compresses and explains the frequencies of traces in an event log, balancing precision and recall in a single score. In [18] Leemans and Polyvyanyy introduced an entropy-based, stochastic-aware precision and recall measures for conformance checking that distinguish frequent from rare deviations between an event log and a stochastic process model.

Trace Likelihood of CTMC. Perkins et al. [25] focused on maximizing the likelihood of CTMCs trajectories, given the initial state, end state, and a time bound. Related research by Grinberg and Perkins [14] and Levin et al. [21] developed methods to find the most likely sequence of states using the idea of *domination of sequences*, but without emphasizing sojourn times.

Temporal Conformance Checking. In [10] Chatain and Rino studied alignments for timestamp sequences and solved the purely timed alignment problem with two different metrics, including the stamp-only distance that we reuse in this paper.

Temporal Stochastic Conformance Checking. Richter et al. [27] introduced a different approach for temporal stochastic conformance checking by using *temporal stochastic conformance fitness*, which is based on estimated probability density functions of activities calculated through a Gaussian kernel, thus avoiding the use of stochastic Petri nets.

2   Preliminaries

Alphabet, Trace, Timed Trace, Log. We let $\Sigma = \{a, b, c, \dots\}$ denote the alphabet of an event log's activities (or actions) and Σ^* the set of traces (words) composed of activities in Σ. We denote $\sigma = \langle (a_1, \theta_1), (a_2, \theta_2), (a_3, \theta_3), \ldots, (a_n, \theta_n) \rangle \in (\Sigma \times \mathbb{R}_{>0})^*$ a timed word where θ_i is the (absolute) time of occurrence of the i-th activity and it is assumed $\forall i \in \mathbb{N}_{>0}$ that $\theta_{i+1} >= \theta_i$. We further denote $\delta_i = \theta_i - \theta_{i-1}$ (with the convention that $\theta_0 = 0$) the relative time of occurrence of action a_i w.r.t. action a_{i-1}. Given a timed word σ we denote $\bar{\sigma} \in \Sigma^*$ its embedded (untimed) word (e.g. for $\sigma = \langle (a, 0.1), (c, 3.7), (b, 15) \rangle$ then $\bar{\sigma} = \langle a, c, b \rangle$.

A *log* is a finite set of recorded (or observed) timed traces. We will usually denote an observed timed trace as $\hat{\sigma}$.

Labelled Petri Net. A labelled Petri net model is a tuple $N = (P, T, F, \Sigma, \lambda, M_0)$, where P is the set of places, T the set of transitions, $F \subseteq (P \times T) \cup (T \times P)$ the set of arcs, M_0 is the initial marking and $\lambda : T \to (\Sigma \cup \{\tau\})$ associates each transition with an activity from alphabet Σ (τ being the silent activity).

The *preset* (resp., *postset*) of a transition $t \in T$ is the set of its input (resp., output) places, i.e., ${}^\bullet t = \{p \in P \mid (p, t) \in F\}$ (resp., $t^\bullet = \{p \in P \mid (t, p) \in F\}$). A state of N consists in the distribution of tokens over its places and it is given by the marking function $M : P \to \mathbb{N}$ with $M(p)$ being the number of tokens in place $p \in P$ in M with $M_0 : P \to \mathbb{N}$ the initial marking. A state change corresponds to the *firing* of a transition. A transition $t \in T$ is enabled (it may be fired) in marking M if all of its input places contain at least one token, formally, if $\forall p \in {}^\bullet t, M(p) > 0$. Accordingly, $en(M) = \{t \in T \mid \forall p \in {}^\bullet t, M(p) > 0\}$ is the set of transitions that are enabled in M. Firing of an enabled transition t in marking M yields a new marking M', written as $M[t\rangle M'$, where M' results from M by removing (resp., adding) a token from (resp., to) each input (resp., output) place, i.e., $M'(p) = M(p) - 1, \forall p \in {}^\bullet t \setminus t^\bullet$, $M'(p) = M(p) + 1, \forall p \in t^\bullet \setminus {}^\bullet t$ and $M'(p) = M(p)$ otherwise. A marking M' is reachable from M if there exists a sequence of n transitions t_i ($1 \le i \le n$) and corresponding markings M_i ($1 \le i \le n+1$) such that $M = M_1[t_1\rangle M_2[t_2\rangle \ldots [t_n\rangle M_{n+1} = M'$ with any $n \ge 1$. We call a PN *run* the firing sequence $\bar{\sigma} = \langle t_1, \ldots, t_n \rangle$ by which M_{n+1} is reached from M_0, through the intermediate markings M_i ($0 < i \le n$), and $\lambda(\bar{\sigma}) = \langle \lambda(t_1), \ldots, \lambda(t_n) \rangle \in \Sigma^*$ the trace corresponding to run $\bar{\sigma}$.

For a PN N, $RS(N)$ is the reachability set of N (i.e., the set of markings reachable from the initial one M_0), and by $RG(N) = (RS(N), A)$ the reachability graph of N where $A \subseteq RS(N) \times RS(N) \times T$ is the set of arcs whose elements $(M, M', t) \in A$ are such that $M[t\rangle M'$.

The language of a labelled PN N, denoted $L(N)$, is the set of sequences of labels corresponding to all runs of the PN, i.e. $L(N) = \{\lambda(\bar{\sigma}) \in \Sigma^* \mid \bar{\sigma} \in Path(RG(N))\}$, where $Path(RG(N))$ is the set of paths in $RG(N)$.

In this paper we consider only *1-safe* Petri nets, i.e. we assume that no place carries more than one token in any reachable marking.

Labelled Timed Stochastic Petri Net. A labelled timed stochastic Petri net (TSPN) is a tuple $N_S = (P, T, F, \Sigma, \lambda, w, M_0)$ where $N = (P, T, F, \Sigma, \lambda, M_0)$ is a PN and $w : T \to \mathbb{R}_{>0}$ gives the rate of exponentially distributed transition delays. The semantics of a TSPN is isomorphic to a continuous-time Markov chain (CTMC) [24], which is subsumed in terms of a reachability graph whose transitions labelled with the rate of the corresponding transition. An execution of a labelled TSPN is a sequence $\rho = M_0 \xrightarrow{t_1,\theta_1} M_1 \xrightarrow{t_2,\theta_2} \ldots \xrightarrow{t_n,\theta_n} M_n$ with $t_i \in T$ and $\theta_i \in \mathbb{R}_{\geq 0}$, the i-th occurred transition, respectively, its (absolute) time of occurrence. For simplicity, in the remainder, we will use $\sigma = \langle(t_1,\theta_1),\ldots,(t_n,\theta_n)\rangle$ to refer to a TSPN *run*, meaning the firing sequence corresponding to an interleaving run (notice that equivalently we may refer to the transition labels to characterise a TSPN run, i.e., $\lambda(\sigma) = \langle(\lambda(t_1),\theta_1),\ldots,(\lambda(t_n),\theta_n)\rangle \in \Sigma^* \times \mathbb{R}_{\geq 0}$. The timed language of a TSPN N_S is defined as $L(N_S) = \{\lambda(\sigma) \in \Sigma^* \times \mathbb{R}_{\geq 0} \mid \lambda(\bar{\sigma}) \in Path(RG(N_S))\}$, where $\overline{(\sigma)} \in \Sigma^*$ is the untimed projection of σ.

Continuous-Time Markov Chain. A (finite-state) continuous-time Markov chain (CTMC) is a tuple $M = (S, Q, \pi_0)$, where $S = s_1,\ldots,s_n$ is a finite set of states, $Q : S \times S \to \mathbb{R}$ is the *infinitesimal generator matrix*, and $\pi_0 : S \to [0,1]$ is the initial probability distribution (i.e., $\sum_{s \in S} \pi_0(s) = 1$).

For $i \neq j$, the entry $q_{ij} = Q(s_i, s_j) \geq 0$ represents the transition rate from state s_i to state s_j. The holding (sojourn) time in state s_i is exponentially distributed with rate $E_i = \sum_{j \neq i} q_{ij}$, called the *exit rate* of state s_i. The diagonal entries satisfy $q_{ii} = -\sum_{j \neq i} q_{ij}$, so that each row of Q sums to zero. In particular, $q_{ii} \leq 0$ for all i. The probability density function that expresses the likelihood of jumping from state s_i to s_j at time t is given by $\ell(\langle s_i, t, s_j \rangle) = q_{ij} \cdot e^{-E_i \cdot t}$ whereas the likelihood of a (timed) path $\langle(s_1,t_1),\ldots,(s_k,t_k)\rangle$ (i.e., $Q(s_i, s_{i+1}) > 0, t_i \in \mathbb{R}_{\geq 0}, 1 \leq i < k$) of a CTMC is given by

$$\ell(\langle(s_1,\theta_1),\ldots,(s_k,\theta_k)\rangle) = \prod_{i=1}^{k-1} \left(q_{i,i+1} \cdot e^{-E_i(\theta_{i+1}-\theta_i)} \right) = \prod_{i=1}^{k-1} q_{i,i+1} e^{-E_i \cdot \delta_i} \quad (1)$$

where $\delta_i = \theta_{i+1} - \theta_i$.

The semantics of a timed stochastic Petri net N_S is usually interpreted as a CTMC whose states correspond to the reachable markings of N_S. This way we can express the likelihood of a run $\sigma = \langle(t_1,\theta_1),\ldots,(t_n,\theta_n)\rangle$ of N_S as

$$\ell(\sigma) = \prod_{i=1}^{n} w_{t_i}\, e^{-W_{t_i}(\theta_i - \theta_{i-1})}, \quad (2)$$

where W_{t_i} is the total firing rate of all transitions enabled in the marking M_i reached after the run $\langle(t_1,\theta_1),\ldots,(t_i,\theta_i)\rangle$: $W_{t_i} = \sum_{j \in en(M_i)} w_{t_j}$.

3   Alignment of Timed Traces to Fully Observable, Perfectly Fitting Timed Stochastic Petri Nets

In order to motivate our approach, let us consider a company *invoice processing* process, which starts when an invoice is received, ends with recording of the received invoice and whose functioning is characterised by the following stages each of which, for simplicity, is represented by a letter of the alphabet $\Sigma = \{a, b, c, c', d\}$:

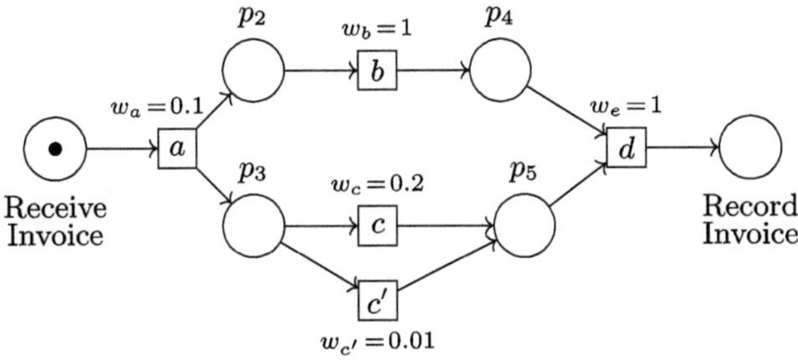

Fig. 1. Timed stochastic Petri net model capturing the invoice processing log L.

- *a: capture invoice data*
- *b: verify purchase order* (PO)
- *c: check fraud (already known client)*
- *c′: check fraud (new client)*
- *d: validate, approve, pay invoice*

Let us imagine that some executions of such process have been observed and stored in the following log L:

$$\langle (a, 10.2), (b, 11.3), (c, 14.1), (d, 14.9) \rangle$$
$$\langle (a, 9.0), (b, 10.2), (c, 15.6), (d, 16.2) \rangle$$
$$\langle (a, 1.1), (b, 10.2), (c, 14.6), (d, 15.5) \rangle$$
$$\langle (a, 8.3), (b, 9.1), (c, 15.1), (d, 16.2) \rangle$$
$$\langle (a, 12.3), (c, 12.5), (b, 13.2), (d, 19.1) \rangle$$
$$\langle (a, 9.2), (b, 10.2), (c', 45.2), (d, 45.9) \rangle$$

where each trace consists of a sequence of pairs (action, timestamp), with the timestamps being absolute occurrence times. Given that the start (a) and end (d) event are common to all traces while the intermediate events (b and c or c') may occur in different orders, it is straightforward to show that the timed stochastic Petri net model in Fig. 1 is capable of reproducing the traces in L.

Notice that the transitions of the model are assumed to be associated with exponential distributions (with corresponding rates $w_a = 0.1$, $w_b = 1$, $w_c = 0.2$, $w_{c'} = 0.01$, $w_d = 1$) characterising the random time of occurrence of each transition.

According to the model, all traces having (a, b, c, d), (a, c, b, d),(a, b, c', d), or (a, c', b, d) as sequence of actions, and increasing timestamps, are possible. Hence, all the sequences above are valid. But this validation is very weak in terms of conformance checking, for two reasons: first, we would rather expect an estimation that takes into account the likelihood of the observed traces; second, in every real log, one must take into account the possibility of erroneous records, like in the classical alignment setting based on model moves and log moves [1]. Now let us look at our example again. The ratio attached to transition a is 0.1, this indicates that the expected firing delay for a is 10 time units, while transition b waits on average only 1 time unit before firing. This is essentially coherent with the observations, except for the third one: $((a, 1.1), (b, 10.2), (c, 13.6), (d, 14.5))$: in this trace, action a occurs surprisingly early, at time 1.1; as such this is absolutely not in contradiction with the model (short dwell times are even the most likely ones in an exponentially distributed law), but what is highly surprising is the long delay $(10.2 - 1.1 = 9.1$ time units) between a and b. Of course, once again, this situation is possible according to the model, but it has very low likelihood. Instead, another explanation needs to be considered: this observation may simply be the result of an error when recording a: it may have actually occurred later, say around 9.5; the corrected trace is then much more likely according to the model.

We will now define the alignment problem between an observed timed trace and a timed stochastic process model as an optimization problem with an objective function that combines the likelihood of a model run with its distance to the observation.

In this section, we focus on timestamps and timed stochastic models, therefore we completely rule out the question of aligning observed traces to an underfitting model, i.e. a model whose language does not contain the observation, usually done by inserting log moves and model moves as in [1]. Hence, in our definition of alignment of an observed trace $\hat{\sigma} = \langle (a_1, \hat{\theta}_1) \ldots (a_n, \hat{\theta}_n) \rangle$, we assume that the model language contains a run $\sigma = \langle (t_1, \hat{\theta}_1) \ldots (t_n, \hat{\theta}_n) \rangle$ such that $\lambda(t_i) = a_i$ for all i, i.e. the model can replay exactly the observed timed trace. Precisely, our point is to show that, even when the model can perfectly replay the observed trace $\hat{\sigma} = \langle (a_1, \hat{\theta}_1) \ldots (a_n, \hat{\theta}_n) \rangle$, this replayed trace may not be the best choice for alignment when one takes into account the likelihood of traces.

Therefore, our alignment problem will combine two dimensions: likelihood and distance to the observation. The importance that the user assigns to these two dimensions is controlled by two parameters α and β.

As a metric for the distance between observed timestamps and those of the aligned model run, we use the *stamp-only distance* from [10], which simply corresponds to the L_1 norm between the vectors of timestamps $\boldsymbol{\theta} = (\theta_1, \ldots, \theta_n)$ and $\hat{\boldsymbol{\theta}} = (\hat{\theta}_1, \ldots, \hat{\theta}_n)$: $\|\boldsymbol{\theta} - \hat{\boldsymbol{\theta}}\|_{L_1} = \sum_i |\theta_i - \hat{\theta}_i|$.

Definition 1 (Alignment of a Timed Trace to a TSPN). *Given an observed trace $\hat{\sigma} = \langle (a_1, \hat{\theta}_1) \ldots (a_n, \hat{\theta}_n) \rangle$ and a TSPN $N_S = (P, T, F, \Sigma, \lambda, w, M_0)$, an alignment of $\hat{\sigma}$ to N_S is a run $\sigma = \langle (t_1, \theta_1) \ldots (t_n, \theta_n) \rangle$ of N_S such that $\bar{\sigma} = \bar{\hat{\sigma}}$ (i.e. $\lambda(t_i) = a_i$ for all i) and the firing times $\theta_1, \ldots, \theta_n$ minimize the objective function*

$$\alpha\left(-\ell(\sigma)\right) + \beta\|\boldsymbol{\theta} - \hat{\boldsymbol{\theta}}\|_{L_1} \tag{3}$$

where $\ell(\sigma)$ denotes the likelihood of σ and $\alpha, \beta \geq 0$ are tunable parameters such that $\alpha + \beta = 1$. The choice of α and β balances the emphasis between the likelihood and the distance to observed times.

In likelihood-maximization problems it is more common to use the log-likelihood; since the log function is monotonically increasing, the likelihood will achieve its maximum value at exactly the same point the log-likelihood achieves its maximum value. The corresponding log-likelihood is

$$\mathcal{L}(\sigma) = \log(\ell(\sigma)) = \sum_{i=1}^{n} \log w_{t_i} - W_{t_i}(\theta_i - \theta_{i-1}) \tag{4}$$

Thus our optimization problem becomes

$$\min_{\theta_1,\ldots,\theta_n} \quad \alpha \sum_{i=1}^{n} W_{t_i}(\theta_i - \theta_{i-1}) + \beta \sum_{i=1}^{n} \left| \theta_i - \hat{\theta}_i \right| \tag{5}$$

$$\text{subject to} \quad \theta_i \le \theta_{i+1}, \quad \forall i = 1,\ldots, n-1$$

$$\theta_n \ge \hat{\theta}_n$$

where we constrain the final timestamp $\theta_n \ge \hat{\theta}_n$ to ensure the optimized trace duration is at least as long as the observed trace, preventing premature termination before all observed events when α it is preponderant.

The solution varies when α, β are tuned and can be found for example applying the *simplex method*, which works by exploring the vertices of the feasible polyhedron.

Example. Consider the invoice workflow depicted in Fig. 1 and suppose we observe the trace $\hat{\sigma} = \langle (a, 1.1), (b, 10.2), (c, 14.6), (d, 15.5) \rangle$ Aligning it to the model N_S amounts to finding a model run $\sigma = \langle (a, \theta_1), (b, \theta_2), (c, \theta_3), (d, \theta_4) \rangle$ with firing times $\boldsymbol{\theta} = (\theta_1, \theta_2, \theta_3, \theta_4)$ which optimize the combination of stamp-only distance to the observed $\hat{\boldsymbol{\theta}} = (1.1, 10.2, 14.6, 15.5)$ and likelihood according to the model, the two criteria being weighted by tunable parameters α (for the likelihood) and β (for the distance). We solved the optimization problem (using log-likelihood) for a dense grid of α and β. In the following table we display the obtained timestamp for some common α values $(0, 0.25, 0.5, 0.75, 1)$ and for the α at which the results change $(0.4760, 0.5560, 0.9520)$, as we will explain later.

α	β	$\boldsymbol{\theta}$
0.0000	1.0000	$[1.1,\ 10.2,\ 14.6,\ 15.5]$
0.4760	0.5240	$[10.2,\ 10.2,\ 14.6,\ 15.5]$
0.5000	0.5000	$[10.2,\ 10.2,\ 14.6,\ 15.5]$
0.5560	0.4440	$[10.2,\ 10.2,\ 15.5,\ 15.5]$
0.9520	0.0480	$[15.5,\ 15.5,\ 15.5,\ 15.5]$
1.0000	0.0000	$[15.5,\ 15.5,\ 15.5,\ 15.5]$

Interpretation of the Results. When $\beta = 1$, the optimizer minimizes only the timestamp distance, thus the resulting transition firing times coincide with the observed timestamps for the corresponding actions. When $\alpha = 0.5$ and $\beta = 0.5$, the optimizer assigns the same weight to the likelihood maximization term and to the timestamp deviation penalty. The resulting timestamps agree with our intuition: since the rate of transition a is low, we expect it to occur later than the observed value 1.1. Indeed, at $\alpha = 0.5$ the model shifts transition a forward in time so that it fires together with transition b (at time 10.2). Because the penalty on timestamp deviation is equally weighted, the remaining transitions retain their observed timestamps.

When α becomes dominant, for example at $\alpha = 0.75$, the likelihood term drives the optimization, and the model also delays transition c, making it fire later in order to increase its contribution to the likelihood maximization.

At the extreme case $\alpha = 1$, the timestamp-distance term is not taken into account; the optimizer maximizes the likelihood by letting all transitions fire at the maximum observed timestamp, resulting in all timestamps being equal to the final observed value. Interestingly, for every α and β between the listed breakpoints, the optimal timestamps remain exactly the same. This means that across the entire interval $\alpha \in [0, 1]$, only four distinct optimal solutions are ever obtained. The existence of these plateaus in the solution space is not accidental, and it is formally explained by the following proposition.

Proposition 1. *For every α and β, there exists a solution of the optimization problem (5) where all the firing times $\theta_i^\star$ are taken in the set of observed timestamps $D = \{\hat{\theta}_1, \ldots, \hat{\theta}_n\}$.*

Proof. The problem (5) can be easily recast in a linear programming (LP) problem (a class of optimization problems in which one minimizes a linear objective function subject to the variables to be contained in a polyhedron), introducing the non-negative slack variables u_i and rewriting the problem as

$$\min_{\theta_1,\ldots,\theta_n} \quad \alpha \sum_{i=1}^{n} W_{t_i}(\theta_i - \theta_{i-1}) + \beta \sum_{i=1}^{n} u_i$$

$$\text{subject to} \quad \theta_i \leq \theta_{i+1}, \quad \forall i = 1, \ldots, n-1$$

$$\theta_n \geq \hat{\theta}_n$$

$$u_i \geq 0, \quad \forall i = 1, \ldots, n \tag{6}$$

$$u_i \geq \theta_i - \hat{\theta}_i, \quad i = 1, \ldots, n,$$

$$u_i \geq -(\theta_i - \hat{\theta}_i), \quad i = 1, \ldots, n.$$

Thus (6) is a linear program in the variables $(\theta_1, \ldots, \theta_n, u_1, \ldots, u_n)$ since, the constraints form a polyhedron, that is the intersection of a finite number of closed half-spaces and hyperplanes in $\mathbb{R}^{2k}$. It is known that, for a linear programming minimization problem where the feasible set is a nonempty polyhedron, either the optimal cost is $-\infty$ or there exists an optimal solution which is an extreme point. In (5) the observed trace can be replayed and the corresponding model-trace is a feasible solution with finite cost and the feasible set is a nonempty polyhedron (since the constraints are linear inequalities); thus we can conclude that it exists an optimal solution which is an extreme point of the feasible set. $\square$

We note that the use of the log-likelihood in the objective function allows us to leverage standard results from linear optimization. If we had formulated the problem using the likelihood itself, the resulting optimization would be nonlinear, and the optimal timestamps would no longer exhibit the piecewise-constant behavior observed here; instead, they would vary continuously between the extreme values.

Likelihood-Only Maximization. If $\alpha = 1$, the distance term is disregarded, and the optimization problem reduces to maximizing the likelihood of the model run. Similarly to what happens in the Perkins' Boundary problem [25], Hence, the minimum is attained when $\delta_j = \hat{\theta}_n,$ and $\delta_i = 0 \quad \forall i \neq j$.

Considering again the observed trace $\hat{\sigma}$ introduced in 3. We have already showed that, when $\alpha = 1$, the optimization problem yields the solution $\sigma = \{(a, 15.5), (b, 15.5), (c, 15.5), (d, 15.5)\}$, in which all timestamps collapse to the final observed time. This behaviour can also be directly justified using the result derived above. The CTMC induced by the TSPN of Fig. 1 has states corresponding to the following markings:

$$M_1 = \{p_1\}, \quad M_2 = \{p_2, p_3\}, \quad M_3 = \{p_3, p_4\}, \quad M_5 = \{p_4, p_5\}, \quad M_6 = \{p_6\}.$$

The exit rates of these CTMC states coincide with the total firing rates of the enabled transitions; the slowest state is therefore M_1, with minimal rate $w_{\min} = w_a = 0.1$. Thus all available time is assigned to M_1:

$$\delta^*_{M_1} = 15.5, \quad \delta^*_{M_2} = 0, \quad \delta^*_{M_3} = 0, \quad \delta^*_{M_5} = 0 \tag{7}$$

this yields exactly the run $\sigma = \{(a, 15.5), (b, 15.5), (c, 15.5), (d, 15.5)\}$, in accordance with the optimal solution obtained earlier.

3.1   Dynamic Programming Solution

By Proposition 1, we can restrict ourselves to solutions of the linear program (5) of the form $(\theta^\star_1, ..., \theta^\star_n)$ where every optimal timestamp belongs to the set of observed ones. Let $D = \{D_0 < D_1 < \cdots < D_k\}$, with $k \leq n$, denote the sorted vector of distinct observed timestamps, where n is the total number of observations. By convention, we set $D_0 = 0$ to represent the initial time point. Thus the continuous problem (5) is equivalent to a finite combinatorial problem in which, for each index i, we select a value θ_i from D. This discrete problem admits an efficient dynamic programming (DP) solution.

Given a feasible partial assignment $(\theta_1, \ldots, \theta_i)$ (with $i \leq n$), we define its *total cost* as

$$C((\theta_1, \ldots, \theta_i)) = \sum_{\ell=1}^{i} \Big(\alpha \, W_\ell \, (\theta_\ell - \theta_{\ell-1}) \; + \; \beta \, |\theta_\ell - \hat{\theta}_\ell| \Big), \tag{8}$$

For each candidate value $D_j \in D$ assigned to θ_i, the *value function* $V(i, j)$ is defined as the minimum total cost of a feasible partial assignment $(\theta_1, \ldots, \theta_i)$ such that $\theta_i = D_j$, namely

$$V(i, j) = \min_{\substack{(\theta_1, \ldots, \theta_i) \text{ feasible} \\ \theta_i = D_j}} C((\theta_1, \ldots, \theta_i)). \tag{9}$$

Given a feasible partial assignment $(\theta_1, \ldots, \theta_{i-1})$, we define the *stage cost* as the cost incurred by adding transition i, that is, by extending the assignment to $(\theta_1, \ldots, \theta_i)$ with $\theta_i = D_j$. This cost is obtained by substituting $\theta_i = D_j$ into the objective of (5):

$$\text{stageCost}(i, j) = \underbrace{\alpha \, W_i \, (D_j - \theta_{i-1})}_{\text{likelihood cost}} + \underbrace{\beta \, |D_j - \hat{\theta}_i|}_{\text{observation cost}} . \tag{10}$$

The *total cost* of a feasible partial assignment $(\theta_1, \ldots, \theta_i)$ with $\theta_i = D_j$ is then given by the sum of the stage cost (10) and the total cost of the previous partial assignment $(\theta_1, \ldots, \theta_{i-1})$. The monotonicity constraint $\theta_{i-1} \leq \theta_i$ implies that if $\theta_i = D_j$, then θ_{i-1} must equal some D_r with $r \leq j$. Thus we can express the value function as

$$V(1, j) = \text{stageCost}(1, j), \tag{11}$$

$$V(i, j) = \text{stageCost}(i, j) + \min_{1 \leq r \leq j} V(i - 1, r), \quad i = 2, \ldots, n, \; j = 1, \ldots, k. \tag{12}$$

The optimal final timestamp must satisfy $\theta_n \geq D_k$, hence the optimal alignment cost of a total assignment is

$$V^\star = \min_{\substack{1 \leq j \leq k \\ D_j \geq D_k}} V(n, j) = \min_{\substack{1 \leq j \leq k \\ D_j \geq D_k}} \min_{\substack{(\theta_1, \ldots, \theta_n) \text{ feasible} \\ \theta_n = D_j}} C((\theta_1, \ldots, \theta_n)). \tag{13}$$

It can be computed by dynamic programming using a table of size $n \times k$. An optimal sequence $(\theta_1^\star, \ldots, \theta_n^\star)$ is than obtained by backtracking. A naïve implementation of the recurrence (12) requires computing a minimum over up to j previous values, resulting in a computational complexity of $O(nk^2)$. However, the recurrence can be reformulated so that each entry is computed in $O(1)$ time using prefix minima, reducing the overall runtime to $O(nk)$.

Specifically, since the *likelihood cost* satisfies $V(i - 1, r) + \alpha W_i(D_j - D_r) = \big[V(i - 1, r) - \alpha W_i D_r\big] + \alpha W_i D_j$, for fixed i, only the first term depends on r. Defining the prefix minimum $M_j = \min_{r \leq j}\big(V(i - 1, r) - \alpha W_i D_r\big)$, the dynamic programming update becomes $V(i, j) = M_j + \alpha W_i D_j + \beta |D_j - \theta_i|$, which can be computed in $O(1)$ time per entry. Maintaining the argmin associated with each prefix minimum ensures correct backtracking, while the total complexity is reduced to $O(nk)$.

4   Handling Silent Transitions and Not Perfectly Fitting Models

In the previous section, we made strong assumptions about the alignment between the model and the observation: in order to isolate the question related to timestamps, we focused on perfectly fitting models and even required that they were fully observable. In this section, we show how to handle more realistic cases, where the model and the observation are not always perfectly aligned. We first show how to handle silent transitions, then we deal with not perfectly fitting models.

Let $\hat{\sigma}$ be an observed trace. Our approach proceeds in two steps:

- First, we consider the corresponding *untimed* trace $\bar{\hat{\sigma}}$ and compute a classical alignment with the Petri net N, obtained from the stochastic Petri net N_s by disregarding temporal and frequency information. Following the approach of [1], we determine a complete run $\bar{\sigma}$ of N that minimizes the Levenshtein distance to $\bar{\hat{\sigma}}$.
- Second, we perform a *timed alignment* between the trace resulting from the untimed alignment and the original stochastic Petri net N_s.

4.1   Silent Transitions

When dealing with real data log, silent transitions are particularly relevant, due to the fact that many process discovery algorithms produce as output a Petri net with silent transitions. We recall that in a *Labelled Stochastic Petri Net N_s* there are two types of timed transitions: *visible* transitions $\lambda(t) \in \Sigma$ and *silent* transitions $\lambda(t) = \tau$. While the firing of a visible transition is explicitly recorded as an event in a trace in an event log, the firing of a silent transition instead indicates the execution of an internal step, not corresponding to any visible activity in the process.

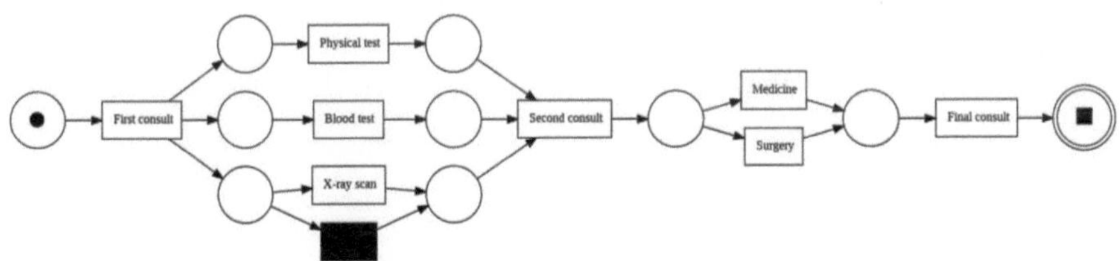

Fig. 2. Labelled Petri Net N from the hospital dataset.

As an example, we consider the labelled Petri net N shown in Fig. 2. Our goal is to show how we handle the silent transitions in the alignment algorithm.

The model N was derived from an artificial hospital event log, publicly available on GitHub. Firstly, we ran the Inductive Miner algorithm [16] as implemented in the PM4Py Python framework [3] to get the untimed Petri net N depicted in Fig. 2, then we assigned exponential firing rates in order to get a Labelled Stochastic Petri Net (LSPN). For observable transition, these rates were computed based on the average firing delays observed in the log. The inductive miner introduces a silent transition in conflict with the *X-ray scan* transition in order to model the fact that this activity may or may not occur. In the literature, silent transitions have often been considered instantaneous, however in our LSPN formalism, they have firing rates like others: the idea is that silent transitions may represent internal steps, which are not observable but still take time. This idea was discussed for instance in [20]. A silent transitions can still be made almost immediate by assigning a high firing rate; however, it is well known that, in a conflict between an immediate and a timed transition, the immediate transition always fires. This behavior would not reflect the observed traces, since the *X-ray scan* activity occurs in 90% of the cases. Conversely, assigning a low firing rate to the silent transition would artificially reduce the waiting time of the subsequent *Second consult* activity, distorting its estimated rate. For these reasons, we inserted an additional silent transition τ_1 before XS. The conflict between the two silent transitions explicitly models the decision of executing the *X-ray scan* or not. These transitions are timed but they are assumed to occur almost immediately and are therefore assigned high firing rates. In particular a higher firing rate is assigned to the silent transition that allows *X-ray scan*, reflecting its higher probability of execution.

The *First consult* activity appears in all traces and is used as a temporal reference point. Specifically, for each process instance, the timestamp of this activity is set to time 0, and all subsequent timestamps are expressed relative to it. Consequently, no exponential rate is associated with this transition, and it is not included in the optimization problem. Its role is exclusively to define a reference time for each process instance, which is necessary to handle different starting times and to make the alignment problem well defined. We rename the activities for clarity as follows: (Physical test $\rightarrow$ PT, Blood test $\rightarrow$ BT, X-ray scan $\rightarrow$ XS, Second consult $\rightarrow$ SC, Surgery $\rightarrow$ S, Medicine $\rightarrow$ M, Final consult $\rightarrow$ FC.)

The corresponding labelled stochastic Petri net (LSPN) N_s is illustrated in Fig. 3.

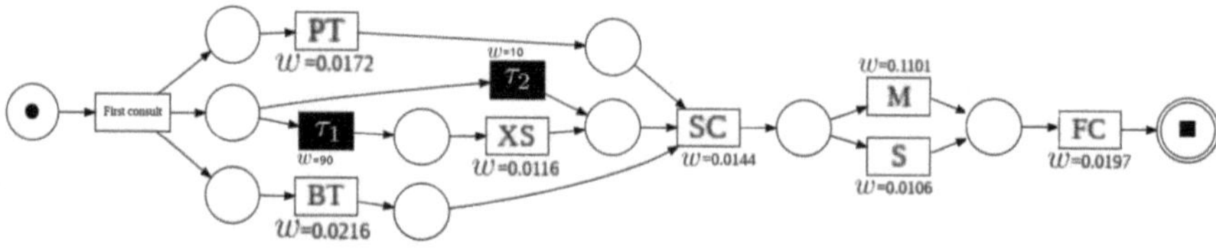

Fig. 3. LSPN N_s from the hospital dataset

Now, we consider the following observed trace and show how to align it with our model N_s:

$$\hat{\sigma} = \langle \, (\textit{First consult, } 2017\text{-}01\text{-}02 \ 11\text{:}40\text{:}11), (\textit{Blood test, } 2017\text{-}01\text{-}02 \ 12\text{:}47\text{:}33),$$
$$(\textit{Physical test, } 2017\text{-}01\text{-}02 \ 12\text{:}53\text{:}50), (\textit{Second consult, } 2017\text{-}01\text{-}02 \ 16\text{:}21\text{:}06),$$
$$(\textit{Surgery, } 2017\text{-}01\text{-}05 \ 13\text{:}23\text{:}09), (\textit{Final consult, } 2017\text{-}01\text{-}09 \ 08\text{:}29\text{:}28) \rangle$$

Our alignment procedure is in two steps.

Step 1: Untimed Alignment. For now we still assume the model to be perfectly fitting, in the sense that, for every observed trace $\hat{\sigma}$ that we consider, the model has a run σ' such that $\lambda(\sigma') = \hat{\sigma}$. Simply, we now allow models which have silent transitions, which means that σ' may be only partially observed. The untimed alignment, with cost 0, yields the model trace $\bar{\sigma}' = \langle \textit{First consult, } \tau_2, \textit{ BT, PT, SC, S, FC, } \rangle$. In this section, our purpose is not to discuss how σ' can be computed (this will come naturally with the next section): what we want to show is how we handle the silent transition in the alignment of timestamps.

Step 2: Timed Alignment. To perform the timed alignment, we construct a corresponding timed trace by assigning observed timestamps only to the transitions that are actually observed. The timestamps are expressed in hours and relatively to the *First consult*, which is excluded from the optimization, since it is common to all traces and serves only as the initial reference point. The resulting (partially) timed trace is therefore

$$\sigma' = \langle (\tau_2, -), (BT, 1.1228), (PT, 1.2275), (SC, 4.6819),$$
$$(S, 73.7161), (FC, 164.8214) \rangle$$

Let $\mathcal{O} \subseteq \{1, \ldots, n\}$ denote the set of indices of transitions in σ' with observed timestamps, in order to allow silent transitions to influence the likelihood of the execution without contributing to the timestamp deviation term, we modify the linear program (5) as follows.

$$\min_{\theta_1, \ldots, \theta_n} \quad \alpha \sum_{i=1}^{n} W_{t_i}(\theta_i - \theta_{i-1}) + \beta \sum_{i \in \mathcal{O}} |\theta_i - \hat{\theta}_i|$$
$$\text{subject to} \quad \theta_i \leq \theta_{i+1}, \quad \forall i = 1, \ldots, n-1 \tag{14}$$
$$\theta_n \geq \max_{i \in \mathcal{O}} \hat{\theta}_i$$

This ensures that only observed transitions are penalized for deviation from the observed timestamps, while all transitions, including silent ones, are considered in the likelihood component of the objective.

The last constraint enforces that the final timestamp θ_n is at least as large as the biggest observed timestamp and can be linearized as the set of constraints $\theta_n \geq \hat{\theta}_i \quad \forall i \in \mathrm{O}$ Thus the problem remains linear and Proposition 1 still holds.

In the DP approach, we adjust the *stage cost* to account for transitions with missing observed timestamps. Formally, the modified stage cost is

$$\text{stageCost}(i,j) = \alpha \, W_i \, (D_j - \theta_{i-1}) \; + \; \begin{cases} \beta \, |D_j - \hat{\theta}_i|, & \hat{\theta}_i \text{ observed,} \\ 0, & \text{otherwise.} \end{cases} \tag{15}$$

This reformulation allows silent transitions to contribute to the likelihood term while not affecting the timestamp deviation.

4.2   Not-Perfectly Aligned Untimed Traces

So far, we have restricted our attention to settings in which the model can perfectly replay the observed trace. We now extend our approach to more realistic scenarios, where the observed traces cannot be replayed exactly by the model. We show that the approach used in the case where the only inserted transitions were the silent ones still applies in this setting. In particular, we consider the model shown in Fig. 1 and assume that the following timed trace is observed:

$$\hat{\sigma} = \langle (a, 10.2), (b, 11.3), (d, 14.1), (c, 14.9) \rangle.$$

Step 1: Untimed Alignment. We first extract the untimed trace $\hat{\bar{\sigma}} = \langle a, b, d, c \rangle$. The optimal untimed alignment consists in swapping the activities c and d. Operationally, this is achieved by deleting d and inserting it after c, resulting in a total alignment cost of 2.

$$\begin{array}{ll} \text{log:} & a \; b \; d \; c \gg \\ \text{model:} & a \; b \gg c \; d \end{array}$$

where $\left(\frac{d}{\gg}\right)$ denotes a log move (deletion), while $\left(\frac{\gg}{d}\right)$ denotes a model move (insertion), each with unit cost.

Step 2: Timed Alignment. The untimed alignment yields the following model trace: $\bar{\sigma}' = \langle a, b, c, d \rangle$. To perform the timed alignment, we construct a corresponding timed trace as in Sect. 4.1, assigning the observed timestamps only to transitions that already align with the model, while inserted transitions (model moves) are considered as unobserved transitions and treated like silent transitions. The resulting (partially) timed trace is therefore

$$\sigma' = \langle (a, 10.2), (b, 11.3), (c, 14.9), (d, -) \rangle,$$

and the timed alignment can be performed using the optimization problem formulation (14).

5 Aligning Partial-Order Traces

The idea of introducing partial orders in the definition of alignments already arised in [28]. Here we show how they can improve our alignment to TSPN models.

Let us come back to our introductory example of Fig. 1, and focus on the log trace $\langle(a, 12.3), (c, 12.5), (b, 13.2), (d, 19.1)\rangle$. We remark that the actions occur in a surprising order (c before b while c waits in average significantly longer than b (rate 1 for b and 0.2 for c). Moreover, the waiting time before d (5.9 time units) is surprisingly long. A reasonable explanation to this is that c actually fired later, around 17.5–18, but was erroneously recorded with the timestamp 12.5. This explains the late firing of d and is perfectly compatible with the expected delay of c. Simply, this modification changes the order of b and c, therefore, our previous method (Sect. 3) will not find it. It will not be able to do better than relating our observed trace to a model run of the type $((a, 12.3), (c, \theta), (b, \theta), (d, 19.1))$ where c fires immediately before b, at some time θ somewhere between 12.5 and 19.1, which is a little bit more likely than the observed trace according to the model, but is certainly not the best explanation.

The goal of this section is to adapt our definition of alignments so that it allows to permute concurrent actions. In the sense of Mazurkiewicz traces [13], we consider two untimed runs σ and σ' of a TSPN N_S as equivalent if they differ only by permutations of transitions which are concurrent according to the model N_S. Two transitions t_1 and t_2 are concurrent if the union of their presets and postsets do not intersect $({}^\bullet t_1 \cup t_1{}^\bullet) \cap ({}^\bullet t_2 \cup t_2{}^\bullet) = \emptyset^1$. This is the case of b and c in our example of Fig. 1, which puts $\langle a, b, c, d\rangle$ and $\langle a, c, b, d\rangle$ in the same equivalence class in the sense of Mazurkiewicz traces.

Now our alignment problem becomes the following.

Definition 2 (Partial-order Alignment of a Timed Trace to a TSPN).
Given an observed trace $\hat{\sigma} = \langle(a_1, \hat{\theta}_1) \ldots (a_n, \hat{\theta}_n)\rangle$ and a TSPN $N_S = (P, T, F, \Sigma, \lambda, w, M_0)$, a partial-order alignment of $\hat{\sigma}$ to N_S is a run $\sigma = \langle(t_1, \theta_1) \ldots (t_n, \theta_n)\rangle$ of N_S such that $\bar{\sigma}$ and $\hat{\bar{\sigma}}$ are equivalent in the sense of Mazurkiwicz traces and the firing times $\theta_1, \ldots, \theta_n$ minimize the objective function

$$\alpha\left(-\ell(\sigma)\right) + \beta\left\|\pi(\boldsymbol{\theta}) - \hat{\boldsymbol{\theta}}\right\|_{L_1} \tag{16}$$

where $\alpha, \beta \geq 0$ are tunable parameters such that $\alpha + \beta = 1$ and π denotes the permutation by which $\pi(\bar{\sigma}) = \hat{\bar{\sigma}}$: the firing times must be permuted accordingly.

Coming back to our example of observed trace $\langle(a, 12.3), (c, 12.5), (b, 13.2), (d, 19.1)\rangle$, our partial-order alignment problem now allows to choose the best run of the model among the runs of the form $\langle(a, \theta_a), (c, \theta_c), (b, \theta_b), (d, \theta_d)\rangle$

---

[1] Notice that such notion of *structural* concurrency is a necessary, but not sufficient, condition for *behavioral* concurrency, in the sense that two transitions can have an empty intersection of presets and postsets and still not be concurrent if they are behaviorally mutually exclusive, such as when they belong to alternative branches of a process.

and $\langle (a, \theta_a), (b, \theta_b), (c, \theta_c), (d, \theta_d) \rangle$ because their untimed support $\langle a, c, b, d \rangle$ and $\langle a, b, c, d \rangle$ are equivalent in the sense of Mazurkiewicz traces. We solve the problem for the two possible orders using the method introduced in Sect. 3. If we consider for example $\alpha = 0.7$, we obtain, for the first trace in which we consider as the order the one that we observed, $\langle (a, 13.2), (c, 13.2), (b, 13.2), (d, 19.1) \rangle$ and the corresponding value of the objective function 5.9223, while for the other sequence, in which we do not trust the observation and consider the right order of the transitions as the most likely one, we find $\langle (a, 13.2), (b, 13.2), (c, 19.1), (d, 19.1) \rangle$ and the corresponding value of the objective function is 4.7235, which is a lower value. This corresponds to our intuition that this is a more satisfactory explanation to the observation as long as one considers the likelihood of the aligned trace. The obvious way to solve our partial-order alignment problem is to solve the optimization problems for all the equivalent traces, trying all permutations of concurrent transitions, but this implies to call the solver several times and compare the results.

5.1 Likelihood of Partial Order Traces

In this section, we show that the objective function for the alignment can be written in a form that is common to all the equivalent traces. This allows one to find the partial-order alignment as the result of one optimization problem for this common formula.

For this, we restrict ourselves to a well known class of Petri nets called extended free choice Petri nets [12].

Two transitions t and t' are in *conflict* if they share an input place. An extended free choice Petri net is a Petri net where every two transitions in conflict, have exactly the same input places.

Definition 3 (Extended Free Choice Petri Net [4,12]). *An* extended free choice *Petri net is a Petri net such that:*

$$\forall t, t' \in T \quad {}^\bullet t \cap {}^\bullet t' \neq \emptyset \implies {}^\bullet t = {}^\bullet t' .$$

Definition 4 (Conflict Cluster[5]). *We define the equivalence relation*

$$t \sim t' \iff {}^\bullet t \cap {}^\bullet t' \neq \emptyset$$

A conflict cluster K is an equivalence class under $\sim$: it is the maximal set of transitions with identical presets ${}^\bullet t$.

For example, in Fig. 1, if we consider transition c, its conflict cluster is made by the transition c' (with which it is in conflict) and itself: $K_{c'} = \{c, c'\}$

For extended free choice Petri nets we can express the likelihood formula without taking into account the order of concurrent transitions.

Definition 5 (Enabling Time). *Let $\sigma = \langle (t_1, \theta_1), \ldots, (t_n, \theta_n) \rangle$ be a run of a timed stochastic Petri net $N_S = (P, T, F, \Sigma, \lambda, w, M_0)$. Denote $M_1, \ldots, M_n$ the markings reached along this run. We assume, for simplicity, that all the*

t_i are distinct. *This allows us to refer to the firing time of transition without mentioning its position in the trace: $\theta_t = \theta_i$ when $t = t_i$.*

We define the enabling time *of transition t_i as the earliest time when t_i got enabled:*

$$e_{t_i} = \min\{\theta_j \mid j \in \{0, \ldots, i-1\},\ t_j \text{ is enabled in marking } M_j\}$$

Let us return again to the example of Fig. 1 and consider the timed trace $\sigma = \langle(a, 12.3), (c, 12.5), (b, 13.2), (d, 19.1)\rangle$ presented at the beginning of Sect. 5; the corresponding enabling times are $e = (e_a = 0, e_c = 12.3, e_b = 12.3, e_d = 13.2)$ In particular, the enabling time of transition d is given by $max(\theta_c, \theta_d) = 13.2$

An important observation for the following is that the enabling time of a transition t can be expressed referring to its causal predecessors, i.e. the set $^{\bullet\bullet}t$ of transitions that produce a token consumed by t:

$$^{\bullet\bullet}t = \{t' \mid t'^{\bullet} \cap {}^{\bullet}t \neq \emptyset\}$$

and these causal predecessors do not depend on the ordering of concurrent transitions.

Proposition 2. *The enabling time of transition t_i can be rewritten as: $e_{t_i} = \max_{t_j \in {}^{\bullet\bullet}t_i} \theta_j$. Now, the indices do not matter any more, and we can simply write $e_t = \max_{t' \in {}^{\bullet\bullet}t} \theta_{t'}$ where $\theta_{t'}$ denotes the firing time of transition t' in the trace.*

The important point is that now the expression for e_t does not depend on the precise ordering of concurrent transitions.

For example, in 1 if we consider two traces $\sigma_1 = \langle(a, 10), (b, 12), (c, 14), (d, 15)\rangle$ and $\sigma_2 = \langle(a, 10), (c, 11), (b, 14), (d, 15)\rangle$ in both cases $e_d = 14$: it is not important in which order b and c happened, but only when the last one fired.

Lemma 1. *For extended free choice Petri nets, the likelihood of every timed trace $\sigma = \langle(t_1, \theta_1), \ldots, (t_n, \theta_n)\rangle$ is, denoting $\Gamma_{t_i} = \sum_{k \in K_{t_i}} w_{t_k}$:*

$$\ell(\sigma) = \prod_{i=1}^{n} w_{t_i} e^{-\Gamma_{t_i}(\theta_{t_i} - e_{t_i})} \tag{17}$$

We omit the proof of this lemma and instead derive the formula in a concrete example to illustrate the underlying reasoning. Let us return to the example of Fig. 1 and consider the two untimed traces $\bar{\sigma}_1 = \langle a, b, c', d\rangle$ and $\bar{\sigma}_2 = \langle a, c', b, d\rangle$. They are equivalent in the sense of Mazurkiewicz traces because they differ only by the permutation of transitions b and c' which are concurrent according to the model. Let us report the likelihood for $\sigma_1 = \langle (a, \theta_1), (b, \theta_2), (c', \theta_3), (d, \theta_4) \rangle$, following (2) (for $\sigma_2 = \langle (a, \theta_1), (c', \theta_2), (b, \theta_3), (d, \theta_4) \rangle$ we use the same approach).

$$\ell(\sigma_1) = w_a e^{-w_a \theta_1} \cdot w_b e^{-(w_b + w_c + w_{c'})(\theta_2 - \theta_1)} \cdot w_{c'} e^{-(w_c + w_{c'})(\theta_3 - \theta_2)} \cdot w_d e^{-w_d(\theta_4 - \theta_3)}$$

$$= w_a e^{-w_a \theta_1} \cdot w_b e^{-w_b(\theta_2 - \theta_1)} \cdot w_{c'} e^{-\Gamma_{c'}(\theta_3 - \theta_1)} \cdot w_d e^{-w_d(\theta_4 - \theta_3)}$$

Knowing that $K_{c'} = \{c, c'\}$ and using the relations: $e_a = \theta_0 = 0, e_b = e_c = \theta_1, e_d = \theta_3, \theta_a = \theta_1, \theta_b = \theta_2, \theta_{c'} = \theta_3, \theta_d = \theta_4$ for trace σ_1, $\theta_a = \theta_1, \theta_b = \theta_3, \theta_{c'} = \theta_2, \theta_d = \theta_4$ for trace σ_2

We obtain the following expressions for the likelihood:

$$\ell(\sigma_1) = w_a e^{-w_a(\theta_a - e_a)} \cdot w_b e^{-w_b(\theta_b - e_b)} \cdot w_{c'} e^{-\Gamma_{c'}(\theta_{c'} - e_{c'})} \cdot w_d e^{-w_d(\theta_d - e_d)} \quad (18)$$

$$\ell(\sigma_2) = w_a e^{-w_a(\theta_a - e_a)} \cdot w_{c'} e^{-\Gamma_{c'}(\theta_{c'} - e_{c'})} \cdot w_b e^{-w_b(\theta_b - e_b)} \cdot w_d e^{-w_d(\theta_d - e_d)} \quad (19)$$

A direct comparison shows that the expressions in (18) and (19) coincide.

5.2 Optimization Problem Formulation

In this section, we adapt the alignment–optimization problem introduced in Sect. 3 to the setting of partial-order traces. From the likelihood expression established in Lemma 1, the log-likelihood of a trace σ is

$$\mathcal{L}(\sigma) = \log(\ell(\sigma)) = \sum_{i=1}^{n} \log(w_{t_i}) - \Gamma_{t_i}(\theta_{t_i} - e_{t_i}). \quad (20)$$

As in the fixed-order case, the terms $\log(w_{t_i})$ do not depend on θ and can therefore be omitted in the optimization. The timestamp-based distance term remains unchanged, but the crucial difference in the partial-order setting is that the enabling times e_{t_i} depend on multiple possible predecessors due to concurrency. We must therefore introduce constraints that encode all admissible linearizations consistent with the partial order. Thus the optimization problem becomes

$$\min_{\boldsymbol{\theta}, \mathbf{e}} \quad \alpha \sum_{i=1}^{n} \Gamma_{t_i}(\theta_{t_i} - e_{t_i}) + \beta \sum_{i=1}^{n} \left| \theta_{t_i} - \hat{\theta}_{t_i} \right|$$

$$\text{s.t.} \quad e_{t_i} = \max_{t_j \in \bullet\bullet t_i} \theta_{t_j}, \qquad \forall i = 1, \ldots, n, \quad (21)$$

$$\theta_{t_i} \geq e_{t_i}, \qquad \forall i = 1, \ldots, n,$$

$$\theta_{t_n} \geq \hat{\theta}_{t_n}.$$

The optimization now involves two groups of variables: $\boldsymbol{\theta}$, $\mathbf{e}$. However, the constraints defining the enabling times are nonlinear due to the presence of the max operator over multiple concurrent predecessors. Consequently, the problem can no longer be reformulated as a linear program, and classical simplex-based LP solvers cannot be applied directly; we can use a *nonlinear solver in MATLAB*, which allows us to directly impose the nonlinear constraints corresponding to the maximum. However, it converges only to *local* minima; different initializations may yield different optima.

6   Experiments

We implemented the functions for solving the Alignment Problem presented in Sect. 3 with both DP approach and linear programming in Matlab. The codes are publicly available on GitHub. Computations were performed on an Apple Mac with an Apple M4 CPU and 16 GB of RAM with 10 cores on *MATLAB_R2025a*. We computed the optimal alignments for the example in Sect. 3 using a grid of 10,000 α values to obtain precise estimates of α and β at which the solution changes. The linear solver converged in 4.6578 s using the dual-simplex method, while the dynamic programming (DP) approach required 0.2094 s with a naïve implementation, reduced to 0.1050 s using prefix minima. The problem from Sect. 4.1 was solved with the same α grid. The LP solver took 6.07 s, while DP required 0.2719 s naïvely, reduced to 0.1005 s with prefix minima. Both LP and DP produced identical aligned timestamps, reported below.

α	β	Optimal Alignment
0.0000	1.0000	$(\tau_2,0)$, (BT,1.123), (PT,1.228), (SC,4.682) (S,73.72), (FC,164.8)
0.9789	0.0211	$(\tau_2,0)$, (BT,0), (PT,1.228), (SC,4.682) (S,73.72), (FC,164.8)
0.9910	0.0090	$(\tau_2,0)$, (BT,0), (PT,1.228), (SC,4.682) (S,164.8), (FC,164.8)
0.9962	0.0038	$(\tau_2,0)$, (BT,0), (PT,1.228), (SC,1.228) (S,164.8), (FC,164.8)
0.9968	0.0032	$(\tau_2,0)$, (BT,0), (PT,0), (SC,0) (S,164.8), (FC,164.8)

These results are fully consistent with Proposition 1: every optimal solution is composed of timestamps drawn from the finite set of observed timestamps, and changes in the optimal solution occur at the five parameter values where two such extreme points exchange optimality. As in the example presented in Sect. 3, for $\alpha = 0$ the optimal timestamps coincide with the observed ones. However, for the silent transition we do not have an observed timestamp. The optimal timestamp assigned to it is 0, which is reasonable: the silent transition has a high rate, making it almost immediate. Moreover, in the objective function for silent transitions, the timestamp-only term is not considered, so the objective reduces to maximizing the likelihood. Since the rate is high, this is achieved by minimizing the difference between the timestamp of the silent transition and that of the previous event (in this case, time 0 of the first consult). The first switch occurs at $\alpha = 0.9789$, indicating that the timestamp distance term dominates the objective function over a wide range of parameter values, while the likelihood term influences the solution only for sufficiently large values of α. This behavior is likely due to the relatively long mean waiting times, which amplify the contribution of the distance term. In addition, expressing timestamps in hours may have an impact in the relative weight of this term in the objective function. When $\alpha = 1$, the objective function reduces to the maximization of the likelihood. This is achieved by allocating the entire waiting time to the transition with the smallest rate, namely *Surgery*.

7   Conclusion

In this paper we have defined the first alignment notion between observed timed traces and timed stochastic process models. Experimental results confirm our intuition of how a trace aligns to a model. We argued that, because of the likelihood of traces, even alignment to a perfectly fitting model is the result of a compromise that we solve using optimization techniques. For non-fitting models, we proposed a simple two-step approach that combines our definitions with the techniques known for untimed models, based on model moves and log moves. We leave as a future perspective the design of more elaborated techniques where the two dimensions of the alignment problem would interact more subtly. Additionally, we are working on heuristics to solve partial-order alignments efficiently in practice.

References

1. Adriansyah, A., van Dongen, B.F., van der Aalst, W.M.: Conformance checking using cost-based fitness analysis. In: 2011 IEEE 15th International Enterprise Distributed Object Computing Conference, pp. 55–64. IEEE (2011)
2. Alkhammash, H., Polyvyanyy, A., Moffat, A., García-Bañuelos, L.: Entropic relevance: a mechanism for measuring stochastic process models discovered from event data. Inf. Syst. **102**, 101927 (2021)
3. Berti, A., Van Zelst, S.J., van der Aalst, W.: Process mining for python (pm4py): bridging the gap between process- and data science. CoRR abs/1905.06169 (2019)
4. Best, E.: Structure theory of petri nets: the free choice hiatus. In: Proceedings of an Advanced Course on Petri Nets: Central Models and Their Properties, Advances in Petri Nets 1986-Part I, pp. 168–205. Springer-Verlag, London, UK (1987)
5. Best, E., Devillers, R., Erofeev, E.: A new property of choice-free petri net systems. In: Application and Theory of Petri Nets and Concurrency: 41st International Conference, PETRI NETS 2020, Paris, France, 24–25 June 2020, Proceedings, volume 12152 of Lecture Notes in Computer Science, pp. 89–108. Springer (2020)
6. Brockhoff, T., Uysal, M.S., van der Aalst, W.M.: Wasserstein weight estimation for stochastic petri nets. In: 2024 6th International Conference on Process Mining (ICPM), pp. 81–88 (2024)
7. Burke, A., Leemans, S.J., Wynn, M.T.: Stochastic process discovery by weight estimation. In: ICPM 2020 Int. Workshops, Padua, Italy, 5–8 October 2020, Revised Selected Papers, volume 406 of Lecture Notes in Business Information Processing, pp. 260–272. Springer (2020)
8. Burke, A., Leemans, S.J., Wynn, M.T.: Discovering stochastic process models by reduction and abstraction. In: 42nd Int. Conf., PETRI NETS 23–25 June 2021, Proceedings, volume 12734 of LNCS, pp. 312–336. Springer (2021)

9. Carmona, J., Van Dongen, B., Solti, A., Weidlich, M.: Conformance Checking - Relating Processes and Models. Springer (2018)

10. Chatain, T., Rino, N.: Timed alignments. In: 2022 4th International Conference on Process Mining (ICPM), pp. 112–119 (2022). hal-03703712

11. Cry, P., Horváth, A., Ballarini, P., Gall, P.L.: A framework for opt imisation based stochastic process discovery. In: Quantitative Evaluation of Systems (QEST) and Formal Modeling and Analysis of Timed Systems (FORMATS), volume 14996 of LNCS, pp. 34–51. Springer Nature Switzerland, Cham (2024)

12. Desel, J., Esparza, J.: Free Choice Petri nets, Cambridge University Press (1995)

13. Diekert, V.: The Book of Traces. World Scientific Publishing Co., Inc, River Edge, NJ, USA (1995)

14. Grinberg, Y., Perkins, T.J.: State sequence analysis in hidden markov models. In: Proceedings of the Thirty-First Conference on Uncertainty in Artificial Intelligence, pp. 336–344, Amsterdam, The Netherlands (2015)

15. Incerto, E., Vandin, A., Ahrabi, S.S.: Stochastic conformance checking based on variable-length markov chains. Inf. Syst. **133**, 102561 (2025)

16. Leemans, S.J.J., Fahland, D., van der Aalst, W.M.P.: Discovering block-structured process models from event logs - a constructive approach. In: Colom, J., Desel, J., (eds.), Application and Theory of Petri Nets and Concurrency, pp. 311–329. Springer, Berlin, Heidelberg (2013)

17. Leemans, S.J.J., Li, T., Montali, M., Polyvyanyy, A.: Stochastic process discovery: can it be done optimally? In: Advanced Information Systems Engineering: 36th Int. Conf., CAiSE 2024, Limassol, Cyprus, 3–7 June 2024, Proceedings, pp. 36–52. Springer-Verlag, Berlin, Heidelberg (2024)

18. Leemans, S.J.J., Polyvyanyy, A.: Stochastic-aware precision and recall measures for conformance checking in process mining. Inf. Syst. **115**, 102197 (2023)

19. Leemans, S.J.J., Syring, A.F., van der Aalst, W.M.: Earth movers' stochastic conformance checking. In: Hildebrandt, T., van Dongen, B.F., Röglinger, M., Mendling, J., (eds.), Business Process Management Forum, pp. 127–143. Springer Int. Publishing, Cham (2019)

20. Leemans, S. J., Maggi, F.M., Montali, M.: Enjoy the silence: analysis of stochastic petri nets with silent transitions. Inf. Syst. **124**, 102383 (2024)

21. Levin, P., Lefebvre, J., Perkins, T.J.: What do molecules do when we are not looking? state sequence analysis for stochastic chemical systems. J. R. Soc. Interface **9**(77), 3411–3425 (2012)

22. Li, T., Polyvyanyy, A., Leemans, S.J.: Stochastic alignments: matching an observed trace to stochastic process models. Bus. Process Manag. (2025)

23. Li, T., Polyvyanyy, A., Leemans, S.J.: Stochastic alignments: matching an observed trace to stochastic process models (2025)

24. Natkin, M., Molloy, J.: A general theory of stochastic petri nets. In: Proceedings of the 3rd International Conference on Application and Theory of Petri Nets, Lecture Notes in Computer Science. Springer (1982)

25. Perkins, T.J.: Maximum likelihood trajectories for continuous-time markov chains. In: Bengio, Y., Schuurmans, D., Lafferty, J., Williams, C., Culotta, A., (eds.), Advances in Neural Information Processing Systems 22, pp. 1–9. Curran Associates Inc. (2009)

26. Polyvyanyy, A., Moffat, A., Garc'ia-Banuelos, L.: An entropic relevance measure for stochastic conformance checking in process mining. In: 2020 2nd Int. Conf. on Process Mining (ICPM), pp. 97–104 (2020)

27. Richter, F., Sontheim, J., Zellner, L., Seidl, T.: TADE: stochastic conformance checking using temporal activity density estimation. In: Fahland, D., Ghidini, C., Becker, J., Dumas, M. (eds.) BPM 2020. LNCS, vol. 12168, pp. 220–236. Springer, Cham (2020). https://doi.org/10.1007/978-3-030-58666-9_13
28. Sommers, D., Sidorova, N., van Dongen, B.F.: In system alignments we trust! explainable alignments via projections. Inf. Syst. **136**, 102631 (2026)

Constructing Weakly Terminating Interface Protocols

Debjyoti Bera[1]([✉])[ID] and Tim A.C. Willemse[2][ID]

[1] TNO-ESI, Eindhoven, The Netherlands
debjyoti.bera@tno.nl
[2] Eindhoven University of Technology, Eindhoven, The Netherlands
t.a.c.willemse@tue.nl

Abstract. Interfaces play a central role in determining compatible component compositions by prescribing permissible interactions between a service provider (server) and its consumers (clients). The high degree of concurrency in asynchronous communicating systems increases the risk of unintentionally introducing deadlocks and livelocks. The weak termination property serves as a basic sanity check to avoid such problems. It assures that in each reachable state, the system has the option to eventually terminate. This paper generalizes existing results that, by construction, guarantee weakly terminating interface compositions. Our generalizations make the theory applicable more broadly in practice. Starting with an interface specification of a server satisfying certain properties, we show how a class of clients modeling different usage contexts can be derived using a partial mirroring relation. Furthermore, we discuss an embedding of our results in an open-source tool to guide modelers in designing weakly terminating interfaces.

Keywords: Interfaces · Asynchronous Systems · Weak Termination

1 Introduction

The first step in the design of a component-based system involves the decomposition of a given system design problem into functionally coherent units of execution called components [15, 20]. Starting from an abstract model, each component is further detailed by means of stepwise refinements. To accomplish a common goal, components may need to interact with each other. In modern distributed systems, these interactions are typically asynchronous via message passing. The set of permissible interactions between pairs of components defines an interface protocol (contract) having two sides: a server side providing the functionality and a client side consuming it. The use of state machine formalisms [9] to specify such interactions is quite prevalent. When interfaces between components have been defined, the resulting system is obtained by composing shared client-server interfaces satisfying some pre-defined notion of compatibility.

A direct consequence of a component-based approach is the high degree of complexity induced by concurrency and asynchronous communication making

J. Desel and A. Kalenkova (Eds.): PETRI NETS 2026, LNCS 16567, pp. 87–108, 2026.
https://doi.org/10.1007/978-3-032-27879-1_5

it difficult to avoid subtle yet critical design errors, resulting in deadlocks or livelocks. It turns out that even if a server side protocol is free of these issues, simply mirroring this protocol to obtain client side protocols to define their usage does not guarantee the absence of errors. Therefore the use of formal modeling and analysis techniques becomes inevitable. Petri nets provide a natural way to model an asynchronous communicating system and are supported by a rich set of verification methods and tools [2]. The graphical representation of a Petri net structure provides an intuitive way of modeling systems and in particular, structural analysis techniques overcome the drawbacks of state space exploration techniques for verifying system behavior.

A basic requirement for any component-based system is that all components in the system should always have the option to reach a desired, final state, and that all messages that have been sent have been received. This property is called *weak termination*. Weak termination does not imply that the system always terminates but that there is always an option to do so. As a consequence of this property the freedom of deadlocks and livelocks is guaranteed.

Many interesting results exist for subclasses of Petri nets such that weak termination is implied by their structure. In [1] a sufficient condition is presented to pairwise verify weak termination for tree-structured compositions. For a subclass of compositions of pairs of components, called ATIS-nets, this condition is implied by their structure [12]. ATIS-nets are constructed from pairs of acyclic marked graphs and isomorphic state machines, and the simultaneous refinement of pairs of places with weakly terminating multi-workflow nets. In [7], an architectural framework based on the notion of components and ports (defining a communication protocol) is proposed. The sufficient conditions for port compatibility, as well as on the acyclic network structure of component compositions, induces the weak termination property on the resulting system (network of asynchronous communicating components). In [6], another architectural framework based on the notion of components and interaction patterns (a parameterized multi-workflow net defining a communication protocol) is proposed. The interaction patterns cover three commonly occurring ones in practice, namely remote procedure call, cancellation (goal-feedback-result) and publish-subscribe. In each communication refinement of the existing system, a new component is added to the system as part of an interaction pattern. In [5], the architectural framework [6] is extended with the mirrored port pattern by lifting the notion of compatible ports from [7] to interaction patterns.

The mirrored port pattern defines a class of weakly terminating single server and multiple client interaction patterns, where each client is an isomorphic mirror of the server state machine satisfying structural properties that enforce communication conditions such as the *choice* and *leg* property. The former guarantees that only one of the parties make a choice whenever there is a conflict, while the latter guarantees that every path between a split and a join always requires the participation of a server and one of the clients. An example of a mirrored port pattern is shown in Fig. 1. It turns out that for practical applications, the mirrored port pattern is quite limiting for at least three reasons:

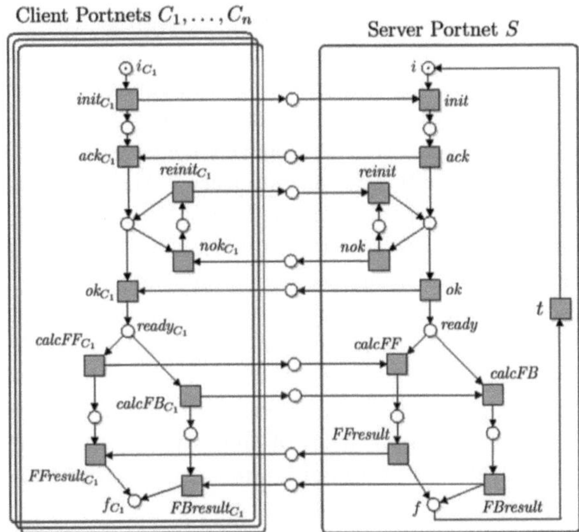

Fig. 1. The Mirrored Portnet Interaction Pattern.

(i) the choice property requires there are no races between a server and client, i.e., the initiative of resolving conflicting transitions is either with the server or one of its clients, but not both. This restriction is quite limiting for practical applications where races are often intensional. For instance, in Fig. 1, it is not possible to model a race between a client firing one of its transitions from the place *ready* and a server that may emit an *error* from *ready*.

(ii) mirrored portnets do not allow multiple transitions in a server/client to be connected to the same interface place. This means that it is not possible to send/receive the same message from more than one state. For instance, in Fig. 1, the transition *reinit* had to be introduced since transition *init* cannot occur twice and share the same interface place.

(iii) the isomorphism requirement implies that client portnets must fully mirror the server portnet even though they may not use all functionality. For instance, in Fig. 1, all clients $C_1, \ldots, C_n$ are required to implement both choices from the place *ready*, even though some of the clients may only use one of them. This is a common requirement since each client may require a different usage of a server depending on the context of the component they occur.

This paper revisits the mirrored port interaction pattern [5] and extends it with a labeling function, relaxes constraints on its structure and weakens the isomorphism requirement on portnets. We call this class of interaction patterns *partial mirrored portnet* and show that the weak termination property holds for these patterns. The results provide a structured method to design and reason about the correctness of a large class of interfaces encountered in practice. Since partial mirrored portnets are an interaction pattern of the component-based framework, we may reuse the stepwise refinements based construction method to design a weakly terminating network of asynchronous communicating components starting from an architectural diagram. A first application of these results has been realized in the open-source ComMA framework [14], widely used by engineers in industry to specify software interfaces. Algorithms have been implemented to automatically detect violations of portnet properties in interface specifications and provide feedback about termination problems at design-time.

This paper is structured as follows: Sect. 2 presents basic mathematical concepts and notations. Section 3 presents labeled portnets, compositions and the partial mirror relation. Section 4 presents restrictions for a labeled portnet to be well-formed. Section 5 proves weak termination in a composition of a well-formed labeled portnet and its partial mirror. Section 6 extends the results to multiple clients. Section 7 shows an application of these results in ComMA. The paper concludes in Sect. 8. Detailed proofs of lemmas, propositions and theorems can be found in [8].

2   Preliminaries

Let S be a set. A *bag* over S is a function $m : S \to \mathbb{N}$, where $\mathbb{N} = \{0, 1, 2, \ldots\}$ denotes the set of natural numbers. For $s \in S$, $m(s)$ denotes the number of occurrences of s in m. The set of all bags over S is denoted by $B(S)$. We use $+$ and $-$ to denote the standard addition and difference of two bags. The relations $=, <,$ and $\leq$ compare bags element-wise. We treat sets as bags in which each element of the set occurs exactly once in the bag.

The set of finite sequences over S is denoted by S^*, with ϵ denoting the empty sequence and $\langle a_0, \ldots, a_n \rangle$ denoting a sequence of length $n + 1$. We use $\circ$ to concatenate two sequences and we denote the length of a sequence σ by $|\sigma|$. A prefix of a sequence σ is a sequence σ' such that $\sigma = \sigma' \circ \sigma''$ for some sequence σ''. We write $x \in \sigma$ when there are sequences σ', σ'' such that $\sigma = \sigma' \circ \langle x \rangle \circ \sigma''$. The projection of a sequence σ onto a set Q, denoted $\sigma_{|Q}$, is defined inductively as follows: $\epsilon_{|Q} = \epsilon$ and $(\langle a \rangle \circ \sigma')_{|Q} = \langle a \rangle \circ (\sigma'_{|Q})$ if $a \in Q$, and $\sigma'_{|Q}$ otherwise. We denote the set obtained by interleaving sequences σ and σ' by $\sigma || \sigma'$.

A *Petri net with labels*, or simply Petri net from hereon, is a mathematical model for describing the dynamics of complex systems.

Definition 1. *A* Petri net *is a tuple* $N = (P, T, F, \mathcal{L}, \mu)$, *where P is the set of places;* T *is a set of* transitions *such that* $P \cap T = \emptyset$, F *is the* flow relation $F \subseteq (P \times T) \cup (T \times P)$, $\mathcal{L}$ *is a set of* labels *and* $\mu : T \to \mathcal{L}$ *is a labeling function.*

Let $N = (P, T, F, \mathcal{L}, \mu)$ be an arbitrary but fixed Petri net. Elements from $P \cup T$ are called *nodes* and elements from F are referred to as *arcs*. For clarity, we occasionally denote the places of net N by P_N, transitions by T_N and similarly for other elements of the tuple. We define the *preset* of a node x as $^\bullet x = \{x' | (x', x) \in F\}$ and the *postset* as $x^\bullet = \{x' | (x, x') \in F\}$ and lift these notations to sets of nodes in the standard way.

The set of paths $PS(N)$ of N is the smallest set satisfying $\epsilon \in PS(N)$, $\langle x \rangle \in PS(N)$ for all $x \in P \cup T$, and for all $(x_0, x_1) \in F$ and finite paths $\langle x_1, \ldots, x_n \rangle \in PS(N)$, also $\langle x_0, x_1, \ldots, x_n \rangle \in PS(N)$. Petri net N is *strongly connected* if for all distinct $x, x' \in P \cup T$ there is a path $\langle x, \ldots, x' \rangle \in PS(N)$.

Definition 2. *We say that N is a* state machine *(S-net) iff for all $t \in T$: $|^\bullet t| \leq 1$ and $|t^\bullet| \leq 1$. It is a* workflow net *(WFN) if there is exactly one place i with $^\bullet i = \emptyset$, called the* initial place, *one place f with $f^\bullet = \emptyset$, called the* final place, *and all nodes $n \in P \cup T$ are on a path from i to f. A place p is called a* split *if $|p^\bullet| > 1$ and it is called a* join *if $|^\bullet p| > 1$.*

Petri nets N and M are disjoint if $(P_N \cup T_N) \cap (P_M \cup T_M) = \emptyset$. They are *isomorphic*, denoted by $N \cong M$ iff there is a bijection $\psi : P_N \cup T_N \to P_M \cup T_M$ such that for all $p \in P_N$ we have $\psi(p) \in P_M$ and for all $t \in T_N$ we have $\psi(t) \in T_M$ and $\mu_N(t) = \mu_M(\psi(t))$, and for all $(x, y) \in (P_N \cup T_N) \times (P_N \cup T_N)$ we have $(x, y) \in F_N$ iff $(\psi(x), \psi(y)) \in F_M$.

The *state* of Petri net $N = (P, T, F, \mathcal{L}, \mu)$ is determined by its *marking* which represents the distribution of tokens over places of the net. Formally, a marking of N is a bag $m \in B(P)$, where $m(p)$ denotes the number of tokens in place $p \in P$. If $m(p) > 0$, place $p \in P$ is called *marked* in marking m. A transition $t \in T$ is *enabled* in m iff $^\bullet t \leq m$, i.e., all the places in its preset are marked. Transition t may *fire* from marking m if it is enabled, resulting in a marking m' that satisfies $m'(p) = m(p) - {}^\bullet t(p) + t^\bullet(p)$ for all $p \in P$. We write $m \xrightarrow{t} m'$ if t is enabled in m and firing t results in m', and lift this notion to sequences of transitions in the expected manner.

A *net system* is a Petri net with an initial and final marking; its semantics is given by a *labeled transition system*.

Definition 3. *Let $N = (P, T, F, \mathcal{L}, \mu)$ be a Petri net. A net system is a 3-tuple $S = (N, m_0, m_f)$, where $m_0 \in B(P)$ is the initial marking and $m_f \in B(P)$ is the final marking.*

Given a net system S, a marking m' of S is *reachable* from m if for some $\sigma \in T^*$ we have $m \xrightarrow{\sigma} m'$; we denote the set of markings reachable from m by $\mathcal{R}(N, m)$. A place $p \in P_N$ in a net system is called *safe* if for all reachable markings $m \in \mathcal{R}(N, m_0)$ we have $m(p) \leq 1$. An important correctness criterion of a net system is whether it is always able to terminate: it implies the absence of (unknown) deadlocks in a system, but also the ability to direct the behavior to a terminal state.

Definition 4. *A net system is* weakly terminating *if from all markings $m \in \mathcal{R}(N, m_0)$ the final marking m_f is reachable.*

3   Labeled Portnets and Their Mirrors

Open Petri nets (OPN) are a well-studied [3,4,16,19] class of Petri nets equipped with open places and visible transitions, i.e., distinguished sets of input/output places and transitions over which a net may interact with its environment, either by exchanging tokens over open places, or by synchronizing on visible transitions. OPNs with open places [21] are ideal for modeling asynchronous communicating systems as a composition of nets obtained by fusing their shared open places, representing interfaces. Portnets [7] are a subclass of OPN that explicitly model the behavior of an interface by defining permissible interactions with its environment, i.e., the order in which tokens may be exchanged over open input and output places. Portnets are ideal to describe interface contracts in the context of component-based systems. An interface of a component providing a service describes expectations from its environment (other components) and obligations towards it by means of a portnet enacting the server side (sell side) of an interface contract. A component that consumes the service provided by another component acts as its environment and defines usage by means of a mirrored portnet enacting the client side (buy side) of an interface contract. The compatibility of a pair of portnets is defined by not only matching input and output labels of open places, but by additionally imposing restrictions on the net structure which guarantees the weak termination property in portnet compositions.

This section introduces the class of *labeled portnets* which extends portnets [7] with a labeling function and a composition operation over sets of labeled portnets. A strict requirement for compatibility of a pair of portnets was for their skeletons to agree on isomorphism. This induces a mirroring relation between a server portnet and its client, which is quite restrictive in practice, since a client may only be interested in using a part of the server portnet. We relax the isomorphism requirement by the *partial mirror* relation between portnet pairs.

An OPN is a Petri net with a distinguished set of interface places, partitioned into input and output places, that represent the interface of the net. The interaction of an OPN with its environment results in tokens (representing messages) being exchanged over interface places. If a transition of an OPN is associated to an interface place, then it is either an input place or an output place, but not both. So a transition either represents a send, or a receive. The net obtained by discarding interface places is called its *skeleton*. By imposing restrictions on the skeleton, many subclasses of OPN are obtained.

Definition 5. *Let P, I, O be pairwise disjoint sets. An* open Petri net (OPN) *is a tuple $(P, I, O, T, F, init, fin, \mathcal{L}, \mu)$, where $(P \cup I \cup O, T, F, \mathcal{L}, \mu)$ is a Petri net, I is the set of input places; we require $^\bullet I = \emptyset$, O is the set of output places; we require $O^\bullet = \emptyset$, for all $t \in T$ we have either $^\bullet t \cap I = \emptyset$ or $t^\bullet \cap O = \emptyset$, init $\subseteq P$ is the set of initial places and fin $\subseteq P$ is the set of final places.*

We refer to the places in I and O as the interface places of the OPN. The *direction of communication* of a transition with respect to a place in an OPN N is a function $\lambda : T \to \{send, receive, \tau\}$ defined as follows:

$$\lambda(t) = \begin{cases} send & \text{if } t^\bullet \cap O \neq \emptyset \\ receive & \text{if } ^\bullet t \cap I \neq \emptyset \\ \tau & \text{otherwise} \end{cases}$$

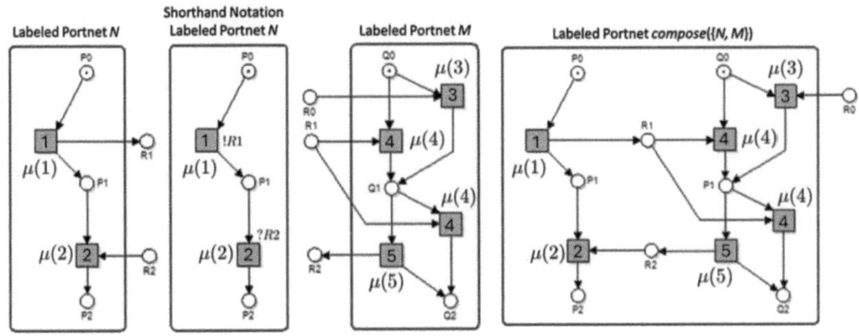

Fig. 2. Labeled Portnets and their Composition.

That is, $\lambda(t) = send$ if t is connected to an output, whereas $\lambda(t) = receive$ if t is connected to an input; $\lambda(t) = \tau$ if neither is the case. We say that t is a *communicating transition* iff $\lambda(t) \neq \tau$. The *skeleton* of N, denoted $\mathcal{S}(N)$, is an OPN $\mathcal{S}(N) = (P, \emptyset, \emptyset, T, F', init, fin, \mathcal{L}, \mu)$ with $F' = F \cap ((P \times T) \cup (T \times P))$. If the skeleton is an S-net then we call it a state machine open Petri net (S-OPN). An OPN N is called an open workflow net (OWN) if its skeleton is a *WFN*. An S-OPN with a skeleton that is a *WFN* is called a state machine open workflow net (S-OWN). The *closure* of a S-OWN N, denoted by *closure(N)*, is an S-OPN obtained by extending N with a fresh transition $t \notin T$ such that $\lambda(t) = \tau$, $^\bullet t = fin$ and $t^\bullet = init$.

A *labeled portnet* is a subclass of OPN, characterized by additional restrictions on their structure.

Definition 6. *An OPN $N = (P, I, O, T, F, init, fin, \mathcal{L}, \mu)$ is a* labeled portnet *if:*

- *N is an S-OWN, and $|init| = |fin| = 1$,*
- *all transitions $t \in T$ satisfy $|(^\bullet t \cup t^\bullet) \cap (I \cup O)| = 1$,*
- *for all $x \in I \cup O$ and all transitions $t, t' \in {}^\bullet x \cup x^\bullet$ we have $\mu(t) = \mu(t')$,*
- *for all transitions $t, t' \in T$ satisfying $\mu(t) = \mu(t')$ we have $(^\bullet t \cup t^\bullet) \cap (I \cup O) = (^\bullet t' \cup t'^\bullet) \cap (I \cup O)$,*

The net system (N, m_0, m_f) of an OPN N has as initial marking $m_0 = init$ and final marking $m_f = fin$, where sets init and fin are interpreted as bags.

Portnets as defined in [7] are instances of labeled portnets where the labeling function μ satisfies $\mu(t) = t$. That is, each transition in a portnet is labeled with its own, unique name. To stress the distinction between portnets and labeled portnets, we sometimes refer to the former as *ordinary portnets*.

Lemma 1. *Let N be a labeled portnet and $t, t' \in T$ such that $\mu(t) = \mu(t')$. Then $\lambda(t) = \lambda(t')$.*

Since the skeleton of a labeled portnet is a state machine workflow net, by Theorem 15 [11], it is weakly terminating, as claimed by the corollary below.

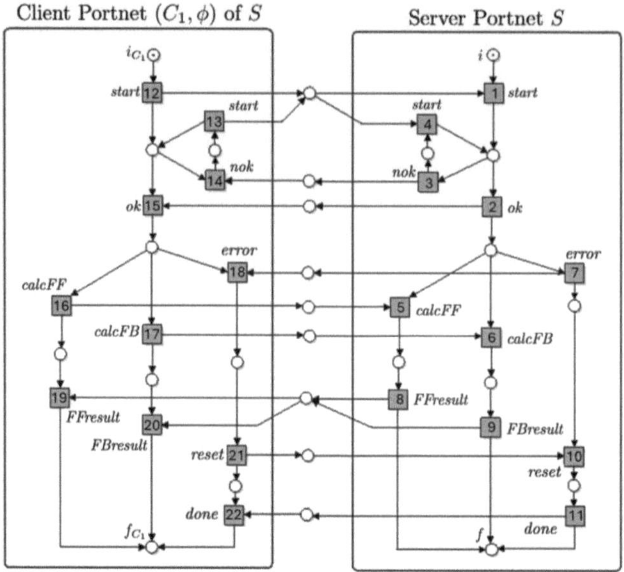

Fig. 3. Deadlock in a Composition of Labeled Portnet and its Mirror.

Corollary 1. *The skeleton of a labeled portnet weakly terminates.*

An example labeled portnet is shown in Fig. 2. As illustrated, the shorthand notation ? (receive) and ! (send) will be used to keep figures readable and when necessary the name of the connected interface places will be prefixed to it. Next to labels on transitions, numbers are used as their identifiers. Depending on the context, we may omit one of them.

Two OPNs are *composable* if the only nodes they share are interface places, and if they share a non-empty set of interface places then either the outputs of one are a subset of the inputs of the other, or their inputs and outputs are disjoint. The *composition* of a set of pairwise composable OPNs is almost a pairwise union of their corresponding tuples except that the shared interface places between OPNs become the internal places of the composition. Figure 2 shows a composition of two composable labeled portnets N and M.

Definition 7. *Two OPNs N and M are* composable *iff*

- $(P_N \cup I_N \cup O_N \cup T_N) \cap (P_M \cup I_M \cup O_M \cup T_M) = ((I_N \cup O_N) \cap (I_M \cup O_M))$,
- *If* $(I_N \cap O_M) \cup (I_M \cap O_N) \neq \emptyset$, *then also:* $O_N \subseteq I_M$ *and* $O_M \subseteq I_N$, *and* $I_N \cap I_M = O_N \cap O_M = \emptyset$.

The composition of a non-empty set S of pairwise composable OPNs is an OPN $compose(S) = (P, I, O, T, F, init, fin, \mathcal{L}, \mu)$*, where*

- $P = (\bigcup_{N \in S} P_N) \cup ((\bigcup_{N \in S} I_N) \cap (\bigcup_{N \in S} O_N))$,
- $I = (\bigcup_{N \in S} I_N) \setminus (\bigcup_{N \in S} O_N)$, $O = (\bigcup_{N \in S} O_N) \setminus (\bigcup_{N \in S} I_N)$,

- $T = (\bigcup_{N \in S} T_N)$, $F = (\bigcup_{N \in S} F_N)$,
- $init = (\bigcup_{N \in S} init_N)$, $fin = (\bigcup_{N \in S} fin_N)$, $\mathcal{L} = (\bigcup_{N \in S} \mathcal{L}_N)$ and $\mu = (\bigcup_{N \in S} \mu_N)$.

We remark that since the shared places of OPNs become internal places when we compose them, in general for sets S_1 and S_2 of pairwise composable OPNs we do not have $compose(S_1 \cup S_2) = compose(S_1 \cup compose(S_2))$.

The *partial mirror* relation relaxes the isomorphism requirement on the skeletons of a pair of portnets [7] by allowing one to drop some transitions with a *send* direction, offering designers the option to exercise only the relevant parts of the other labeled portnet.

Definition 8. *A* partial mirror *of a labeled portnet N is a pair (M, ϕ) where M is a labeled portnet and $\phi : M \to N$ an injective mapping satisfying:*

- $\phi(P_M) \subseteq P_N$, $\phi(T_M) \subseteq T_N$, $\phi(I_M) \subseteq O_N$ and $\phi(O_M) \subseteq I_N$,
- F_M *satisfies that for all $(x, y) \in F_M$:*
  - *if $x \notin I_M$ and $y \notin O_M$, then $(\phi(x), \phi(y)) \in F_N$,*
  - *if $x \in I_M$ or $y \in O_M$, then $(\phi(y), \phi(x)) \in F_N$,*
- $\phi(init_M) = init_N$ *and* $\phi(fin_M) = fin_N$,
- *for all $t \in T_M$ we have $\mu_M(t) = \mu_N(\phi(t))$,*
- *for all places $p \in P_M$ and all transitions $t \in \phi(p)^{\bullet}$ in N, if $\lambda(t) = send$ then there is some $t' \in p^{\bullet}$ such that $\phi(t') = t$.*

We say that (M, ϕ) is a *mirrored labeled portnet* in case ϕ is a bijection. The labeled portnet M is referred to as the *client portnet* of the *server portnet N*.

A partial mirror relation by itself does not guarantee weak termination of the composition of a labeled portnet and its (partial) mirror. Figure 3 shows an example of such a composition w.r.t. ϕ, where one of the deadlocks can be reached by firing the sequence of transitions $\langle 12, 1, 2, 15, 7, 16 \rangle$.

4   Well-Formed Labeled Portnets

In [7], a notion of compatibility between pairs of ordinary portnets was introduced that guarantees weak termination of their composition. This notion of compatibility requires the skeletons of the ordinary portnets to satisfy the *choice* and *leg* property; for the sake of completeness, we repeat these below. These are structural properties of the ordinary portnets that can be checked efficiently.

The *choice property* does not allow transitions in the postset of a place to present a race between an input and an output: either all such transitions perform a send, or receive, but not a mix of both. As a result it is not possible to model a race between a client and server portnet. The *leg property* prevents a client and server portnet from getting confused and going down different paths. By requiring every path from a split (a place with postset greater than one, including the initial place) to a join (a place with preset greater than one, including the final place) to have at least two transitions with different communication direction, confusions are prevented.

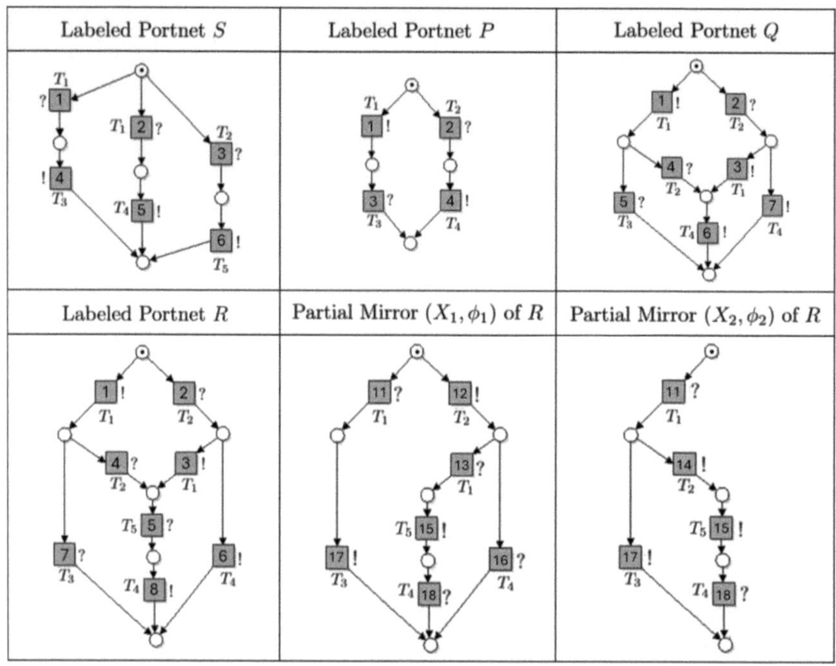

Fig. 4. Portnet Properties and Partial Mirrors

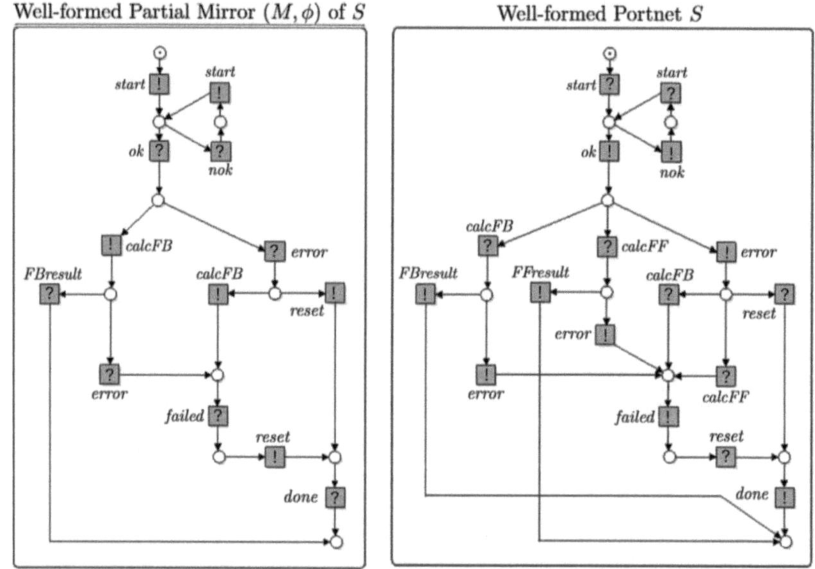

Fig. 5. Well-formed portnet and its well-formed partial mirror portnet.

Definition 9 (Choice and Leg Property). *A portnet N satisfies the* choice *property if $\lambda(t) = \lambda(t')$ for all $t, t' \in p^\bullet$ and all $p \in P_N$. It satisfies the* leg *property if $\forall \beta = \langle p_1, t_1, \ldots, t_{n-1}, p_n \rangle \in PS(N) : ((|p_1{}^\bullet| > 1 \vee p_1 = init_N) \wedge (|{}^\bullet p_n| > 1 \vee p_n = fin_N)) \implies \exists t, t' \in \beta : \lambda(t) \neq \tau \wedge \lambda(t') \neq \tau \wedge \lambda(t) \neq \lambda(t').*

This section presents a relaxation of these properties by introducing the *observable choices*, the *diamond* and *loop* properties. Moreover, we consider these properties in the more general setting of labeled portnets. As a consequence, the class of labeled portnets satisfying these properties encompasses a larger variety of interaction protocols, commonly encountered in practice.

A labeled portnet satisfies the *observable choices* property if distinct transitions in the postset of a place have different labels. This enables one to decide, based on observing labels only, exactly which non-interface place of a labeled portnet becomes marked upon firing a sequence of transitions. For instance, portnet S in Fig. 4 violates this property, while the others satisfy it.

Definition 10 (Observable Choices). *A labeled portnet satisfies the observable choices property if all $p \in P$, and all distinct $t, t' \in p^\bullet$ we have $\mu(t) \neq \mu(t')$.*

The *diamond property*, stated below, relaxes the aforementioned choice property by allowing a race between a client and server to initiate sending of messages to each other from a split place. Intuitively, such a race is benign if, whichever way the race is resolved, the option to handle the interaction that lost the race persists and a subsequent handling of that interaction still has the option to lead to an aligned view on each-other's state. This intuition is formalized below.

Definition 11 (Diamond Property). *A labeled portnet satisfies the diamond property if all its places $p \in P$ and all $t, t' \in p^\bullet$ with $\lambda(t) \neq \lambda(t')$ satisfy*

$$\forall q \in t^\bullet, q' \in t'^\bullet : \exists u \in q^\bullet, u' \in q'^\bullet : u^\bullet \cap u'^\bullet \neq \emptyset \wedge \mu(t) = \mu(u') \wedge \mu(t') = \mu(u)$$

Property 1. If a labeled portnet satisfies the choice property then it also satisfies the diamond property.

Labeled portnet S, depicted in Fig. 4, satisfies the choice property, while the other labeled portnets violate it. Labeled portnet P violates the diamond property, while labeled portnets Q and R satisfy it.

Definition 12. *For a given labeled portnet N with place $p \in P$, transition t and label $l \in \mathcal{L}$, we define pathset(p, t, l) as follows:*

$$\{\langle p \rangle \circ \pi \in PS(N) \mid \exists \sigma \in T^*, t'' \in T : \pi_{|T} = \langle t \rangle \circ \sigma \circ \langle t'' \rangle \wedge$$
$$\mu(t'') = l \wedge \forall u \in T : u \in \sigma \Rightarrow \mu(u) \neq l\}$$

The *loop property*, formalized below, requires any path that starts with resolving competing receiving (resp. sending) interactions by handling one of them, to also handle the interaction with the other one, but not before a send (resp. receive) interaction occurred. The goal is to prevent confusion between client and server, causing both to go down different paths. For instance, consider send transitions 3 and 7 in the postset of a split in portnet Q (Fig. 4). Suppose Q has a mirror $(\overline{Q}, \phi)$ with transition t such that $\phi(t) = 7$. If Q fires the sequence of *send* transitions $\langle 3, 6 \rangle$, then $\overline{Q}$ may choose to fire t (since 6 and 7 have the same label T_4, they are connected to the same interface place), resulting in leftover tokens in the interface place of transition 3. Portnet R resolves this by forcing a synchronization with *receive* transition 5.

Definition 13 (Loop Property). *A labeled portnet satisfies the loop property if all places $p \in P$ satisfy that all distinct transitions $t, t' \in p^\bullet$ with $\lambda(t) = \lambda(t')$ and all paths $\sigma \in pathset(p, t, \mu(t'))$ have a transition $t'' \in \sigma$ with $\lambda(t'') \neq \lambda(t)$.*

Definition 14 (Well-Formed Portnet). *A labeled portnet is* well-formed *if it satisfies the observable choices, the diamond and the loop properties.*

The portnet S depicted in Fig. 5 is a modification of portnet S, depicted in Fig. 3; the modifications ensure the portnet is well-formed. Its well-formed partial mirror M drops the transition labeled *calcFF* and as a consequence, also the conflict transitions *FFResult* and *error* to satisfy that the skeleton of M remains a WFN. We remark that although a mirrored labeled portnet of a well-formed labeled portnet is well-formed, well-formedness is not preserved by partial mirroring. To see this, consider the well-formed portnet R, depicted in Fig. 4 and its partial mirrors, where $\phi_1(11) = \phi_2(11) = 1$, $\phi_1(12) = 2$, $\phi_1(13) = 3$, $\phi_2(14) = 4$, $\phi_1(15) = \phi_2(15) = 5$, $\phi_1(16) = 6$, $\phi_1(17) = \phi_2(17) = 7$, and $\phi_1(18) = \phi_2(18) = 8$. The partial mirror (X_1, ϕ_1) is not well-formed since it violates the diamond property. On the other hand, the partial mirror (X_2, ϕ_2) is well-formed, and it remains so when additionally, transition 17 is dropped.

5   Weak Termination of Composition of Labeled Portnets

As demonstrated by the example in Fig. 3, the composition of a labeled portnet and its partial mirror does not necessarily lead to a weakly terminating OPN. In this section, we show that weak termination of a composition of two labeled portnets N and M *is* guaranteed when both are well-formed and M is a partial mirror of N. Indeed, the main theorem of this section is the following:

Theorem 1. *Let N and M be well-formed labeled portnets and let (M, ϕ) be a partial mirror of N. Let $S = compose(\{N, M\})$ and $i_S = i_N + i_M$ be the initial marking of S and let $f_S = f_N + f_M$ be the final marking of S. Then the net system (S, i_S, f_S) weakly terminates.*

In the remainder of this section we break down the correctness argument of the theorem before ultimately presenting its proof.

Fix a well-formed labeled portnet N—from hereon the *server*—and a well-formed partial mirror (M, ϕ) of N—from hereon the *client*. Suppose that in their interactions, at some point the server wishes to execute a send transition. The client may still be catching up with previously executed send transitions, due to the asynchronous nature of their communication. However, due to the diamond property, it does not matter when exactly the server fires its send transition, since it will have the same effect on the overall state of the system. The following lemma formalizes this phenomenon.

Lemma 2. *Let N and M be well-formed labeled portnets and let (M, ϕ) be a partial mirror of N. Let $S = compose(\{N, M\})$ and $i_S = i_N + i_M$ be the initial marking of S and $f_S = f_N + f_M$ be the final marking of S. For any marking $m, m', m_0, \ldots, m_n \in \mathcal{R}(S, i_S)$, and all transitions $t, t_1, \ldots, t_n$ of N such that $m = m_0 \xrightarrow{t_1} \ldots \xrightarrow{t_n} m_n$, where each $\lambda(t_i) = receive$, and $m \xrightarrow{t} m'$ with $\lambda(t) = send$, there exists markings $m'_0, \ldots, m'_n \in \mathcal{R}(S, i_S)$ and transitions $\bar{t}, t'_1, \ldots, t'_n$ of N such that $m' = m'_0 \xrightarrow{t'_1} \ldots \xrightarrow{t'_n} m'_n$ and $m_n \xrightarrow{\bar{t}} m'_n$ and*

1. *$\mu(t_i) = \mu(t'_i)$ and $\lambda(t_i) = \lambda(t'_i)$;*
2. *$\mu(t) = \mu(\bar{t})$ and $\lambda(t) = \lambda(\bar{t})$.*

Similar to the exchange lemma (see [10]), we show that an enabled transition can be fired anywhere in a firing sequence, as long as the transition does not share any place with the preset of transitions in that executable firing sequence.

Lemma 3. *Let (N, m_0, m_f) be a net system, with N a labeled OPN. Suppose $t \in T$ and $\sigma \in T^*$ are such that for all $t' \in \sigma$, $^\bullet t'$ is disjoint with $^\bullet t \cup t^\bullet$. Then for every $m \in \mathcal{R}(N, m_0, m_f)$ satisfying $m \xrightarrow{t}$ and $m \xrightarrow{\sigma}$ there is some m' such that for every $\sigma' \in (\langle t \rangle \| \sigma)$ we have $m \xrightarrow{\sigma'} m'$.*

A receive transition of a labeled portnet and a receive transition of its partial mirror do not share an interface place. So in a composition of a well-formed labeled portnet and its well-formed partial mirror, a sequence of receive transitions of the former may be shuffled with the latter, leading to the same marking.

Lemma 4. *Let N and M be well-formed labeled portnets and let (M, ϕ) be a partial mirror of N. Let $S = compose(\{N, M\})$ and $i_S = i_N + i_M$ be the initial marking of S and $f_S = f_N + f_M$ be the final marking of S. For any reachable marking $m_0, \ldots, m_n$, and all transitions $t_1, \ldots, t_n \in T_N$ and $u_1, \ldots, u_k \in T_M$ such that $m_0 \xrightarrow{t_1} \ldots \xrightarrow{t_n} m_n \xrightarrow{u_1} \ldots \xrightarrow{u_k} m_{n+k}$ and $\lambda(t_i) = \lambda(u_j) = receive$ for all relevant i, j, there are markings $m'_0, \ldots, m'_{n+k}$ such that $m_0 = m'_0 \xrightarrow{u_1} \ldots \xrightarrow{u_k} m'_k \xrightarrow{t_1} \ldots \xrightarrow{t_n} m'_{n+k} = m_{n+k}$.*

From a reachable marking of a composition of a well-formed server portnet and its well-formed partial mirror client, firing an enabled sequence of receive transitions of the server cannot disable an enabled send transition of a client since the latter does not consume from an interface place.

Lemma 5. *Let N and M be well-formed labeled portnets and let (M, ϕ) be a partial mirror of N. Let $S = compose(\{N, M\})$ and $i_S = i_N + i_M$ be the initial marking of S and $f_S = f_N + f_M$ be the final marking of S. For any reachable marking $m_0, \ldots, m_n, m'_0$, and all transitions $t_1, \ldots, t_n \in T_N$ with $\lambda(t_i) = receive$, and*

$u \in T_M$ with $\lambda(u) = send$ such that $m_0 \xrightarrow{t_1} \ldots \xrightarrow{t_n} m_n$ and $m_0 \xrightarrow{u} m'_0$, there are markings $m'_1, \ldots, m'_n$ such that $m'_0 \xrightarrow{t_1} \ldots \xrightarrow{t_n} m'_n$ and $m_n \xrightarrow{u} m'_n$.

Building on the previous lemmas, the next proposition shows that for any reachable marking in a composition of a well-formed server portnet and its well-formed partial mirror client, there exists a sequence of receive transitions that can be executed first in the server, then in the client, leading to a marking where a place in the client and its mirror are marked, and all other places are empty. This means, informally, that server and client can catch up with each other and align their view on each other's state by processing receive transitions only.

Proposition 1. *Let N and M be well-formed labeled portnets and let (M, ϕ) be a partial mirror of N. Let $S = compose(\{N, M\})$ and $i_S = i_N + i_M$ be the initial marking of S and $f_S = f_N + f_M$ be the final marking of S. For any marking $m_0, \ldots, m_n$, and all transitions $t_1, \ldots, t_n$ such that $i_S = m_0 \xrightarrow{t_1} \ldots \xrightarrow{t_n} m_n$, there are markings $m_{n+1}, \ldots, m_{n+k}$ and transitions $t_{n+1}, \ldots, t_{n+k}$ such that $m_n \xrightarrow{t_{n+1}} \ldots \xrightarrow{t_{n+k}} m_{n+k}$ and:*

1. *for all $i \geq n + 1$, we have $\lambda(t_i) = receive$;*
2. *$t_{n+1} \ldots t_{n+k} \in T_N^* T_M^*$; i.e. the transitions first execute in N and then in M;*
3. *there is a place $p \in P_M$ such that $m_{n+k}(p) = m_{n+k}(\phi(p)) = 1$ and $m_{n+k}(p') = 0$ for all $p' \notin P_N \cup P_M$; i.e. all interface places are empty and there is a token in both place p of M and its mirror $\phi(p)$ of N.*

From a reachable marking in a composition of a well-formed server portnet and its well-formed partial mirror client such that a place in the client and its mirror are marked, and all other places empty, an alternating sequence of send and mirrored receive transitions can be fired leading to the final marking. The proof relies on weak termination of portnet skeletons (Corollary 1).

Proposition 2. *Let N and M be well-formed labeled portnets and let (M, ϕ) be a partial mirror of N. Let $S = compose(\{N, M\})$ and $i_S = i_N + i_M$ be the initial marking of S and $f_S = f_N + f_M$ be the final marking of S. Assume the skeleton of M is weakly terminating. Let $m \in \mathcal{R}(S, i_S, f_S)$ such that for some $p \in P_M$, $m(p) = m(\phi(p)) = 1$ and $m(q) = 0$ for all $q \notin \{p, \phi(p)\}$. Then marking f_S is reachable from marking m.*

We are now in a position to formally prove Theorem 1:

Proof (Theorem 1). Let N, M and S be as stated. Let $m \in \mathcal{R}(S, i_S, f_S)$.

By Proposition 1, there must be a sequence of transitions $t_1, \ldots, t_n$ such that $m \xrightarrow{t_1} \ldots \xrightarrow{t_n} m'$ such that $\lambda(t_i) = receive$ for all $i \in 1 \ldots n$, and there is a place $p \in P_M$ such that $m'(p) = m'(\phi(p)) = 1$ and all other places of S are empty. By Proposition 2 it follows that f_S is reachable. Hence, S is weakly terminating. $\square$

The previous result is strengthened by showing that if the final places of a composition of well-formed labeled portnet and its well-formed partial mirror are marked, then no other places of the composition are marked.

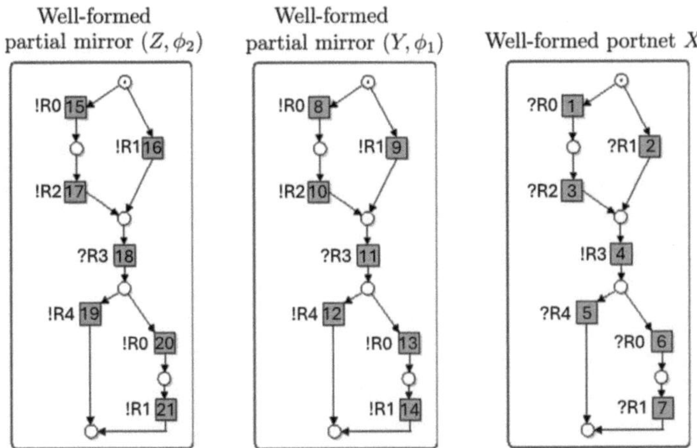

Fig. 6. Deadlock in composition: $compose(\{X, Y, Z\})$.

Theorem 2 (Proper Completion). *Let N be a well-formed labeled portnet and let (M, ϕ) be a partial mirror of N. Let $S = compose(\{N, M\})$ and $i_S = i_N + i_M$ be the initial marking of S and $f_S = f_N + f_M$ be the final marking of S. For all markings reachable in the net system (S, i_S, f_S), if the final places of N and M are marked then all other places are empty.*

Proof. Suppose that the final places of N and M are marked and there are leftover tokens in other places of net S. There cannot be any leftover tokens in the internal places of N and M, since they are state machine nets having a single token (initial place) each in their initial marking. If there are leftover tokens in interface places, then by Theorem 1, there must exist a firing sequence that leads to the final marking f_S. This is not possible since the final places of N and M have empty postsets, which is a contradiction.   □

6 Weak Termination of Composition of Multiple Portnets

In practice, multiple clients may wish to consume the same service provided by a server, repeatedly. The results we established in the previous section do not generalize to such a setting unchanged. To see this, consider the well-formed labeled portnets X, Y and Z depicted in Fig. 6. Observe that (Y, ϕ_1) and (Z, ϕ_2) are partial mirrors of X. While the compositions $compose(\{X, Y\})$ and $compose(\{X, Z\})$ are deadlock free, the firing sequence $\langle 8, 10, 16, 2, 4, 6, 18, 19 \rangle$ in the composition $compose(\{X, Y, Z\})$ leads to a deadlock. The main reason is that clients may interfere with each other. It turns out that in order to achieve the desired result we must embed our results in a synchronization pattern.

A synchronization pattern requires that a server first makes a selection among the clients with which to cooperate. Such negotiation can be modeled by means of the pattern shown in Fig. 7. Starting from the initial marking, one or more clients may send their requests, one of which is picked up by the server, followed by a response back. The response is picked up by exactly one client, which guarantees that exactly one client may progress with the server to eventually mark their respective final places without interference from others. Once the final place of the server is marked a closure transition t returns the token back to the initial place, and from thereon the server can pick the next client. This process repeats until all clients have been handled and the final marking is reached.

Definition 15. *The pattern depicted in Fig. 7 (left) is a synchronization pattern* $(\mathcal{C}, S, Q)$ *where* S *is a closure of a well-formed labeled portnet* N, $\mathcal{C} = \{C_1, C_2, \ldots C_n\}$ *is a set of* n *($n \in \mathbb{N}$) well-formed mirrored clients of* N, *and* Q *is the set of refinable places* $\{q, q_1, q_2, \ldots q_n\}$. *The net system is defined as* $(compose(\mathcal{C} \cup \{S\}), m_0, m_f)$, *where* m_0 *is the initial marking given by* $\{i\} \cup \{i_j \mid j \leq n\}$, *and final marking* m_f *given by* $\{i\} \cup \{r_j \mid j \leq n\}$.

Lemma 6. *The synchronization pattern is weakly terminating.*

The structure of a synchronization pattern leads to the following observations. First, since the skeletons of client and server are S-nets with a single token in *init*, there can be at most one token in their internal places. Second, exactly one token produced by one of the clients can be consumed by the server. The server can then produce exactly one token back towards its clients. Third, at most one client consumes a token produced by the server and it becomes the chosen client. when a client is chosen the mirror places of that client and server are marked (first server then client) and all interface tokens produced by one were consumed by the other. From here the chosen client produces a token and marks its final place, enabling the server to pick this token and mark its final place.

Lemma 7. *Let* m *be a marking of a synchronization pattern with* $n \in \mathbb{N}$ *clients satisfying the following properties*

1. $m(i) + m(p) + m(q) + m(r) = 1$
2. $\forall j \in \{1 \ldots n\} : m(i_j) + m(p_j) + m(q_j) + m(r_j) = 1$
3. $\sum_{j \in \{1 \ldots n\}} m(p_j) = m(a_1) + m(p) + m(a_2)$,
4. $(\sum_{k \in \{1 \ldots n\}} m(q_k)) + m(a_2) \leq 1$,
5. $m(a_2) + m(a_3) \leq 1$ *and* $m(a_2) + m(a_3) = m(q) - \sum_{k \in \{1 \ldots n\}} m(q_k)$

Then firing any enabled transition t *from* m *does not violate these properties.*

Corollary 2. *The net system of a synchronization pattern with* $n \in \mathbb{N}$ *clients, satisfies*

- $\forall m \in \mathcal{R}(N, m_0) : m(q) = 1 \implies \sum_{j \in \{1 \ldots n\}} m(q_j) \leq 1$
- $\forall m \in \mathcal{R}(N, m_0) : \sum_{j \in \{1 \ldots n\}} m(q_j) = 1 \implies m(q) = 1$

Refinements of places and transitions [17] are a well-known strategy to incrementally elaborate the behavior of a Petri net while preserving soundness. We extend the classical place refinement operation to labeled OPNs.

Definition 16. *Given two disjoint labeled OPNs N and M. The refinement of a place $p \in P_N$ by M, denoted by $N \odot_p M$, is a labeled OPN $(P, I, O, T, F, init, fin, \mu)$ with $P = (P_N \setminus \{p\}) \cup P_M$, $I = I_N \cup I_M$, $O = O_N \cup O_M$, $T = T_N \cup T_M$, $F = (F_N \setminus ((^\bullet p \times \{p\}) \cup (\{p\} \times p^\bullet))) \cup F_M \cup (^\bullet p \times init_M) \cup (fin_M \times p^\bullet)$, $init = init_N$, $fin = fin_N$, $\mu = \mu_N \cup \mu_M$.*

We formally introduce the notion of a *partial mirrored portnet pattern*. Such a pattern captures the essence of a server communicating with multiple clients, obtained by (partial) mirroring, embedded inside a synchronization pattern by means of place refinements. An example is shown in Fig. 7.

Definition 17 (Partial Mirrored Portnet Pattern). *Let N be a weakly terminating well-formed labeled portnet and let $(M_1, \phi_1), \ldots, (M_n, \phi_n)$, for $n \in \mathbb{N}$, be a set of disjoint partial mirrors of N. Let $R = (C, S, \{q, q_1, \ldots q_n\})$ be a synchronization pattern with n clients such that $M_1, \ldots, M_n, N, C, S$ have disjoint skeletons. A partial mirrored portnet pattern is a composition $L = compose(C' \cup \{S'\})$, where $C' = \{C_1 \odot_{q_1} M_1, \ldots, C_n \odot_{q_n} M_n\}$ and $S' = S \odot_q N$ with initial marking $m_0 = i_S + i_{C_1} + \ldots + i_{C_n}$, final marking $m_f = i_S + f_{C_1} + \ldots + f_{C_n}$ and a net system (L, m_0, m_f). If $M_1, \ldots, M_n$ are full mirrors of N, then we call L a mirrored portnet pattern.*

Theorem 3. *The partial mirrored portnet pattern weakly terminates.*

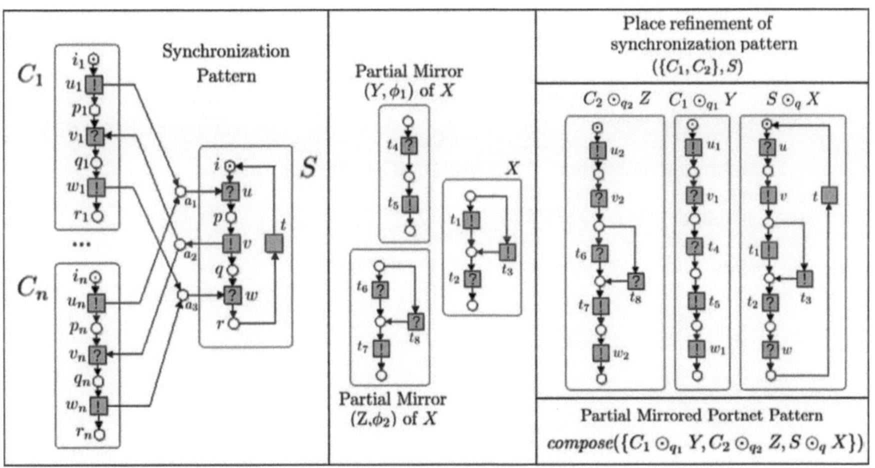

Fig. 7. Synchronization Pattern and Partial Mirrored Portnet Pattern.

Proof. By Lemma. 6 we know that a synchronization pattern consisting of $n \in \mathbb{N}$ clients weakly terminates. By Corollary 2. we know that first a server needs to mark a place q and only then exactly one client C_j, for some $j \in \{1 \ldots n\}$, can mark its place q_j. Furthermore, the server cannot remove a token from q until q_j is marked. In the skeleton of the server, the place q is always on a path from the initial place i to final place r. So all firing sequences from the initial marking to the final marking of a synchronized portnet, first marks place q with a token, followed by place q_j, while q remains marked.

By Definition 16 and Definition 17, only the initial and final place of a well-formed labeled portnet N and its partial mirrors $(M_1, \phi_1), \ldots, (M_n, \phi_n)$ have nodes (transitions) of the synchronization pattern in their preset and postset, respectively. So a marking of a synchronization pattern R containing one token in q and q_j, corresponds to a marking in the partial mirrored portnet pattern L with a token in initial places of N and M_j, respectively. By Theorem 1, we know that from all reachable markings from such a marking, there exist a firing sequence to the final marking, i.e., final places of N and M_j. Furthermore, by Theorem 2, we know that when the final places of N and M_j are marked, then all other places of N and M_j are empty. By Definition 16 and Definition 17,

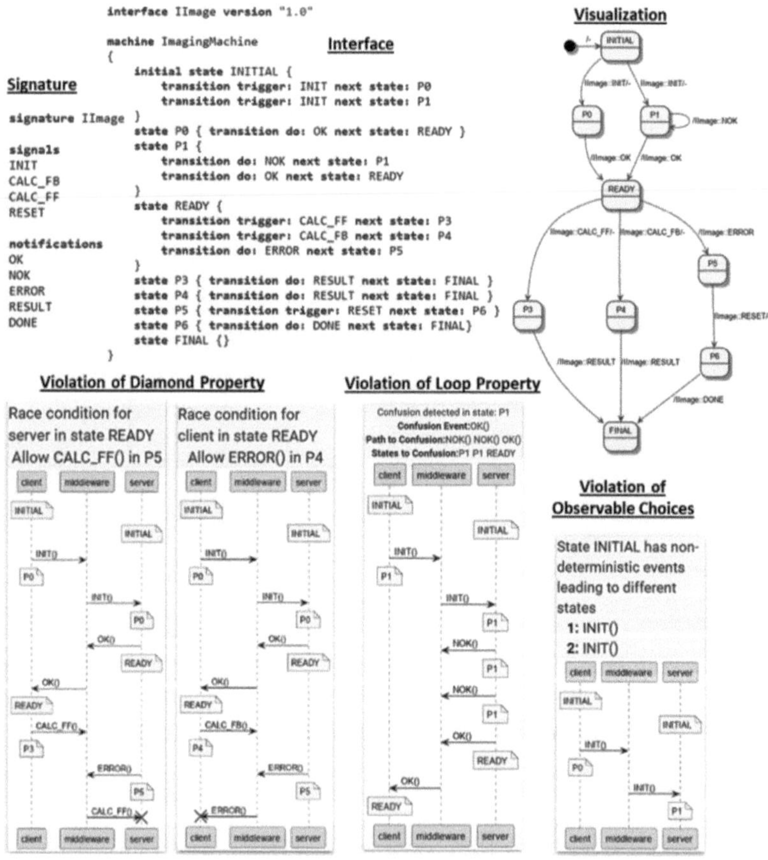

Fig. 8. Application of the results to ComMA.

a marking of L containing a token in the final places of N and M_j (and no other tokens in places of S), enables exactly the same set of transitions as that of the net R with a token in the places q and q_j. Hence, the refinement with a multi-client portnet does not violate the weak termination property of the synchronization pattern.   □

7 Application

The growing recognition that a lack of explicit and precise specifications of component interfaces potentially leads to problems during system integration have led to the development of an open-source interface modeling tool ComMA [14]. The goal is to capture both static and dynamic aspects of an interface as a sound basis for a formal contract between service providers and consumers. ComMA specifications are used to validate whether implementations conform to their specifications using run-time monitoring [13] and more recently, model-based testing techniques. The tool is developed in collaboration with industry and is used by software engineers at Philips and Thales as part of their development processes. To fully realize the benefits and detect design errors earlier, the need for correct interface specifications was recognized. The results of this paper have been implemented in ComMA as algorithms to check well-formedness of interface specifications. The feature has proven to be successful in raising awareness about the pitfalls of designing asynchronous systems and serve as a guide for engineers to create weakly terminating interface protocols by construction.

ComMA provides a family of domain specific languages that includes types, signature, interface and input parameters. *Types* capture data definitions built of basic types (int, real, string, etc.), enumerations, vectors, records and maps. *Signature* captures event definitions (i.e., their labels and data parameter types) of type: command-reply (client sends a command and is blocked until server reply), signal (client sends a non-blocking event) and notification (server sends a non-blocking event). *Interface* captures allowed order of input and output events as transitions of a protocol state machine, defined from the perspective of a service provider (server), also known as a *provided interface*. The state machine may additionally define global variables and manipulate them using an expressions language. Output events containing data arguments refer to global variables, while data arguments of input events are provided by clients. When defining usage of a provided interface by another component, it is common practice to simply mirror the protocol of the server. So an explicit model of a client protocol is omitted. For standalone simulation of the server state machine, *input parameters* specify possible data values of input event arguments (in each possible state). Analysis and simulation of server state machine is made possible by transformation to colored Petri nets using SNAKES [18]. A formal treatment of this transformation is out of scope. The remainder of this section will omit data aspects and informally discuss the mapping to labeled portnets.

A ComMA specification of the server portnet S in Fig. 3 is shown in Fig. 8. The signature defines signals and notifications with names corresponding transition labels of S. The syntax of an interface state machine comprises of state labels (with an explicit initial state) and transitions are directed edges between states. A generated visualization of the state machine in UML syntax is shown in Fig. 8. Each state maps to a place in the underlying portnet with a name corresponding its label. Each state defines a set of possible transitions and each transition defines a target *next state*. Two kinds of transitions are distinguished. A *non-triggered transition* defines an initiative of a server in a *do* section containing a notification which maps to a *send* transition with a preset corresponding the state (place) it is defined within and a postset corresponding the state (place) referred to by *next state*. A *triggered transition* defines an initiative of a client in a *trigger* section containing a signal which maps to a *receive* transition with a preset corresponding the state (place) it is defined within and a postset corresponding the state (place) referred to by *next state*. With these mapping rules, a labeled portnet is derived from a state machine syntax. The underlying translation discards interface places since we are only interested in analyzing the skeleton. For compact specifications, many shorthand notations are available, e.g., triggered transitions with multiple *do* sections, each having a sequence of notifications, *for all states* prevents duplication of same transition in multiple states, etc. All constructs can be expressed in terms of concepts presented so far.

To support a modeler in creating weakly terminating specifications, algorithms are implemented in ComMA to first validate whether the skeleton is weakly terminating and then to validate whether the observable choices, diamond and loop properties hold. Whenever a violation is detected, a UML sequence diagram is generated (see Fig. 8) illustrating the problem in a composition of a server with an assumed client that fully mirrors the protocol.

8   Conclusion

This paper presented sufficient conditions to guarantee weak termination in a composition of a server and multiple *partial* mirrored client protocols. These conditions impose structural restrictions characterized by well-formedness, relaxing the strict restrictions imposed by existing results, thereby encompassing a larger class of communication protocols encountered in practice. By embedding well-formedness checks on the server protocol and assuming a symmetric mirrored client in the ComMA tool, modelers are made aware about potential hidden issues in component interactions, early on during design-time. As future work, we aim to (i) further relax the requirement for partial mirrors to be well-formed. The example in Fig. 4, where composition $compose(\{Q, X_1\})$ weakly terminates even through X_1 is not well-formed, seems to suggest such a relaxation is possible, (ii) generalize the synchronization pattern which was introduced to preserve weak termination in a composition of multiple clients, and (iii) extend our results to the component modeling language of ComMA, where functional constraints define restrictions over events of provided and required interfaces.

Acknowledgements. The research is carried out as part of the TechFlex and Matala program under the responsibility of TNO-ESI in cooperation with Thales and ASML.

The research activities are co-funded by Holland High Tech | TKI HTSM via the PPP Innovation Scheme (PPP-I) for public-private partnerships.

References

1. van der Aalst, W.M., van Hee, K.M., Massuthe, P., Sidorova, N., van der Werf, J.M.: Compositional service trees. In: Applications and Theory of Petri Nets, 30th International Conference, PETRI NETS 2009, Paris, France, 22–26 June 2009. Proceedings. Lecture Notes in Computer Science, vol. 5606, pp. 283–302. Springer (2009). https://doi.org/10.1007/978-3-642-02424-5_17
2. Amat, N.: Behind the scene of the model checking contest, analysis of results from 2018 to 2023. In: Beyer, D., Hartmanns, A., Kordon, F. (eds.) TOOLympics Challenge 2023, pp. 52–89. Springer Nature Switzerland (2025)
3. Baldan, P., Bonchi, F., Gadducci, F., Monreale, G.V.: Modular encoding of synchronous and asynchronous interactions using open petri nets. Sci. Comput. Program. **109**, 96–124 (2015). https://doi.org/10.1016/J.SCICO.2014.11.019
4. Baldan, P., Corradini, A., Ehrig, H., Heckel, R.: Compositional semantics for open petri nets based on deterministic processes. Math. Struct. Comput. Sci. **15**(1), 1–35 (2005). https://doi.org/10.1017/S0960129504004311
5. Bera, D.: Petri nets for modeling robots. Phd thesis, Technische Universiteit Eindhoven (2014). https://doi.org/10.6100/IR780942
6. Bera, D., van Hee, K.M., van der Werf, J.M.: Designing weakly terminating ROS systems. In: Application and Theory of Petri Nets - 33rd International Conference, PETRI NETS 2012, Hamburg, Germany, 25–29 June 2012. Proceedings. Lecture Notes in Computer Science, vol. 7347, pp. 328–347. Springer (2012). https://doi.org/10.1007/978-3-642-31131-4_18
7. Bera, D., Van Hee, K.M., Van Osch, M., van der Werf, J.M.E., et al.: A component framework where port compatibility implies weak termination. In: Duvigneau, M., Moldt, D., Hiraishi, K. (eds.) Proceedings of the International Workshop on Petri Nets and Software Engineering, Newcastle upon Tyne, UK, 20–21 June 2011. CEUR Workshop Proceedings, vol. 723, pp. 152–166. CEUR-WS.org (2011). https://ceur-ws.org/Vol-723/paper11.pdf
8. Bera, D., Willemse, T.A.C.: Constructing weakly terminating interface protocols (2026). https://arxiv.org/abs/2603.15675
9. De Alfaro, L., Henzinger, T.A.: Interface automata. In: Tjoa, A.M., Gruhn, V. (eds.) Proceedings of the 8th European Software Engineering Conference held jointly with 9th ACM SIGSOFT International Symposium on Foundations of Software Engineering 2001, Vienna, Austria, 10–14 September 2001, pp. 109–120. ACM (2001). https://doi.org/10.1145/503209.503226
10. Desel, J., Esparza, J.: Free Choice Petri Nets. Cambridge Tracts in Theoretical Computer Science, Cambridge University Press (1995). https://doi.org/10.1017/CBO9780511526558
11. van Hee, K.M., Sidorova, N., Voorhoeve, M.: Soundness and separability of workflow nets in the stepwise refinement approach. In: van der Aalst, W.M.P., Best, E. (eds.) Applications and Theory of Petri Nets 2003, 24th International Conference, ICATPN 2003, Eindhoven, The Netherlands, 23–27 June 2003, Proceedings. Lecture Notes in Computer Science, vol. 2679, pp. 337–356. Springer (2003). https://doi.org/10.1007/3-540-44919-1_22

12. van Hee, K.M., Sidorova, N., van der Werf, J.M.: Construction of asynchronous communicating systems: weak termination guaranteed! In: Software Composition - 9th International Conference, SC@TOOLS 2010, Malaga, Spain, 1–2 July 2010. Proceedings. Lecture Notes in Computer Science, vol. 6144, pp. 106–121. Springer (2010). https://doi.org/10.1007/978-3-642-14046-4_8

13. Kurtev, I., Hooman, J., Schuts, M.: Runtime monitoring based on interface specifications. In: ModelEd, TestEd, TrustEd - Essays Dedicated to Ed Brinksma on the Occasion of His 60th Birthday, Lecture Notes in Computer Science, vol. 10500, pp. 335–356. Springer (2017). https://doi.org/10.1007/978-3-319-68270-9_17

14. Kurtev, I., Hooman, J., Schuts, M., van der Munnik, D.: Model based component development and analysis with ComMA. Sci. Comput. Program. **233**, 103067 (2024). https://doi.org/10.1016/J.SCICO.2023.103067

15. McIlroy, M.D., Buxton, J., Naur, P., Randell, B.: Mass-produced software components. In: Proceedings of the 1st International Conference on Software Engineering, Garmisch Pattenkirchen, Germany, pp. 88–98 (1968)

16. Milner, R.: Bigraphs for petri nets. In: Desel, J., Reisig, W., Rozenberg, G. (eds.) Lectures on Concurrency and Petri Nets, Advances in Petri Nets [This tutorial volume originates from the 4th Advanced Course on Petri Nets, ACPN 2003, held in Eichstätt, Germany in September 2003.]. Lecture Notes in Computer Science, vol. 3098, pp. 686–701. Springer (2003). https://doi.org/10.1007/978-3-540-27755-2_19

17. Murata, T.: Petri nets: properties, analysis and applications. Proc. IEEE **77**(4), 541–580 (1989). https://doi.org/10.1109/5.24143

18. Pommereau, F.: SNAKES: a flexible high-level petri nets library (tool paper). In: Devillers, R., Valmari, A. (eds.) Application and Theory of Petri Nets and Concurrency - 36th International Conference, PETRI NETS 2015, Brussels, Belgium, 21–26 June 2015, Proceedings. Lecture Notes in Computer Science, vol. 9115, pp. 254–265. Springer (2015). https://doi.org/10.1007/978-3-319-19488-2_13

19. Sassone, V., Sobocinski, P.: A congruence for Petri nets. In: Ehrig, H., Padberg, J., Rozenberg, G. (eds.) Proceedings of the Workshop on Petri Nets and Graph Transformations, PNGT@ICGT 2004, Rome, Italy, 2 October 2004. Electronic Notes in Theoretical Computer Science, vol. 127, pp. 107–120. Elsevier (2004). https://doi.org/10.1016/J.ENTCS.2005.02.008

20. Szyperski, C.A., Gruntz, D., Murer, S.: Component software - beyond object-oriented programming, 2nd edn. Addison-Wesley component software series, Addison-Wesley (2002). https://www.worldcat.org/oclc/248041840

21. Wolf, K.: Does my service have partners? Trans. Petri Nets Other Models Concurr. II Spec. Issue Concurr. Process-Aware Inf. Syst. **2**, 152–171 (2009). https://doi.org/10.1007/978-3-642-00899-3_9

Persistent Permutability Implies Persistence for Pure Dissymmetric Choice Petri Nets

Eike Best[1]([✉]) [ID] and Raymond Devillers[2] [ID]

[1] Department of Computer Science, Carl von Ossietzky Universität Oldenburg,
26111 Oldenburg, Germany
`eike.best@informatik.uni-oldenburg.de`
[2] Département d'Informatique, Université Libre de Bruxelles, Brussels, Belgium
`raymond.devillers@ulb.be`

Abstract. Persistence is a strong, global, behavioural property of a Petri net, meaning that no activity (represented by a transition of the net) can disable a different transition. Persistent permutability is a comparatively local, weaker, property pertaining only to the individual interleavings of a Petri net. We prove that in pure, dissymmetric choice, Petri nets, persistent permutability already suffices to imply persistence. Dissymmetric choice Petri nets generalise free-choice Petri nets and are related to Petri's notion of "asymmetric confusion".

Keywords: Dissymmetric choice · embedding · Parikh equivalence · permutation equivalence · persistence · Petri nets

1 Introduction

The notion of persistence has been well-established in Petri net theory since [17] (Karp and Miller), [19] (Keller), and especially [20] (Landweber and Robertson who proved that persistent Petri nets have semilinear reachability sets). Persistence is usually viewed as a property of a marked Petri net as a whole. In [22], however, Ochmański extended the concept to the interleavings (i.e., the firing sequences) of a Petri net. He examined a persistent permutability property which states that one can deduce, from the existence of a given *non-persistent* interleaving, the existence of a *persistent* interleaving into which the given one can be permuted equivalently (in terms of available concurrent permutations).

If a Petri net is persistent, then it is also persistently permutable, for the simple reason that the identity permutation can be applied to any firing sequence. Our paper is concerned with the converse question:

$$\text{Does persistent permutability imply persistence?} \tag{1}$$

In general, the answer to (1) is "no", but in [5], it was shown to be "yes" within the class of pure and safe dissymmetric choice (DC) nets as well as in the class

J. Desel and A. Kalenkova (Eds.): PETRI NETS 2026, LNCS 16567, pp. 109–129, 2026.
https://doi.org/10.1007/978-3-032-27879-1_6

of safe free-choice (FC) Petri nets [10,13].[1] DC nets generalise FC nets and exclude a class of asymmetrically confused nets.[2] The class of pure DC nets properly includes the class of pure FC nets (but not the class of safe FC nets). Thus, the result in the present paper extends the first (but not the second) result in [5].

The structure of the paper is as follows. Section 2 consists of two parts. In the first part, all necessary basic concepts, in particular pureness, safeness, dissymmetric choice nets, and persistence, are introduced. The second part contains the definition of a pattern and how a pattern can be embedded in the reachability graph of a Petri net. In Sect. 3, persistent permutations and the permutability property are defined and explained. In Sect. 4, the main result suggested in the title is stated and proved. The proof is indirect, showing that under some assumptions, a pattern which is not DC can be embedded in the reachability graph of a net. Section 5 presents a few remarks on context and on possible future work. Appendix A contains a brief historical account relating to the class of dissymmetric choice Petri nets and their relationship to Petri's asymmetric confusion.

2   Labelled Transition Systems, Petri Nets, and Patterns

2.1   Labelled Transition Systems and Petri Nets

The reachability graph of a Petri net is a labelled transition system (LTS). Conversely, a labelled transition system may, or may not, be isomorphic to the reachability graph of some Petri net. In that sense, labelled transition systems are more general and will be defined first. In the correspondence between Petri net reachability graphs and labelled transition systems, the set of labels of an LTS and the set of transitions of a Petri net correspond uniquely to each other. This is why we use the same letter, T, for both.

Definition 1. LABELLED TRANSITION SYSTEMS

A *labelled transition system* with initial state, abbreviated LTS, is a quadruple $TS = (S, \rightarrow, T, s_0)$ where S is a set of *states*, T is a set of *labels*, $\rightarrow \subseteq (S \times T \times S)$ is the *transition relation*, and $s_0 \in S$ is an *initial state* (so that S may not be empty). A label t is *enabled* in a state $s \in S$, denoted by $s \xrightarrow{t}$, if there is some state $s' \in S$ such that $(s, t, s') \in \rightarrow$, and *disabled at* s if there is *no* state $s' \in S$ such that $(s, t, s') \in \rightarrow$. For $t \in T$, we write $s \xrightarrow{t} s'$ iff $(s, t, s') \in \rightarrow$, meaning that s' is *reachable* from s through the execution of t. The definitions of enabledness and of the reachability relation are extended to finite (firing) sequences $\sigma \in T^*$ as follows:

---

[1] In [5], we called DC nets asymmetric choice, but in the present paper we shall clarify and disambiguate this terminology, as explained below in Appendix A.
[2] This terminology is due to Petri [24].

$s \xrightarrow{\varepsilon}$ and $s \xrightarrow{\varepsilon} s$ are always true, by definition, and

$s \xrightarrow{\sigma t} (s \xrightarrow{\sigma t} s')$ if $\exists q \in S$ with $s \xrightarrow{\sigma} q \xrightarrow{t} (s \xrightarrow{\sigma} q \xrightarrow{t} s'$, respectively)

For a state $s \in S$, $[s\rangle$ denotes the set of states reachable from s, i.e., the set of states s' such that $s \xrightarrow{\sigma} s'$ for some $\sigma \in T^*$. Two LTS with the same label set, $(S, \rightarrow, T, s_0)$ and $(S', \rightarrow', T, s_0')$, will be called *isomorphic* if there is a bijection $\beta \colon S \to S'$ such that $s_0' = \beta(s_0)$ and $(r, t, s) \in \rightarrow$ iff $(\beta(r), t, \beta(s)) \in \rightarrow'$.

For a finite sequence $\sigma \in T^*$ of labels, the *Parikh vector* $\Psi(\sigma)$ is a T-vector (i.e., a vector of natural numbers with index set T), where $\Psi(\sigma)(t)$ denotes the number of occurrences of t in σ. Two sequences $\sigma, \tau \in T^*$ are called *Parikh-equivalent* if $\Psi(\sigma) = \Psi(\tau)$.

A labelled transition system $(S, \rightarrow, T, s_0)$ is called *persistent* [20] if for all reachable states $s, s', s'' \in [s_0\rangle$, and labels $t, u \in T$, if $s \xrightarrow{t} s'$ and $s \xrightarrow{u} s''$ with $t \neq u$, there is some (reachable) state $r \in S$ such that both $s' \xrightarrow{u} r$ and $s'' \xrightarrow{t} r$ (i.e., once two different labels are both enabled at some state, none of them can lead to a disabling of the other, and executing both, in any order, leads to the same state).   □ D1

Definition 2. PETRI NETS

A (finite, initially marked, place-transition, arc-weighted) Petri net is a quadruple $N = (P, T, F, M_0)$ such that P is a finite set of *places*, T is a finite set of *transitions*, with $P \cap T = \emptyset$, F is a *flow* function $F \colon ((P \times T) \cup (T \times P)) \to \mathbb{N}$, and M_0 is the *initial marking*, where a *marking* is a mapping $M \colon P \to \mathbb{N}$ (indicating a number of *tokens* in each place). A transition $t \in T$ is *enabled by* a marking M, denoted by $M \xrightarrow{t}$, if for all places $p \in P$, $M(p) \geq F(p, t)$. If t is enabled at M, then t can *occur* (or *fire*) in M, leading to the marking M' defined by $M'(p) = M(p) - F(p, t) + F(t, p)$ (denoted by $M \xrightarrow{t} M'$). The set of markings reachable from M is denoted $[M\rangle$. The *reachability graph of* N, $RG(N)$, is the labelled transition system with the set of vertices $[M_0\rangle$, the set of edges $\{(M, t, M') \mid M, M' \in [M_0\rangle \wedge M \xrightarrow{t} M'\}$, and initial state M_0.

For a place p and a transition t of a Petri net $N = (P, T, F, M_0)$, let ${}^\bullet t = \{p \in P \mid F(p, t) > 0\}$ be the *preset of t* containing *pre-places*, $t^\bullet = \{p \in P \mid F(t, p) > 0\}$ its *postset* containing *post-places*, ${}^\bullet p = \{t \in T \mid F(t, p) > 0\}$ the *preset of p*, and $p^\bullet = \{t \in T \mid F(p, t) > 0\}$ its *postset*. N is called *plain* if $cod(F) \subseteq \{0, 1\}$ (i.e., there are no multiple arcs);[3] *pure* or *side-condition free* if $p^\bullet \cap {}^\bullet p = \emptyset$ for all places $p \in P$; *free-choice* (FC) [10,13] if N is plain and for all $t, t' \in T$, $({}^\bullet t \cap {}^\bullet t' \neq \emptyset)$ entails $({}^\bullet t = {}^\bullet t')$; *dissymmetric choice* (DC) if N is plain and for all $t, t' \in T$, $({}^\bullet t \cap {}^\bullet t' \neq \emptyset)$ entails $({}^\bullet t \subseteq {}^\bullet t') \vee ({}^\bullet t' \subseteq {}^\bullet t)$; and *asymmetric choice* (AC) [6,13] if N is plain and for all $p, p' \in P$, $(p^\bullet \cap p'^\bullet \neq \emptyset)$ entails $(p^\bullet \subseteq p'^\bullet) \vee (p'^\bullet \subseteq p^\bullet)$.

---

[3] In this paper, only plain nets will be considered. $cod(F)$ refers to the codomain (i.e., the set of images) of the function F.

112   E. Best and R. Devillers

A Petri net $N = (P, T, F, M_0)$ is *k-bounded*, for some $k \in \mathbb{N}$, if $\forall M \in [M_0\rangle\colon \forall p \in P\colon M(p) \leq k$ (i.e., the number of tokens on any place never exceeds k); *safe* if it is 1-bounded; *bounded* if $\exists k \in \mathbb{N}\colon N$ is k-bounded; and *persistent* if so is its reachability graph.

We can also omit the initial marking and write $N = (P, T, F)$. The *reverse dual* of an unmarked net is obtained by exchanging the roles of places and transitions and reversing all the arcs: $\mathcal{R}D(N) = (T, P, F')$, where $F'(x, y) = F(y, x)$.

□D2

We shall use letters $a, b, c, \ldots \in T$ (but also $t, u \in T$) for the labels of a transition system (or, respectively, for the corresponding transitions of a Petri net); $\alpha, \beta, \sigma, \tau \in T^*$ for sequences of transitions; r, s for the states of an LTS; p, q for the places of a net; and M, J, K, L for the markings of a net; those denotations may also be "decorated", such as in a_1, σ', $\widetilde{p}$, and $\widehat{M}$.

Definition 3. Solvability

A Petri net N *solves* (or *generates*) a transition system TS if $RG(N)$ and TS are isomorphic.).

□D3

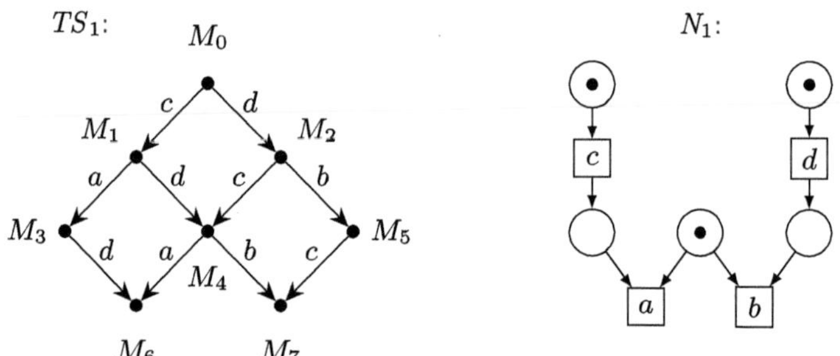

Fig. 1. An LTS TS_1 and a Petri net solution N_1. The net N_1 is plain, pure, and safe (and AC), but not dissymmetric choice because ${}^\bullet a \cap {}^\bullet b \neq \emptyset$, $\neg({}^\bullet a \subseteq {}^\bullet b)$, and $\neg({}^\bullet b \subseteq {}^\bullet a)$

Example 1. Figure 1: an LTS TS_1 (L.H.S.) and a solution N_1 (R.H.S.)

In line with common practice, labelled transition systems such as TS_1 and Petri nets such as N_1 are represented by two different kinds of directed graphs. The states in TS_1 are already named in such a way that they correspond uniquely to the reachable markings of N_1, with M_0 being the initial state. Formally,

$$TS_1 = \Big(\{M_0, \ldots, M_7\}, \{(M_0, c, M_1), \ldots, (M_5, c, M_7)\}, \{a, b, c, d\}, M_0\Big)$$

but such a description is somewhat unwieldy and we shall not also specify it for the net (the places would have to be named first anyway).).

□E1

Remark 1. FREE-CHOICE, DISSYMMETRIC, AND ASYMMETRIC PETRI NETS

In Definition 2, the free-choice property has been defined as

$$\forall t, t' \in T : {}^\bullet t \cap {}^\bullet t' \neq \emptyset \Rightarrow {}^\bullet t = {}^\bullet t' \tag{2}$$

which is equivalent to its reverse dual

$$\forall p, p' \in P : p^\bullet \cap p'^\bullet \neq \emptyset \Rightarrow p^\bullet = p'^\bullet \tag{3}$$

The dissymmetric choice property arises from (2) by weakening the right-hand side to $({}^\bullet t \subseteq {}^\bullet t') \vee ({}^\bullet t \subseteq {}^\bullet t')$. The asymmetric choice property, as in [6,10] and other works, arises from (3) by weakening the right-hand side to $(p^\bullet \subseteq p'^\bullet) \vee (p^\bullet \subseteq p'^\bullet)$. The AC and DC properties are no longer equivalent, but they are still reverse duals of each other. The excluded structures are shown in Fig. 2. We refer to Appendix A for a fuller historical synopsis.   □R1

Fig. 2. DC excludes the Petri net pattern shown on the left-hand side whereas AC excludes the pattern shown on the right-hand side. An arc with a cross on top means "no such arc".

Remark 2. REACHABILITY GRAPHS ARE SPECIAL LTS

By Definition 3, the set of Petri net reachability graphs corresponds to a subset of labelled transition systems. This subset is proper because it has the following special properties (and more). A labelled transition system $(S, \rightarrow, T, s_0)$ is called *totally reachable* if $[s_0\rangle = S$. Petri net reachability graphs are totally reachable. A labelled transition system is called *deterministic* if $\forall s, s', s'' \in S \ \forall \sigma, \tau \in S^* :$ $s \xrightarrow{\sigma} s' \wedge s \xrightarrow{\tau} s'' \wedge \Psi(\sigma) = \Psi(\tau) \Rightarrow s' = s''$. Petri net reachability graphs are deterministic. Region theory [1,6,11,12] provides a full characterisation of the set of Petri net solvable LTS.   □R2

2.2 Patterns and Embeddings

In the following, we shall capture interesting substructures of an LTS by means of patterns. In general, patterns specify that some features are mandatory and other ones are excluded. In the present paper, we shall restrict our attention to rather simple patterns, where the mandatory features are labelled arcs and the excluded ones are enabling relations.

114   E. Best and R. Devillers

Definition 4. PATTERN OF AN LTS

A *pattern* is a quadruple $PT = (S, \rightarrow, T, D)$ where S is a set of states, T a set of labels, $\rightarrow \subseteq (S \times T \times S)$ (represented as usual), and $D \subseteq (S \times T)$ (represented by labelled arcs without specified endpoint and a red cross).   □ D4

This is illustrated in Fig. 3. The pattern shown in this figure is designed to characterise a non-persistent part of an LTS.

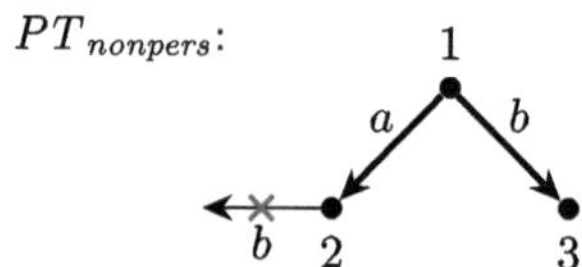

Fig. 3. The pattern $PT_{nonpers} = (\{1, 2, 3\}, \{(1, a, 2), (1, b, 3)\}, \{a, b\}, \{(2, b)\})$ mirrors a part of TS_1 and highlights the violation of persistence at M_4 which stems from the disabling of b by a. The set $\{(2, b)\}$ specifies an excluded enabling.

The fact that a pattern is present in a transition system will be captured by an embedding notion.

Definition 5. EMBEDDING OF A PATTERN IN AN LTS

Let $PT = (S_1, \rightarrow_1, T_1, D_1)$ be a pattern and $TS = (S, \rightarrow, T, s_0)$ an LTS. Then PT is *embedded in* TS if there is a function $f = f_1 \cup f_2$, with $f_1 \colon (S_1 \rightarrow S)$ and $f_2 \colon (T_1 \rightarrow T)$, such that $\forall (s, a, s') \in \rightarrow_1 \colon (f(s), f(a), f(s')) \in \rightarrow$, and $\forall (s, a) \in D_1 \colon \neg (f(s) \xrightarrow{f(a)})$. Moreover, we shall assume that f_2 is injective, meaning that it does not "fuse" labels: $\forall t, t' \in T_1 \colon t \neq t' \Rightarrow f_2(t) \neq f_2(t')$.   □ D5

Example 2. EMBEDDING $PT_{nonpers}$ INTO TS_1 AND INTO TS_2.

A possible embedding of the pattern $PT_{nonpers}$ into the LTS TS_1 is given by the function f such that $f(1) = M_4$, $f(2) = M_6$, $f(3) = M_7$, $f(a) = a$ and $f(b) = b$. Here the embedding is also injective on the states, but Fig. 4 exemplifies a situation where some states are fused: $f(1) = s_0$, $f(2) = s_1 = f(3)$, $f(a) = x$, and $f(b) = y$. This still captures a non-persistent situation (s_0 enables x and y but executing x disables y). We require injectivity on labels, since if $f(a) = f(b)$ for $a \neq b$, we lose the fact that, in the definition of persistence, the two considered labels are different (for example, in Fig. 4, if $f(a) = x = f(b)$ in $PT_{nonpers}$, we could drop y and we would simply have two x's in a row, which would have nothing to do with a situation of non-persistence). The embedding of $PT_{nonpers}$ proves that neither TS_1 nor TS_2 can be solved by a persistent Petri net. As it happens, TS_2 cannot be realised by any pure or plain net, but it could occur as the reachability graph of a non-pure, non-plain, Petri net.   □ E2

Fig. 4. An LTS TS_2 containing the pattern $PT_{nonpers}$, fusing states 2 and 3 into s_1 and witnessing the non-persistence of this LTS. The Petri net N_2 solves TS_2.

3 Persistent Equivalent Permutations, and SPE

The transition system TS_1 shown in Fig. 1 is not persistent, nor is, therefore, its generating Petri net N_1. The offending state is M_4 where both a and b are enabled in such a way that executing a disables b (and vice versa). This is witnessed by the pattern $PT_{nonpers}$ that is embedded two times in TS_1. M_4 might be called a "proper choice state", as opposed to the "fake choice states" M_0, M_1 and M_2. Those latter states denote a choice in the order of subsequent execution of two concurrent, independent, transitions, whereas M_4 does not allow the concurrent execution of a and b. One tends to say that, from M_0, executing cd and executing dc really denote the same (concurrent) computation, differing only in the order of the occurrence of c and d. Formally, $M_0 \xrightarrow{cd}$ and $M_0 \xrightarrow{dc}$ are only "a permutation of two transitions apart" from each other. More generally, we call two sequences, say $M_0 \xrightarrow{dca}$ and $M_0 \xrightarrow{cad}$, to be in relation $\equiv$ if they are a number of transition permutations apart; e.g.,

$$M_0 \xrightarrow{dca} \quad \equiv \quad M_0 \xrightarrow{cda} \quad \equiv \quad M_0 \xrightarrow{cad}$$

with single permutations in between, and two permutations between the first and the last. Yet there is still a difference between these sequences in terms of persistence: it may happen that one of them goes through an arc which switches an enabled transition into disabled while this does not happen for the other sequence:

$$M_0 \xrightarrow{d} M_1 \xrightarrow{c} M_4 \xrightarrow{a} M_6 \quad \equiv \quad M_0 \xrightarrow{c} M_1 \xrightarrow{a} M_3 \xrightarrow{d} M_6 \tag{4}$$

$M_4 \xrightarrow{a} M_6$ where $M_4 \xrightarrow{b}$ but not $M_6 \xrightarrow{b}$. The first sequence is then called nonpersistent while the other one is called persistent.

This paper is concerned with circumstances in which nonpersistent firing sequences have equivalent persistent permutations, as in (4) where the nonpersistent sequence $M_0 \xrightarrow{dca}$ can be permuted equivalently to $M_0 \xrightarrow{cad}$.

First, let us define the notion of a persistent firing sequence formally, and then, the notion of permuting sequences. In the following, let $N = (P, T, F, M_0)$ be a Petri net with some (initial) marking M_0.

Definition 6. PERSISTENCE OF FIRING SEQUENCES [22,23]

A finite firing sequence $M_0 \xrightarrow{a_1} \ldots \xrightarrow{a_n} M_n$ of N is called *persistent* if for every $i \in \{1, \ldots, n\}$ and every $t \in T$ with $t \neq a_i$, if $M_{i-1} \xrightarrow{t}$ then also $M_i \xrightarrow{t}$.   □ D6

The persistence of a sequence $M_0 \xrightarrow{a_1} M_1 \xrightarrow{a_2} \ldots$ means that no single step $M_{i-1} \xrightarrow{a_i} M_i$ of the sequence may switch some transition $t \neq a_i$ from enabled at M_{i-1} to disabled at M_i. This is similar to the persistence of a net, but restricted to a single firing sequence. The following corollaries are straightforward consequences of the definition:

Corollary 1. PERSISTENCE FACTORISATION

A firing sequence $M \xrightarrow{\alpha} M' \xrightarrow{\beta}$ is persistent iff so are $M \xrightarrow{\alpha} M'$ and $M' \xrightarrow{\beta}$. Also, $M_0 \xrightarrow{a_1} M_1 \xrightarrow{a_2} \ldots$ is persistent iff the singleton sequences $M_{i-1} \xrightarrow{a_i} M_i$ are persistent for every used index $i > 0$.   □ C1

Corollary 2. PERSISTENT NETS AND PERSISTENT FIRING SEQUENCES

A Petri net (N, M_0) is persistent if and only if all of its firing sequences $M_0 \xrightarrow{\sigma}$ are persistent.   □ C2

Permutation equivalence is described by a relation $\equiv$ between firing sequences starting from the same state, which is normally the initial state M_0. The relation $\equiv$ is based on a relation $\equiv_0$ which refers to a single permutation; then $\equiv$ is defined as the reflexive and transitive closure $\equiv_0^*$.[4]

Definition 7. PERMUTATION-EQUIVALENCE OF FINITE FIRING SEQUENCES

Whenever $M_0 \xrightarrow{\alpha t_i t_{i+1} \beta}$ and $M_0 \xrightarrow{\alpha t_{i+1} t_i \beta}$ for $\alpha, \beta \in T^*$, then

$$M_0 \xrightarrow{\alpha t_i t_{i+1} \beta} \equiv_0 M_0 \xrightarrow{\alpha t_{i+1} t_i \beta}$$

Moreover, $\equiv \; = \; \equiv_0^*$ (by definition).   □ C 7

In the case of Fig. 1, the two sequences $M_0 \xrightarrow{cad}$ and $M_0 \xrightarrow{cda}$ are permutation-equivalent ($M_0 \xrightarrow{cad} \equiv_0 M_0 \xrightarrow{cda}$, hence $M_0 \xrightarrow{cad} \equiv_0^* M_0 \xrightarrow{cda}$ and $M_0 \xrightarrow{cad} \equiv M_0 \xrightarrow{cda}$), with only one permutation $ad \rightsquigarrow da$ between them.
 Again with an immediate proof, we have:

Corollary 3. PERMUTATION-EQUIVALENCE IMPLIES PARIKH-EQUIVALENCE

If $(M_0 \xrightarrow{\sigma_1} \equiv M_0 \xrightarrow{\sigma_2})$, then $(\Psi(\sigma_1) = \Psi(\sigma_2))$.   □ C3

But having the same Parikh vector does not imply permutation equivalence, even if the two sequences are firable from the same marking, as illustrated by Fig. 5.

---

[4] This definition, and also the previous one, would need to be generalised for infinite sequences [5,7]. But in the present paper, no infinite sequences are considered.

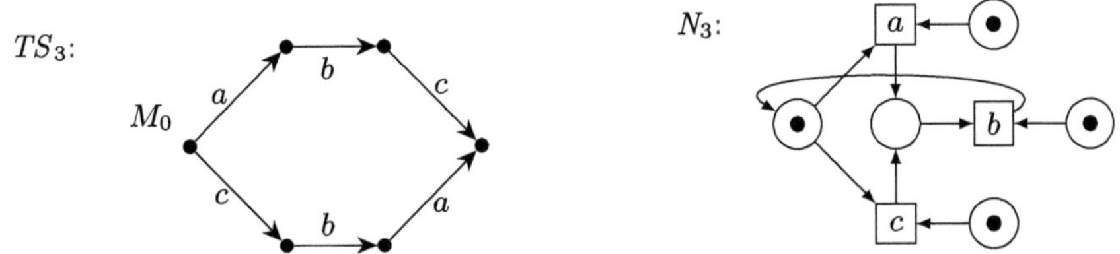

Fig. 5. A plain, pure, and safe, net N_3 on the right and its reachability graph TS_3 on the left: we have $\Psi(abc)=(1,1,1)=\Psi(cba)$ but not $M_0 \xrightarrow{abc} \equiv M_0 \xrightarrow{cba}$, since neither $M_0 \xrightarrow{bac}$ nor $M_0 \xrightarrow{acb}$.

Definition 8. PERSISTENT EQUIVALENTS

A marked net (N, M_0) is called SPE (for "short persistent equivalents") if every finite firing sequence starting from M_0 has a persistent permutation equivalent. It is called $\widetilde{\text{SPE}}$ if every finite firing sequence starting from M_0 has a persistent Parikh equivalent.   □ C8

Corollary 3 implies that SPE is stronger than $\widetilde{\text{SPE}}$.

The name SPE comes from [5,22,23] where SPE is contrasted with various other persistent equivalence notions such as "fair sequences have persistent permutation equivalents", called FPE. These other notions play no relevant role in the present paper.[5]

The plain, pure, and safe, net N_1, M_0) in Fig. 1 satisfies SPE (and thus $\widetilde{\text{SPE}}$). For example, the nonpersistent sequence $M_0 \xrightarrow{cda}$ has the persistent permutation equivalent $M_0 \xrightarrow{cad}$.

Remark 3. NON-INHERITANCE OF PERSISTENT EQUIVALENTS

As opposed to persistence *per se*, SPE is a property which is not inherited in forward direction. It may happen that a marked net (N, M) satisfies SPE and that $M \xrightarrow{t} M'$ but (N, M') does not satisfy SPE, nor $\widetilde{\text{SPE}}$. For example, in Fig. 1, the sequence $M_0 \xrightarrow{c} M_1 \xrightarrow{db}$ has a persistent permutation equivalent in (N_1, M_0), namely $M_0 \xrightarrow{dbc}$. However, after firing c first, i.e., in (N_1, M_1), which is N_1 with initial marking M_1, the sequence $M_1 \xrightarrow{db}$ has no persistent Parikh equivalent, and *a fortiori*, no persistent permutation equivalent.   □ R3

4   Pure DC Petri Nets Satisfying $\widetilde{\text{SPE}}$ Are Persistent

In this section, we prove our main theorem: if a pure, dissymmetric choice, Petri net satisfies $\widetilde{\text{SPE}}$, then it is necessarily persistent (and satisfies SPE). To this end, we first argue that no LTS that contains the pattern shown in Fig. 6 can be realised by a DC net.

---

[5] But see Sect. 5 (the concluding section).

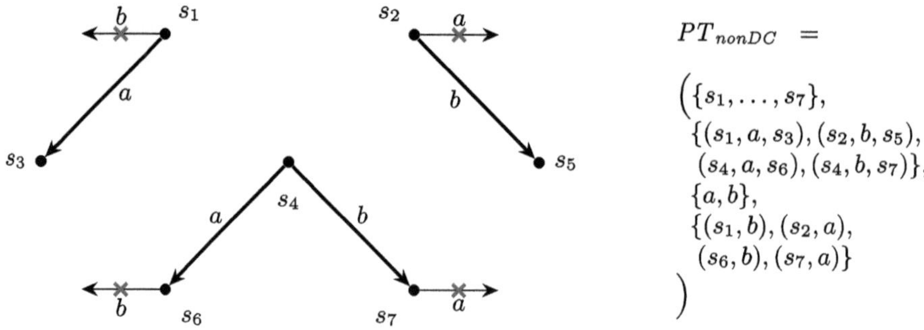

Fig. 6. The pattern PT_{nonDC} indicating that the DC property is violated.

Lemma 1. PT_{nonDC} CAN RECOGNISE NON-DC NETS

Suppose that $N = (P, T, F, M_0)$ is a Petri net and $TS = (S, \rightarrow, T, s_0)$ is (isomorphic to) its reachability graph. Suppose that PT_{nonDC} is embedded in TS. Then N is not dissymmetric choice.

Proof: Let f be the embedding function from PT_{nonDC} to TS. Then:

- The marking (corresponding to) $f(s_4)$ enables both $f(a)$ and $f(b)$. The two non-enablings (s_6, b) and (s_7, a) imply that $f(a)$ and $f(b)$ have at least one common preplace, that is, ${}^\bullet f(a) \cap {}^\bullet f(b) \neq \emptyset$ in N.
- The marking corresponding to $f(s_1)$ enables $f(a)$ but, by $(s_1, b) \in D$ and the definition of embedding, not $f(b)$. This entails $\neg({}^\bullet f(b) \subseteq {}^\bullet f(a))$ in N.
- Similarly, the marking corresponding to $f(s_2)$ enables $f(b)$ but, by $(s_2, a) \in D$, not $f(a)$, entailing $\neg({}^\bullet f(a) \subseteq {}^\bullet f(b))$ in N.

Taken together, this implies that N is not DC. ∎

Now we turn our attention to the claim that if a pure, dissymmetric choice, Petri net satisfies $\widetilde{SPE}$, then it is necessarily persistent. This claim is proved in the following logically equivalent way: Let a given net be pure and assume it satisfies $\widetilde{SPE}$ and is nonpersistent. Then it cannot be dissymmetric choice. The proof is done in three steps as follows:

1. (Section 4.1.) By pureness, any choice state in the reachability graph of a plain net either starts a half-diamond (proper choice) or a full diamond (proper concurrency). There are no 3/4-diamonds.
2. (Section 4.2.) Any two persistent sequences from M_0 having the same Parikh vectors and coming together at some M can be permuted such that they go through some state just two labels before M (unless they are very short, and provided all shorter sequences are also persistent).
3. (Section 4.3.) Nonpersistence of N means that some nonpersistent state M can be reached, which we can assume to be reachable by some shortest possible sequence γ (so that all sequences of lengths $< |\gamma|$ are also persistent).

By nonpersistence and Step 1., there is a half-diamond at M, which is symmetric. $\widetilde{\text{SPE}}$ implies that the two ends of the half-diamond can be reached by persistent sequences α, β from M_0 which are (almost) Parikh-equivalent to γ (except for the two legs of the half-diamond). Using Step 2. twice, we can reconstruct the pattern shown in Fig. 6 and apply Lemma 1.

4.1 There Are No 3/4-Diamonds

Lemma 2. 3/4-DIAMONDS CAN BE COMPLETED TO 4/4-DIAMONDS

Let $N = (P, T, F)$ be a pure, plain, net. Let M be a marking of N and let $x, y \in T$. Suppose that $M \xrightarrow{y} M'' \xrightarrow{x} \widehat{M}$ is a firing sequence and that $M \xrightarrow{x} M'$. Then $M \xrightarrow{x} M' \xrightarrow{y} \widehat{M}$ is also a firing sequence.

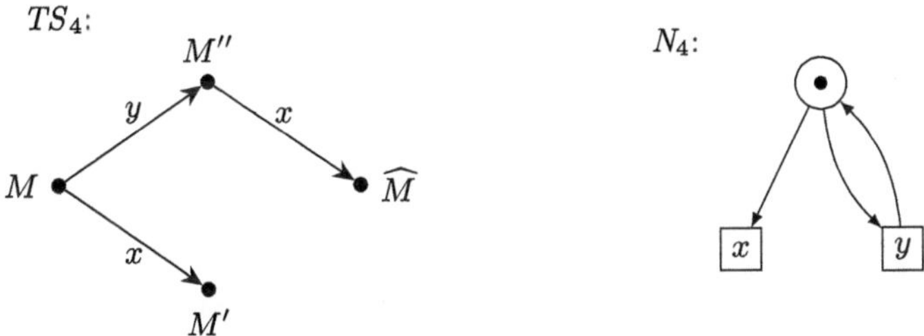

Fig. 7. A 3/4-diamond TS_4. The net N_4 proves that the 3/4-diamond has a Petri net solution which is impure but plain, safe, and FC (with $M = M''$ and $M' = \widehat{M}$).

Proof: We show that $M \xrightarrow{xy}$ is a firing sequence:

By $M \xrightarrow{yx}$, pureness and plainness, the places $p \in {}^\bullet x \cap {}^\bullet y$ satisfy $M(p) \geq 2$, and by $M \xrightarrow{x}$, the places $q \in ({}^\bullet x \setminus {}^\bullet y) \cup ({}^\bullet y \setminus {}^\bullet x)$ satisfy $M(q) \geq 1$. Hence, after firing x from M by $M \xrightarrow{x} M'$, the places $r \in {}^\bullet y$ satisfy $M'(r) \geq 1$, and therefore, y is firable from M'. Of course, $\widehat{M}$ is reached after $M \xrightarrow{xy}$, from determinism, because $\Psi(xy) = \Psi(yx)$. ∎

Note that if N is not safe, it may be the case that $x = y$, but the proof is unaffected. Informally, Lemma 2 means that in a pure and plain net, there may only be 1/2 (half; meaning proper choice) or full (1/1; indicating proper concurrency) diamonds at choice points of the reachability graph, and no 3/4-diamonds. The pureness requirement is essential for this property, as exhibited by the net N_4 on the right-hand side of Fig. 7.

4.2 Unifying Persistent Firing Sequences

Proposition 1. Persistent sequences can be unified

Suppose that $N = (P, T, F, M_0)$ is a pure, plain, Petri net with initial marking M_0 and that $\widetilde{M}$ is a marking that can be reached by two equally long sequences $\alpha = a_1 \ldots a_n$ and $\beta = b_1 \ldots b_n$ as follows:

$$M_0 \xrightarrow{a_1 \ldots a_{n-1}} \widetilde{M_1} \xrightarrow{a_n} \widetilde{M} \quad and \quad M_0 \xrightarrow{b_1 \ldots b_{n-1}} \widetilde{M_2} \xrightarrow{b_n} \widetilde{M}$$

such that $a_n \neq b_n$, i.e., the last two transitions are different, and $\Psi(\alpha) = \Psi(\beta)$. Further, assume that all firing sequences $M_0 \xrightarrow{\tau}$ of length $|\tau| \leq n-1$, in particular $M_0 \xrightarrow{a_1 \ldots a_{n-1}}$ and $M_0 \xrightarrow{b_1 \ldots b_{n-1}}$, are persistent.
Then there is a marking J which can be reached by a persistent sequence $M_0 \xrightarrow{\sigma} J$ satisfying $J \xrightarrow{a_n} \widetilde{M_2}$, $J \xrightarrow{b_n} \widetilde{M_1}$, and $\Psi(\sigma a_n b_n) \, (= \Psi(\sigma b_n a_n)) = \Psi(\alpha) \, (= \Psi(\beta))$.

Remark: The proposition states that there is a full $1/1$-diamond with starting point J, legs a_n, b_n, and end point $\widetilde{M}$ (see the right-hand side of Fig. 8).

Proof: Suppose that all premises are satisfied for N, $\widetilde{M}$, $\alpha = a_1 \ldots a_n$, and $\beta = b_1 \ldots b_n$. We shall prove the existence of J.

Let $\gamma = a_1 \ldots a_m = b_1 \ldots b_m$ be the longest common prefix of α and β (see the left-hand side of Fig. 8). The length of γ can be zero (if $a_1 \neq b_1$), and it can be at most $n - 2$, since lengths $n - 1$ and n are prohibited by $\Psi(\alpha) = \Psi(\beta)$ in combination with $a_n \neq b_n$. Thus, $0 \leq m \leq n - 2$, and $a_{m+1} \neq b_{m+1}$ since γ is defined as the longest common prefix. The proof proceeds by downward induction on $n - 2 - m \geq 0$ (i.e., by upward induction on m).

Base case: $n - 2 - m = 0$, that is, $m = n - 2$.

Then the marking just before $\widetilde{M_1}$ in α must, by definition of γ, be the same as the marking just before $\widetilde{M_2}$ in β. Call it J. By $\Psi(\alpha) = \Psi(\beta)$, it follows that $a_{n-1} = b_n$ and $b_{n-1} = a_n$, yielding the desired diamond.

Inductive step: Let us assume that the claim has been proved when two Parikh-equivalent firing sequences of length n have a common prefix of length smaller than m; we now prove it for γ of length m (with $0 \leq m < n - 2$).

Figure 8 depicts the general setup. From M_0 up to K, α and β agree with each other, and γ is their common prefix. Because of $a_{m+1} \neq b_{m+1}$, they deviate from point K onwards. The sequence $a_{m+1} \ldots a_{n-1}$ leads from K to $\widetilde{M_1}$ via a_{m+1}, M_1, a_{m+2}, M_1', and α' in the lower part of the figure. The sequence $b_{m+1} \ldots b_{n-1}$ leads from K to $\widetilde{M_2}$ via b_{m+1}, K_1, β', K_2, b_k, K_3, b_{k+1}, K_4, and β'' in the upper part of the figure. Moreover, both sequences are persistent because they are tails of the persistent sequences $a_1 \ldots a_{n-1}$ and $b_1 \ldots b_{n-1}$, respectively. (Compare Corollary 1.)

Abbreviate $a = a_{m+1}$. Because $a \neq b_{m+1}$ and $\Psi(a_{m+1} \ldots a_n) = \Psi(b_{m+1} \ldots b_n)$, a also occurs amongst the $b_{m+2}, \ldots, b_n$; we shall assume that

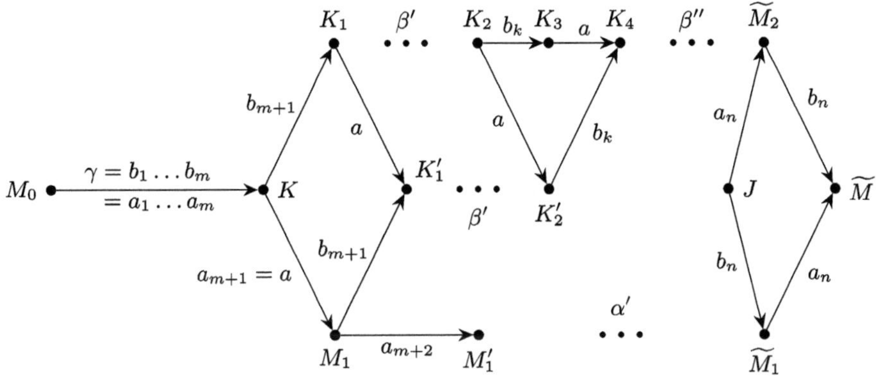

Fig. 8. Sketch of γ (left), $\alpha = a_1 \ldots a_n$ (bottom), and $\beta = b_1 \ldots b_n$ (top) with $\Psi(\alpha) = \Psi(\beta)$. Node J results from the construction in the proposition. Eventually, J can be reached from M_0 by a (persistent) sequence σ which satisfies $\Psi(\sigma a_n b_n) = \Psi(\alpha) = \Psi(\beta)$.

the first such occurrence is just after b_k $(m + 2 \leq k \leq n - 1)$, i.e., $a = b_{k+1}$. Because $a \neq b_{m+1}$ and because $b_{m+1} \ldots b_{n-1}$ is persistent, K_1 enables a. In fact, by the same reason and because $b_{m+1} \ldots b_k$ do not contain a, all markings up to and including K_2 (and K_3, of course) enable a. At K_2, there is thus a 3/4 diamond formed by K_2, K_3, K_2', K_4 and legs a, b_k, which can be "closed" by a b_k-arrow from K_2' to K_4 according to Lemma 2.

Now replace β by the sequence $\widetilde{\beta}$ which leads from M_0 to $\widetilde{M}$ via γ, a_{m+1}, M_1, b_{m+1}, K_1', β', K_2', b_k, K_4, β'', $\widetilde{M}_2$, and b_n. The pair $\alpha, \widetilde{\beta}$ has the same salient properties as the pair α, β before, but it has a longer common initial prefix.

Hence the induction hypothesis can be applied. Eventually, the base case takes effect, proving the existence of a marking J, as was claimed. ∎

Remark 4. ON PROPOSITION 1 AND ITS PROOF

Various special cases may arise. For example, the new initial prefix may be much longer than the previous one (not just by one transition). Essentially, we have permuted a backwards in front of $b_{m+1} \ldots b_k$. Also, the markings $\widetilde{M}_1$ and $\widetilde{M}_2$ may coincide if a_n and b_n are duplicated transitions. It may even be the case that the whole diamond collapses to a single point.

Note that Proposition 1 provides a special case in which there is a kind of reverse of Corollary 3, in the sense that we construct a permutation from two Parikh-equivalent sequences. Such a construction does not work in general (for example, it does not in Fig. 5) because the sequences do not need to be persistent. The proof bears some similarity with the proof of Keller's theorem [3, 19]. □ R4

4.3   Constructing a Non-DC Pattern from a Nonpersistent State

Proposition 2. USING A NONPERSISTENT STATE TO DERIVE FIG. 6

Suppose that $N = (P, T, F, M_0)$ is a pure, plain, nonpersistent, Petri net satisfying the $\widetilde{SPE}$ property. Then in its reachability graph, the pattern PT_{nonDC} shown in Figure 6 is embedded.

Proof: Suppose that N satisfies all premises. We shall prove that PT_{nonDC} (Fig. 6) can be embedded into $RG(N)$.

Since N is nonpersistent, there is some nonpersistent reachable state M. Pick one of them that is nearest to M_0, that is, we have $M_0 \xrightarrow{\delta} M$ (central horizontal line in Fig. 9), and whenever $M_0 \xrightarrow{\tau} M'$ for another nonpersistent state M', then $|\tau| \geq |\gamma|$. In particular, no nonpersistent state is reachable by a sequence that is shorter than δ, and δ is itself persistent. Also, let M_1 and M_2 with $M \xrightarrow{a} M_1$ and $M \xrightarrow{b} M_2$ be the corners of the 1/2-diamond starting at M (right-hand side of Fig. 9). Of course, $a \neq b$, since otherwise, there is no proper choice at M.

By the $\widetilde{SPE}$ property, there is some persistent sequence, Parikh-equivalent to $M_0 \xrightarrow{\delta a} M_1$, by which M_1 can be reached from M_0. This sequence cannot lead through M, since it would not be persistent in this case. (As a consequence, $\delta \neq \varepsilon$.) Hence there is a marking $K_1 \neq M$ such that this persistent sequence has the form $M_0 \xrightarrow{\alpha'} K_1 \xrightarrow{x} M_1$, for some transition x, as shown in the lower part of Fig. 9. By $K_1 \neq M$ and determinism, also $a \neq x$. Similarly, there is a reachable marking K_2 and a persistent sequence $M_0 \xrightarrow{\beta'} K_2 \xrightarrow{y} M_2$ for some transition y, as shown in the upper part of the figure. By $K_2 \neq M$, $b \neq y$.

We can now invoke Proposition 1 two times. For the lower part of the figure, set $\alpha = \alpha'x$ and $\beta = \delta a$ (or the other way round). This results in state J_1, reachable by γ_1, and the backward diamond completions $J_1 \xrightarrow{a} K_1$ and $J_1 \xrightarrow{x} M$. For the upper part of the figure, set $\alpha = \delta b$ and $\beta = \beta'y$ (or the other way round). This results in state J_2 and the backward diamond completions $J_2 \xrightarrow{b} K_2$ and $J_2 \xrightarrow{y} M$.

We shall now check the failed enablings induced by the failed enablings of b at M_1 and a at M_2. K_1 does not enable b because the sequence $\alpha'x$ is persistent since it stems from the $\widetilde{SPE}$ property. This implies that J_1 also does not enable b since otherwise, we would have a nonpersistent state (J_1) that is properly nearer to M_0 than M, contradicting the choice of M. Similarly, in the upper part of the figure, we have $\neg(K_2 \xrightarrow{a})$ and $\neg(J_2 \xrightarrow{a})$. The pattern PT_{nonDC} is embedded by a function $f = f_1 \cup f_2$ (referring to Definition 5) as follows:

$$f_1(s_1) = J_1, \; f_1(s_2) = J_2, \; f_1(s_3) = K_1, \; f_1(s_4) = M,$$
$$f_1(s_5) = K_2, \; f_1(s_6) = M_1, \; f_1(s_7) = M_2 \tag{5}$$
$$\text{and} \quad f_2 = \text{identity on } \{a, b\}$$

This completes the proof. $\blacksquare$

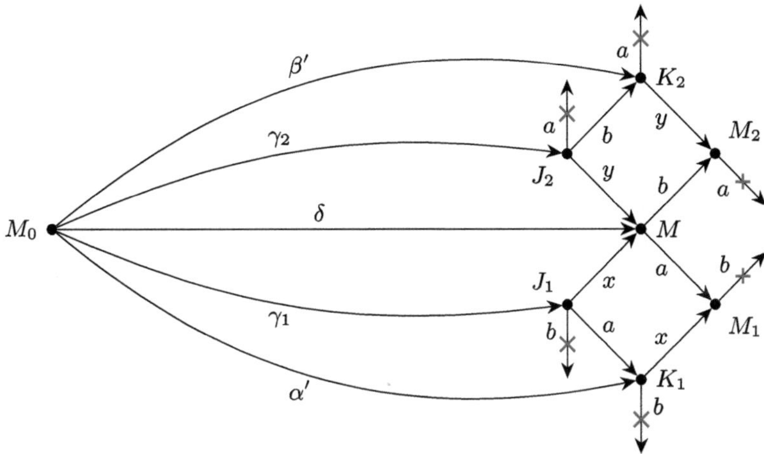

Fig. 9. There is a proper choice (1/2-diamond with legs a, b, $a \neq b$, and $\neg(M_2 \xrightarrow{a})$, $\neg(M_1 \xrightarrow{b})$) at state M. By $\widetilde{\text{SPE}}$, we find a persistent sequence $M_0 \xrightarrow{\alpha' x} M_1$ and another persistent sequence $M_0 \xrightarrow{\beta' y} M_2$. Besides $a \neq b$, we also have $a \neq x$ and $b \neq y$. The non-enablings at K_1, K_2, J_1, J_2 follow as described in the proof of Proposition 2.

Theorem 1. PURE, PLAIN, NONPERSISTENT NETS SATISFYING $\widetilde{\text{S PE}}$ ARE NOT DC

Let $N = (P, T, F, M_0)$ be a pure, plain and nonpersistent, Petri net satisfying the $\widetilde{\text{SPE}}$ property. Then N is not a dissymmetric choice net.

Proof: Referring to Fig. 9, the pattern PT_{nonDC}, defined in Fig. 6, can be embedded in the reachability graph of N as defined above, in (5). We can now simply invoke Lemma 1. This completes the proof. $\blacksquare$

Actually, the embedding (5) is almost injective: We cannot have $M = M_1$ because the former enables b while the latter does not. We cannot have $J_1 = K_1$ because this would imply that $M = M_1$. We cannot have $J_1 = M$ because the former does not enable b while the latter does. We cannot have $K_1 = M_1$ because that would imply $J_1 = M$. Also, we cannot have $M = K_1$ because the former enables b while the latter does not. We cannot have $J_1 = M_1$ by pureness and plainness because in the pure and plain case, it is not possible that, from M, a occurs twice in a row but (still from M) after only one b, a is disabled (in M, the places in ${}^\bullet a$ must have at least two tokens, and b may only take one of them). (The other part of the figure, referring to the diamond J_2, K_2, M, M_2, is symmetric.) However, we might have $K_1 = K_2$.

It is presently unknown whether or not the statement of the main theorem withstands dropping pureness from the set of premises. Lemma 2 goes out of the window in that case, and it is doubtful whether it can be compensated easily.

Corollary 4. PURE, PLAIN, NONPERSISTENT NETS SATISFYING SPE ARE NOT DC

Let $N = (P, T, F, M_0)$ be a pure, plain and nonpersistent, Petri net satisfying the SPE property. Then N is not a dissymmetric choice net.   □ C4

Said differently, with the aid of Corollary 2, we have that pure DC Petri nets satisfying $\widetilde{\text{SPE}}$ or SPE are persistent (as suggested by the title of this paper).

5   Conclusion

As promised in the title, we have proved the correctness of SPE $\Rightarrow$ persistent and of $\widetilde{\text{SPE}} \Rightarrow$ persistent for the class of pure dissymmetric choice nets in Sect. 4. Appendix A describes some historical context motivating the class of dissymmetric choice nets. The implication $\widetilde{\text{SPE}} \Rightarrow$ persistent is valid in other Petri net classes, such as (extended, and possibly non-pure) free-choice nets [4], but not, of course, in larger contexts such as exemplified by Fig. 1.

Several questions are open. A first, immediate, question is whether the premise "pureness" can be dropped from Theorem 1 and Corollary 4 of this paper. Plainness can of course not be dropped because the DC property is undefined for non-plain nets. However, in the same way as *equal conflict* [26] extends choice freeness to the non-plain context, and *homogeneous asymmetric choiceness* [14, 15] extends asymmetric choiceness to the same context, we may also consider extension(s) of dissymmetric choiceness to the non-plain context. For instance, we could say that a Petri net (P, T, F) has *dissymmetric bounds* if, for all $t, t' \in T$, $({}^{\bullet}t \cap {}^{\bullet}t' \neq \emptyset)$ entails $(\forall p \in P : F(p, t) \leq F(p, t')) \vee (\forall p \in P : F(p, t') \leq F(p, t))$, meaning that whenever t and t' may be in conflict, enabling t also enables t', or enabling t' also enables t.

A second open question concerns the computational complexity analysis of the properties SPE and $\widetilde{\text{SPE}}$. If a net is either pure dissymmetric choice or (extended) free-choice, the results presented in this paper and in [4] show that we simply have to check persistence, for which there are known terminating algorithms. In general, these algorithms are very complex, since persistence checking belongs to the class of hard Petri net analysis problems [16, 27]. Specific and more efficient algorithms could possibly be found for the net subclasses mentioned above. For more general subclasses of nets, the problem seems to be much more delicate.

A third question is whether or not

$$\text{SPE} \Rightarrow \text{FPE} \quad \text{for} \quad \text{pure, plain, and safe, Petri nets} \tag{6}$$

which is a slight strengthening of Ochmański's original conjecture [22, 23].[6] From our various results, including the ones in [5], we can argue that if (6) turns out

---

[6] An (infinite) firing sequence σ is fair if no transition is enabled infinitely often along σ but occurs only finitely often in σ. FPE means that every fair firing sequence has a persistent equivalent.

to be true, then it is a very tight result. We hope that the proof techniques we have developed in these papers ([5] and the present one) can be re-used in order to contribute to a resolution of the open questions.

Acknowledgement. The authors gratefully acknowledge the remarks by the reviewers of this paper.

A Historical Remarks on (Dis-/a-)symmetric Choice Nets

The class of asymmetric choice Petri nets, and by (reverse) duality, also the class of dissymmetric choice nets, has at least two different roots. Firstly, both classes of nets are pretty immediate extensions of the class of free-choice nets. In fact, Michel Hack, the originator of FC net theory, already saw this [13] and proposed to investigate the class of "simple nets" as a promising candidate to which the free-choice net theory could be extended.[7] As it turns out, simple nets are asymmetric choice, and conversely, every asymmetric choice net can be simulated in a strong sense by simple nets [6]. Some structural analysis results such as Fred Commoner's liveness theorem [8] can partly be extended to asymmetric choice nets [2,10]. We know of no such research on DC nets, however. We are not even aware of a name given to that class, which is why we borrowed a suitable moniker from [9].

A second line of heritage of asymmetric and dissymmetric choice nets can be traced back to none other than C.A. Petri. In several of his works, he was trying to sort out different ways in which concurrent agents can interact in a distributed setting. He identified various constellations such as "contact", "conflict", and "confusion". By "confusion", he meant, broadly speaking, interactions characterised by a mixture of concurrency and conflict. In [24], for instance, we find the two prototypical confusion patterns shown in Fig. 10.

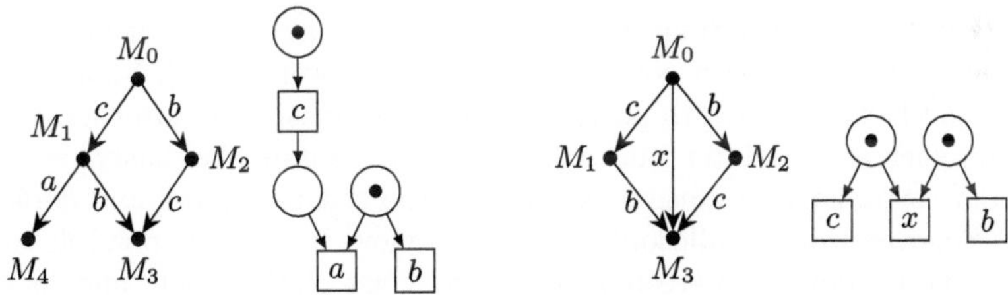

Fig. 10. Asymmetric confusion (left-hand side) and symmetric confusion (right-hand side), according to Carl Adam Petri [24].

In asymmetric confusion as in Fig. 10, if c occurs before b, there is a transient conflict between a and b, but if b occurs before c, there is no such conflict, and yet,

---

[7] Hack's original definition of "simple nets" requires that every transition has at most one shared pre-place, i.e., $\forall p, p' \in P, p \neq p' : (p^\bullet \cap p'^\bullet = \emptyset) \vee |p^\bullet| \leq 1 \vee |p'^\bullet| \leq 1$.

b and c are completely independent. Petri argues that since in a truly concurrent and distributed setting, it is "physically", so to speak, not possible to determine a specific order between b and c, there is some fuzziness about the conflict between a and b, which is what the term "confusion" aims at expressing.

In symmetric confusion as in Fig. 10, there is a related kind of fuzziness: b and c are independent, and they could be physically very far away from each other. The conflict between the two transitions b (or c) and x is resolved not just by one of them occurring, but also by a likely uncontrollable occurrence of c (resp. b). This indicates that it could be quite hard to implement b and c in a distributed manner such that the conflicts between c, x and x, b are respected as specified by the "M" shaped Petri net on the right-hand side of Fig. 10, preferably not using devices such as "busy wait".

Actually, the dissymmetric choice restriction can accommodate both of Petri's confusion patterns shown in Fig. 10. None of them violates the DC condition

$$(^\bullet t \cap {}^\bullet t' = \emptyset) \ \vee \ (^\bullet t \subseteq {}^\bullet t') \ \vee \ (^\bullet t' \subseteq {}^\bullet t)$$

and both sorts of confusion could occur in a DC net. What DC does prohibit, though, is a particularly nasty form of dissymmetric choice when the pattern shown on the left-hand side of Fig. 10 is redoubled by adding a d branch in parallel to the c branch, leading into b. This symmetricised version of asymmetric choice leads straight to the net N_1 in Fig. 1. As we have seen, the absence of this pattern in DC nets is one of the cruxes of our arguments in Sect. 4. (And it is its presence which complicates very considerably the proof of Ochmański's conjecture for plain, pure, and safe, nets.)

Originally, Petri was aiming at finding rules for orderly means of communication, and for some time, he was actually trying to argue that all kinds of confusion should be avoided (if at all possible). This led to a lot of discussion because many real-life use cases seem to involve confusion of some sort quite unavoidably. If one follows Einar Smith in his function (if there is such a function) as Petri's "scientific executor" [25], then it should be advisable to come to terms with confused situations, rather than trying to avoid them from the outset, an opinion which has been corroborated by other work ([18], by Joost-Pieter Katoen and Doron Peled). In particular, Smith argues, there are some pieces of hardware such as an arbiter that will necessarily contain the "nasty" symmetric version of asymmetric confusion sketched in the last paragraph and depicted in Fig. 1.[8] So, even from a philosophical point of view, it is hardly possible to argue that we should not be interested in the DC class of Petri nets and in solving Ochmański's original conjecture one way or the other.

It is possible to prohibit asymmetric confusion by a structural condition which is akin to the conditions defining AC and DC nets. To this end, let us call a Petri net $N = (P, T, F)$ $\widetilde{\text{DC}}$ if

$$\forall p, p' \in P: p^\bullet \cap p'^\bullet \neq \emptyset \Rightarrow (p^\bullet \subseteq p'^\bullet) \vee (^\bullet p' \subseteq {}^\bullet p) \tag{7}$$

---

[8] To appreciate this, compare Fig. 8 of [25] with Fig. 1.

As it was noted in [9], (7) excludes Petri's asymmetric confusion (Fig. 10), and it was in fact designed to do so. This indicates that (7) has some intuitive appeal, but not much seems to have resulted in terms of any closer investigation of $\widetilde{\text{DC}}$ nets. Besides, (7) is neither dual nor reverse dual to the AC or DC conditions, so that we feel justified in proposing a more fitting change of names.[9]

Like AC and DC, the $\widetilde{\text{DC}}$ property (7) generalises the FC property, but unlike DC, it does not come close to characterising the pattern in Fig. 1 whose absence is a relevant ingredient of our main result. For the sake of completeness, we give an example of a non-free-choice net which satisfies all of the other three properties (Fig. 11). This example indicates that it could be interesting, in future work, to investigate the class of Petri nets which simultaneously satisfy AC, DC, and $\widetilde{\text{DC}}$, and are thus confusion-free in the sense of Petri, without necessarily already being free-choice.

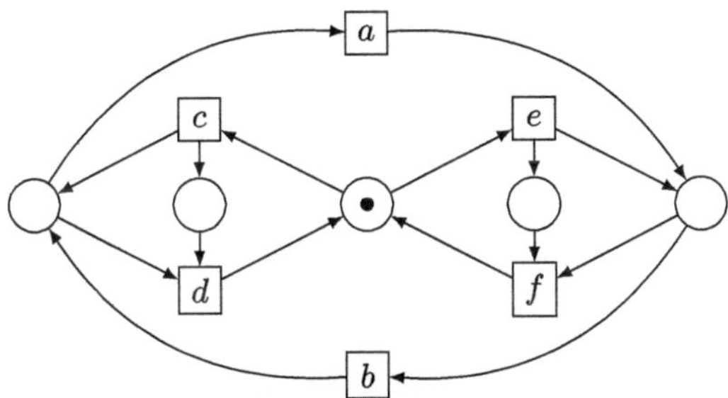

Fig. 11. A net, taken from [9], which is AC, DC, and $\widetilde{\text{DC}}$, but not FC. In [9], it is argued informally that no FC net can simulate it in any strong sense. The argument is that after firing c, a is enabled in conflict with d, whereas after firing eb, a is the only enabled transition, which is very much unlike the behaviour of an FC net. A formal proof of this, given some notion of simulation, can be challenging [21].

References

1. Badouel, E., Darondeau, P.: Theory of regions. In: Lectures on Petri Nets I: Basic Models, Advances in Petri Nets, the volumes are based on the Advanced Course on Petri Nets, held in Dagstuhl, September 1996. Reisig, W., Rozenberg, G., (eds.) vol. 1491, pp. 529–586. Springer, Heidelberg (1998). https://doi.org/10.1007/3-540-65306-6_22

2. Barkaoui, K., Minoux, M.: A polynomial-time graph algorithm to decide liveness of some basic classes of bounded Petri nets. In: Application and Theory of Petri Nets 1992, 13th International Conference, Sheffield, UK, June 22-26, 1992, Proceedings. Ed. by Kurt Jensen.Vol. 616. Lecture Notes in Computer Science. Springer, 1992, pp. 62–75. https://doi.org/10.1007/3-540-55676-1_4

---

[9] It may perhaps be amusing to consider the net class defined by the reverse dual of (7), but in our estimation, this class of nets might not be quite so interesting.

3. Best, E., Darondeau, P.: A decomposition theorem for finite persistent transition systems. Acta Inf. **46**(3), pp. 237–254 (2009). https://doi.org/10.1007/s00236-009-0095-6

4. Best, E., Devillers, R.: Persistent permutability in choice petri nets (2026). Submitted for publication

5. Best, E., Devillers, R.: Persistent permutations, fairness, asymmetric choice petri nets, and ochmański's conjecture. In: Application and Theory of Petri Nets and Concurrency - 46th International Conference, PETRI NETS 2025, Paris, France, 22–27 June 2025, Proceedings. Ed. by Elvio Gilberto Amparore and Lukasz Mikulski. Vol. 15714. Lecture Notes in Computer Science. Springer, pp. 155–173 (2025). https://doi.org/10.1007/978-3-031-94634-9_8

6. Best, E., Devillers, R.: Petri net primer - a compendium on the core model, analysis, and synthesis. Springer (2024). https://doi.org/10.1007/978-3-031-48278-6

7. Best, E., Devillers, R.: Sequential and concurrent behaviour in petri net theory. Theor. Comput. Sci. **55**(1), 87–136 (1987). https://doi.org/10.1016/0304-3975(87)90090-9

8. Fred, G.: Commoner. deadlocks in petri nets. Technical Report CA-7206-2311 (apparently not available online). Applied Data Research, Wakefield, Mass (1972)

9. Desel, J., Best, E.: AC/DC-systems. In: Petri Net Newsletters Herzog, O., Reisig, W., Valk, R., (eds.), Vol. 33 : Gesellschaft für Informatik, 1989, pp. 3–8 (2025). https://www.inf.uni-hamburg.de/inst/bib/loan/members/petri-net-newsletter-33.pdf

10. Desel, J., Esparza, J.: free choice petri nets. Cambridge Tracts Theoret. Comput. Sci. **40** (1995). Cambridge University Press

11. Desel, J., Reisig, W.: The synthesis problem of petri nets. Acta Inf. **33**(4), 297–315 (1996). https://doi.org/10.1007/s002360050046

12. Ehrenfeucht, A., Rozenberg, G.: Partial (Set) 2-structures. part ii: state spaces of concurrent systems. Acta Inf. **27**(4), 343–368 (1990). https://doi.org/10.1007/BF00264612

13. Michel, H., Hack, T.H.: Analysis of production schemata by petri nets. Tech. rep. Massachussetts Institute of Technology, MAC TR-94, 1974 (based on his MSc thesis 1972)

14. Hujsa, T., Devillers, R.: On deadlockability, liveness and reversibility in subclasses of weighted petri nets. Fundam. Inf. **161**(4), 383–421 (2018). https://doi.org/10.3233/FI-2018-1708

15. Jiao, L., Cheung, T., Lu, W.: On liveness and boundedness of asymmetric choice nets. Theor. Comput. Sci. **311**(1–3), 165–197 (2004). https://doi.org/10.1016/S0304-3975(03)00359-1

16. Jones, D, N., Landweber, L.H., Lien, Y.E.: Complexity of some problems in petri nets. Theor. Comput. Sci. **4**(3), 277–299 (1977). https://doi.org/10.1016/0304-3975(77)90014-7

17. Karp, R.M., Miller, R.E.: Parallel program schemata. J. Comput. Syst. Sci. **3**(2), 147–195 (1969). https://doi.org/10.1016/S0022-0000(69)80011-5

18. Katoen, J.-P., Peled, D.: Taming confusion for modeling and implementing probabilistic concurrent systems. In: Programming Languages and Systems - 22nd European Symposium on Programming, ESOP 2013, Held as Part of the European Joint Conferences on Theory and Practice of Software, ETAPS 2013, Rome, Italy, 16–24 March 2013. Proceedings. Ed. by Matthias Felleisen and Philippa Gardner. Vol. 7792. Lecture Notes in Computer Science. Springer, 2013, pp. 411–430. https://doi.org/10.1007/978-3-642-37036-6_23

19. Keller, R.M.: A fundamental theorem of asynchronous parallel computation. In: Parallel Processing, Proceedings of the Sagamore Computer Conference, Sagamore, Adirondack Mountains, NY, USA, 20–23 August 1974. Feng, T (ed.) vol. 24, pp. 102–112. Springer, Heidelberg (1975). https://doi.org/10.1007/3-540-07135-0_113
20. Landweber, L.H., Robertson, E.L.: Properties of conflict-free and persistent petri nets. J. ACM **25**(3), 352–364 (1978). https://doi.org/10.1145/322077.322079
21. Linde-Göers, H.: Free choice simulation of petri nets. In: Proceedings of the Fourth International Workshop on Petri Nets and Performance Models, PNPM 1991, Melbourne, Victoria, Australia, 2–5 December 1991. IEEE Computer Society, pp. 236–245 (1991). https://doi.org/10.1109/PNPM.1991.238796
22. Ochmański, E.: On conflict-free executions of elementary nets. Syst. Sci. Wydawca Oficyna Wydawnicza Politechniki Wrocławskiej **27**(2), 89–105 (2001)
23. Ochmański, E.: Persistent runs in elementary nets. FoLCo (Formal Languages and Concurrency) Memorandum, pp. 2–6. University of Toruń (2014)
24. Petri, C, A.: General net theory. In: On the Teaching of Computing Science. Ed. by Brian Randell. Newcastle International Seminars, pp. 131–169 (1976). http://homepages.cs.ncl.ac.uk/brian.randell/Seminars/ (last checked December 2025): University of Newcastle upon Tyne
25. Smith, E.: On the border of causality: contact and confusion. Theor. Comput. Sci. **153**(1–2), 245–270 (1996). https://doi.org/10.1016/0304-3975(95)00123-9
26. Teruel, E., Silva, M.: Liveness and home states in equal conflict systems. In: Application and Theory of Petri Nets 1993, 14th International Conference, Chicago, Illinois, USA, 21–25 June 1993, Proceedings, pp. 415–432 (1993). https://doi.org/10.1007/3-540-56863-8_59
27. Wimmel, H.: Entscheidbarkeit bei Petrinetzen, Springer-Verlag (2008). Textbook. ISBN: 978-3-540-85471-5

ANIMATE: Automated Framework for Scalable Design of Tsetlin Machines Using 1-Safe Petri Nets

Alex Chan$^{(\boxtimes)}$, Mohamed Tarraf , Rishad Shafik , and Alex Yakovlev

Newcastle University, Newcastle upon Tyne NE1 7RU, UK
`{alex.chan,m.tarraf2,rishad.shafik,alex.yakovlev}@newcastle.ac.uk`

Abstract. The evergrowing demand for state-of-the-art machine learning (ML) models has introduced systems with extremely large, often combinatorial, state spaces, which severely impact existing validation and verification techniques needed for ensuring the ML model's behaviour is guaranteed, safe and dependable. Tsetlin Machines (TMs) are a promising ML model, as their discrete event-based behaviour is naturally explainable and aligns with formal modelling techniques, where prior work shows how they can be captured using ordinary 1-safe Petri nets. However, those TM-Petri nets lack automated construction support, are only validated on a toy-sized dataset, do not cover the assembling process, and use Cartesian product-based operations that scale poorly to the TM's hyperparameters and larger datasets. In this paper, we present ANIMATE as a holistic data-driven framework that supports automatic construction of our improved TM-Petri nets, which replace the Cartesian product-based operations with parametric mechanisms that scale to any binarised dataset and TM hyperparameter configuration. ANIMATE integrates existing Petri net tools like Workcraft and TINA, and introduces a tool called MaestroPN that supports data-driven simulation. Our experimental results demonstrate the scalability of our new TM-Petri net design, and report the complete net's assembling and simulation times.

Keywords: Petri nets · Tsetlin Machine · Learning Automata · Machine Learning · Data-driven Simulation · Parametric Mechanism

1 Introduction

Over the last few years, Artificial Intelligence (AI) has played an important role for analysing, optimising and implementing state-of-the-art systems for many different areas of application, where we have seen a huge influx of new, yet highly sophisticated, machine learning (ML) models that address key challenges such as cancer diagnosis, financial risk analysis and safety-assured automobiles.

With the introduction of generative AI models like ChatGPT and DeepSeek, it is now evermore important for the industry to create smarter, more energy-efficient, models to reduce the enormous amount of resources that they require

© The Author(s), under exclusive license to Springer Nature Switzerland AG 2026
J. Desel and A. Kalenkova (Eds.): PETRI NETS 2026, LNCS 16567, pp. 130–153, 2026.
https://doi.org/10.1007/978-3-032-27879-1_7

for training and generating on-demand solutions. It is therefore crucial that ML designers do not only focus on the operations of these models, but also the underlying algorithm used to generate their outputs and make their decision.

However, despite recent advancements in ML research, many ML models still remain as *black box* models, where their internal workings and decisions cannot be easily observed. This poses significant challenges for ML designers, who must validate and verify the behaviour of these models to ensure that their behaviours are guaranteed, safe and dependable.

While there are many existing verification and explainability techniques for ML models, these tend to come with some small caveats. On one hand, verification techniques ensure that the ML model's behaviour is guaranteed, but comes with extremely expensive computational costs that are not feasible for general-purpose workstations. On the other hand, explainability techniques ensure that the ML model's behaviour can be explained, but can only provide feature-level insights that explain how features may contribute to the model's prediction and not what may cause it. This motivates the use of automata-based learners, as they naturally support both verification and explainability, where they can be verified using reachability-based approaches and their discrete event-based structure can be described using symbolic semantics [40,43].

Tsetlin Machines (TMs) [17] stand out as a promising automata-based learner, as they are naturally explainable due to their *white box* architecture. Their decision logic is built from a combination of the observed datapoints and their collection of Tsetlin Automata (TAs) [38] that form propositional clauses, which help build rule-based expressions and help describe how input features affect the TM's decision making [32]. In recent works, TMs have been used to solve several large classification problems [7,9,10], while demonstrating real-world use cases in critical domains like microbiome analysis [36] and identification of cancer recurrence risk factors [5]. TMs have also seen significant use in practical settings like hardware accelerators [22,29] and financial forecasting [16], with a plethora of supported tools that visualise and explain the TM's behaviour [27,28].

In fact, due to the TM's discrete event-based behaviour and high levels of concurrency, they can even be captured using ordinary 1-safe Petri nets to create so-called "TM-Petri nets" [13], which allow their inference and learning dynamics to be validated and verified. Unfortunately, despite these TM-Petri nets bridging the gap between ML and Petri nets, they have several key limitations:

1. **Feasibility**: The original TM-Petri net was based on a TM learning a two-input XOR gate using 2 classes (1 per output), 2 clauses and 6 TA states [13], which took around 2 months to design. In a real world setting, this is clearly unreasonable for designers, especially if they are a non-Petri net expert, as designing an equivalent TM in software requires significantly less time.
2. **Scalability**: As the original TM-Petri net only considered a toy-sized dataset, it is unclear how it would scale up to larger TM hyperparameters, e.g. more classes, clauses and literals. In fact, the dataset's number of features and rows also play a significant role, as explicitating each datapoint as a transition like the original dataloader would exponentially increase the net size.

3. **Assembling the complete TM-Petri net**: Despite the blueprint designs
   for each TM-Petri net component were shown, the original approach did not
   cover how they are used to assemble the complete TM-Petri net. This limitation is noticeable as we scale up these TM-Petri nets with different hyperparameters and datasets, where some components cannot be easily built due
   to the use of Cartesian product-based constructions.
4. **Behavioural correctness**: While the original TM-Petri net was validated
   to be 1-safe and deadlock-free via simulation, and their components verified
   for deadlock-freeness via MPSAT [19], there is no guarantee that these results
   will hold for different hyperparameters and datasets. The probability-based
   operations used during the TM's feedback was also not covered, as it was
   assumed to be handled by the environment, which is currently not possible for
   most, if not all, Petri net simulators as they do not support such operations.

In this paper, we propose a new holistic data-driven framework called
`ANIMATE` that enables automatic construction of our improved TM-Petri nets,
where we replaced the Cartesian product-based operations found in several components of the original TM-Petri net design [13] with parametric mechanisms,
which scale better for any arbitrary binarised dataset and TM hyperparameter
configuration.

Like in [13], our improved TM-Petri nets are still modelled as ordinary 1-safe
Petri nets to ensure that they are structurally aligned with the TM's core design,
where they can capture all logical operations required by the TM including
logical AND/OR expressions and positive/negative clauses, and be verified for
1-safeness and deadlock-freeness using MPSAT. To ensure that our complete TM-
Petri net can be assembled, while preserving each block's verified properties, we
leverage the concept of distributed places [20] by treating each TM-Petri net
component block as a 'box' and combine them through a merged place that acts
as an input-output communication port between each block.

Beyond the net designs, `ANIMATE` also integrates support for several existing
Petri net tools like `Workcraft` [2], `GraphRack` [33] and `TINA` [1,39], and offers
a new data-driven simulator called `MaestroPN` for automated transition-specific
firing of data instantiation and probability-based operations.

To evaluate `ANIMATE`, our experimental results demonstrate the scalability
of our improved TM-Petri nets with respect to different TM hyperparameter
configurations and datasets, and report both the assembling and simulation times
of our fully constructed TM-Petri net.

Thus, the main contribution of this paper is `ANIMATE`, which is our proposed
holistic data-driven framework (Sect. 3) that includes:

- Automated support for construction and verification of TM-Petri nets using
  any binarised dataset and TM hyperparameter configuration, and complements its supported Petri net tools[1] (Sect. 3.1).
- A new data-driven simulation tool called `MaestroPN` that demonstrates
  the ability to control transition firing of runtime data instantiation and
  probability-based operations during simulation (Sect. 3.2).

---

[1] We currently have plans to add support for more tools for this work in the future.

- An assembling method of the complete TM-Petri net using distributed places that act as each component's input-output ports (Sect. 3.3).
- Improved scalable designs for the TM-Petri net components including data loader, clauses, TM classes, argmax and all feedback modules (Sect. 4).

2   Preliminaries

In this section, we cover the necessary background in both TAs and TMs, as well as several existing Petri net tools that we used or were inspired by, to help understand our work shown in the later sections.

2.1   Tsetlin Automata

TAs are a class of reinforcement learning automata [38], where it changes between different states upon receiving a penalty action or a reward action. As shown in Fig. 1, the states to the left of the midpoint represent an *exclude state*, while the states to the right of the midpoint represent an *include state*, such that they produce an output of 0 and 1 respectively. Note that continued reward actions at end states 1 and $2v$ cause the TA to saturate, and penalty actions at midstates v and $v+1$ cause the TA to cross the decision boundary and invert its output.

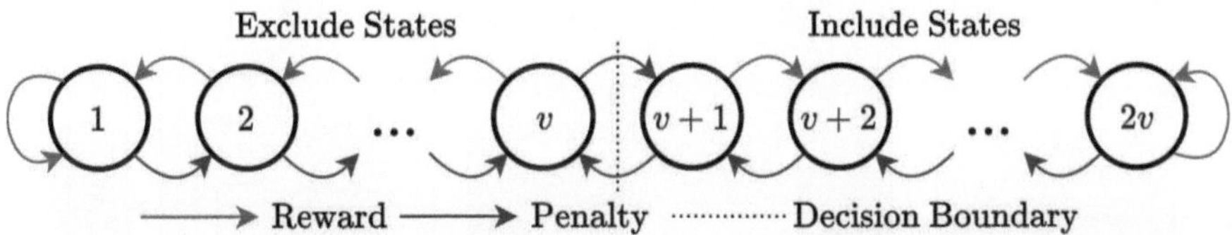

Fig. 1. Overview of a TA.

2.2   Tsetlin Machines

TM [17] is a ML model that uses a collection of TAs to learn new patterns through propositional logic. Like other ML models, TMs use both inference and feedback to contribute to their learning, where inference involves running data points to calculate some output, while feedback involves learning from the output by comparing it to the expected outcome and adjusting its weighting. To help understand how TMs work, Fig. 2 illustrates the whole TM pipeline.

Firstly, every TA in a team of TAs (Fig. 2a) becomes associated with a Boolean literal ℓ generated from the input vector loading process of K literals (Fig. 2b) and are randomly initialised to either midstates v or $v+1$.

Next, a clause m (Fig. 2c) uses its team of K TAs to compute its output through a logical AND expression, which comprises every OR expression required

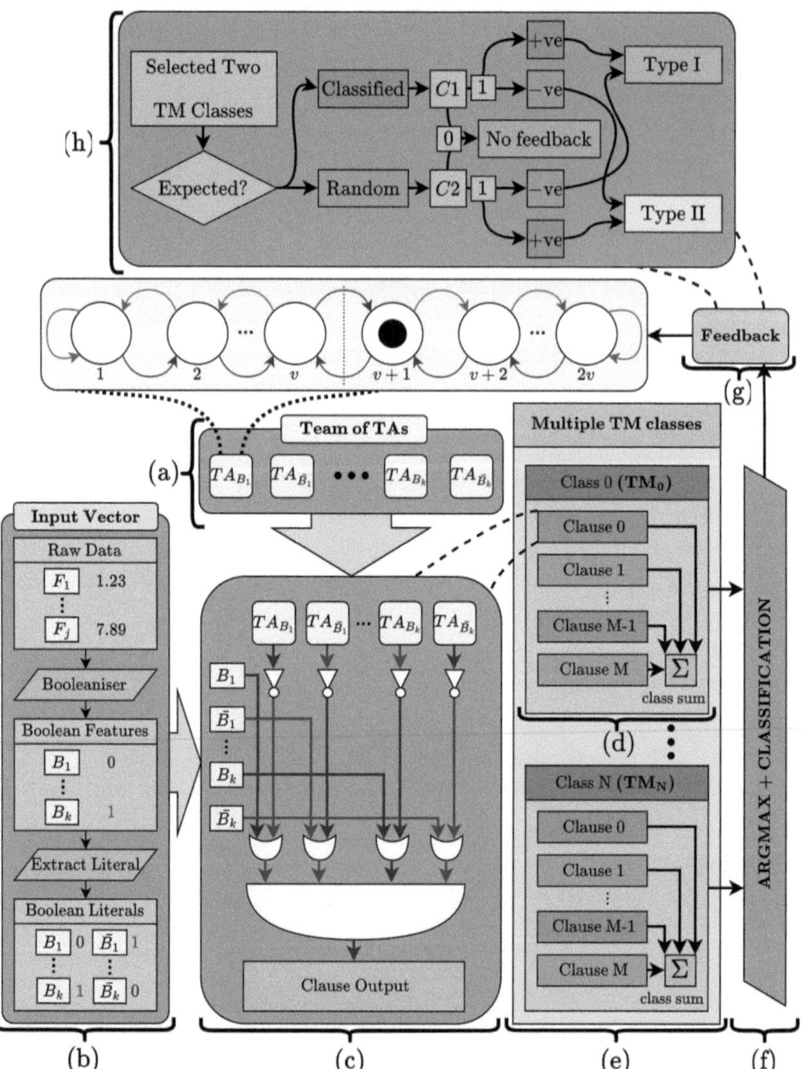

Fig. 2. Overview of TM pipeline. (a) Team of TAs. (b) Booleanised input vector. (c) Clause computation. (d) Single TM class. (e) Multiple TM classes. (f) Classification via Argmax. (g) Feedback component. (h) Feedback type calculation.

between each Boolean literal ℓ and the inverted value of the ℓ-associated TA, where this TA produces an output of 1 (0) if it is in the exclude (include) state.

Once all M clause outputs are calculated, each TM (Fig. 2d) calculates its class sum, i.e., *vote count*, where the polarity in clauses alternate, such that a clause with a positive (negative) polarity means that they vote for (against)

Table 1. TM's Payoff matrix

FB2	none	T1	T1	T1	T1	T1	T1	T1	T1	T1	T1	T2	T2	T2
include	×	1	1	1	1	0	0	0	0	0	0	1	0	0
clause	×	0	0	1	1	0	0	1	1	1	1	×	1	0
literal	×	×	×	×	×	×	×	0	0	1	1	×	0	×
p_s	×	0	1	0	1	0	1	1	0	0	1	×	×	×
FB3	I	P	I	I	R	R	I	I	R	I	P	I	P	I

the class. Note that there are up to N-number of TM classes depending on the number of unique expected outputs provided in the dataset (Fig. 2e).

The TM's predicted output, i.e., *classification*, is then calculated using an argmax (Fig. 2f), i.e., *arguments of the maxima*, before this result is forwarded to the TM's feedback mechanism (Fig. 2g). From there, the classified TM and a random TM are selected, and each of their clauses are checked consecutively to determine whether they should receive feedback or not. If a clause is said to receive feedback, then the feedback type given to its team of TAs is then determined by its polarity and if its associating class is classified or not (Fig. 2h).

The exact feedback action given to the TAs can be calculated using the TM's payoff matrix [17,41]. Here, the TM's feedback logic can be split into three stages: 1) FB1 that focuses on TM-level feedback, 2) FB2 that focuses on clause-level feedback, and 3) FB3 that focuses on TA-level feedback. Note that each stage's output acts as the next stage's input, e.g. the output of FB1 is the input of FB2, until FB3's output is reached, which the corresponding TA receives as its input.

Stages FB1 and FB2 can be assigned a value of none, T1 or T2, where none represents no feedback, T1 represents Type I feedback that combats *false negatives* and T2 represents Type II feedback that combats *false positives*. Stage FB3 can be assigned a value of penalty, reward or inaction, which correspond to the TA's received feedback action.

These particular details of the TM's payoff matrix can be formalised as a truth table, as shown in Table 1 where the indices 'FB2' represents the feedback type received, 'include' represents the TA's current state such that 0 (1) means exclude (include), 'clause' represents the clauses output value, 'literal' represents the literal value, 'p_s' represents the probability of FB3's output using the TM's hyperparameter s such that 0 (1) means a probability of $\frac{1}{s}$ $\left(\frac{s-1}{s}\right)$, and 'FB3' represents the action given to the TA such that R, P and I represent reward, penalty and inaction respectively. Note that any cell with a × means that the value can be either 0 or 1, as they have no effect on FB3's output.

2.3 Petri Net Toolsets

To help understand our work in the later sections, we introduce `Workcraft`, `GraphRack` and `TINA`, which were used to simulate the TM-Petri nets generated

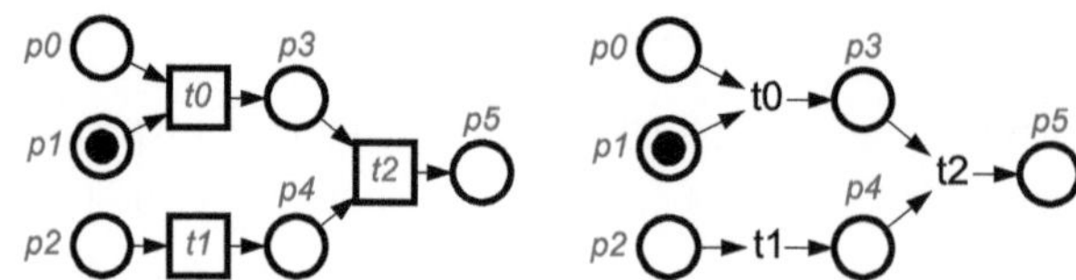

(a) General design of a Petri net (left) and Workcraft's simplified design (right).

(b) Explicit read arcs (left) and simplified read arc (right).

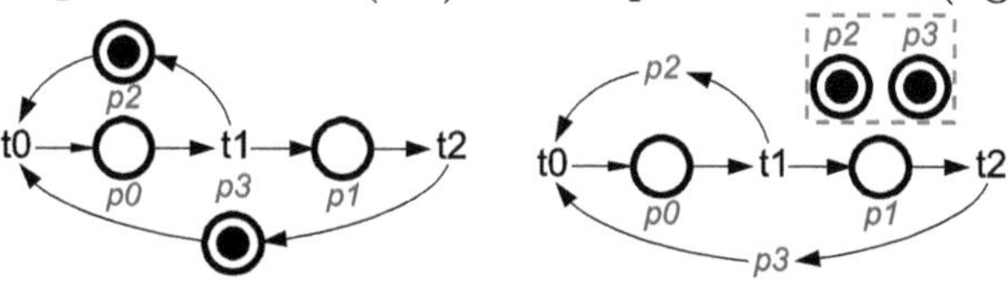

(c) Non-proxies (left) and proxies (right).

Fig. 3. Visual comparisons of general Petri net designs with Workcraft's features.

from ANIMATE that is later presented in Sect. 3, before we later validate and analyse their scalability across different simulation environments.

Workcraft [2] is a visual framework that supports the design automation of many interpreted graph models including ordinary Petri nets [26] and Signal Transition Graphs [30], with an established back-end of tools including PETRIFY [15] and MPSAT [19] for verification and circuit synthesis. In our previous work [13], Workcraft was used both to manually design and visualise the TM-Petri nets. In this work, however, the TM-Petri net designs are fully automated independently of Workcraft, which is used solely to visualise the resulting nets as later shown in Sect. 4. In particular, we simplify the visualisation by making use of Workcraft's simplified Petri net design shown on the right in Fig. 3aand its simplified read arc also shown on the right in Fig. 3b. Like in [13], we also used *proxy places* to graphically collapse the 'physical' place and its connections without changing the Petri net's behaviour as shown in Fig. 3c.

GraphRack [33] is an extensible C++ library that supports the design, simulation and optimisation of digital and analogue circuits. It has been designed from the ground-up to address the difficulties that are encountered, when using Petri nets to design large-scale ML implementations. By opening the problem to a wider scope of graph-network based computations, GraphRack provides the necessary tools to build, simulate, optimise and seamlessly convert these networks, and becomes a powerful asset in the design process. GraphRack also makes large-scale experimentation possible with Petri net models feasible by establishing programmatic interfaces into its internal models. As such, GraphRack played a vital role for analysing the scalability of the original TM-Petri net, and helped us optimise our designs to become more scalable.

TINA (TIme petri Net Analyzer) [39] is a well-established framework that supports editing, simulation and verification of Petri Nets and their time-based variants [23]. TINA offers a wide range of frontend and backend tools [1] including NETDRAW for graphical and textual editing, TINA and SIFT for building and checking reachability graphs, and PLAY and WALK for simulation. Here, we took specific interest on WALK due to its high-speed simulation and support of specifying a list of transitions to 'favour' or 'avoid', which are desirable features for Petri nets specifying ML application and have inspired the implementation of our MaestroPN tool in Sect. 3.2. Moreover, its support for Petri net extensions like Sleptsov nets [8] and Salwicki nets [12], i.e., Petri nets with 'step-based' firing semantics [24,25], allow for greater support of larger Petri net models.

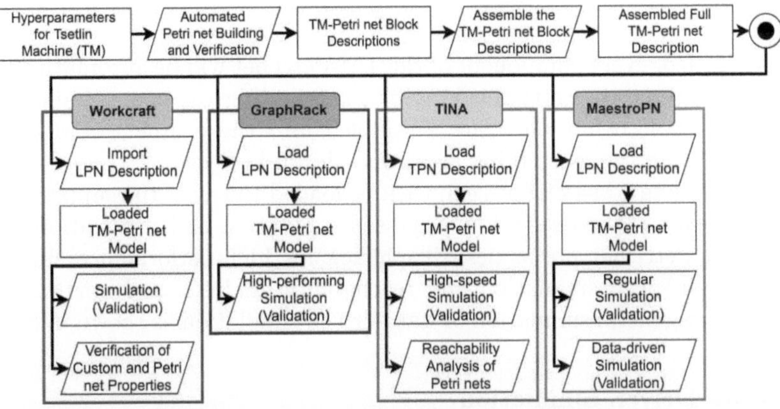

Fig. 4. Design workflow of the ANIMATE.

3 ANIMATE Framework

In this section, we cover the details of ANIMATE, where we highlight its design workflow in Sect. 3.1, our MaestroPN tool in Sect. 3.2 and our algorithm to assemble the complete TM-Petri net in Sect. 3.3.

3.1 Design Workflow

The design workflow of ANIMATE is shown in Fig. 4, where we can specify some TM hyperparameters like the number of clauses, number of TA states and its dataset. Note that the input features and expected output features determine the number of literal TAs and TM classes respectively.

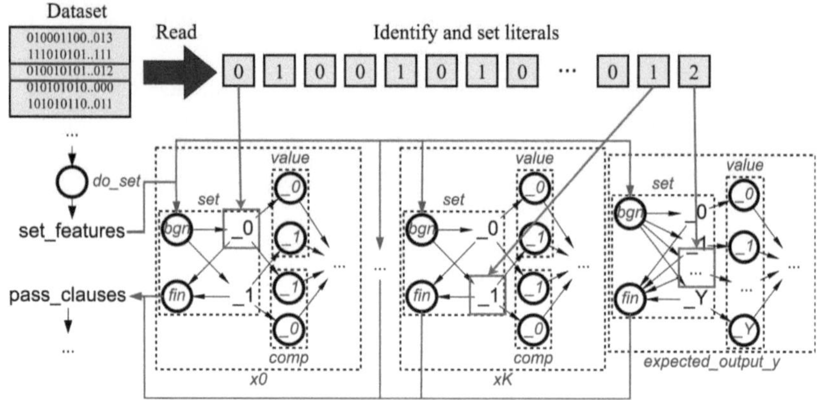

Fig. 5. Simulation paradigm of `MaestroPN` tool.

Once loaded, every component of the TM-Petri net will be automatically constructed into separate Petri net descriptions following the new scalable designs shown in Sect. 4, before they are assembled into the complete TM-Petri net description using the assembling algorithm shown in Sect. 3.3.

This complete TM-Petri net can then be loaded to any of the supported Petri net tools such as `Workcraft` for visualised simulation and verification of both custom and Petri net properties, `GraphRack` for high-performance simulation, `TINA` for high-speed simulation and reachability analysis, and our `MaestroPN` tool for either general simulation or data-driven simulation as shown in Sect. 3.2.

3.2 Data-Driven Simulation

`MaestroPN` is a data-driven simulation tool that was implemented with the ability to automatically detect and fire transitions for enabling data instantiation and probability-based operations during runtime. Unlike most general Petri net simulators, `MaestroPN` has built-in support for loading arbitrary binarised datasets to fire specific transitions, which set the given datapoint's expected value. This feature is essential to support the training of Petri net-specified ML models, where datapoints must be temporally read row-by-row for each *epoch* (i.e. learning cycle) in order to accurately capture the training dynamics of the ML model. An example is shown in Fig. 5, which illustrates how the dataset is read before configuring `MaestroPN` to fire the exact transitions of the corresponding literals.

While simulators like `TINA`'s WALK tool offer support to read in favoured transitions, they are fixed to the exact transition names that are provided before simulation is executed. This limitation prevents temporal firing of transitions associated to a given literal, where they must be fired at different time intervals to correctly set the literal's value based on the dataset's current row.

This feature also allows support of firing transitions that capture probability-based operations, which is essential for capturing the stochastic decisions made during TM training. For example, the feedback decision processes $C1$ and $C2$ that depend on $rand() \leq \frac{T-Class\_sum}{2 \times T}$ and $rand() \leq \frac{T+Class\_sum}{2 \times T}$ respectively, and the TA feedback actions that depend on $rand() \leq \frac{1}{s}$ and $rand() \leq \frac{s-1}{s}$, where T thresholds the TM's feedback frequency, s determines the TM's learning specificity, and $Class\_sum = clip(class\_sum(TM_n), [-T, T])$.

Although guards in Petri nets can be used to mimic this behaviour, it cannot fully capture the stochastic nature of these probability-based operations. Alternatively, one can capture this behaviour by using greatest common denominator (GCD)-based transition rationing or using the IEEE-754 representation. Unfortunately, GCD-based operations tend to be computationally expensive especially for sufficiently large nominators and small denominators, while IEEE-754 likely cause state explosion due to its fine-grain precision.

3.3 Assembling the Tsetlin Machine-Petri Net

In [13], it was not shown how the TM-Petri net components were used to assemble complete TM-Petri net. This is coupled by the fact that the assembling was only done for a TM using 2 classes, 2 clauses and 4 TAs with 6 six states each to learn a 2-input XOR dataset, which makes assembling larger TM-Petri nets non-trivial.

For this work, we used the concept of distributed places [20] by treating each TM-Petri net component block as a 'box' and combine them through a merged place, which acts as an input-output communication port between each block.

Here, we chose distributed places because our TM-Petri net component blocks are designed using ordinary 1-safe Petri nets, as later shown in Sect. 4, to ensure that they are structurally aligned with the TM's core design, and to ensure that they can be verified for 1-safeness and deadlock-freeness using MPSAT.

Our assembling process is formalised as listed in Algorithm 1, where it generates every component blocks' transitions, places, arcs and markings and checks if each place is a merge place or not. Note that this assembling process is only executed after all TM-Petri net components are constructed.

4 Scalable Design of Petri Net-Based Tsetlin Machine

In this section, we illustrate the designs of our improved TM-Petri net components, which can be automatically constructed using ANIMATE described in Sect. 3. To understand why we redesigned our TM-Petri net, we first provide our reasoning in Sect. 4.1, before showing the design of each TM-Petri net component in the following subsections.

To ensure that our improved TM-Petri net components can be verified during automatic construction, we also add an optional ADD_INIT flag that creates the necessary transitions and places to mimic the component's interaction with the environment, where MPSAT can then be used to automatically verify that they are 1-safe and deadlock-free. Note that using this flag also means composition cannot take place, and therefore both must be used independently.

Algorithm 1. Assembling the complete TM Petri Net

Require: Set of TM-Petri net blocks $\mathcal{B}$
Ensure: Composed Sets $\mathcal{P}, \mathcal{T}, \mathcal{A}, \mathcal{M}$
 1: $\mathcal{P} \leftarrow \emptyset,\ \mathcal{T} \leftarrow \emptyset,\ \mathcal{A} \leftarrow \emptyset,\ \mathcal{M} \leftarrow \emptyset$
 2: **function** GETAFFIXEDPLACE(p)   ▷ Affixes a place with block name or merge
 3:     **if** p.name.starts_with("input") $\vee$ p.name.starts_with("output") **then**
 4:         $p_{new} \leftarrow$ PLACE("merge." $+p$.name)
 5:     **else**
 6:         $p_{new} \leftarrow$ PLACE(b.name $+$ "." $+p$.name)
 7:     **end if**
 8:     **return** p_{new}
 9: **end function**
10: **for all** $b \in \mathcal{B}$ **do**
11:     **for all** $p \in b$.places **do**
12:         $\mathcal{P} \leftarrow \mathcal{P}\ \cup\ \{$GETAFFIXEDPLACE($p$)$\}$
13:     **end for**
14:     **for all** $t \in b$.transitions **do**
15:         $t_{new} \leftarrow$ TRANSITION(b.name $+$ "." $+p$.name)
16:         $\mathcal{T} \leftarrow \mathcal{T}\ \cup\ \{t_{new}\}$
17:     **end for**
18:     **for all** $a \in b$.arcs **do**
19:         $s \leftarrow a$.source, $d \leftarrow a$.destination
20:         **if** $s \in b$.places **then**
21:             $s \leftarrow$ GETAFFIXEDPLACE(s), $d \leftarrow$ TRANSITION(b.name $+$ "." $+d$.name)
22:             $\mathcal{A} \leftarrow \mathcal{A}\ \cup\ \{$ARC($s, d$)$\}$
23:         **else if** $d \in b$.places **then**
24:             $s \leftarrow$ TRANSITION(b.name $+$ "." $+s$.name), $d \leftarrow$ GETAFFIXEDPLACE(d)
25:             $\mathcal{A} \leftarrow \mathcal{A}\ \cup\ \{$ARC($s, d$)$\}$
26:         **end if**
27:     **end for**
28:     **for all** $p \in b$.marking **do**
29:         $\mathcal{M} \leftarrow \mathcal{M}\ \cup\ \{$GETAFFIXEDPLACE($p$)$\}$
30:     **end for**
31: **end for**
32: **return** $\mathcal{P}, \mathcal{T}, \mathcal{A}, \mathcal{M}$

4.1 From Cartesian Products to Parametric Mechanisms

As briefly discussed in Sect. 1, our original TM-Petri net [13] used Cartesian product-based operations to design the clause output computation, TM's class summation, argmax's classification, and feedback's action calculation. However, this design choice inherently limits the scalability of the TM-Petri net's design, as every possible combination must then be captured, which is computationally intensive and can lead to incomplete program executions.

In particular, a Cartesian product over i sets $A_1\ \cdot\ A_2\ \cdot\ ...\ \cdot\ A_i$ produces a combinatorial explosion of $|A_1| \times |A_2| \times ... \times |A_i|$ transitions, which quickly becomes infeasible to construct even for modest TM configurations. To exemplify this, let us consider the following scenarios below:

1. **Clause output computation**: A clause output is calculated based on the logical AND expression of every OR expression between a literal ℓ and its associated TA from a set of K literals. As each expression produces either a 0 or 1, the logical AND expression requires 2^K transitions to calculate its result based on every possible value of each logical OR expression.
2. **TM Class summation**: A TM consists of M clauses, which are used to determine its class sum. As clauses either increment or decrement the sum, this requires $2^M + 1$ transitions to capture the sum range of $-\frac{M}{2}$ to $\frac{M}{2}$.
3. **Argmax classification**: Given a dataset with N unique outputs, this requires N TM classes that the argmax must compare based on their vote count. As there can only be one winner, this requires 2^N transitions for each possible class sum value $v \in \{-\frac{M}{2}, ..., \frac{M}{2}\}$ per TM.
4. **Feedback calculation**: TMs have four different possible feedbacks being Type I Clause 0 (T1C0), Type I Clause 1 (T1C1), Type II Clause 0 (T2C0) and Type II Clause 1 (T2C1). T2C0 scales linearly as it only produces inactions. But, T1C0, T1C1 and T2C1 must read each TA's state, and both T1C1 and T2C1 must also read each literal's value. Since feedback can give different actions and is repeated for M clauses per TM class $n \in N$, this results up to $2^{N \times M \times K \times K}$ transitions.

Now, with these scenarios in mind, if we try to implement a realistic TM implementation for MNIST, which has 784 input features (i.e. 1568 literals) and 10 outputs (i.e. 10 classes), this requires a minimum of 100 clauses for sufficient performance and will result in $2^{10 \times 100 \times 1568 \times 1568}$ (i.e. $2^{2458624000}$) transitions for just scenario 4. This is obviously infeasible, as such a generation is impossible to complete, even with unlimited computational resources.

On top of the above, the original dataloader further compounds this problem, as it explicitates the loading of every datapoint in each row. Again, for the MNIST dataset: there are also 60000 rows meaning the original would essentially produce 60000 transitions with 1568×60000 connections to every literal.

This motivates our shift towards designing TM-Petri net components with 'parametric mechanisms', where we redesign the components to scale more proportionally to the parameters and avoid producing combinatorial connections. For example, let us consider the redesigned clause computation shown in Sect. 4.4: We heuristically optimise the connections, where instead of connecting every OR-expression to an AND-expression, we can connect each OR-expression to the next OR-expression until the last one, which connects to the AND-expression. This reduces the combinatorial generation of $|A_1| \times |A_2| \times ... \times |A_i|$ transitions to $|A_1| + |A_2| + ... + |A_i|$ transitions, as we sequentialise the connections between the OR-expressions and the AND-expression. Since this approach adds inherent latency, we use an early exit mechanism [11] to mitigate this, as shown in the following subsections that include redesigns of other components.

4.2   Data Loader

Our new dataloader is shown in Fig. 6, which now scales linearly to the number of total features and outputs of a binarised dataset, by creating set and reset

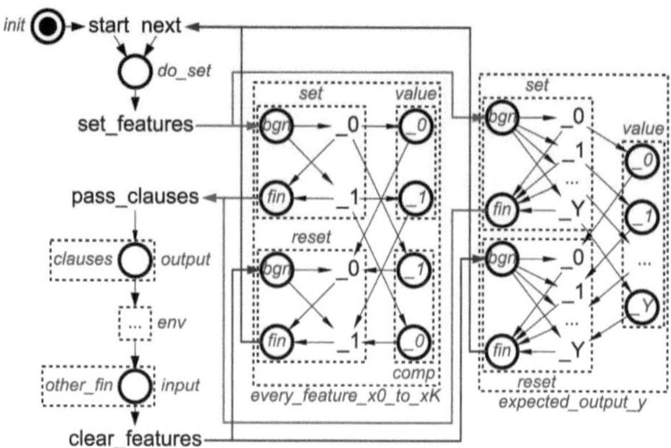

Fig. 6. Design of the new dataloader.

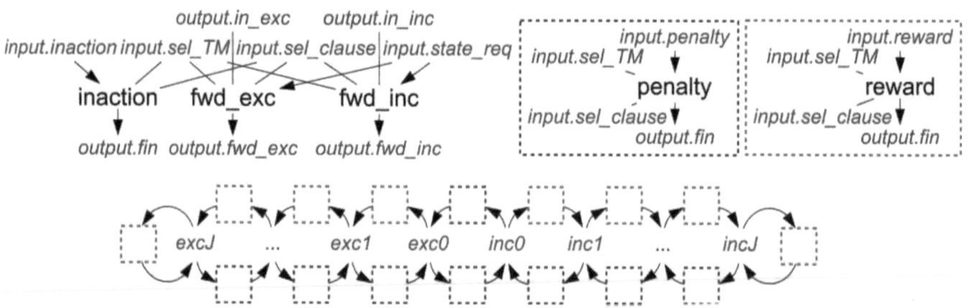

Fig. 7. Design of the 'self-aware' TA that determines when the TA should receive a reward, penalty or inaction from the TM's feedback calculation.

transitions for each literal and expected output instead of the explicit loading. For example, 784 set and reset transitions for 1568 literals instead of 60000 explicit transitions for 1568 literals. Note that `MaestroPN` especially leverages off this design, as it can automatically fire row-specific transitions temporally.

4.3   Tsetlin Automata

Our modified TA component is shown in Fig. 7, where the core TA logic is kept. Like in the original TM-Petri net, we use the input port places to determine when the TA receives a penalty, reward or inaction, and output port places to determine the TA's current state. A key difference, however, is that our new TA component is made '*self-aware*', where every TA now knows when it should receive feedback through a pair of new input port places that detect when the TA's corresponding TM and clause have been selected for feedback. This design was done to simplify the feedback logic shown in Sect. 4.8, as the original design was combinatorial to N TM classes, M classes, K TAs and K literals.

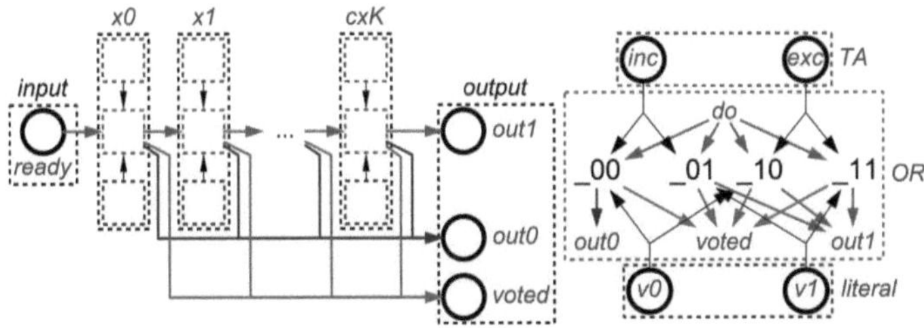

Fig. 8. Incremental clause computation for $2K$ literals with an early exit mechanism: If OR computation outputs 0, exit. If OR computation outputs 1, continue.

4.4 Clause

Our modified clause component is shown in Fig. 8, where the original Cartesian product-based operation for calculating the clause output is now replaced with an incremental OR-calculation towards the final AND-calculation, which scales more linearly but at the cost sequentialising the operation. To counteract this, we enable an *early exit* mechanism, where the clause computation stops as soon as there is an OR calculation of 0. If an OR calculation of 1 is detected, the calculation will continue to the next OR calculation until either a 0 is detected early or the last OR calculation is reached.

4.5 Tsetlin Machine

Our modified TM class component is shown in Fig. 9, where the original Cartesian product-based operation that calculates the class sum is replaced with an incremental vote counter, where it counts from clause 0 until clause M is reached. Note that positive clauses (top-left and top-middle blocks) increment the vote count, while negative clauses (top-right block) decrement the vote count.

4.6 Argmax

Our modified argmax component, which selects the class with the highest vote count, is shown in Fig. 10, where the original Cartesian product-based operation is replaced with an incremental vote checker that includes an early exit mechanism based on a summation-wise elimination process inspired by the FUTURE-BUS arbitration protocol [11]. The process starts from the maximum possible class sum $\frac{M}{2}$ to the minimum possible class sum $-\frac{M}{2}$, where, at each vote count $v_c \in [\frac{M}{2}, ..., -\frac{M}{2}]$, the vote counts of every TMs are checked. If one TM is matched, it is the classified TM. If multiple TMs are matched then one TM is selected at random (or from the leftmost position) to be classified.

4.7 Feedback Type Calculation

Our original feedback component is now split into several components with the first being the feedback type calculator shown in Fig. 11, where it determines

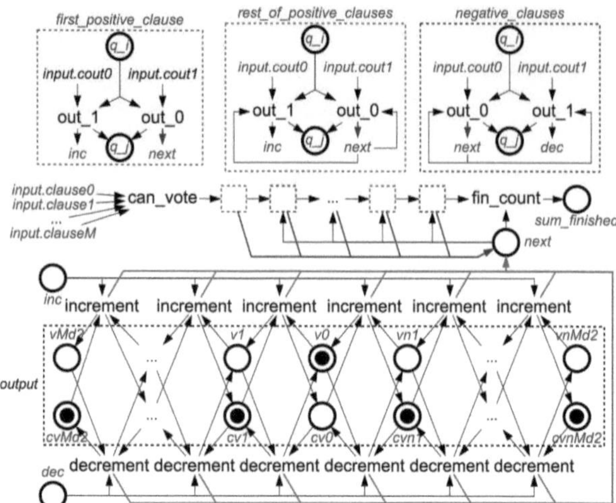

Fig. 9. Incremental TM class summation from clause 0 to clause M.

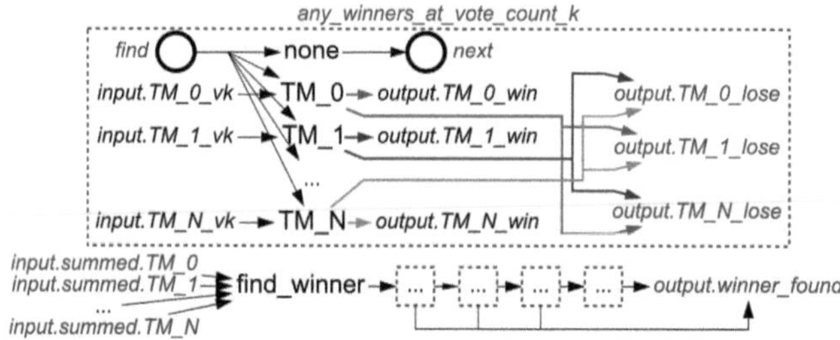

Fig. 10. Summation-based elimination argmax with an early exit mechanism.

which TM receives feedback and the feedback type given to each of its clauses and their respective TAs. The transition flow for both classified and random TMs are the same, with the exception of flags *is_clsfd* and *is_rnd* determining if the TM is classified or random, and $C1$ and $C2$ checks determining the feedback type given to the TM's clauses. Note that flags *selected_TM_n* and *selected_$clause_m$* are used to help TAs identify when it should receive feedback, while flags *pos* and *neg* are used to determine the clause's polarity for feedback calculation.

4.8   Feedback Action Calculation

The design of our feedback action calculator is shown in Fig. 12, which calculates the exact feedback type block that should be called for the TAs of the selected TM-clause to receive the correct feedback actions.

Here, the four feedback blocks can be summarised in Fig. 13, where Fig. 13a shows the common flow for all feedback blocks, Fig. 13b shows the possible

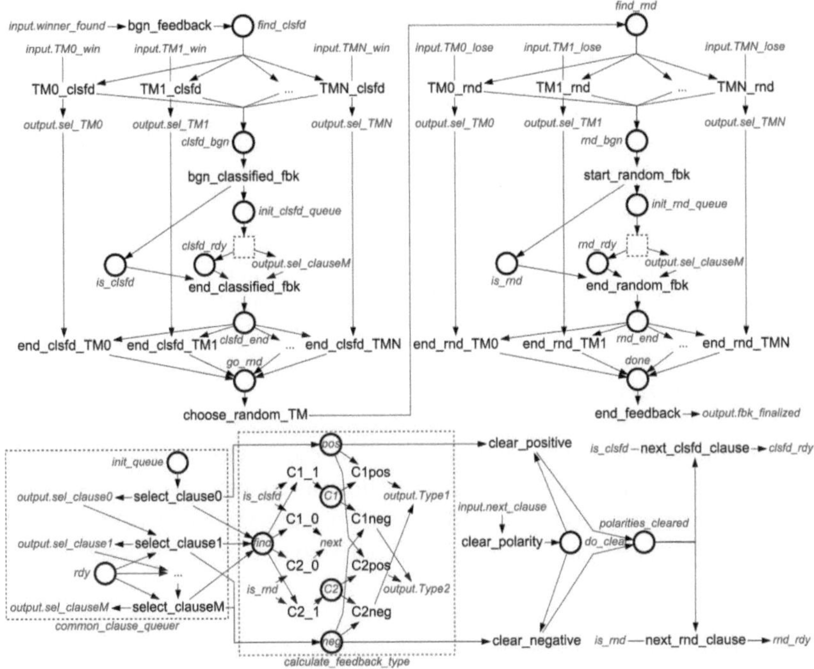

Fig. 11. Feedback type calculator that determines which TM receives feedback, before sequentially calculating the feedback type for each of its clauses and TAs.

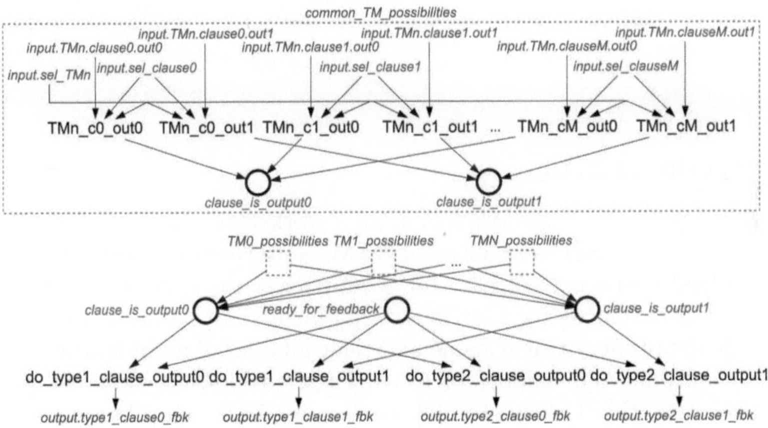

Fig. 12. Feedback action calculator that calculates what feedback action the TAs should receive, before forwarding it to the expected feedback type block.

actions for Type 1 Clause output 0, Fig. 13c for Type 1 Clause output 1, Fig. 13d for Type 2 Clause output 0, and Fig. 13e for Type 2 Clause output 1.

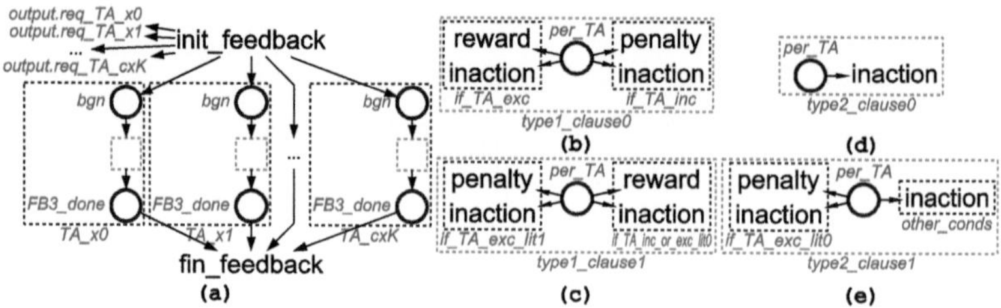

Fig. 13. Feedback blocks. (a) Common transitions and places. (b) Type 1 Clause 0. (c) Type 1 Clause 1. (d) Type 2 Clause 0. (e) Type 2 Clause 1.

5   Experimental Results

In this section, we cover three experiments using ANIMATE. Section 5.2 illustrates the scalability of our new TM-Petri net designs using different datasets and TM hyperparameters. Section 5.3 highlights the automated construction times for each TM-Petri net component. Section 5.4 covers both the assembling and simulation times of our complete TM-Petri net for each dataset.

Table 2. Dataset details of number of input features, unique outputs and rows.

Number of	SimpleXOR	Tic-Tac-Toe	NoisyXOR	2DNoisyXOR	Heart	BreastCancer	IRIS	Income	Avila	MNIST	GTSRB
Inputs	2	9	12	16	22	30	16	108	10	784	1024
Outputs	2	2	2	2	2	2	3	2	12	10	43
Rows	4	958	5000	2500	80	568	150	32561	10430	60000	39209

5.1   Experiment Setup

All of our experiments were conducted on a workstation with an Intel i7-12700 CPU, a 1TB NVMe SSD and 32GB of RAM. Note that GPU details were omitted, as ANIMATE does not yet have GPU support at the time of writing.

The datasets that we used for these experiments are shown in Table 2, where we provide details about their name and their number of input features, number of unique outputs and number of rows. These include the same simple XOR dataset used in [13], Tic-Tac-Toe [6], both noisy XOR and 2D noisy XOR [4], Heart [14], Breast Cancer [42], IRIS [3], Income [21], Avila [35], MNIST [31], and German Traffic Sign Recognition Benchmark (GTSRB) [34].

5.2   Design Scalability

In our first experiment, we benchmarked the scalability of our TM-Petri net designs against the specified datasets and different TM configurations of clauses

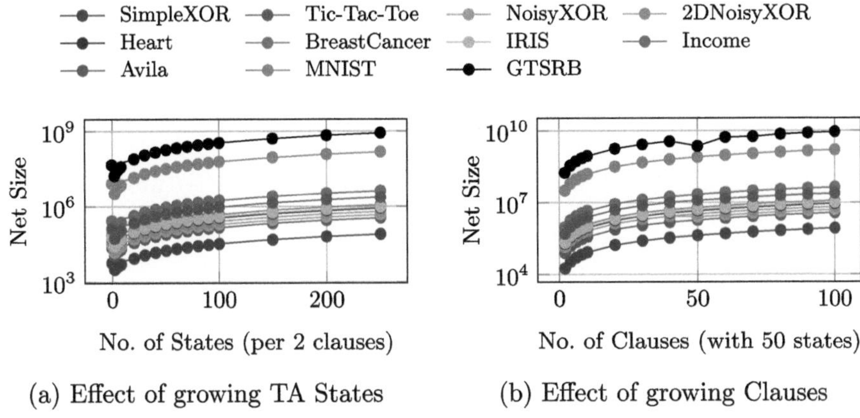

(a) Effect of growing TA States   (b) Effect of growing Clauses

Fig. 14. Scalability of TM-Petri net Constructions.

Table 3. Baseline total net sizes of each TM-Petri net component per dataset.

Component	SimpleXOR	Tic-Tac-Toe	NoisyXOR	2DNoisyXOR	Heart	BreastCancer	IRIS	Income	Avila	MNIST	GTSRB
Loader	117	341	565	565	757	1013	582	3509	543	25277	33518
	+8C	+16C	+16C	+16C	+16C	+16C	+24C	+16C	+96C	+80C	+344C
TM	108	108	108	108	108	108	108	108	108	108	108
	+88C	+88C	+88C	+88C	+88C	+88C	+88C	+88C	+88C	+88C	+88C
Clause	156	632	836	1108	1516	2060	1108	7364	700	53332	69652
TA	60	60	60	60	60	60	60	60	60	60	60
	+11J	+11J	+11J	+11J	+11J	+11J	+11J	+11J	+11J	+11J	+11J
Argmax	117	117	117	117	117	117	175	117	967	377	1136
	+52C	+52C	+52C	+52C	+52C	+52C	+80C	+52C	+512C	+388C	+4480C
Type	193	193	193	193	193	193	216	193	423	377	1136
	+26C	+26C	+26C	+26C	+26C	+26C	+26C	+26C	+26C	+26C	+26C
Action	127	127	127	127	127	127	172	127	577	487	1972
	+90C	+90C	+90C	+90C	+90C	+90C	+134C	+90C	+530C	+442C	+1894C
T1C0	178	780	1038	1382	1898	2586	1382	9294	866	67430	88070
T1C1	346	1536	2046	2726	3746	5106	2726	18366	1706	133286	174086
T2C0	66	276	366	486	666	906	486	3246	306	23526	30726
T2C1	214	942	1254	1670	2294	3126	1670	11238	1046	81542	106502

and TA states. As shown in Fig. 14, our TM-Petri net components scale linearly with respect to the number of clauses C and TA states J, where Table 3 details each component's total net size with respect to each dataset. From this, we establish a baseline net size using the smallest possible hyperparameter configuration (i.e. 1 clause and 2 TA states), where a scaling factor can be applied for each component affected by C and/or J.

5.3   Automated Construction

In our second experiment, we benchmarked the execution time and total net size of ANIMATE's automated construction of the TM-Petri net components for each dataset under configurations, which better reflect real-world implementations,

Table 4. Automated construction times of TM-Petri net Components

Dataset Name	Time (s)	Tsetlin Machine			Petri net ($\times 10^6$)			
		Classes	Clauses	States	Places	Transitions	Arcs	Total
SimpleXOR	0.19	2	2	6	0.0008	0.0006	0.004	0.005
Tic-Tac-Toe	1.52	2	20	50	0.050	0.079	0.613	0.741
NoisyXOR	7.78	2	50	100	0.282	0.499	3.955	4.735
2DNoisyXOR	11.63	2	60	120	0.527	0.951	7.552	9.029
Heart	14.93	2	80	80	0.682	1.179	9.333	11.193
BreastCancer	3.38	2	20	50	0.163	0.259	2.036	2.456
IRIS	16.59	3	80	80	0.744	1.287	10.184	12.214
Income	102.08	2	80	120	4.714	8.544	67.885	81.142
Avila	31.56	12	60	80	1.398	2.414	19.114	22.925
MNIST	9979.38	10	120	200	406.6	765.9	6102.4	7274.8
GTSRB	61043.50	43	200	200	3805.9	7169.2	57121.7	68096.8

as shown in Table 4. Compared to the original TM-Petri net design that took roughly 2 months to manually model, `ANIMATE` seamlessly constructs the TM-Petri net for any given dataset in significantly less time.

5.4   Assembling and Simulation

In our third experiment, we benchmarked the assembling and simulation times of the complete TM-Petri net using `Workcraft`, `GraphRack`, `TINA`'s WALK and `MaestroPN` as shown in Table 5. The simulation times are averaged over a single datapoint's reading, where DNL means did not load and GTSRB was excluded.

Notably, the assembling times are near-equivalent to their respective component construction times across most datasets, which demonstrate a minimal overhead of assembling the complete TM-Petri net. This is attributed to a template-based string substitution approach used in `ANIMATE`, where each template follows the generic component designs described in Sect. 4 and are assembled according to Algorithm 1 to construct the complete TM-Petri net efficiently.

The simulation times reveal distinct performance profiles across the four tools, which reflect their differing design purposes. `TINA`'s WALK is the fastest simulator as it is optimised for reachability analysis, but it does not natively support data-driven simulation since replicating this behaviour requires external support, which inherently reduces its performance. `MaestroPN`, while slower, natively supports data-driven simulation for ML applications without additional overhead, making it the most capable tool for this specific use case. `GraphRack` exhibits the slowest simulation times, which is likely attributed to its relative immaturity and broader focus on generic graph models beyond Petri nets. `Workcraft`, being primarily designed for asynchronous circuit design and design automation of interpreted graph models, was unable to load any net beyond simple XOR.

Table 5. Assembling and Simulation times of complete TM-Petri nets

Dataset	Assembling Time (s)	Average Simulation Time (s)			
		`Workcraft`	`GraphRack`	`MaestroPN`	`TINA` walk
SimpleXOR	0.016	0.22	0.012	0.01	0.001
Tic-Tac-Toe	1.345	DNL	101.08	15.14	0.64
NoisyXOR	6.785	DNL	3027.47	443.53	23.84
2DNoisyXOR	12.598	DNL	11201.60	1252.27	70.72
Heart	16.320	DNL	17276.50	2602.61	140.857
BreastCancer	3.639	DNL	1512.45	198.64	7.75
IRIS	17.747	DNL	14039.50	1644.99	121.94
Income	116.134	DNL	DNL	DNL	4190.28
Avila	34.001	DNL	DNL	DNL	208.55
MNIST	10261.315	DNL	DNL	DNL	DNL

6   Conclusion

In this paper, we proposed `ANIMATE`, which is a data-driven framework for automated construction of our improved TM-Petri nets. While TMs offer explainable architectures that are verifiable through reachability-based methods, the original TM-Petri net designs suffered from poor scalability due to its Cartesian product-based constructions, lack of automation, and limited dataset coverage.

`ANIMATE` addresses these limitations by replacing the Cartesian product-based operations with parametric mechanisms, which scale to the TM's hyperparameter configurations and dataset, and uses distributed places to assemble the complete TM-Petri net and preserve its 1-safe and deadlock-free properties.

To support runtime data instantiation and stochastic dynamics, we developed a data-driven simulator called `MaestroPN`, which enables transition-specific firing and probability-based operations to mirror the training of software-based TMs.

Our experimental results demonstrate that `ANIMATE`'s template-based string substitution approach achieves minimal assembling overhead, and show that our improved TM-Petri net components scale linearly with respect to the TM's clauses and TA states, which highlight the effectiveness of our parametric design approach across diverse binarised datasets and hyperparameter configurations.

6.1   Discussion

As shown throughout this paper, `ANIMATE` represents a significant step towards designing and verifying large-scale TM-Petri nets. However, our experimental results also reveal a "scalability paradox": While we can generate and verify larger-scale TM-Petri nets, critical bottlenecks remain in simulation times for larger nets and storage requirements. In Table 5, the completed TM-Petri net implementation for MNIST execeeded 200GB, which is completely infeasible in practice, especially when there are more complex datasets like GTSRB, where the full-system representation is rendered intractable.

To alleviate these issues, we considered three alternative Petri net extensions, though each presented underlying caveats. Sleptsov Nets [8] support multi-firing of transitions, but may struggle with the TAs' individual, data-dependent feedback updates. Salwicki/Step-based Nets [12,24,25] offer parallel firing of concurrent transitions, but unfolding their semantics across thousands of TAs may introduce risks to mutual exclusion and model complexity. Coloured Petri nets [18] reduce visual complexity using labelled tokens, but their identity-dependent nature may complicate the verification and structure of our TM-Petri net designs.

These limitations suggest that the bottleneck is not merely a software engineering challenge, but a representational one. As discussed in [33], regardless of a tool's capability, the sheer size of monolithic Petri nets will severely impact simulation times as shown in Table 5. Our findings point towards the need for a novel Petri net extension, where transition firing semantics are generalised to operate over high-dimensional structures, and treat large blocks of concurrent token changes as multi-dimensional operations rather than individual firings. This effectively allows large-scale nets like TM-Petri nets to retain their fine-granular representation, while leveraging the computational efficiency of modern tensor processing, for modelling and verifying real-world, high-dimensional ML systems.

6.2   Future Work

Following from our discussion, we first consider the development of a Petri net extension that supports a new data structure to alleviate some of the limitations discussed in Sect. 6.1, and improve ANIMATE's design flow. Next, we plan to support more Petri net-based designs of other ML models for ANIMATE including our recent Petri net-based Binary Neural Networks [37], and extend ANIMATE's support for GPUs and ML libraries such as PYTORCH and KERAS. To improve our formal verification capabilities, we consider backwards reachability analysis to observe what combinations trigger the ML model's decision-making processes.

Acknowledgments. We would like to thank the anonymous reviewers for their feedback. This work was supported by EPSRC EP/X036006/1 Scalability Oriented Novel Network of Event Triggered Systems (SONNETS) and EPSRC EP/X039943/1 UKRI-RCN: Exploiting the dynamics of self-timed machine learning hardware (ESTEEM).

Disclosure of Interests. The authors have no competing interests to declare that are relevant to the content of this article.

References

1. TIme petri Net Analyzer (TINA) (2006). https://projects.laas.fr/tina/home.php. Accessed 12 Feb 2025

2. Workcraft. https://workcraft.org/. Accessed 12 Feb 2025

3. Binarized IRIS (2018). https://github.com/cair/TsetlinMachine/blob/master/BinaryIrisData.txt. Accessed 17 Nov 2025

4. pyTsetlinMachine Examples (2019). https://github.com/cair/pyTsetlinMachine/tree/master/examples/. Accessed 17 Nov 2025

5. Abbas, S., Soomro, N., Granmo, O.C., Shafik, R., Heer, R., Adhikari, K.: AI-Based Clinical Rule Discovery for NMIBC Recurrence through Tsetlin Machines. In: International Symposium Tsetlin Machine (ISTM), pp. 24–29 (2025). https://doi.org/10.1109/ISTM67926.2025.00012

6. Aha, D.: Tic-Tac-Toe Endgame. UCI Machine Learning Repository (1991). https://doi.org/10.24432/C5688J

7. Berge, G.T., Granmo, O.C., Tveit, T.O., Goodwin, M., Jiao, L., Matheussen, B.V.: Using the tsetlin machine to learn human-interpretable rules for high-accuracy text categorization with medical applications. IEEE Access **7**, 115134–115146 (2019). https://doi.org/10.1109/ACCESS.2019.2935416

8. Berthomieu, B., Zaitsev, D.A.: Sleptsov nets are turing-complete. Theoret. Comput. Sci. **986**, 114346 (2024). https://doi.org/10.1016/j.tcs.2023.114346, https://www.sciencedirect.com/science/article/pii/S030439752300659X

9. Bhattarai., B., Granmo., O., Jiao., L.: Measuring the novelty of natural language text using the conjunctive clauses of a tsetlin machine text classifier. In: International Conference on Agents and Artificial Intelligence (ICAART), pp. 410–417. INSTICC, SciTePress (2021). https://doi.org/10.5220/0010382204100417

10. Blakely, C.D., Granmo, O.-C.: Closed-Form Expressions for Global and Local Interpretation of Tsetlin Machines. In: Fujita, H., Selamat, A., Lin, J.C.-W., Ali, M. (eds.) IEA/AIE 2021. LNCS (LNAI), vol. 12798, pp. 158–172. Springer, Cham (2021). https://doi.org/10.1007/978-3-030-79457-6_14

11. Boussinot, F., Ramesh, S., Shyamasundar, R.K., De Simone, R.: Validation and analysis of the futurebus arbitration protocol: a case study. Sadhana **21**(2), 185–211 (1996). https://doi.org/10.1007/BF02745519

12. Burhard, H.D.: On priorities of parallelism: Petri nets under the maximum firing strategy. In: Salwicki, A. (ed.) Logics Programs Their Appl., pp. 86–97. Springer, Berlin Heidelberg, Berlin, Heidelberg (1983)

13. Chan, A., Wheeldon, A., Shafik, R., Yakovlev, A.: Design of Event-Driven Tsetlin Machines Using Safe Petri Nets. In: Kristensen, L.M., van der Werf, J.M. (eds.) Application and Theory of Petri Nets and Concurrency, pp. 357–378. Springer Nature Switzerland, Cham (2024)

14. Cios, K., Kurgan, L., Goodenday, L.: SPECT Heart. UCI Machine Learning Repository (2001). https://doi.org/10.24432/C5P304

15. Cortadella, J., Kishinevsky, M., Kondratyev, A., Lavagno, L., Yakovlev, A.: Petrify: A Tool for Manipulating Concurrent Specifications and Synthesis of Asynchronous Controllers. IEICE Trans. Inf. Syst. **E80-D**, 315–325 (1997)

16. Erkus, E.C., Chan, A., Distaso, W., Thomas, D., Yakovlev, A., Shafik, R.: Latent-Graph: From Latent States to Rule-based Expressions for Explainable Financial Forecasting, pp. 745–752. Association for Computing Machinery, New York, NY, USA (2025). https://doi.org/10.1145/3768292.3770428

17. Granmo, O.C.: The Tsetlin Machine - A Game Theoretic Bandit Driven Approach to Optimal Pattern Recognition with Propositional Logic. arXiv preprint arXiv:1804.01508 (2018). https://arxiv.org/abs/1804.01508

18. Jensen, K.: Coloured Petri Nets: Basic Concepts, Analysis Methods and Practical Use, vol. 2. Springer-Verlag, Berlin, Heidelberg (1995)

19. Khomenko, V., Koutny, M., Yakovlev, A.: Logic synthesis for asynchronous circuits based on petri net unfoldings and incremental SAT. In: International Confernce on Application of Concurrency to System Design (ACSD), pp. 16–25 (2004)
20. Khomenko, V., Koutny, M., Yakovlev, A.: Distributed Places and Safe Net Reduction. In: Amparore, E., Mikulski, Ł (eds.) Application and Theory of Petri Nets and Concurrency, pp. 265–286. Springer Nature Switzerland, Cham (2025)
21. Kohavi, R.: Census Income. UCI Machine Learning Repository (1996). https://doi.org/10.24432/C5GP7S
22. Mao, G., et al.: Dynamic tsetlin machine accelerators for on-chip training using FPGAs. IEEE Trans. . Circu. Syst. Regul. Papers **72**(11), 6962–6975 (2025). https://doi.org/10.1109/TCSI.2025.3564875
23. Merlin, P., Farber, D.: Recoverability of communication protocols - implications of a theoretical study. IEEE Trans. Commun. **24**(9), 1036–1043 (1976). https://doi.org/10.1109/TCOM.1976.1093424
24. Nielsen, M., Plotkin, G., Winskel, G.: Petri nets, event structures and domains, part I. Theor. Comput. Sci. **13**(1), 85–108 (1981). https://doi.org/10.1016/0304-3975(81)90112-2, https://www.sciencedirect.com/science/article/pii/0304397581901122
25. Mukund, M.: Petri nets and step transition systems. Int. J. Found. Comput. Sci. **03**(04), 443–478 (1992). https://doi.org/10.1142/S0129054192000231
26. Murata, T.: Petri nets: properties, analysis and applications. Proc. IEEE **77**(4), 541–580 (1989). https://doi.org/10.1109/5.24143
27. Pattison, B., et al.: TMAtlas: an interactive visual analytics framework for explaining tsetlin machine outputs. In: International Symposium on Tsetlin Machine (ISTM), pp. 116–123 (2025). https://doi.org/10.1109/ISTM67926.2025.00025
28. Rafiev, A., et al.: Visualization of machine learning dynamics in tsetlin machines. In: International Symposium on Tsetlin Machine (ISTM), pp. 81–88 (2022). https://doi.org/10.1109/ISTM54910.2022.00020
29. Rahman, T., et al.: Runtime tunable tsetlin machines for edge inference on eFPGAs. In: IEEE Sensors Applications Symposium (SAS), pp. 1–6 (2025). https://doi.org/10.1109/SAS65169.2025.11105163
30. Rosenblum, L., Yakovlev, A.: Signal graphs: from self-timed to timed ones. In: International Workshop on Timed Petri Nets, pp. 199–206 (1985)
31. Salakhutdinov, R., Murray, I.: On the quantitative analysis of deep belief networks. In: Proceedings of International Conference on Machine Learning (ICML), pp. 872–879. ACM (2008)
32. Shafik, R., Wheeldon, A., Yakovlev, A.: Explainability and dependability analysis of learning automata based AI hardware. In: IEEE International Symposium on On-Line Testing and Robust System Design (IOLTS), pp. 1–4 (2020). https://doi.org/10.1109/IOLTS50870.2020.9159725
33. Squires-Parkin, H., Chan, A., Shafik, R., Wheeldon, A., Yakovlev, A.: Asynchronous design of a bitwise elimination argmax via high-level modeling in GraphRack. In: IEEE International Symposium on Asynchronous Circuits and Systems (ASYNC), pp. 89–98 (2025). https://doi.org/10.1109/ASYNC65240.2025.00021
34. Stallkamp, J., Schlipsing, M., Salmen, J., Igel, C.: The German Traffic Sign Recognition Benchmark: A multi-class classification competition. In: IEEE International Joint Conference on Neural Networks (IJCNN), pp. 1453–1460 (2011)
35. Stefano, C., Fontanella, F., Maniaci, M., Freca, A.: Avila. UCI Machine Learning Repository (2018). https://doi.org/10.24432/C5K02X

36. Tarasyuk, O., et al.: Prediction of the infecting organism in peritoneal dialysis patients with acute peritonitis using interpretable Tsetlin machines. Bioinform. Adv. **5**(1), vbaf140 (2025)
37. Tarraf, M., Chan, A., Yakovlev, A., Shafik, R.: Eventizing Traditionally Opaque Binary Neural Networks as 1-safe Petri net Models (2026). https://arxiv.org/abs/2602.13128
38. Tsetlin, M.L.: On behaviour of finite automata in random medium. Avtomat. i Telemekh **22**(10), 1345–1354 (1961)
39. Vernadat, F., Berthomieu, B.: Time petri nets analysis with TINA. In: International Conference on Quantitative Evaluation of Systems (QEST), pp. 123–124 (2006). https://doi.org/10.1109/QEST.2006.56
40. Weiss, G., Goldberg, Y., Yahav, E.: Extracting automata from recurrent neural networks using queries and counterexamples. In: Dy, J., Krause, A. (eds.) Proceedings of International Conference on Machine Learning (ICML), vol. 80, pp. 5247–5256. PMLR (2018). https://proceedings.mlr.press/v80/weiss18a.html
41. Wheeldon, A., Yakovlev, A., Shafik, R.: Self-timed reinforcement learning using tsetlin machine. In: International Symposium on Asynchronous Circuits and Systems (ASYNC), pp. 40–47 (2021). https://doi.org/10.1109/ASYNC48570.2021.00014
42. Wolberg, W., Mangasarian, O., Street, N., Street, W.: Breast Cancer Wisconsin (Diagnostic). UCI Machine Learning Repository (1993). https://doi.org/10.24432/C5DW2B
43. Xu, Z., Wen, C., Qin, S., He, M.: Extracting automata from neural networks using active learning. PeerJ Comput. Sci. **7**, e436 (2021). https://doi.org/10.7717/peerj-cs.436

Supporting Modularity by (De-)Composition of Distributed P/T Nets Based on Karger's Algorithm for Distributed Execution

Laif-Oke Clasen[(✉)], Justus Middendorf, Felix Gorke, Jens Brüggemann, Lukas Eshun, Simon Bott, Marie Chevalier, Efe Nayci, and Daniel Moldt

Faculty of Mathematics, Informatics and Natural Sciences, Department of Informatics, University of Hamburg, Hamburg, Germany
`laif-oke.clasen@uni-hamburg.de`
`http://www.paose.de`

Abstract. Systems grow in complexity, inflating model size and structural density. Place/Transition (P/T) nets are a standard formalism, but large instances strain editing, simulation, and verification.

A modular perspective treats distributed P/T nets as composable units with a module interface. The research question is how to define and operationalize a concept of interface-guided modularity via Decomposition and Composition so that large P/T net models become more tractable while preserving interface compatibility.

Decomposition and composition are defined over interfaces via distributed synchronous channels and realized as a proof of concept (PoC) in RENEW using constructivist prototyping. Decomposition applies Karger's randomized contraction algorithm to an undirected graph derived from the net, yielding a cut that induces subnet boundaries and interfaces. The PoC selects the minimum cut to minimize cross-partition coupling and distributed communication. Composition reconciles interfaces by aligning and merging required interface elements. Runtime is analyzed with respect to the interface-based definitions, and benchmarks are evaluated by wall-clock time, the throughput, and the runtime-to-net-size ratio.

Asymptotic bounds are derived for both operations. NETSPLIT scales as $\mathcal{O}(n^2 + n \cdot t)$, where n is the number of nodes and t the number of transitions. NETJOIN is bounded by $\mathcal{O}\left(\sum_{i=1}^{n} m_i \cdot l\right)$, where m_i is the number of net elements in the i-th net and l the number of distributed synchronous channels to be reconciled during composition.

Interface-guided modularization improves operational efficiency for complex Petri net models, makes larger nets more tractable, and supports dynamic repartitioning in distributed simulation and verification.

Keywords: Composition · Decomposition · Distributed P/T Nets · Synchronous Channels · Kargers Algorithm · Distributed Systems · Modularization

1   Introduction

Systems and their formal models continue to grow in size and structural density, increasing the effort required for model construction, maintenance, simulation, and verification. Place/Transition (P/T) nets are a widely used formalism for concurrent systems, yet large nets can exceed practical limits of modeling, simulation, and verification.

A common strategy to regain tractability is modularization: decomposing a large net into smaller, separately comprehensible subnets and recombining them when needed. For modeling, a decomposition can improve readability and local reasoning provided that inter-module coupling remains bounded. For simulation and verification, decomposition can enable horizontal scaling by distributing workloads across compute resources. However, this benefit depends on the induced interface size and the synchronization and communication overhead between partitions. In contrast, purely vertical scaling often exhibits disproportionate cost increases at higher performance levels, which motivates distributed approaches in scenarios where a single-machine execution becomes impractical. [28,39,44]

To make modularization precise and machine-checkable, we adopt a module-and-interface perspective for distributed P/T nets: a module is a subnet equipped with an explicit interface that covers its externally visible interaction points. The central objective is to introduce a concept for interface-guided composition that (i) specifies when modules are compatible, (ii) provides composition and decomposition operators that preserve a well-defined execution semantics, and (iii) algorithms for these two operations.

Accordingly, this work addresses the following research question: how to establish and operationalize a concept of interface-guided modularity via Decomposition and Composition so that large P/T net models become more tractable while preserving interface compatibility? Here, "tractable" is treated operationally in terms of measurable criteria such as wall-clock time, the throughput, and the runtime-to-net-size ratio.

We introduce a concept for decomposition and composition as interface-driven operators for distributed P/T nets, where inter-module interaction is mediated by distributed synchronous channels with rendezvous synchronization. Following a constructivist prototyping methodology [2,6,43,54], which primarily evaluates through the construction of a proof of concept (PoC), we derive operational procedures for both operators and instantiate them in a PoC implementation in RENEW, using the executable prototype to iteratively validate the adequacy of the concepts against concrete model executions. We then evaluate the approach along the tractability criteria introduced above.

This work proceeds from foundations (Sect. 2) and the problem statement (Sect. 3) to the proposed conceptual solution (Sect. 4). Its realization is presented via the prototypes NETSPLIT (Sects. 5) and NETJOIN (Sects. 6), followed by evaluation (Sects. 7). Finally, the results are discussed (Sects. 8), related to prior work (Sects. 9, and future work (Sects. 10).

2   Foundations

2.1   Distributed Place/Transition Nets

Distributed Place/Transition Nets (DPTNs) were introduced by Clasen et al. [9] for distributed simulation of classical Place/Transition (P/T) nets and extend P/T nets with synchronous channels [53]. A synchronous channel enforces rendez-vous-style synchronization between two transitions. While Voß et al. [53] restrict synchronization to transitions within a single simulator, Clasen et al. [9] generalize it to distributed synchronous channels that connect transitions across simulator boundaries. This extension is limited to transition synchronization, preserves the fundamental execution semantics of P/T nets, and enables communication between concurrently executed simulation components.

In this formalism, distributed synchronous channels are the primary coordination mechanism between simulators: a participating transition fires only as part of a joint, synchronized firing with its remote counterpart, coupling execution steps across distributed subnets. The induced communication events encode both causal ordering and synchronization dependencies among distributed transitions.

Clasen et al. implement these semantics using an event-driven communication layer based on Apache Kafka [26,36]. Kafka provides distributed event streaming and real-time data processing. This abstraction aligns with inter-simulator synchronization requirements, including persistent event logs, ordered message delivery, and scalable publish-subscribe semantics. Each synchronous channel corresponds to a Kafka topic through which transition events are transmitted and consumed, ensuring consistent propagation of firing states across simulators.

Figure 1 illustrates Clasen et al.'s [9] distributed producerâĂŞstorageâĂŞconsumer configuration as an example of inter-simulator synchronization via synchronous channels. Distributed transitions exchange control tokens across simulator boundaries while maintaining local causality. Each simulator may expose multiple communication endpoints – uplinks (DU) and downlinks (DD) – corresponding to participation in one or more synchronous channels. Together, these channels realize a single logical Petri net whose execution is distributed across multiple simulator environments.

2.2   Petri Net Modules and Composition

A Petri net module is a subnet equipped with an explicit interface that defines its externally observable interaction points [35]. Composition and decomposition are inverse structural transformations over such interfaces: composition combines compatible modules into a unified net, while decomposition partitions a net into modules that reconstruct it under composition. Both operations must preserve the execution semantics of the original net.

Two formal foundations underpin this work. For composition, we adopt Reisig's associative composition calculus [24,45,46], in which modules expose typed interface elements - gates - paired via a matching relation; matched gates

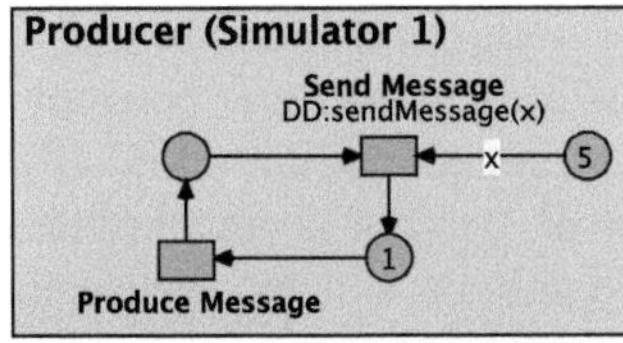

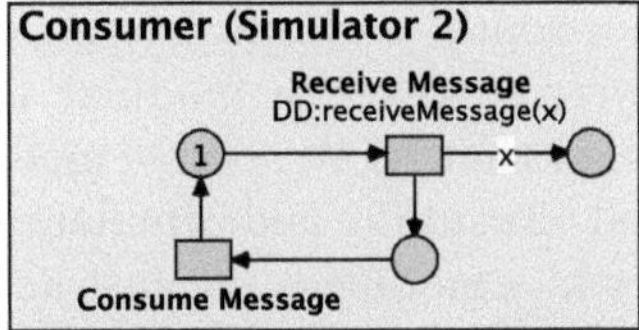

(a) Producer and Consumer Net

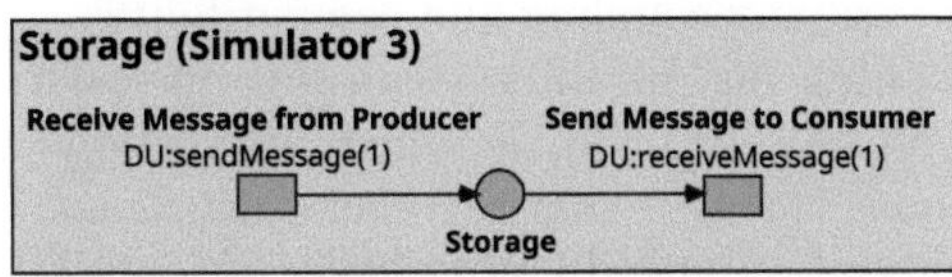

(b) Storage Net

Fig. 1. Components of the producer storage consumer example [9]

are internalized upon composition while unmatched gates form the resulting module's interface. For interfaces, we build on the synchronous channel model of Voß et al. [53], in which transitions serve as interface points via rendezvous-style synchronous channels. This combination provides both the semantic framework and the concrete interface primitive required for the definitions in Sect. 4. Alternative composition frameworks and interface concepts are surveyed in Sect. 9.

2.3 RENEW

RENEW[1] [38] is an open-source modeling, simulation, and verification environment for Petri nets. Originally designed for Reference Nets [37], it has evolved into a multi-formalism platform [9,12,37,53] with a plugin system for integrating formalisms such as P/T nets [20]. These formalisms support configurations ranging from simulation in a single local instance to distributed, cloud-native environments [9,12].

3 Problem Description

The structural and functional complexity of modern systems is steadily increasing, requiring correspondingly complex models. Place/Transition nets (P/T nets) provide a well-established way to model such systems. However, monolithic tools increasingly reach their limits in resource utilization, technical constraints, and cognitive load. Since the costs of vertical scaling tend to rise disproportionately – particularly at higher performance tiers – it is not a practical option in all scenarios. [28,39,44]

---

[1] **Reference Net W**orkshop, available at http://renew.de,.

For horizontal scaling of large P/T net models in distributed simulation and analysis contexts, a modular modeling and operations concept is therefore required that (i) structures models into manageable units and (ii) supports distributed execution and simulation. Central to this is an explicit interface specification: a module is not just an isolated subnet, but has well-defined interaction points through which it can be coupled with other modules. This interface syntactically specifies which interactions are possible and defines an observable external view for verifying coupling and compatibility.

This work focuses on an interface-driven modularity concept for DISTRIBUTED P/T NETS (DPTNs), in which interactions are expressed via distributed synchronous channels that serve as synchronization primitives. It aims to define the composition and decomposition of large DPTNs as clear, traceable, and repeatable operations. This ensures that the resulting modules and their interfaces remain precisely compatible. Methodologically, this work (1) uses or adapts a formal foundation based on Reisig's established composition calculus, including explicit compatibility conditions, and (2) specifies concrete algorithms that execute composition and decomposition as Proof of Concept. The algorithmic implementation is evaluated using transparent metrics that measure manageability and distributed coupling effort.

The following research questions arise from this problem context: Main Research Question: How to establish and operationalize a concept of interface-guided modularity via Decomposition and Composition so that large P/T net models become more tractable while preserving interface compatibility?

RQ1: Which perspective of modules and interfaces is suitable for DPTNs?

RQ2: Which algorithmic methods can practically implement composition and decomposition of DPTNs under the assumptions in RQ1, and how can transparent metrics compare the results' quality and required effort?

The primary objective of the results presented here is to investigate the distribution of simulations of distributed P/T nets [9, 10, 13]. In particular, we study the dynamic scaling of these simulations for applications such as adaptive load balancing in large-scale distributed simulations. [13] Moreover, the concepts of dynamic scaling for distributed simulations are expected to be transferable to distributed verification, for example, when verifying distributed protocols with high communication traffic. Finally, the results can also support collaborative editing of large P/T nets: the nets are first decomposed and then modified concurrently by different modelers. Afterwards, the resulting subnets are merged again, for instance, in scenarios where several modelers work in parallel on different submodels.

4   Conceptual Solution

A DPTN is understood as a finite combination of multiple modules. A module comprises (i) its internal net structure, meaning the associated places, transitions, and edges, as well as (ii) an explicitly designated interface. The internal net structure behaves in the same way as with P/T nets (Sect. 2.1). The marking is understood as the state of a concrete module instance and thus as an attribute that determines the module's execution in a given simulation. A module description is organized as a net template that combines structure, interface, and the initial-state information into a uniform representation. The specific storage format is an implementation decision and not a model-theoretical component. A module can be instantiated and simulated independently of other modules, provided that an execution environment is available that supports the synchronization semantics described below via the interface.

The interaction between modules is based on rendezvous synchronization and is implemented using distributed synchronous channels. Rendezvous synchronization [5,30,47] is a communication style where the synchronization is a bidirectional, synchronous blocking mechanism. The sender and receiver establish a mutual synchronization point and engage in a mutual exchange. Each distributed synchronous channel has a downlink or uplink in each participating module. A channel describes a synchronous coupling relationship between the modules involved. A synchronization step via a channel occurs precisely when all module sides involved in the channel are ready to synchronize simultaneously. Synchronization readiness results from the activation of the local transitions of the distributed synchronous channel to which it is bound. Synchronization is thus modeled as a joint, atomic action that coordinates the modules involved and allows interaction exclusively via the interface.

The interface of a module is determined by the set of its distributed synchronous channels. It thus encompasses the interaction possibilities through which the module can be synchronously coupled with other modules. Distributed synchronous channels are assigned a minimal signature that defines the synchronization-relevant properties. A channel signature contains a channel name, a type, and a parameter specification. The channel name must be unique within the specified naming scope. The type distinguishes the two endpoint roles: Downlink (active) and Uplink (passive), consistent with the model's synchronization semantics. In this sense, interface compatibility means that modules use consistent signatures across all channels used in common. In the PoC, compatibility is operationalized accordingly as a primarily syntactic consistency condition. Syntactic compatibility guarantees wellformedness of the coupling, but not progress properties such as deadlock freedom.

On this basis, the operations of decomposition and composition are defined. Decomposition generates new modules, along with newly formed interface channels that precisely represent the structural couplings that span the new module boundary. Decomposition partitions a given DPTN into multiple modules and can be applied hierarchically by recursively subdividing the already-formed modules, treating each as input again. The resulting module boundaries and the interfaces derived from them are chosen so that all interactions between the resulting modules are conducted exclusively via distributed synchronous channels. In particular, inter-module dependencies should be evident at the interfaces, and no additional couplings outside the interface channels are introduced.

Composition, on the other hand, combines a set of modules into an overall net, provided that their interfaces are compatible. Compatibility in this context means that the channels used jointly match in terms of their synchronization-relevant properties, so that the coupling can be implemented entirely via the agreed channels. A central design principle is that composition does not create any additional couplings between modules that are not mediated by interface channels. Inter-modular dependencies should be visible only in the interface and thus become explicit as the subject of compatibility checks.

We use Reisig's interface-based composition calculus [25,45,46] as a formal basis for the composition and decomposition. A module has a set of interface elements (hereinafter, "gates"), each of which is assigned a label. A matching relation on their gates defines the composition of two modules: Only gates with mutually compatible labels are matched. Labels that cannot be matched remain unchanged. The gates involved are then considered internalized, i.e., they are removed from the exported interface set of the new module and are only effective within the composite module's internal structure. All gates without a compatible match remain as exported interface elements and form the composite module's interface.

In this contribution, gates are interpreted as transitions of distributed synchronous channels. For a match, the channel name and parameter list must match, in particular, the same number and order, as well as consistent parameter types, if typed. In addition, matching is only permitted between complementary types. We distinguish between up- and downlink for this purpose. This ensures that the composition connects only channels that match as sending and receiving counterparts for rendezvous synchronization.

Both operations require algorithms that can pursue different strategies. Possible strategies include recognizing and exploiting recurring structural patterns on the one hand, and transforming the underlying net structure into a graph problem on the other hand, to leverage known optimization methods. The appropriate strategy depends on the properties of the input, such as size, density, and the presence of symmetries or repetitive structures. Accordingly, strategies may involve different trade-offs between result quality and runtime. Typical examples are the number or weight of inter-modular coupling points, which become visible as interface channels in the resulting modularization. In addition, constraints may become relevant, such as requirements for a specific module granularity or a balanced distribution.

A PoC in the context of RENEW demonstrates how this concept can be put into practice. The PoC initially considers paired channels, but multi-party channels are conceptually possible. For decomposition, a graph representation is derived from the net structure, expressing the net's structural coupling in a form suitable for cut optimization. The decomposition is then formulated as a cutting problem, with minimizing the cut as the optimization goal to reduce inter-modular coupling. Karger's randomized contraction algorithm is used as the specific method, which aims to identify structures with a minimal cut in undirected graphs and, at the same time, allows the underlying quality criterion to be explicitly formulated. Due to randomization, the decomposition is to be understood as a parametrizable process determined, for example, by the number of runs and a selection rule for the final cut. The selected cut is then chosen based on the specified quality criterion.

In the PoC, a minimum cut is selected as the quality criterion. This minimizes distributed communication and inter-modular coupling, as the cut size captures them in the graph representation. At the same time, the concept deliberately remains open to alternative quality criteria, such as other weightings, balance constraints, or additional restrictions that control the permissible granularity and form of the resulting modules.

5   Prototype: NETSPLIT

5.1   Requirements

The NETSPLIT [13] plugin must decompose DPTNs efficiently to support distributed simulations. For complex and large-scale DPTNs, efficient splitting is critical for simulation scalability and performance.

To address these challenges, NETSPLIT must be redesigned through several key improvements, as detailed below. First, the input size for the partitioning algorithm must be reduced via a compact net abstraction. This is critical because the contraction algorithm, which merges nodes to simplify the net, scales quadratically with the number of nodes, rendering literal (explicit) representations of large nets impractical.

Second, to further enhance efficiency, the algorithmic complexity of decomposition should be reduced by adopting a direct strategy. This means using a method that computes partitions in a single step, thereby eliminating the computational overhead caused by the repeated splitting cycles of the previous implementation.

Third, to maintain scalability for large models, the reconstruction of subnets and the assignment of DISTRIBUTED SYNCHRONOUS CHANNELS must be improved using optimized data structures and mapping strategies. Collectively, these improvements are expected to yield a reduction in overall decomposition time compared to the initial implementation, creating a more effective solution.

Beyond algorithmic efficiency, seamless integration into the modern RENEW environment (Sect. 2.3) is also essential for the plugin. RENEW supports operation in containerized environments such as Kubernetes clusters. The NETSPLIT plugin must leverage the event-driven infrastructure based on Apache Kafka introduced in Sect. 2.1 to log splits, suitable for distributed environments where multiple simulators operate concurrently.

5.2   Specification

First, to improve runtime performance on large nets, the NETSPLIT plugin must provide an input optimization that significantly reduces the graph size and complexity processed by the contraction algorithm. This representation should focus on cut-relevant connections by eliminating transition nodes and self-loops.

Second, the NETSPLIT plugin should compute multiple partitions in a single run of the decomposition algorithm. Instead of repeatedly applying 2-cuts – iteratively splitting the largest remaining component into exactly two parts as in the previous version [13] – the redesigned version utilizes a direct k-cut strategy, terminating as soon as the desired number of subnets is reached.

Third, the NETSPLIT plugin should provide efficient reconstruction of partitioned DPTNs from the contracted graph structure. It must be possible to derive the necessary DISTRIBUTED SYNCHRONOUS CHANNELS directly from the cut edges and assign affected transitions to the corresponding subnets efficiently.

Fourth, the plugin should accelerate contraction operations by using a disjoint-set union data structure with path compression and union-by-size heuristics to manage node groups during edge contractions. This addresses algorithmic efficiency, since union operations have near-constant amortized time complexity [17], drastically reducing computation time during the contraction of large models.

Finally, the NETSPLIT plugin should be integrable into modern distributed execution setups based on RENEW CloudNative infrastructure and Kafka. This integration is necessary because RENEW supports operation in containerized environments such as Kubernetes clusters [12]. As described by [49], this infrastructure uses Spring Framework for Java to provide HTTP endpoints for remote control and simulation initiation. It should be possible to invoke decomposition operations on simulator instances via these REST endpoints, and the plugin should emit structured events about performed decompositions to Kafka topics for logging split histories.

5.3   Design

To reduce input complexity, the new approach transforms the original DPTNs into an undirected graph with places as the only nodes. For each transition t, one undirected edge per pair in $\mathrm{pre}(t) \times \mathrm{post}(t)$ is added, where multiple transitions with identical pre/post sets each produce their own parallel edge rather than a merged one, preserving all structural information. Arc multiplicities are not mapped to edge weights, so the cut weight equals the number of transitions crossed and thereby directly minimizes the number of introduced DISTRIBUTED SYNCHRONOUS CHANNELS, which all cut edges directly represent, simplifying reconstruction and subnet assignment.

The core decomposition process is outlined in Algorithm 1, a modified version of Karger's randomized contraction algorithm redesigned for a k-cut strategy. In a trial, the process runs a configurable number of independent trials r to identify the best partition with minimum cut weight. At the start of each trial, the edge contraction order is determined once via a shuffle operation to eliminate overhead from repeated random selections. Contractions proceed until exactly k super-nodes remain, representing the desired number of subnets; across all r trials, the design compares results and selects the partition with minimum total cut weight. The probability guarantee of the original algorithm is established for the 2-cut case; formally extending it to the k-cut strategy remains an open theoretical question. The multi-trial design partially compensates by empirically increasing the likelihood of finding a near-minimum cut.

Algorithm 1: Modified k-cut Karger's Algorithm

Input: Graph G, target k supernodes, r repetitions
Output: Best partition (set of $k + 1$ subnets)
$bestCut \leftarrow \emptyset$;
$bestValue \leftarrow \infty$;
for $i \leftarrow 1$ **to** r **do**
    $H \leftarrow$ copy of G;
    $edges \leftarrow ShuffleEdges(H)$;   `// optional fixed edge order`
    while H *has more than* k *supernodes* **do**
        $e \leftarrow$ next entry from $edges$ or random edge;
        $Contract(H, e)$;   `// merge nodes (Union-Find)`
    $value \leftarrow EvaluateCut(H)$;
    if $value < bestValue$ **then**
        $bestValue \leftarrow value$;
        $bestCut \leftarrow ExtractCut(H)$;
return $bestCut$

After contraction, graph-to-net reconstruction analyzes the identified cut edges to determine their corresponding original transitions. These transitions are reconstructed, assigned to subnets, and coupled via DISTRIBUTED SYN-CHRONOUS CHANNELS. Arc directionality is preserved from the original net: for a cut transition t, the pre(t) side determines the Uplink and the post(t) side the Downlink of the resulting distributed synchronous channel. For internal execution efficiency, the plugin uses a Disjoint Set Union (Union-Find) data structure with path compression and union-by-size heuristics to manage node groups during edge contractions, enabling near-constant amortized merge time [17]. The design also replaces standard list structures with HashMap implementations in multiple components to ensure constant access for recurrent searches during mapping.

To enable integration into modern containerized environments, NETSPLIT exposes a Representational State Transfer (REST) interface through the CLOUDNATIVE [48,49] plugin at the `/netsplit` endpoint. This interface preserves the functionality of the command-line *netsplit* command and communicates outcomes through standard HTTP status codes, including 200 OK and 400 Bad Request. To support traceability and monitoring, NETSPLIT asynchronously publishes structured JSON events with per-operation metadata to an Apache Kafka broker via the `NetSplitTopic`.

Phase 1 (graph transformation) implements a place-only abstraction: DPTN places are nodes and transitions are implicit edges. Because each net element is processed once in a linear traversal, this phase remains $\mathcal{O}(n)$.

Phase 2 (algorithmic partitioning) is the most significant optimization. Karger's randomized contraction method [31] yields $\mathcal{O}(n^2)$ per trial for an undirected graph with n nodes, traditionally producing exactly two parts. By terminating contraction as soon as exactly k supernodes remain, multiple sequential cycles of Karger's algorithm are eliminated, reducing the theoretical single-trial runtime from $\mathcal{O}(k \cdot n^2)$ to $\mathcal{O}(n^2)$. Result quality is maintained via a configurable number of independent trials r, which is not fixed at one and is set to Karger's optimal value by default, while the user can also manually set the number of trials.

Phase 3 reconstructs subnets and assigns DISTRIBUTED SYNCHRONOUS CHANNELS by analyzing cut edges to reconstruct affected transitions within their new subnets. Using high-performance data structures and decoupling previously nested processing loops reduces the effective reconstruction complexity to $\mathcal{O}(n \cdot t)$, where n is the number of graph nodes and t the number of transitions in the original net. Overall, the per-trial complexity of Algorithm 1 is $\mathcal{O}(n^2 + n \cdot t)$, fulfilling the reduced-runtime requirement.

5.4   Implementation

The redesigned plugin extends and optimizes the initial NETSPLIT approach [13]. Following RENEW architectural standards and development practices (Sect. 2.3), the implementation uses Java 17 [18] and integrates via the Java Platform Module System (JPMS) [11,40].

GUI components use Java Swing [19] to ensure visual and functional consistency with the established RENEW environment. Testing continues to use JUnit 5.9 [50] with Mockito 4.8 [51] to mock complex interactions and dependencies. Validation of Kafka event logic is enhanced using org.json [15] for parsing and verifying structured JSON events, and the suite uses Testcontainers for Apache Kafka 1.19.4 [16] to simulate real broker interactions. This setup verifies NETSPLIT event logging via the KAFKACLIENT, an internal RENEW plugin that provides a generic interface for asynchronous communication and event exchange via an Apache Kafka broker in containerized environments.

Integration into distributed environments is realized through the CLOUDNATIVE [48,49] plugin, which uses Spring Boot [52] to provide a REST interface. A central `@RestController` manages the `/netsplit` endpoint, enabling remote initiation of decomposition operations from other simulator instances.

Empirical validation is conducted using the Java Microbenchmark Harness [14]. The benchmark measures throughput, average execution time per operation, and memory consumption, providing a granular view of how the redesigned k-cut strategy and optimized data structures affect NETSPLIT's efficiency.

5.5   Evaluation

All three algorithmic requirements are satisfied: the place-only abstraction achieves linear transformation complexity $O(n)$; the direct k-cut strategy eliminates iterative splitting overhead, reducing single-trial complexity to $O(n^2)$; and decoupled reconstruction loops with optimized data structures yield $O(n \cdot t)$ reconstruction complexity. Integration into distributed environments is verified through successful remote initiation of decomposition operations via the /netsplit REST endpoint based on the CLOUDNATIVE [48,49] RENEW plugin and structured JSON event logging to an Apache Kafka broker via the NetSplitTopic, fulfilling the fourth requirement.

Table 1. Performance evaluation of DPTN subnet decomposition across diverse topologies(All benchmarks were performed on a system with Ubuntu 25.10, an Intel Core i9-13900HX (24 cores, 2 threads), 30 GiB RAM, and a 1 TB NVMe SSD (WD Blue SN570))

Model	Implementation	Average	Throughput	Speedup	Min. Subnet
8 Phil	Original	0.464 ms/op	2.169 ops/ms	58.0x	3
	Redesigned	0.008 ms/op	123.023 ops/ms		
80 Phil	Original	18.488 ms/op	0.053 ops/ms	176.1x	3
	Redesigned	0.105 ms/op	9.936 ops/ms		
800 Phil	Original	1909.185 ms/op	0.001 ops/ms	1322.2x	3
	Redesigned	1.444 ms/op	0.698 ops/ms		
Disconnected Net	Original	0.035 ms/op	28.797 ops/ms	17.5x	4
	Redesigned	0.002 ms/op	644.873 ops/ms		
Dumbbell	Original	0.229 ms/op	4.407 ops/ms	16.4x	9
	Redesigned	0.014 ms/op	71.276 ops/ms		
Fully Connected	Original	0.174 ms/op	5.841 ops/ms	17.4x	5
	Redesigned	0.010 ms/op	104.657 ops/ms		
Service Org.	Original	0.195 ms/op	5.092 ops/ms	39.0x	6
	Redesigned	0.005 ms/op	199.807 ops/ms		

Table 1 summarizes NETSPLIT performance across various topologies, measured in average execution time (avgt), throughput (thrpt), speedup over the previous implementation, and minimum result subnet size. The Dining Philosophers model at scales 8, 80, and 800 provides a controlled scaling benchmark over a regular, symmetric topology. The redesigned implementation achieves speedups of 58.0x, 176.1x, and 1322.2x, respectively, with a consistent minimum subnet size of 3. This trend of increasing speedup aligns with the superlinear growth of the previous implementation's per-trial complexity $O(n^2 + n \cdot t)$, highlighting how the redesigned contraction and data structures scale more favorably at larger inputs.

Transitioning from scale to structure, we next evaluate structural generalizability by benchmarking four topologies with distinct structural properties. The Disconnected Net consists of two isolated subgraphs of 4 elements each, providing a baseline for structurally separable inputs; NETSPLIT achieves a 17.5x speedup with a minimum subnet of 4. The Fully Connected net comprises nine elements, representing a densely coupled structure in which all places share common transitions; the redesign yields a 17.4x speedup with a minimum subnet of 5. The Dumbbell topology connects two fully connected nets with a single bridge, creating an asymmetric bottleneck; it yields a 16.4x speedup and a minimum subnet of 9. The Service Organisation net, a DPTN adaptation of a structured service system [23], yields a 39.0x speedup with a minimum subnet size of 6.

Across all benchmarks, the redesigned NETSPLIT consistently outperforms the previous implementation. Notably, the speedup factors for the four additional topologies—ranging from 16.4x to 39.0x—are lower than those observed for large Philosopher instances. This difference is expected given their smaller absolute input sizes. Taken together, these results indicate that the algorithmic improvements are effective across structurally diverse inputs, supporting the generalizability of the redesign within the scope of this PoC.

6   Prototype: NETJOIN

6.1   Requirements

In contrast to the NETSPLIT (Sect. 5) plugin, the new plugin NETJOIN has been developed, which provides the functionality for composing DPTNs (Sect. 2.1). Although NETJOIN is meant to complement NETSPLIT, their functionality is not symmetric: using NETSPLIT on DPTNs that were previously merged with NETJOIN does not guarantee restoration of the original DPTNs. Therefore, NETSPLIT cannot be regarded as the inverse of NETJOIN. To achieve this objective, the prototype adopts a principle that composes multiple DPTNs into a unified model while preserving their individual identities. The design is expected to achieve linear-time complexity with respect to the input size, which is essential for practical applicability to large-scale networks; long processing times would reduce usability and may negatively affect user acceptance.

A further requirement is integration into RENEW (Sect. 2.3). Compatibility is ensured via existing interfaces and data formats, enabling smooth adoption by users familiar with the environment. Beyond the graphical user interface, the prototype shall provide a REST endpoint that allows all core NETJOIN operations to be triggered via defined paths, enabling automation and external system integration.

Moreover, the prototype must remain consistent with the NETSPLIT plugin by providing the same fundamental, non-decomposition-specific features, in particular the user interfaces and their interaction concepts. Accordingly, all relevant NETJOIN operations shall support Kafka-based logging in the same manner as NETSPLIT to ensure traceability, reproducibility, and consistent process monitoring across both plugins. This preserves a familiar user experience, eases switching between tools, and maintains interoperability for future extensions.

6.2   Specification

NETJOIN composes two or more DPTNs into a single resulting DPTN. The user explicitly selects the nets to be joined, either by specifying a file path or via RENEW's graphical user interface. Runtime requirements impose linear complexity relative to the size of the input nets. In the worst case, runtime is bounded by $(|P|+|T|+|F|)$, where P, T, and F are the sets of places, transitions, and arcs in the composed DPTNs. This ensures predictable scalability as the DPTN grows.

The plugin must follow RENEW's architectural constraints and interfaces to integrate seamlessly with the existing infrastructure. The NETJOIN plugin must also maintain consistency with the NETSPLIT plugin. NETJOIN offers the same core mechanisms, including a user-friendly graphical interface, a command-line interface, and a REST interface. User-friendliness is supported by features such as auto-save, Kafka-based activity logging, and clear documentation.

6.3   Design

The NETJOIN design integrates multiple DPTNs into a single consistent model in two stages, reversing the decomposition performed by NETSPLIT: (1) structural merging of net elements and (2) functional composition by reconciling synchronous channels. Algorithm 2 formalizes this composition process.

Algorithm 2: Automated Composition Algorithm

Input: Set of DPTNs $\mathcal{N} = \{N_1, N_2, \ldots, N_n\}$
Output: Composed DPTN $N_{composed}$
$D \leftarrow$ new empty Drawing;
// **Phase 1: Structural Merging**
foreach $N_i \in \mathcal{N}$ **do**
    foreach *element* $e \in N_i$ **do**
        add prefixed deep copy of e to D;

// **Phase 2: Channel Reconciliation**
foreach *channel identifier* c *with exactly 1 uplink* u *and 1 downlink* d **do**
    redirect all arcs from u to d;
    aggregate duplicate arcs;
    remove u and channel inscriptions from d;
return D

Architectural consistency is realized via a dual-module structure: an API module for external interfaces (GUI, CLI, REST) and an implementation module for core logic. As in Algorithm 2, composition starts by creating a new **Drawing** as a container. Then, each element of every selected subnet is deep-copied into this container, one at a time. Next, to prevent naming collisions, prefix each element ID with the name of its source net.

After combining the elements, scan the unified drawing for transitions with identical channel identifiers to reconcile synchronous channels. Then analyze the inscriptions and group them into pairs, ensuring each pair contains exactly one uplink and one downlink transition. In accordance with the composition calculus, discard incomplete pairs that violate the interface requirement (exactly two occurrences per interface) to ensure model correctness.

Pairs are merged by restructuring the topology. All input and output arcs of the uplink transition are redirected to the matching downlink transition. Afterwards, the uplink transition and explicit channel inscriptions are removed. If multiple arcs between the same places and transitions result, aggregate their arc weights following Renew conventions. This reduction of interfaces links previously separated subnets through a single consistent transition, completing the composition process.

For integration into containerized environments, NETJOIN provides a REST interface consistent with the NETSPLIT architecture. For portability and security, the REST endpoint primarily processes nets currently active in the RENEW environment. Design correctness is validated by an automated test suite in the build pipeline. This suite tests incomplete channel pairs, multi-arcs, and various net structures to prevent regression.

NETJOIN complexity follows the two-step design (Sect. 6.3). Structural merging of n subnets requires a linear traversal for deep-copying and prefixing, yielding $O(\sum_{i=1}^{n} m_i)$ where m_i is the number of elements (places, transitions, arcs) in subnet i. Channel reconciliation scans the unified drawing to identify, group, and restructure channel inscriptions; for l synchronous channels, this is $O(\sum_{i=1}^{n} m_i \cdot l)$. Since l is typically smaller than the total number of net elements, the algorithm remains efficient for large-scale models.

6.4   Implementation

The implementation was realized as an independent extension with strict adherence to RENEW implementation details for seamless integration. To achieve this, a new `Drawing` is created, and the selected net elements are iteratively deep-copied into it, ensuring independence from the original nets. This copying process is encapsulated in the implementation layer and leverages existing RENEW APIs, allowing for low complexity and tight integration. Furthermore, the design (see Sect. 6.3) is implemented in Java 17. The graphical user interface, in turn, uses the Java Swing Framework. Analogous to NETSPLIT, integration into modern distributed environments is provided via the CLOUDNATIVE [48,49] plugin. This plugin uses the Spring Boot framework to expose a REST interface. To support remote operations, a central `@RestController` manages the `/netjoin` endpoint, enabling remote initiation of composition operations from other simulation instances.

6.5   Evaluation

NETJOIN fulfils all specified requirements. Functional correctness is confirmed: structural merging and channel reconciliation compose multiple DPTNs into a unified model while preserving composition semantics, as validated by automated regression tests covering incomplete channel pairs and multi-arc aggregation. The combined complexity of $O(\sum_{i=1}^{n} m_i \cdot l)$ confirms the linear-time requirement. Integration into RENEW is achieved via existing interfaces, and consistency with NETSPLIT is ensured through a shared interaction model comprising GUI, REST endpoint, Kafka-based logging, and autosave.

Table 2 evaluates NETJOIN across topologies of varying size and structural character.

Table 2. Performance evaluation of DPTN subnet composition across diverse topologies([2] All benchmarks were performed on a system with Ubuntu 25.10, an Intel Core i9-13900HX (24 cores, 2 threads), 30 GiB RAM, and a 1 TB NVMe SSD (WD Blue SN570)

Configuration	N	S/C	Average	Throughput	Ratio
8 Philosophers	8	2/2	30.85 ± 5.38 ms/op	0.035 ops/ms	—
80 Phil. (baseline)	80	2/2	93.33 ± 7.19 ms/op	0.011 ops/ms	0.303
80 Phil. (4 sync)	80	2/4	91.80 ± 7.39 ms/op	0.011 ops/ms	—
80 Phil. (4 subnets)	80	4/4	91.76 ± 7.47 ms/op	0.011 ops/ms	—
800 Philosophers	800	2/2	581.49 ± 17.28 ms/op	0.002 ops/ms	0.623
Disconnected Net	2	2/0	6.657 ± 0.514 ms/op	0.148 ops/ms	—
Dumbbell	2	2/1	22.431 ± 1.751 ms/op	0.046 ops/ms	—
Fully Connected	2	2/4	16.908 ± 1.239 ms/op	0.058 ops/ms	—
Service Org.	2	2/2	0.939 ± 0.016 ms/op	1.059 ops/ms	—

The Dining Philosophers model at scales 8, 80, and 800 provides a controlled scaling benchmark. The complexity ratios of 0.303 (8 to 80 philosophers) and 0.623 (80 to 800 philosophers) are consistent with the theoretical bound $O(\sum_{i=1}^{n} m_i \cdot l)$, indicating that runtime grows more slowly than input size across these configurations. The three 80-philosopher configurations vary in subnet and channel count; their runtimes are statistically indistinguishable (91.76–93.33 ms/op), suggesting that the l factor does not dominate runtime at this scale.

To assess structural generalizability, four additional topologies derived from the NETSPLIT benchmarks were evaluated, each composed of two subnets. Comparing these topologies reveals that runtime correlates more strongly with net size than with channel count: the Dumbbell and Fully Connected nets share similar structural complexity yet differ in channel count (1 vs. 4), while their runtimes remain in the same order of magnitude (22.431 vs. 16.908 ms/op). The Disconnected Net, containing no synchronous channels, isolates the baseline cost of structural merging at 6.657 ms/op. The Service Organisation net, the most compact of the four, achieves the lowest runtime at 0.939 ms/op. Together, these results are consistent with the dominance of the m_i factor in the complexity bound.

Across all benchmarks, NETJOIN preserves composition semantics while delivering stable, practically usable performance across structurally diverse topologies. The higher absolute runtimes compared to NetSplit are attributable to deep-copy overhead during structural merging and channel reconciliation; both remain practical for the intended use cases.

7 Evaluation

This section evaluates how to define and operationalize interface-driven modularity for Distributed P/T Nets (DPTNs). It builds on the research questions from the problem description (Sect. 3).

RQ1: Which perspective of modules and interfaces is suitable for DPTNs? We propose a modular perspective in which a DPTN consists of a set of net modules. Each net module is itself a DPTN. This enables hierarchical modeling. The modules are distributed across multiple simulator instances. Subsequently, communication between net modules occurs via explicit interfaces implemented via distributed synchronous channels. To facilitate composition and decomposition, we use Reisig's associative composition calculus [25, 45, 46] as the formal framework.

RQ2: Which algorithmic methods can practically implement composition and decomposition of DPTNs under the assumptions in RQ1, and how can transparent metrics compare the results' quality and required effort? Karger's algorithm forms the basis for decomposition. Its cutting criteria can be swapped. To apply it to Petri nets rather than graphs, we first transform the Petri nets. An asymptotic upper bound is $O(n^2 + n \cdot t)$, where n is the number of net elements and t is the number of transitions in the original net. For composition, a matching procedure is applied to the distributed synchronous channels. The up- and down-links of a channel are paired. This assigns corresponding interface connections between the nets to be composed. We specify an asymptotic upper bound of $O(\sum_{i=1}^{n} m_i \cdot l)$. Here, n is the number of nets to be composed. m_i is the number of net elements in the i-th net, and l is the number of distributed synchronous channels.

We evaluate the PoC using transparent metrics. These include wall-clock time of the operation, throughput (number of complete algorithm runs per millisecond), and runtime-to-net-size ratio. Additionally, we consider the runtime improvement for decomposition compared to the version described in [13], thereby providing a comparative perspective.

Main Research Question: How can a concept of interface-driven modularity be defined and operationalized via decomposition and composition so that large P/T net models become more manageable without violating formally defined interface compatibility? To address this, RQ1 establishes a modular view of DPTNs, including explicit interfaces and the formal composition framework used. Building upon this foundation, we operationalize (de)composition using the methods from RQ2 so that modules can be automatically distributed, decomposed, and reassembled.

In accordance with the constructivist prototyping methodology, these concepts are made comprehensible through a PoC implementation: decomposition and composition are implemented using the RENEW plugins NETSPLIT and NETJOIN, respectively. Measurements of the metrics are reported in the evaluations of NETSPLIT (Sect. 5.5) and NETJOIN (Sect. 6.5). Due to current use cases focusing on distributed simulation and dynamic scaling [13], decomposition and composition are therefore provided exclusively automatically and not as manual modeling operations.

8   Discussion

The provided operations allow existing DPTNs to be decomposed into modules along their interfaces, or modules to be composed via their interfaces. The well-defined interface specification consists of distributed synchronous channels, enabling automatic modularization based on these specifications.

Decomposing large DPTNs into smaller net modules reduces local complexity for modelers. Instead of the entire net, only the current module and its external interface must be understood.

The implementation of RENEW plugins enables direct embedding into existing modeling, simulation, and analysis workflows. In particular, other RENEW plugins can be used in the same context, e.g., the plugins established in the DPTN environment [9,10,13]. This yields a consistent tool stack for modeling, transformation, simulation, and analysis.

Compared to the NETSPLIT plugin version in [13], optimizations have been implemented in the algorithm and the data structures. In the evaluation (Sect. 7), these changes show a shorter runtime or a speedup. Metrics, baseline, and measurement setup are documented there.

The decomposition criteria are implemented as an interchangeable strategy. This facilitates adaptation to different modeling objectives while making explicit that modularization quality is not absolute, since it depends on the selected criterion and the context.

As a PoC, NETSPLIT uses a randomized min-cut algorithm, namely Karger's algorithm. Hence, results generally vary across runs; with a fixed seed, a run is reproducible deterministically. Moreover, the transformation is generally not invertible. If already split modules are merged and then split again, the original modulation is not expected to be exactly restored. This behavior is acceptable when decomposition is understood as heuristic partitioning rather than as a canonical normal form; however, defining explicit splitting constraints – such as user-annotated cut boundaries or decomposition provenance preserved as metadata during composition – could mitigate this asymmetry while preserving Karger's randomized nature.

For example, using a minimal cut as an optimization goal can yield cuts that are unfavorable in specific contexts. In particular, structuring properties – such as invariants or net-theoretical patterns (e.g., traps/siphons) – may be distributed across module boundaries or compromised in interpretability. A robust evaluation, therefore, requires context-specific quality measures beyond cut sizes, such as communication effort, balance, or property preservation.

Currently, the split in NETSPLIT is determined exclusively automatically. For modeling practice, a manual or interactive mode may also help preserve existing structures and incorporate domain knowledge into the partitioning.

Both NETSPLIT and NETJOIN offer potential for further algorithmic and data structure optimizations. For NETSPLIT, a recursive min-cut variant such as Karger-Stein [32] is an obvious option and can offer expected runtime advantages over a non-recursive implementation. Similarly, for NETJOIN, alternative representations and more efficient update operations can improve runtime.

The concept and its PoC implementation currently refer to DPTNs. Transferring it to other net classes requires explicit clarification of which semantic and syntactic features of the class in question are essential and how decomposition and composition correctly map these features.

9   Related Work

Decomposition and composition of Petri nets are well-established techniques for improving modularity and verification. Classical approaches introduce interfaces at structural cut boundaries to preserve behavioral properties [1], or use place and transition fusion for modular analysis [8]. Kindler [35] introduces place-based interfaces with a compositional partial order semantics, enabling modular verification via temporal logic. However, these approaches assume monolithic execution and do not address distributed operation.

Formal composition has been studied extensively, for example, in the classification by Gomes and Barros [29] and in Reisig's composition calculus, which defines precise compatibility conditions for Petri net modules. Algebraic foundations for component composition are further developed in Heraklit's component calculus [24]. These works provide a rigorous semantic basis but do not consider operational criteria, such as synchronization or communication overhead, in distributed execution.

Algorithmic decomposition often relies on general graph partitioning. Multi-level partitioners like METIS [33] optimize fixed objectives, usually minimizing cut size. They provide limited flexibility for domain-specific metrics. Alternatively, decomposition can be framed as an optimization problem using integer linear programming (ILP) [21,22]. This allows expressive objectives but does not scale well to large nets. Gaede et al. [27] propose automatic modularization of P/T nets using pattern recognition and structural analysis. They identify recurring substructures as potential module boundaries. Their method complements graph-based partitioning by exploiting domain-specific knowledge in net structures. However, it does not explicitly address distributed execution or interface-driven composition.

Modularity metrics provide quantitative criteria for evaluating decomposition quality. Briand et al. [4] establish a property-based framework for software engineering measurement. This framework includes principles for assessing coupling and cohesion in modular systems. Morasca [41] adapts these concepts to Petri nets. He introduces metrics for structural attributes such as complexity, coupling, and modularity in concurrent specifications. Bott et al. [3] define coupling and cohesion measures for P-invariant calculation in modular P/T nets. These metrics quantify the impact of module boundaries on verification complexity. While this work uses cut size as a proxy for inter-module coupling, these metrics suggest additional quality dimensions, such as intra-module cohesion, balance constraints, and preservation of structural properties, such as invariants.

Distributed P/T nets enable execution via synchronous channels across simulator boundaries [9]. The synchronous channel concept originates from Christensen and Hansen [7], who introduce it for Colored Petri Nets as an interaction primitive between modules. Existing tool support focuses on infrastructure rather than interface-guided decomposition and composition.

In response, this work defines decomposition and composition as operational, interface-driven operators for distributed P/T nets, combining adaptable graph-based partitioning with Petri-net-specific interface semantics. The approach operationalizes quality criteria through empirically validated metrics. It remains open to integrating additional modularity metrics to improve decomposition strategies.

10   Future Work

Several directions can extend the composition and decomposition of DISTRIBUTED P/T NETS. For NETSPLIT, a key avenue is faster, deterministic splitting criteria and algorithms. Beyond random contraction, alternative cut criteria—classical static metrics, dynamic runtime metrics, and graph-structural metrics—may further improve scalability on very large nets; in particular, multilevel graph partitioning (e.g., METIS [34]) and spectral clustering [42] could further reduce runtime. A modular architecture also supports swapping the splitting algorithm by use case.

Further enhancements include manual split-boundary selection (free choice of edges to cut), allowing users to apply domain knowledge. Moreover, defining explicit splitting constraints would restrict the algorithm to cuts that preserve logical units, which mitigates the asymmetry between composition and decomposition. Similarly, structure-aware splitting could detect common net motifs (modules, pipelines) to keep splits aligned with semantic boundaries.

Finally, it remains to be investigated to what extent these concepts transfer to other classes of Petri nets by identifying formal conditions under which underlying interface semantics and (de-)composition operations remain valid, and characterizing required adaptations when additional net features are present.

References

1. Berthelot, G.: Transformations and decompositions of nets. In: Brauer, W., Reisig, W., Rozenberg, G. (eds.) Petri Nets: Central Models and Their Properties, Advances in Petri Nets 1986, Part I, Proceedings of an Advanced Course, Bad Honnef, Germany, 8-19 September 1986. Lecture Notes in Computer Science, vol. 254, pp. 359–376. Springer, Cham (1986). https://doi.org/10.1007/BFB0046845

2. Bischofberger, W.R., Pomberger, G.: Prototyping-Oriented Software Development - Concepts and Tools. Texts and Monographs in Computer Science, Springer, Cham (1992). https://doi.org/10.1007/978-3-642-84760-8

3. Bott, S., Clasen, L.O., Hansson, M., Levens, N., Moldt, D.: Coupling and cohesion measures for calculating p-invariants for modular p/t nets (2025)

4. Briand, L.C., Morasca, S., Basili, V.R.: Property-based software engineering measurement. IEEE Trans. Softw. Eng. **22**(1), 68–86 (2002)

5. Brookes, S.D., Hoare, C.A.R., Roscoe, A.W.: A theory of communicating sequential processes. J. ACM **31**(3), 560–599 (1984). https://doi.org/10.1145/828.833

6. Budde, R., Kautz, K., Kuhlenkamp, K., Züllighoven, H.: What is prototyping? Inf. Technol. People **6**(2/3), 89–95 (1990)

7. Christensen, S., Damgaard Hansen, N.: Coloured petri nets extended with channels for synchronous communication. In: Valette, R. (ed.) Application and Theory of Petri Nets 1994, pp. 159–178. Springer, Berlin Heidelberg, Berlin, Heidelberg (1994)

8. Christensen, S., Petrucci, L.: Modular analysis of petri nets. Comput. J. **43**(3), 224–242 (2000). https://doi.org/10.1093/COMJNL/43.3.224

9. Clasen, L., Bartelt, S., Stahl, Y., Moldt, D.: Distributed P/T net simulation prototypes based on event streaming. In: Köhler-Bußmeier, M., Moldt, D., Rölke, H. (eds.) Proceedings of the International Workshop on Petri Nets and Software Engineering 2024 co-located with the 45th International Conference on Application and Theory of Petri Nets and Concurrency (PETRI NETS 2024), June 24–25, 2024, Geneva, Switzerland. CEUR Workshop Proceedings, vol. 3730, pp. 192–216. CEUR-WS.org (2024). https://ceur-ws.org/Vol-3730

10. Clasen, L., Leonhardt, P., Wichelmann, L.: Resilient distributed P/T net simulators. In: Köhler-Bußmeier, M., Moldt, D., Rölke, H. (eds.) Proceedings of the International Workshop on Petri Nets and Software Engineering 2025 co-located with the 46th International Conference on Application and Theory of Petri Nets and Concurrency (PETRI NETS 2025), June 22–27, 2025, Paris, France. CEUR Workshop Proceedings, vol. 3998. CEUR-WS.org (2025). https://ceur-ws.org/Vol-3998

11. Clasen, L., Moldt, D., Hansson, M., Willrodt, S., Voß, L.: Enhancement of Renew to version 4.0 using JPMS. In: Köhler-Bußmeier, M., Moldt, D., Rölke, H. (eds.) Proceedings of the International Workshop on Petri Nets and Software Engineering 2022 co-located with the 43rd International Conference on Application and Theory of Petri Nets and Concurrency (PETRI NETS 2022), Bergen, Norway, June 20th, 2022. CEUR Workshop Proceedings, vol. 3170, pp. 165–176. CEUR-WS.org (2022), https://ceur-ws.org/Vol-3170

12. Clasen, L., Nayci, C., Moldt, D.: Distributed reference net simulation based on event streaming. In: Amparore, E.G., Mikulski, L. (eds.) Application and Theory of Petri Nets and Concurrency - 46th International Conference, PETRI NETS 2025, Paris, France, June 22–27, 2025, Proceedings. Lecture Notes in Computer Science, vol. 15714, pp. 130–154. Springer, Cham (2025). https://doi.org/10.1007/978-3-031-94634-9_7

13. Clasen, L., Nayci, C., Nacyi, E., Middendorf, J., Mack, T.: Investigations Towards Dynamic Scaling of Distributed P/T Nets. In: Köhler-Bußmeier, M., Moldt, D., Rölke, H. (eds.) Proceedings of the International Workshop on Petri Nets and Software Engineering 2025 co-located with the 46th International Conference on Application and Theory of Petri Nets and Concurrency (PETRI NETS 2025), June 22–27, 2025, Paris, France. CEUR Workshop Proceedings, vol. 3998. CEUR-WS.org (2025). https://ceur-ws.org/Vol-3998

14. Community, O.: Java microbenchmark harness (jmh) (2009). https://github.com/openjdk/jmh
15. Contributors, J.: Json in java – org.json:json:20240303 (2024). https://www.javadoc.io/doc/org.json/json/20240303/index.html
16. Contributors, T.: Testcontainers kafka module – kafkacontainer 1.19.4 (2024). https://www.javadoc.io/doc/org.testcontainers/kafka/1.19.4/org/testcontainers/containers/KafkaContainer.html
17. Cormen, T.H., Leiserson, C.E., Rivest, R.L., Stein, C.: Introduction to Algorithms, fourth edition. MIT Press (2022). https://mitpress.mit.edu/9780262046305/introduction-to-algorithms/
18. Corporation, O.: Java platform, standard edition 17 documentation (2022). https://docs.oracle.com/en/java/javase/17/
19. Corporation, O.: javax.swing (java se 17) (2022). https://docs.oracle.com/en/java/javase/17/docs/api/java.desktop/javax/swing/package-summary.html
20. Duvigneau, M.: Konzeptionelle Modellierung von Plugin-Systemen mit Petrinetzen, Agent Technology – Theory and Applications, vol. 4. Logos Verlag, Berlin (2010). http://www.logos-verlag.de/cgi-bin/engbuchmid?isbn=2561&lng=eng&id=
21. Ferreira, C.E., Martin, A., de Souza, C.C., Weismantel, R., Wolsey, L.A.: Formulations and valid inequalities for the node capacitated graph partitioning problem. Math. Program. **74**(3), 247–266 (1996)
22. Ferreira, C.E., Martin, A., de Souza, C.C., Weismantel, R., Wolsey, L.A.: The node capacitated graph partitioning problem: a computational study. Math. Program. **81**(2), 229–256 (1998)
23. Fettke, P., Reisig, W.: Heraklit-fallstudie: Service-system (2020). https://doi.org/10.13140/RG.2.2.27956.27521
24. Fettke, P., Reisig, W.: Modularization, composition, and hierarchization of petri nets with heraklit. arXiv preprint arXiv:2202.01830 (2022)
25. Fettke, P., Reisig, W.: Once and for all: how to compose modules–the composition calculus. In: International Symposium on Leveraging Applications of Formal Methods, pp. 173–190. Springer, Cham (2024)
26. Foundation, A.S.: Apache kafka 2.0 documentation (2025). https://kafka.apache.org/20/documentation.html, abgerufen am 10.10.2025
27. Gaede, J., Overath, J.H., Wallner, S.: Automatic modularization of place/transition nets. In: Köhler-Bußmeier, M., Moldt, D., Rölke, H. (eds.) Proceedings of the International Workshop on Petri Nets and Software Engineering 2024 co-located with the 45th International Conference on Application and Theory of Petri Nets and Concurrency (PETRI NETS 2024), June 24–25, 2024, Geneva, Switzerland. CEUR Workshop Proceedings, vol. 3730, pp. 53–73. CEUR-WS.org (2024). https://ceur-ws.org/Vol-3730
28. Gandhi, A., Dube, P., Karve, A., Kochut, A., Zhang, L.: Modeling the impact of workload on cloud resource scaling. In: 2014 IEEE 26th International Symposium on Computer Architecture and High Performance Computing, pp. 310–317 (2014). https://doi.org/10.1109/SBAC-PAD.2014.16
29. Gomes, L., Barros, J.P.: Structuring and composability issues in petri nets modeling. IEEE Trans. Industr. Inf. **1**(2), 112–123 (2005). https://doi.org/10.1109/TII.2005.844433
30. Hoare, C.A.R.: Communicating sequential processes. Commun. ACM **21**(8), 666–677 (1978). https://doi.org/10.1145/359576.359585

31. Karger, D.R.: Global min-cuts in rnc, and other ramifications of a simple min-cut algorithm. In: Ramachandran, V. (ed.) Proceedings of the Fourth Annual ACM/SIGACT-SIAM Symposium on Discrete Algorithms, pp. 21–30. 25–27 January 1993, Austin, Texas, USA. ACM/SIAM (1993), http://dl.acm.org/citation.cfm?id=313559.313605

32. Karger, D.R., Stein, C.: An o (n^2) algorithm for minimum cuts. In: Kosaraju, S.R., Johnson, D.S., Aggarwal, A. (eds.) Proceedings of the Twenty-Fifth Annual ACM Symposium on Theory of Computing, pp. 757–765. May 16-18, 1993, San Diego, CA, USA. ACM (1993). https://doi.org/10.1145/167088.167281

33. Karypis, G., Kumar, V.: Metis: A software package for partitioning unstructured graphs, partitioning meshes, and computing fill-reducing orderings of sparse matrices (1997)

34. Karypis, G., Kumar, V.: A fast and high quality multilevel scheme for partitioning irregular graphs. SIAM J. Sci. Comput. **20**(1), 359–392 (1998). https://doi.org/10.1137/S1064827595287997

35. Kindler, E.: A compositional partial order semantics for Petri net components. In: Azéma, P., Balbo, G. (eds.) ICATPN 1997. LNCS, vol. 1248, pp. 235–252. Springer, Heidelberg (1997). https://doi.org/10.1007/3-540-63139-9_39

36. Kreps, J., Narkhede, N., Rao, J., et al.: Kafka: A distributed messaging system for log processing. In: Proceedings of the NetDB, vol. 11, pp. 1–7. Athens, Greece (2011)

37. Kummer, O.: Referenznetze. Logos Verlag, Berlin (2002). http://www.logos-verlag.de/cgi-bin/engbuchmid?isbn=0035&lng=eng&id=

38. Kummer, O., et al.: Renew – the Reference Net Workshop (2025). http://www.renew.de/, release 4.2

39. Manne, U., Scholar Y, R.: Horizontal vs. vertical scaling in modern database systems: a comparative analysis of performance and cost trade-offs. Int. J. Comput. Eng. Technol. **15**, 514–524 (2024). https://doi.org/10.5281/zenodo.13851050

40. Moldt, D., et al.: RENEW: modularized architecture and new features. In: Gomes, L., Lorenz, R. (eds.) Application and Theory of Petri Nets and Concurrency - 44th International Conference, PETRI NETS 2023, Lisbon, Portugal, June 25–30, 2023, Proceedings. Lecture Notes in Computer Science, vol. 13929, pp. 217–228. Springer, Cham, Switzerland (2023). https://doi.org/10.1007/978-3-031-33620-1_12

41. Morasca, S.: Measuring attributes of concurrent software specifications in petrinets. In: Proceedings Sixth International Software Metrics Symposium (Cat. No. PR00403), pp. 100–110. IEEE (1999). https://doi.org/10.1109/METRIC.1999.809731

42. Ng, A.Y., Jordan, M.I., Weiss, Y.: On spectral clustering: Analysis and an algorithm. In: Advances in Neural Information Processing Systems, vol. 14, pp. 849–856 (2001). https://papers.nips.cc/paper_files/paper/2001/hash/801272ee79cfde7fa5960571fee36b9b-Abstract.html

43. Ralph, P., et al.: ACM SIGSOFT empirical standards. CoRR abs/2010.03525 (2020). https://arxiv.org/abs/2010.03525

44. Razavi, K., Salmani, M., Mühlhäuser, M., Koldehofe, B., Wang, L.: A tale of two scales: Reconciling horizontal and vertical scaling for inference serving systems. CoRR abs/2407.14843 (2024). https://doi.org/10.48550/arXiv.2407.14843

45. Reisig, W.: Associative composition of components with double-sided interfaces. Acta Informatica **56**(3), 229–253 (2019). https://doi.org/10.1007/s00236-018-0328-7

46. Reisig, W.: Composition of Component Models - A Key to Construct Big Systems. In: Margaria, T., Steffen, B. (eds.) ISoLA 2020. LNCS, vol. 12477, pp. 171–188. Springer, Cham (2020). https://doi.org/10.1007/978-3-030-61470-6_11
47. Roscoe, A.W.: The Theory and Practice of Concurrency. Prentice Hall International Series in Computer Science, Prentice Hall (1997)
48. Röwekamp, J.H.: Skalierung von nebenläufigen und verteilten Simulationssystemen für interagierende Agenten. Ph.D. thesis, University of Hamburg, Department of Informatics, Vogt-Kölln Str. 30, D-22527 Hamburg (2023). https://ediss.sub.uni-hamburg.de/handle/ediss/10040
49. Röwekamp, J.H., Taube, M., Mohr, P., Moldt, D.: Cloud native simulation of reference nets. In: Köhler-Bußmeier, M., Kindler, E., Rölke, H. (eds.) Proceedings of the International Workshop on Petri Nets and Software Engineering 2021 co-located with the 42nd International Conference on Application and Theory of Petri Nets and Concurrency (PETRI NETS 2021), Paris, France, June 25th, 2021 (due to COVID-19: virtual conference). CEUR Workshop Proceedings, vol. 2907, pp. 85–104. CEUR-WS.org (2021). http://ceur-ws.org/Vol-2907
50. Team, J.: JUnit 5 User Guide (2025). https://junit.org/junit5/docs/current/user-guide/
51. Team, M.: Mockito core 4.8.0 api (2022). https://www.javadoc.io/doc/org.mockito/mockito-core/4.8.0/org/mockito/Mockito.html
52. Team, S.B.: Spring boot 2.4.2 reference documentation (2021). https://docs.spring.io/spring-boot/docs/2.4.2/reference/htmlsingle/
53. Voß, L., Willrodt, S., Moldt, D., Haustermann, M.: Between expressiveness and verifiability: P/T-nets with synchronous channels and modular structure. In: Köhler-Bußmeier, M., Moldt, D., Rölke, H. (eds.) Proceedings of the International Workshop on Petri Nets and Software Engineering 2022 co-located with the 43rd International Conference on Application and Theory of Petri Nets and Concurrency (PETRI NETS 2022), Bergen, Norway, June 20th, 2022. CEUR Workshop Proceedings, vol. 3170, pp. 40–59. CEUR-WS.org (2022). https://ceur-ws.org/Vol-3170
54. Wilde, T., Hess, T.: Forschungsmethoden der Wirtschaftsinformatik. Wirtschaftsinformatik 4(49), 280–287 (2007)

Preserving LTL Properties in Sweep-Line State Space Exploration with Partial-Order Reduction

Sami Evangelista[1], Lars M. Kristensen[2(✉)], and Laure Petrucci[1]

[1] LIPN, CNRS UMR 7030, Université Sorbonne Paris Nord, Villetaneuse, France
`sami.evangelista@univ-paris13.fr`, `laure.petrucci@lipn.univ-paris13.fr`
[2] Department of Computer Science, Electrical Engineering, and Mathematical Sciences, Western Norway University of Applied Sciences, Bergen, Norway
`lmkr@hvl.no`

Abstract. Model checking often requires the combined use of multiple reduction techniques to mitigate the state explosion problem. The contribution of this paper is a generalised sweep-line algorithm that enforces the conditions necessary for partial-order reduction in the form of stubborn sets to preserve LTL_{-X} properties. The core idea in our algorithm is to compute strongly connected components and detect cycles within each state space layer explored by the sweep-line method, while leveraging persistent states as anchor states for enforcing cross-layer partial-order reduction conditions. We have implemented our new algorithm and conducted an experimental evaluation on set of benchmarks from the Petri Nets Model Checking Contest. The results show our LTL_{-X} preserving combination of sweep-line exploration with partial-order reduction even in a conservative implementation may substantially reduce peak memory usage. Furthermore, the overhead is mostly similar to plain partial-order preserving LTL_{-X} reduction.

Keywords: Model checking · Verification · Temporal Logic · State explosion

1 Introduction

The sweep-line method [3,10] and partial-order reduction [11] have been developed to alleviate the inherent state explosion problem in model checking. The sweep-line method exploits a progress measure on states to partition the state space into layers and delete states from memory on-the-fly during state space exploration. This reduces the number of states that must be stored simultaneously in memory. Partial-order reduction analyses dependencies between transitions (actions/events) and explores in each encountered state only a subset of the enabled transitions. This reduces the interleaved executions considered and implies that only a *reduced state space*, i.e., a subset of the full state space is explored. It has been demonstrated [14] that the sweep-line method and

J. Desel and A. Kalenkova (Eds.): PETRI NETS 2026, LNCS 16567, pp. 178–199, 2026.
https://doi.org/10.1007/978-3-032-27879-1_9

partial-order reduction are complementary, i.e., using both methods simultaneously yields better reduction than when either method is used alone. In many cases, the sweep-line method must be combined with partial-order reduction to effectively alleviate state explosion.

Linear-time Temporal Logic (LTL) is being widely used for expressing behavioural properties in model checking [1] and is often combined with partial-order reduction to improve efficiency. Model checking of safety properties with the sweep-line method was described in [10]. To the best of our knowledge, earlier research [14] on the combination of the sweep-line method and partial-order reduction has not considered the preservation of LTL_{-X} properties (linear-time temporal logic without the next operator) for the generalized sweep-line method [10], i.e., covering both monotonic and non-monotonic progress measures. In earlier work [5], we developed an automata-based algorithm for LTL model checking with the sweep-line method combining the MAP algorithm [2] with nested depth-first to detect acceptance cycles [8]. The combined use of the sweep-line method and partial-order reduction for LTL_{-X} properties was, however, not investigated in [5].

Partial-order reduction generally requires certain conditions to be fulfilled by the set of transitions selected for exploration in each encountered state. The conditions to be enforced depend upon the class of properties considered and must ensure that the property under verification is *preserved*, i.e., that it holds in the reduced state space if and only if it holds in the full state space. In the context of this paper, we are concerned with how partial-order reduction conditions for LTL_{-X} can be enforced with sweep-line state space exploration.

The sweep-line method when not combined with other reduction methods explores the full state space. Hence, when combined with partial-order reduction it will explore the same reduced state space as the partial-order method prescribes as per the conditions being enforced. This means that the sweep-line method does not impose specific additional conditions in terms of the property being preserved. The primary aspect is therefore how to ensure that the conditions on the selected transitions of the partial-order reduction are being enforced. This is the focus of this paper, where we consider partial-order reduction in the framework of stubborn sets [15,16] and the conditions imposed on stubborn sets for the preservation of LTL_{-X} properties. A secondary aspect is how properties can be verified (assuming that they are preserved) given that the sweep-line method: 1) only stores a subset of the state space being explored in memory at a time; and 2) prescribes a least-progress first exploration order. For LTL_{-X}, we addressed this secondary aspect already in [5] by developing a full LTL model checking algorithm compatible with the sweep-line exploration. In the context of the present paper, the algorithm of [5] would be used on the sweep-line explored space space reduced using stuborn sets preserving LTL_{-X}.

The rest of this paper is organised as follows. Section 2 introduces basic state space notations and the example state space that we use throughout the paper for illustration purposes. Section 3 introduces the sweep-line method, and Sect. 4 introduces the stubborn set method. In Sect. 5 and Sect. 6 we develop our new sweep-line algorithm enforcing stubborn set conditions for preserving LTL_{-X} properties. Section 7 introduces our initial prototype implementation and presents results from our experimental evaluation. Finally, in Sect. 8 we sum up conclusions and discuss future work. We assume familiarity with the basic principles of explicit-state model checking.

2   Background

To maintain independence from any particular modeling language for concurrent systems, our algorithms are presented at the level of state spaces.

Definition 1. *A* **state space** *is a tuple* (S, T, Δ, ι)*, where* S *is a finite set of* **states***,* T *is a finite set of* **transitions***,* $\Delta \subseteq S \times T \times S$ *is a* **transition relation***, and* $\iota \in S$ *is the* **initial state***.*

By Definition 1, we consider state spaces to be finite. Furthermore, we assume transitions to be deterministic to simplify the presentation of the stubborn set method. Our algorithm can, however, be adapted to accomodate also non-deterministic transitions.

A transition $t \in T$ is said to be *enabled* in a state $s \in S$ if there exists a state $s' \in S$ such that $(s, t, s') \in \Delta$. We denote by $\mathsf{enab} : S \to 2^T$ the mapping that associates with each state s the set of enabled transitions in s, i.e. for a state $s \in S$, $\mathsf{enab}(s) = \{t \in T \mid \exists s' \in S \wedge (s, t, s') \in \Delta\}$. Similarly, we denote by $\mathsf{succ} : S \times T \to S$ the mapping that associates with each state s and enabled transition t in s, the successor state s' resulting from executing t in s. Formally, for $(s, t, s') \in \Delta$, $\mathsf{succ}(s, t) = s'$. If $t \in \mathsf{enab}(s)$ we write $s - t \to$, and when $s' = \mathsf{succ}(s, t)$ we write $s - t \to s'$. When the transition is not important, we may omit the label and write $s \to s'$. A *transition sequence* $s_0 - t_1 \to s_1 - t_2 \to s_2 \cdots s_{n-1} - t_n \to s_n$ is a sequence of states s_i and transitions t_i such that $s_{i-1} - t_i \to s_i$ for all $1 \leq i \leq n$. If a state s' can be obtained from a state s by the execution of a (possibly empty) transition sequence, we say that s' is *reachable* from s and write this as $s - t_1 t_2 \cdots t_n \to s'$ Basic state space exploration starts from the initial state and successively *processes* states by exploring the enabled transitions and computing successor states. A set of *visited states* is stored in memory to ensure that each encountered state is only processed once.

The state space of a system can be viewed as a directed graph, and Fig. 1 depicts the state space that we will use for illustration purposes. Each node (state) is identified by an integer written inside the state. Node 0 corresponds to the initial state ι. Cycles and *strongly connected components (SCCs)* play a key role in our approach to enforcing the conditions on stubborn sets to preserve LTL_{-X} properties. The SCCs of the state space in Fig. 1 containing more than one node are indicated using light gray boxes and are maximal subgraphs such that two states s_1 and s_2 are in the same SCC if and only if s_1 is reachable from s_2 and vice versa. A *terminal SCC S* is an SCC such that no state contained in S has outgoing edges to states contained in other SCCs. The SCCs marked with

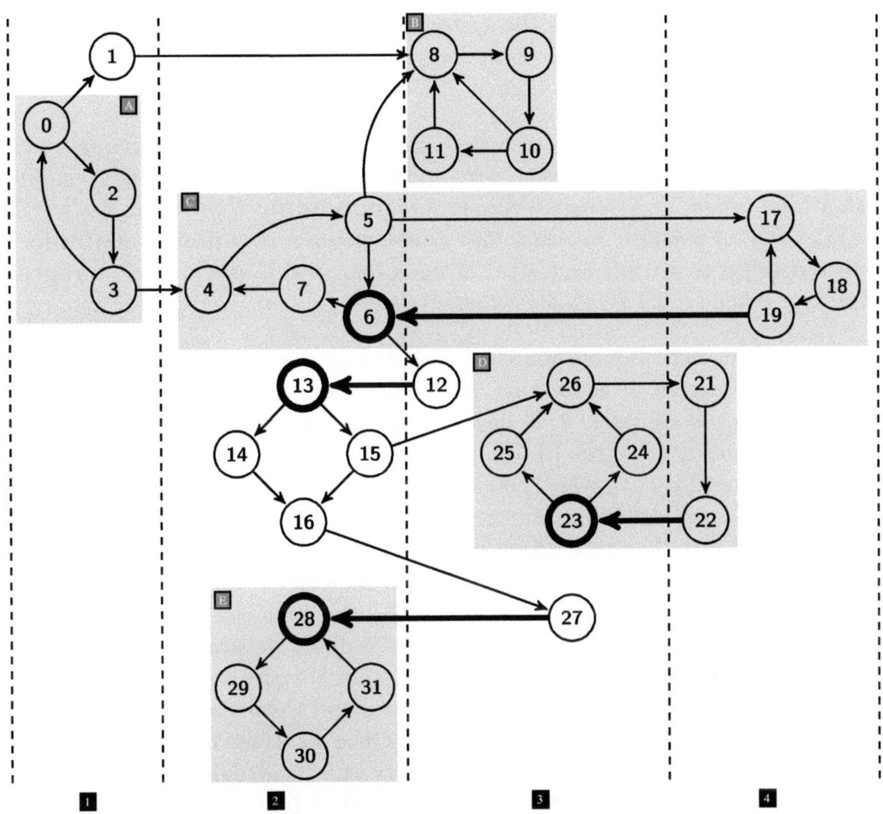

Fig. 1. Example state space and strongly connected components (light gray boxes).

B, D, and E are terminal SCCs. The dashed vertical lines and the numbered black squares will be explained in the next section.

An LTL property to be verified is composed of atomic propositions and temporal operators, where an atomic proposition $\phi : S \mapsto \{T, F\}$ maps each state into a Boolean value specifying whether the atomic proposition is true (T) or false (F) in the state. Given an LTL property Φ to be verified, a transition t is *visible* if there exists states s and s' with $(s, t, s') \in \Delta$ and an atomic proposition ϕ of Φ such that $\phi(s) \neq \phi(s')$, i.e., the execution of t in s changes the truth value of the atomic proposition. A transition that is not visible is *invisible*.

3   Sweep-Line State Space Exploration

The sweep-line method [3,10] relies on a *progress measure* intended to quantify the system progression that a state represents. The basic intuition of a progress measure is that it must capture some inherent progress in the system such that progress generally increases as the system is being executed. The progress measure partitions the state space into *layers* where each layer consists of states mapped to the same progress value.

Definition 2. *Let (S, T, Δ, ι) be a state space. A* **progress measure** *is a mapping $\psi : S \mapsto O$, where O is a non-empty set of* **progress values** *equipped with a total order $\sqsubseteq$. A progress measure is* **monotonic** *if for all $(s, t, s') \in \Delta$: $\psi(s) \sqsubseteq \psi(s')$. A progress measure that is not monotonic is* **non-monotonic.** *A* **forward edge** *is a triple $(s, t, s') \in \Delta$ such that $\psi(s) \sqsubseteq \psi(s')$, while a* **regress edge** *is a triple $(s, t, s') \in \Delta$ such that $\psi(s') \sqsubset \psi(s)$.*

The dashed vertical lines and the numbered black squares in Fig. 1 illustrate an example progress measure that partitions the state space into four layers based on progress values 1, 2, 3, and 4. Layer 1 consists of states 0–3. Layer 2 consists of states 4–7, states 13–16, and states 28–31. Layer 3 consists of states 8–12, and states 23–27. Layer 4 consists of states 17–19 and 21–22.

The sweep-line algorithm starts with the layer containing the initial state and then explores the state space one layer at a time. When all states in a given layer have been processed, the states in the current layer are deleted since the assumption is that they are not anymore needed for comparison with newly encountered states, i.e., determining whether a newly generated successor state is already contained in the set of visited states. For the state space in Fig. 1, the sweep-line method would start from state 0 and then compute the successor states 1 and 2 which will then be processed. Once all states in layer 1 have been visited, the states will be deleted from memory and the exploration will continue with states 4–7 in layer 2. States in layer 2 will then be deleted, and states 8–11 and state 12 in layer 3 will be processed. When the algorithm processes state 12, it will discover the *regress edge* to state 13 and mark state 13 as *persistent* which means that it can no longer be deleted from memory. Non-persistent states in layer 2 will then be deleted, and the algorithm will explore states 17–19 where it will discover the regress edge from state 19 to state 6 making state 6 persistent. States 6 and 13 will now be used as *roots* for a second iteration in the sweep-line algorithm. From state 6, states 4–7 will be (re)explored followed by state 12, states 8–11, and states 17–19. When the regress edge from state 12 to state 13 and the regress edge from state 19 to state 6 are explored for the second time, states 6 and 13 will be present in memory (as they became persistent in the previous iteration). From state 13, states 13–16 will be explored before the exploration will continue from state 26 and state 27 in layer 3 which will eventually lead to the full state space having been explored.

Algorithms 1 and 2 specify the generalised sweep-line algorithm [10]. The sweep-line algorithm starts with the initial state ι as the root state for the first sweep (l. 6–8). The algorithm iterates as long as new roots exists (l. 9–13) using the current roots $\mathcal{R}$ as unprocessed states $\mathcal{Q}$. The procedure SWEEPITERATION (l. 16–18) explores the states layer by layer in a least progress first order as $\mathcal{Q}$ is a priority queue on progress values. The procedures EXPLORELAYER and EXPLORESTATE work basically as a standard state space exploration, where states are expanded and their successors that have not been visited so far are put in $\mathcal{Q}$ to be later processed. Unlike standard state space exploration, the algorithm maintains states of the current layer in $\mathcal{L}$ (l. 3 and 21 in Algorithm 2) such that we can efficiently delete non-persistent states in the current layer when we have explored the layer (l. 8 in Algorithm 2). Furthermore, persistent states, regress edges, and new roots are detected l. 14–16 in Algorithm 2.

Algorithm 1. The sweep-line algorithm of [10]

1: Set of states $\mathcal{S}$ ▷ Visited states currently stored in memory
2: Set of root states $\mathcal{R} \subseteq \mathcal{S}$ ▷ States that will serve as roots in next sweep iteration
3: Progress measure ψ
4: Priority queue $\mathcal{Q} \subseteq \mathcal{S}$ ▷ States to be processed, priority based on progress value

5: **procedure** SWEEP()
6:     $\iota.pers \leftarrow$ false
7:     $\mathcal{R} \leftarrow \{\iota\}$
8:     $\mathcal{S} \leftarrow \{\iota\}$

9:     **while** $\mathcal{R} \neq \emptyset$ **do**   ▷ Explore starting from current set of roots
10:        $\mathcal{Q} \leftarrow \mathcal{R}$
11:        $\mathcal{R} \leftarrow \emptyset$
12:        SWEEPITERATION()
13:     **end while**

14: **end procedure**

15: **procedure** SWEEPITERATION()

16:     **while** $\mathcal{Q} \neq \emptyset$ **do**   ▷ Explore as long as there are unprocessed states
17:        $\psi_l \leftarrow \mathcal{Q}.\text{MINPROGRESS}()$
18:        EXPLORELAYER(ψ_l)
19:     **end while**

20: **end procedure**

It can be seen from Fig. 1 that SCCs may span multiple layers. A (global) SCC is called an *inter-layer SCC* if it contains states from multiple layers, while a (global) SCC is said to be an *intra-layer SCC* if all its states are confined within a single layer. SCC C is an example of an inter-layer SCC, while SCC B is an example of an intra-layer SCC. Similarly, an *inter-layer cycle* is a cycle in the state space containing states from at least two different layers, and an *intra-layer cycle* is a cycle containing states only from a single layer. Throughout this paper, we use the following fundamental properties of sweep-line state space exploration.

Algorithm 2. The sweep-line ExploreLayer and ExploreState procedures

---

1: States $\mathcal{L} \subseteq S$ ▷ states in the current layer

2: **procedure** ExploreLayer(ψ_l)
3:     $\mathcal{L} \leftarrow \emptyset$

4:     **while** $\mathcal{Q}$.minProgress() $= \psi_l$ **do** ▷ Explore as long as we are in current layer
5:        $s \leftarrow \mathcal{Q}$.dequeue()
6:        ExploreState(s)
7:     **end while**

8:     $S \leftarrow S \setminus \{\, s \in S \mid s \in \mathcal{L} \wedge \neg s.pers \,\}$ ▷ Delete non-persistent states in current layer
9: **end procedure**

10: **procedure** ExploreState(s)

11:     **for** $(s, t, s') \in \Delta$ **do**

12:        **if** $s' \notin S$ **then**
13:           $S \leftarrow S \cup \{s'\}$
14:           $s'.pers \leftarrow \psi(s') \sqsubset \psi(s)$

15:           **if** $\psi(s') \sqsubset \psi(s)$ **then**   ▷ Regress edge discovered
16:              $\mathcal{R} \leftarrow \mathcal{R} \cup \{s'\}$ ▷ s' will be root for a subsequent sweep
17:           **else**
18:              $\mathcal{Q}$.Enqueue(s')
19:           **end if**

20:           **if** $\psi(s) = \psi(s')$ **then**Âă   ▷ new state within current layer
21:              $\mathcal{L} \leftarrow \mathcal{L} \cup \{s'\}$
22:           **end if**
23:        **end if**
24:     **end for**
25: **end procedure**

Theorem 1. *Let (S, T, Δ, ι) be a state space and let $\psi : S \mapsto O$ be a progress measure. Then the following holds:*

1. *The sweep-line state space exploration algorithm in Algorithm 1–2 terminates after having explored all states reachable from ι at least once [10].*
2. *An inter-layer cycle and an inter-layer SCC contain at least one regress edge and at least one persistent state.*
3. *If ψ is a monotonic progress measure, then all cycles are intra-layer cycles and all SCCs are intra-layer SCCs.*

Proof. Item 1 on completeness and termination of the sweep-line algorithm was already proved in the original paper [10] that introduced the generalized sweep-line method. For item 2, let $s = s_0 - t_1 \rightarrow s_1 - t_2 \rightarrow \cdots s_{n-1} t_n - s_n = s \rightarrow$ be a cycle. Since it is an inter-layer cycle, there must be at least two states s_i and s_j with $i \neq j$ such that $\psi(s_i) \neq \psi(s_j)$. Hence, there must be at least one regress edge and associated persistent state on the cycle. The second part of item 2 follows as an inter-layer SCC spans multiple layers and hence must contain at least one inter-layer cycle. Item 3 follows from the fact that when the progress measure is monotonic, regress edges cannot exist. Hence it is impossible to have inter-layer cycles since states belonging to two different layers cannot be mutually reachable without the existence of a regress edge. From this it also follows that it is impossible to have inter-layer SCCs with monotonic progress measures.

$\square$

4   Stubborn Set State Space Exploration

The stubborn set method is based on computing in each state s encountered during state space exploration, a set of transitions $\mathsf{stub}(s) \subseteq T$ called the stubborn set. The actual computation of stubborn sets [16] is based on a structural (static) analysis of enabling and disabling dependencies between transitions for the model under consideration. The method then only explores the enabled transitions in the stubborn set. This results in the exploration of a reduced state space which is a subset of the full state space. Hence, the very basic integration with the sweep-line algorithm is to replace the loop over all enabled transitions in l. 11–24 of Algorithm 2 with a loop over the enabled transitions in the stubborn set of s. In the following we use *stubborn set transitions* to refer to transitions contained in the stubborn set and *outside transitions* to refer to transitions that are not stubborn set transitions. In its basic form, the stubborn sets computed must imply the following semantic requirements reformulated from [16] in each encountered state s.

[Condition D0 (At least one enabled stubborn set transition, if it exists)]
If $\mathsf{enab}(s) \neq \emptyset$, then $\mathsf{stub}(s) \cap \mathsf{enab}(s) \neq \emptyset$.

[Condition D1 (Outside transitions cannot enable stubborn set transitions)]
If $t \in \mathsf{stub}(s)$, $t_1, t_2, \ldots, t_n \notin \mathsf{stub}(s)$, $s = s_0 - t_1 t_2 \cdots t_n \rightarrow s_n$ and $s_n - t \rightarrow s_n'$ then there is an s_0' such that $s_0 - t \rightarrow s_0'$ and $s_0' - t_1 t_2 \cdots t_n \rightarrow s_n'$.

[Condition D2 (Outside transitions cannot disable stubborn set transitions)]
If $t \in \mathsf{stub}(s) \cap \mathsf{enab}(s)$, $t_1, t_2, \ldots, t_n \notin \mathsf{stub}(s)$ and $s = s_0 - t_1 t_2 \cdots t_n \rightarrow s_n$, then $t \in \mathsf{enab}(s_n)$.

Stubborn sets satisfying conditions **D0**, **D1**, and **D2** preserve in the reduced state space all terminal states and the existence of an infinite execution [16]. In order to preserve additional properties in the reduced state space, further dynamic requirements must be enforced on the stubborn sets used. Specifically, for a property in LTL_{-X}, the following conditions reformulated from [16] must be fulfilled by the stubborn sets used in states s encountered during exploration of the reduced state space:

[Condition S (no enabled transition is ignored)] If $t \in \text{enab}(s)$, then there is an execution $s = s_0 - t_1 t_2 \cdots t_n \rightarrow s_n$ such that $t \in stub(s_n)$ and $t_i \in \text{stub}(s_{i-1})$ for $1 \leq i \leq n$.

[Condition V (no enabled visible transitions or all visible transitions)]
If $t \in \text{stub}(s) \cap \text{enab}(s)$ and t is visible, then all visible transitions t' are in stub(s).

[Condition L1 (at least one enabled invisible stubborn set transition, if it exists)]
If there exists a transition $t \in \text{enab}(s)$ and t is invisible, then there is at least one invisible transition $t' \in \text{stub}(s) \cap \text{enab}(s)$.

[Condition L2 (cycles considers all visible transitions)] If $s = s_0 - t_1 t_2 \cdots t_n \rightarrow s_n = s$ is a cycle of the stubborn set reduced state space, then for each visible transition t there is some state s_i on the cycle such that $t \in \text{stub}(s_i)$.

Condition **S** can be implemented by ensuring that no transitions enabled in the states of terminal SCCs are ignored [15]. This requirement is compatible with the sweep-line method when using a monotonic progress measure since any SCC is confined within a single layer cf. Theorem 1(item 3). Hence, the states of the SCCs are in memory at the same time and SCCs can be computed layer by layer. It is an open problem how to enforce condition **S** when using non-monotonic progress measures. The reason is that non-monotonic progress measures do not guarantee that states belonging to an SCC of the (reduced) state space are confined within a layer. This is for instance the case with the terminal SCC D with states 21-26 in Fig. 1. It implies that the states of the SCC, including terminal SCCs may not be in memory at the same time.

The conditions **V** and **L1** do not induce particular complications for the sweep-line method as these are local properties that can be checked directly from the current state being processed. Condition **L2** can be checked using nested-depth first search (or similar) within each layer in the case of monotonic progress measures since states on a cycle are confined within a single layer and hence in memory at the same time by Theorem 1 (item 3). In this paper we address the open problem of how to enforce **L2** in the case of non-monotonic progress measures where cycles may span multiple layers such as the cycle 5, 17, 18, 19, 6, 7, 4, 5 in Fig. 1. Table 1 summarises the presentation above in terms of where there are open problems to be explored in the combination of partial-order reduction and the sweep-line method. For the sweep-line method, the cases of monotonic and non-monotonic progress measures are distinguished.

Table 1. Open problems related to implementation of partial-order conditions for LTL_{-X} when combined with sweep-line state space exploration.

Stubborn Set Condition	Progress Measure	
	Monotonic	Non-monotonic
S	✓ (single layer property)	Open
V	✓ (single state property)	✓ (single state property)
L1	✓ (single state property)	✓ (single state property)
L2	✓ (single layer property)	Open

5   Enforcing LTL_{-X} Conditions

Preservation of LTL_{-X} requires that conditions **D0**, **D1**, **D2**, **S**, **V**, **L1**, and **L2** are enforced in the stubborn set reduced state space explored with the sweep-line method. It follows from the previous section as summarised in Table 1 that the two central research questions concern how to enforce and implement conditions **S** and **L2** for the generalised sweep-line method, i.e., in the presence of non-monotonic progress measures. We address these questions in the following two subsections.

5.1   Enforcing Condition S

Condition **S** has traditionally been enforced and implemented by on-the-fly computation of the SCCs during state space exploration, and ensuring that no transition is ignored in any of the terminal SCCs. It is sufficient to consider terminal SCCs since from any reachable state it is possible to reach the states contained in at least one terminal SCC. If a transition is detected as being ignored in a terminal SCC, then the stubborn set in one of the states in the SCC is augmented with an ignored transition, the stubborn set recomputed, and the exploration of the reduced state space is resumed, including the on-the-fly computation of the SCCs. This approach can be carried forward in the case of monotonic progress measures as SCCs are confined within a single layer (Theorem 1, item 3) and as a consequence in memory at the same time. In the case of non-monotonic progress measures, some terminal SCCs may span multiple layers. SCC D in Fig. 1 illustrates this.

Our approach to resolving ignoring with non-monotonic progress measures is based on computing SCCs within each layer as they are being explored combined with additional checks. If we encounter a SCC in which no state has outgoing edges to states in other layers or to another SCC in the same layer, then this is intra-layer SCC is also a terminal SCC. We therefore check whether any transition is ignored in the SCC. If there are ignored transitions, then we augment the stubborn set in one of the states of the SCC with an ignored transition, recompute the stubborn set, and resume the state space exploration, including the computations of the SCCs. If we encounter a locally computed SCC with outgoing edges to other layers such as the local SCC comprised of states 4,5, 6, and 7 in Fig. 1, then we cannot at that stage know whether the states contained in this SCC are part of an inter-layer terminal SCC. The key observation is, however, that an inter-layer SCC must contain at least one regress edge and hence a persistent state (Theorem 1, item 2). This means that if the states of the SCC are part of an inter-layer terminal SCC, then a persistent state of the SCC will eventually be encountered again, and we can then ensure that no transition is ignored in such a state. Hence, the idea is to use persistent states (destination states of regress edges) as *anchor states* for ensuring that no transition is ignored in a terminal SCC. Clearly, this becomes a safe over-approximation as not all persistent states may belong to an inter-layer terminal SCC.

An *eager* approach to resolving ignoring in inter-layer (terminal) SCCs is therefore to include all enabled transitions in the stubborn sets of persistent states the first time that they are discovered via a regress edge. That would, however, potentially be wasteful as we may not know at the time that it becomes persistent (l. 14 of Algorithm 2), whether the persistent state is part of an inter-layer SCCs. The persistent state 13 in Fig. 1 for instance is not part of an inter-layer terminal SCC and it may be sufficient in terms of the other stubborn set conditions to only explore one of its enabled transitions. It is, however, possible to apply a *lazy* approach that relaxes the eager resolving of ignoring by exploiting the following proposition.

Proposition 1. *Let* (S, T, Δ, ι) *be the state space explored by the sweep-line algorithm using* $\psi : S \mapsto O$ *as a progress measure. Let be* $s_0 - t_1 \rightarrow s_1 - t_2 \rightarrow \cdots s_{i-1} - t_{i-1} \rightarrow s_i - t_i \rightarrow \cdots s_{n-1} - t_n \rightarrow s_n = s_0$ *be an inter-layer cycle of the explored state space. There exists a persistent state* s_i *on the cycle that has been encountered once via a regress edge and after that at least once again either via a regress edge or a forward edge.*

Proof. The existence of a persistent state s_i follows from Theorem 1 (item 2). Since s_i has been made persistent it must have been encountered via a regress edge and used as root in a subsequent sweep iteration. Since s_i is on a cycle it is reachable from itself. Hence if we do not encounter s_i again in the sweep iteration that uses s_i as root (or in a subsequent sweep iteration), it must be because the cycle contains another state s_j that was made persistent in an earlier sweep iteration. This means that s_j (which is also on the cycle) has been encountered twice: once when it was made persistent and now again. Hence, we can choose s_j as the persistent state on the cycle that is encountered twice.   □

An inter-layer SCC will contain at least one inter-layer cycle and hence we can postpone the resolving of ignoring in a persistent state until the second time the persistent state is encountered. Formulated differently, it is only persistent states that are encountered twice that we need to fully expand to guarantee that ignoring is resolved. For the state space in Fig. 1 this implies that state 6 (explored twice via a regress and a forward edge), state 23 (explored twice via a regress edge) and state 28 (explored twice via a regress and a forward edge) will be fully expanded, but not state 13. The potential advantage of the lazy approach is that it may result in a smaller stubborn set reduced state space. The disadvantage is that it may cause additional re-exploration of states. It is the opposite for the eager approach.

5.2   Enforcing Condition L2

For the implementation of condition **L2** it can be observed that the exploration with the sweep-line method induces two categories of cycles. Intra-layer cycles which are confined within one layer and inter-layer cycles which span multiple layers. In the case of intra-layer cycles, all states of the cycle will be present in memory simultaneously and conventional methods, e.g. based on depth-first search can be used to ensure that condition **L2** is enforced. In the case of inter-layer cycles, then each such cycle contains at least one persistent state as per Theorem 1(item 2). Hence, we can use a similar approach as in the case of condition **S** on inter-layer SCCs using persistent states as anchor states for enforcing **L2**. We may again consider a lazy and eager approach, where the lazy approach exploits that at least one of the persistent states on the cycle will be encountered a second time after becoming persistent.

6   Stubborn Set Sweep-Line Algorithm for LTL_{-X}

We now present our new algorithm for enforcing conditions **S**, **V**, **L1**, and **L2** when combining sweep-line state space exploration with partial-order reduction in the form of stubborn sets. As conditions **V** and **L1** can be enforced based upon the state for which a stubborn set is being computed, we denote by $\mathsf{stub}_{\{V,L1\}}$: $S \mapsto T$ a function which provides for a state s, a stubborn set $\mathsf{stub}_{\{V,L1\}}(s)$ satisfying **V** and **L1** (in addition to **D0**, **D1**, and **D2**). By $\mathsf{stub}^E_{\{V,L1\}} : S \mapsto T$ we denote a function which returns the enabled transitions contained in the stubborn set provided by $\mathsf{stub}_{\{V,L1\}}$. We assume similar functions $\mathsf{stub}_{\{S,V,L\}}$ and $\mathsf{stub}^E_{\{S,V,L\}}$ for stubborn sets satisfying **S**, **V**, **L1**, and **L2** by including all enabled transitions (for condition S) and visible transitions (for condition L2) in the stubborn set. We also assume that the anchor attribute of a state is set to 0 when this state is created.

Algorithm 3 shows the adapted SWEEP and SWEEPITERATION procedures. The priority queue Q in l. 4 controlling the exploration order now contains pairs of states and enabled transitions to be explored. This is done so that we can dynamically augment the enabled transitions to be explored in a state when possibly augmenting the stubborn sets in order to enforce conditions **V** and **L1**. Priority is still based on the progress measure of the state. The main difference is in ll. 10–16 where we now ensure that if a state has been marked as an anchor state by having been discovered a second time via a regress edge, then we ensure that the stubborn set in such a state enforces conditions **S** and **L2**. There are no changes to the SWEEPITERATION procedure.

Algorithm 4 provides the EXPLORELAYER procedure. As we now need to compute SCCs and cycles in the current layer, we store in $\mathcal{L}$ all edges in the

---

Algorithm 3. The stubborn set sweep-line Sweep and SweepIteration procedures

---

1: Set of states $\mathcal{S}$ ▷ States currently stored in memory
2: Set of states $\mathcal{R} \subseteq \mathcal{S}$ ▷ States that will serve as roots in next sweep iteration
3: Progress measure ψ
4: Priority queue $\mathcal{Q} \subseteq \mathcal{S} \times T$ ▷ States to be processed, priority based on progress value

5: **procedure** Sweep()
6:     $\iota.pers \leftarrow$ false
7:     $\mathcal{R} \leftarrow \{\iota\}$
8:     $\mathcal{S} \leftarrow \{\iota\}$

9:     **while** $\mathcal{R} \neq \emptyset$ **do**

10:       **for** $s \in \mathcal{R}$ **do**

11:         **if** $s.anchor = 2$ **then**
12:           $stub^E \leftarrow \mathsf{stub}^E_{\{S,V,L\}}(s)$ ▷ Expand anchor discovered twice via regress edge
13:         **else**
14:           $stub^E \leftarrow \mathsf{stub}^E_{\{V,L1\}}(s)$
15:         **end if**

16:         $\mathcal{Q}.$Enqueue($\{\,(s,t) \mid t \in stub^E\,\}$)

17:       **end for**

18:       $\mathcal{R} \leftarrow \emptyset$
19:       SweepIteration()

20:     **end while**

21: **end procedure**

22: **procedure** SweepIteration()

23:     **while** $\mathcal{Q} \neq \emptyset$ **do**   ▷ Explore as long as there are unprocessed states
24:       $\psi_l \leftarrow \mathcal{Q}.$minProgress()   ▷ Progress value for the layer to be explored
25:       ExploreLayer(ψ_l)
26:     **end while**

27: **end procedure**

---

current layer. ll. 7–13 now make sure that when we are about to move into the next layer, we perform a check to ensure that conditions **S** and **L2** are enforced by checking the local terminal SCCs and the local cycle of the layer. If needed, we augment the stubborn set used in the current layer and add additional state transition pairs to be processed. Any new state and transition pairs added will effectively postpone moving into the next layer until we again are about to leave the current layer at which point the same checks will be performed again by ll. 7–13. We assume that procedure TERMINALLOCALSCCS in l. 8 computes the terminal local SCCs in the current layer, and that ENFORCECONDITIONS_S given a terminal SCC computes stubborn sets in the states of the SCC required to ensure condition **S** and return the corresponding pairs of states and enabled transitions. We use $stub_S^E$ to collect the additional state and enabled transitions pairs that may have to be added to ensure that the stubborn sets satisfies condition **S**. Similarly, we assume that procedure LOCALCYCLE in l. 10 computes the local cycles of the current layer, and that ENFORCECONDITION_L2 given a cycle computes stubborn sets in the states of the cycle required to ensure condition **L2** and returns the corresponding pairs of states and enabled transitions. We use $stub_{L2}^E$ to collect any additional states and enabled transitions pairs required by the stubborn sets to satisfy condition **L2**. We provide details on how these can be implemented in practice in the next section.

Algorithm 5 specifies the EXPLORETRANSITION procedure which is a transition-oriented variant of the EXPLORESTATE procedure in Algorithm 2. The algorithm implements the lazy approach to ensure conditions **S** and **L2**. The first time we discover a persistent state s' via a regress edge, we therefore set the *anchor* attribute of s' to 1 in ll. 8–11.

If we discover a persistent state for the second time via a regress edge ll. 16–19, then we mark it as an anchor state and add it to the set of roots for the iteration such that it will be expanded in ll. 11–12 of Algorithm 3 such that conditions **S**, **L1** and **L2** are enforced. In case of discovering a persistent state the second time via a forward edge (ll. 20–23), we insert the s' and the enabled transitions in the stubborn set into the priority queue for processing as they need to be handled in the current sweep iteration.

The correctness of our new algorithm in terms of preserving LTL_{-X} properties is stated in the theorem below. It should be noted that the theorem only concerns preservation of the LTL_{-X} properties and not how the property can be checked. To model check an LTL_{-X} property based upon the reduced state space explored by our new algorithm, one would have to employ for instance the MAP-based algorithm from [5] to perform the actual checking of the property under a sweep-line state space exploration.

Theorem 2. *Let $\mathcal{S} = (S, T, \Delta, \iota)$ be a full state space and let $\psi : S \mapsto O$ be a progress measure. The sweep-line algorithm in specified in Algorithms 3–5 terminates after having explored a reduced state space $\mathcal{S}' = (S', T', \Delta', \iota)$ with $S' \subseteq S$, $T' \subseteq T$, $\Delta' \subseteq \Delta$, and an LTL_{-X} property ϕ holds in $\mathcal{S}$ if and only if ϕ holds in $\mathcal{S}'$.*

Proof. We assume termination of the procedures TERMINALLOCALSCCS, ENFORCECONDITION_S, LOCALCYCLES and ENFORCECONDITION_L2, and the computation of stubborn sets by the functions $\mathsf{stub}_{\{V,L1\}}$ and $\mathsf{stub}_{\{S,V,L\}}$

Algorithm 4. The stubborn set sweep-line ExploreLayer procedure

---

1: States and edges $\mathcal{L} \subseteq S \times T \times S$ ▷ current layer

2: **procedure** ExploreLayer(ψ_l)

3:     $\mathcal{L} \leftarrow \emptyset$

4:     **while** $\mathcal{Q}$.minProgress() $= \psi_l$ **do**   ▷ Explore as long as we are in the current layer

5:        $(s, t) \leftarrow \mathcal{Q}$.dequeue()
6:        ExploreTransition(s, t)

7:        **if** $\psi_l \sqsubset \mathcal{Q}$.minProgress() $\vee \mathcal{Q}$.empty() **then**   ▷ About to complete current layer

8:           $tscc \leftarrow$ TerminalLocalSccs($\mathcal{L}$)   ▷ Check S on Terminal SCCs
9:           $stub_S^E \leftarrow$ EnforceCondition_S($tccs$)

10:          $cycs \leftarrow$ LocalCycles($\mathcal{L}$)   ▷ Check L2 on cycles
11:          $stub_{L2}^E \leftarrow$ EnforceCondition_L2($cycs$)

12:          $\mathcal{Q}$.enqueue($stub_S^E \cup stub_{L2}^E$)   ▷ Explore additional transitions

13:        **end if**

14:     **end while**

15:     $S \leftarrow S \setminus \{\, s \in S \mid (s, t, s') \in \mathcal{L} \wedge \neg s.pers \,\}$ ▷ Delete non-persistent states in layer

16: **end procedure**

---

for which we have not provided a detailed implementation. Termination of the algorithm then follows from termination of the generalised sweep-line algorithm (cf. Theorem 1(1)) and the fact that a persistent state can only be added as an anchor state into roots in ll. 16–19 or be expanded via ll. 20–23 of Algorithm 5 once. A reduced state space is explored since the enabled transitions in the stubborn set in a given state is always a subset of the enabled transition in the state. For preservation of LTL_{-X} we need to argue that conditions **D0-D2, S, V, L1,** and **L2** are satisfied by the reduced state space. The computation of stubborn sets in l. 12 and l. 14 of Algorithm 3, and l. 13 and ll. 21–22 of Algorithm 5 ensures that conditions **D0-D2, V,** and **L1** are guaranteed. Condition **S** is satisfied for intra-layer terminal SCCs via ll. 8–9 in Algorithm 4, and for inter-layer terminal SCCs via detection of anchor states in ll. 16–19 of Algorithm 5 and the computation of stubborn set for anchor states in l. 12 of Algorithm 3. Condition

---

Algorithm 5. The stubborn set sweep-line EXPLORETRANSITION procedure

---

1: **procedure** EXPLORETRANSITION(s, t)

2: $\quad s' \leftarrow \mathsf{succ}(s, t)$

3: $\quad$ **if** $\psi(s) = \psi(s')$ **then** $\qquad\qquad\qquad\qquad\qquad\qquad$ $\triangleright$ edge within this layer
4: $\quad\qquad \mathcal{L} \leftarrow \mathcal{L} \cup \{(s, t, s')\}$
5: $\quad$ **end if**

6: $\quad$ **if** $s' \notin S$ **then**

7: $\quad\qquad S \leftarrow S \cup \{s'\}$

8: $\quad\qquad$ **if** $\psi(s') \sqsubset \psi(s)$ **then** $\qquad\qquad\qquad$ $\triangleright$ Regress edge discovered
9: $\quad\qquad\qquad s'.pers \leftarrow$ true $\qquad\qquad\qquad\qquad$ $\triangleright$ Make s' persistent
10: $\quad\qquad\qquad s'.anchor \leftarrow 1$ $\quad$ $\triangleright$ Potential anchor s' discovered once via a regress edge
11: $\quad\qquad\qquad \mathcal{R} \leftarrow \mathcal{R} \cup \{s'\}$ $\qquad\qquad$ $\triangleright$ s' will be root for a subsequent sweep
12: $\quad\qquad$ **else**
13: $\quad\qquad\qquad \mathcal{Q}.\text{ENQUEUE}(\{(s', t) \mid t \in \mathsf{stub}^E_{\{V, L1\}}(s')\})$
14: $\quad\qquad$ **end if**

15: $\quad$ **else**

16: $\quad\qquad$ **if** $\psi(s') \sqsubset \psi(s) \wedge s'.pers \wedge s'.anchor = 1 \wedge s' \notin \mathcal{R}$ **then**
17: $\quad\qquad\qquad s'.anchor \leftarrow 2$ $\quad$ $\triangleright$ Persistent state s' discovered twice via regress edges
18: $\quad\qquad\qquad \mathcal{R} \leftarrow \mathcal{R} \cup \{s'\}$ $\quad$ $\triangleright$ s' anchor state will be root for a subsequent sweep
19: $\quad\qquad$ **end if**

20: $\quad\qquad$ **if** $\psi(s) \sqsubseteq \psi(s') \wedge s'.pers \wedge s'.anchor = 1$ **then**
21: $\quad\qquad\qquad stub^E \leftarrow \mathsf{stub}^E_{\{S, V, L\}}(s)$ $\qquad\qquad$ $\triangleright$ Persistent state s' discovered twice via forward edges
22: $\quad\qquad\qquad \mathcal{Q}.\text{ENQUEUE}(\{(s, t) \mid t \in stub^E\})$
23: $\quad\qquad$ **end if**

24: $\quad$ **end if**
25: **end procedure**

L2 is satisfied for intra-layer cycles via ll. 10–11 in Algorithm 4, and for inter-layer cycles via detection of anchor states in ll. 16–19 of Algorithm 5 and the computation of stubborn set for anchor states in l. 12 of Algorithm 3.

$\qquad\qquad\qquad\qquad\qquad\qquad\qquad\qquad\qquad\qquad\qquad\qquad\qquad\qquad\qquad\qquad\qquad$ $\square$

The variants of Algorithms 3–5 implementing the suggested eager approach can be obtained by eliminating the use of the anchor attribute on states, and removing the if-then-else statement in ll. 11–15 of Algorithm 3 and keep only the statement in the if-branch such that persistent states are always expanded according to conditions **S,V, L1,** and **L2**.

7   Implementation and Initial Evaluation

We report in this section on results from preliminary experiments undertaken with the Helena verification tool [6] where we have implemented our new algorithm.

Input Models and Progress Measures. We experimented with 14 Petri net instances from the Model Checking Contest [9] database. We focused on instances modeling real life protocols (e.g., coming from the industry) or algorithms (e.g., mutual exclusion algorithms). Progress measures were automatically generated by the Lola tool [17]. These are based on transition invariants following the method described in [14].

Partial Order Reduction Implementation. We focused on the evaluation of the **L2** condition (required for the verification of LTL_{-X} properties). We did not check actual properties, but implemented our partial-order reduction algorithms in such a way that, in the reduced state space, it is guaranteed that at least one fully expanded state lies on each cycle – hence enforcing conditions **D0**, **D1**, **D2** and **L2**. Such a reduction mechanism is denoted by POR^{L2} hereafter. For comparison purposes (i.e., in order to evaluate the cost of cycle detection and transition ignoring prevention) we also evaluated a reduction mechanism, denoted by POR, that only preserves conditions **D0**, **D1**, **D2** (i.e., conditions that can be checked on individual states) for terminal states detection.

It should be noted that guaranteeing that any cycle in the reduced graph contains a fully expanded state also trivially ensures that the **S** condition (i.e., any enabled transition is explored is some future) is enforced. Therefore, our implementation could also be used for the verification of safety properties, although it is clearly not optimal in that context as it could perform undesired full state expansions. Moreover, having each cycle contain a fully expanded state also implicitly assumes all transitions to be visible. Therefore our experiments evaluate our algorithm independently of any particular property and as such can be considered as a worst-case evaluation from the perspective of the **S** condition and the **V** condition.

Our implementation of the algorithm described in this paper combines the priority-first search induced by the sweep-line algorithm (SL) with local depth-first searches (DFS) within the same layer to detect intra-layer cycles. To implement POR^{L2}, we used a method described in [4] for intra-layer cycle detection—more precisely, the ColoredDest algorithm based on the full expansion of destination states of cycle-closing edges.

Moreover, we experimented with two implementations of the algorithm described in this paper. In the first one, POR_e^{L2}, a persistent state is systematicaly fully expanded (the eager approach), while the second, POR_l^{L2}, makes use of the anchor attribute to fully expand persistent states at their second visit only (the lazy approach).

Overall, this means that we have 6 different algorithms implemented in our evaluation: depth-first exploration (DFS) equipped with POR or $\mathsf{POR}^{\mathsf{L2}}$ reduction; and sweep-line exploration without partial-order reduction or equipped with POR, $\mathsf{POR}^{\mathsf{L2}}_{\mathsf{e}}$ (eager approach) or $\mathsf{POR}^{\mathsf{L2}}_{\mathsf{l}}$ (lazy approach) reduction.

Experimental Results. Table 2 summarises our results. For the two runs using DFS, the table gives the number of explored states which coincides with the number of stored states. For each of the four runs using the sweep-line exploration, two numbers are provided: the peak number of stored states and the number of explored states. *OOT* means that the execution used more than the 30 min we used as a timeout.

Table 2. Experimental results for 14 Petri net instances.

Instance	DFS		Sweep-line exploration							
	POR	$\mathsf{POR}^{\mathsf{L2}}$	No POR		POR		$\mathsf{POR}^{\mathsf{L2}}_{\mathsf{e}}$		$\mathsf{POR}^{\mathsf{L2}}_{\mathsf{l}}$	
	explored	explored	peak	explored	peak	explored	peak	explored	peak	explored
aslink(1,a)	999,852	1,023,574	OOT		37,482	2,041,736	40,366	2,216,812	43,877	3,350,000
anderson(4)	12,621	12,621	7,541	72,652	2,355	33,402	2,355	33,568	2,399	46,188
deploy(3,a)	50,593	72,608	127,322	288,471	17,866	75,446	33,261	157,658	31,466	185,449
des(2,a)	71,666	81,238	OOT		2,803	128,601	3,221	183,868	4,275	253,184
discovery(6,a)	68	72,298	1,445,467	2,740,646	22	106	27,902	286,490	22,848	225,564
eisenmcguire(4)	304,178	305,183	998,480	3,048,866	69,836	547,990	92,022	782,193	88,223	1,086,742
firewire(7)	1,662,462	1,715,905	7,185,255	12,559,256	690,547	2,674,127	885,644	3,068,237	837,410	3,946,358
gpufp(4,b)	3,432	4,312	1,422,350	23,026,207	353	5,234	511	8,538	532	13,382
peterson(3)	109,360	133,311	1,724,805	7,852,439	20,567	313,428	32,892	467,491	33,086	636,291
shield(t,iip,1,b)	1,784	1,997	1,423,708	10,136,661	294	5,342	859	18,250	633	12,438
stigcomm(2,b)	377,605	2,781,703	OOT		14,003	548,850	671,234	5,924,879	641,220	7,323,018
stigelec(3,b)	52,513	407,183	2,988,578	51,968,497	4,087	118,857	84,959	1,030,966	84,679	1,374,618
szymanski(2)	28,779	59,203	39,263	162,383	12,113	73,758	21,477	160,284	21,420	254,722
tcp(5)	635,357	806,500	1,478,519	6,930,857	133,580	1,430,837	164,019	1,493,184	157,576	2,734,743

We first observe that the two methods (partial-order and sweep-line) combine nicely. A comparison of algorithms DFS + POR and Sweep-Line + POR reveals that the sweep-line method could further reduce the peak number of stored states produced by partial-order reduction by a factor of 9.88 on the average of the 14 model instances while increasing the number of explored states by a factor of 2.06 on the average.

In order to have a more precise understanding of the cost of transition ignoring prevention required for the verification of LTL_{-X} properties, we give in Table 3 a comparison of reduction POR with respect to $\mathsf{POR}^{\mathsf{L2}}$ for DFS, and with respect to $\mathsf{POR}^{\mathsf{L2}}_{\mathsf{e}}$ and $\mathsf{POR}^{\mathsf{L2}}_{\mathsf{l}}$ for the sweep-line algorithm. Each number is obtained by dividing the number of states (from Table 2) of a run with $\mathsf{POR}^{\mathsf{L2}}$, $\mathsf{POR}^{\mathsf{L2}}_{\mathsf{e}}$ or $\mathsf{POR}^{\mathsf{L2}}_{\mathsf{l}}$ by the corresponding number with the POR setting.

Generally speaking, we see that the cost of transition ignoring prevention with the sweep-line method is comparable to this cost when DFS (which is the standard algorithm when checking LTL_{-X} properties) is used. However, we see three instances for which this assertion does not hold: stigcomm(2,b), stigelec(3,b) and, to a lesser extent, shield(t,iip,1,b). It would be interesting to study the structure of these nets and try to determine what could cause this behaviour. Nevertheless, the trend in these results is generally that although not as efficient as in depth-first search (that revealed to be especially efficient as reported in [7]), the ignoring prevention mechanism proposed in this paper performs quite well and does not cancel the benefits of combining these two methods. A comparison of algorithms DFS + POR^{L2} and Sweep-Line + POR_e^{L2} reveals that, even when transition ignoring is prevented, the sweep-line method can further reduce the memory consumption (i.e., peak number of stored states) produced by partial-order reduction by a factor 6.95 on the average at the cost of an increase of the number of explored states by a factor 2.95 on the average (compared respectively to 9.88 and 2.06 when ignoring is not taken care of).

Table 3. Measuring the cost of transition ignoring prevention

Instance	DFS	Sweep-Line			
	POR^{L2} vs POR	POR_e^{L2} vs POR		POR_I^{L2} vs POR	
	explored	peak	explored	peak	explored
aslink(1,a)	× 1.024	× 1.077	× 1.086	× 1.171	× 1.641
anderson(4)	× 1.000	× 1.000	× 1.005	× 1.019	× 1.383
deploy(3,a)	× 1.435	× 1.862	× 2.090	× 1.761	× 2.458
des(2,a)	× 1.134	× 1.149	× 1.430	× 1.525	× 1.969
discovery(6,a)	× 1063	× 1268	× 2703	× 1039	× 2128
eisenmcguire(4)	× 1.003	× 1.318	× 1.427	× 1.263	× 1.983
firewire(7)	× 1.032	× 1.283	× 1.147	× 1.213	× 1.476
gpufp(4,b)	× 1.256	× 1.448	× 1.631	× 1.507	× 2.557
peterson(3)	× 1.219	× 1.599	× 1.492	× 1.609	× 2.030
shield(t,iip,1,b)	× 1.119	× 2.922	× 3.416	× 2.153	× 2.328
stigcomm(2,b)	× 7.367	× 47.94	× 10.80	× 45.79	× 13.34
stigelec(3,b)	× 7.754	× 20.79	× 8.674	× 20.72	× 11.57
szymanski(2)	× 2.057	× 1.773	× 2.173	× 1.768	× 3.453
tcp(5)	× 1.269	× 1.228	× 1.044	× 1.180	× 1.911

We conclude this section by some observations regarding the benefits of the lazy approach (i.e., relying on the use of the anchor attribute). A comparison of columns POR_e^{L2} and POR_I^{L2} in Table 2 shows that this attribute does unfortunately not bring a significant improvement. It can even, for some instances (e.g., aslink(1,a)), increase the memory consumption while, in general, it tends to increase to number of explored states as it was expected (see Sect. 5.1). We conjecture that this is due to the design of the progress measures we experimented with. The principle of the lazy approache is to avoid the full expansion of persistent states that are not part of a cycle. Since our progress measures are based on transition invariants it is likely that a destination state of a regress edge indeed lies on a cycle meaning that it must anyway be fully expanded (which is done by the eager approach, i.e., POR_e^{L2}, at the first visit). It is therefore worthwhile investigating our algorithm with other progress measures or with models specified in other formalisms that do not have structural analysis tools such as invariants computation.

8   Conclusions and Future Work

We have developed a variant of the sweep-line method such that it can be combined with the LTL$_{-X}$ preserving stubborn set method. The core idea was the local check of terminal SCCs and cycles with each layer combined with using persistent states as anchor states for ensuring the conditions for terminal SCCs and cycles that span multiple layers. We have considered partial-order reduction in stubborn set framework, but our results apply also to other partial-order frameworks such as ample sets [12]. Our algorithm does not address how LTL$_{-X}$ would be model checked during the combined stubborn set and sweep-line state space exploration. To perform model checking, one would have to run a sweep-line compatible LTL model checking algorithm such as the MAP-based algorithm of [5] on top of our proposed algorithm.

A direct perspective is to perform a more thorough experimentation of our algorithm. First, although we have seen that our implementation could be used for the verification of safety properties (as it ensures both the **L2** and **S** conditions), it is not optimal in that context. We thus plan to experiment with a less conservative approach specifically tailored for the **S** condition. Second, we have seen that our transition ignoring mechanism is not suited for some model instances. We plan to further investigate these to identify where this inadequacy comes from. Along these lines, we plan to evaluate our algorithm with other progress measures or modelling languages to identify whether our conjecture regarding the anchor attribute is valid and identify the benefit of using this attribute in that context.

We have not considered the use of persistent predicates in our approach. This is an addition to the sweep-line method where a predicate is used to mark states as persistent the first time they are discovered. A persistence predicate can be used to reduce the re-exploration of states. To use persistent predicates in the context of our algorithm, we would have to make it an anchor state the first time that a persistent state is rediscovered via a regress edge. It can also be observed that some cycles in the state space may contain multiple persistent states and that it is not required to fully expand all in order to enforce conditions **S** and **L2**. As part of future work, it may be investigated whether the concept of maximal persistent predecessor inspired from [5] can be used to address this aspect.

The proposed LTL$_{-X}$ stubborn set sweep-line method has addressed an open problem within the sweep-line state space exploration framework. While we are now able to perform LTL model checking, it remains an open problem how to perform full CTL and CTL$_{-X}$ with the sweep-line method. The challenge here is that the CTL model checking would have to be done following the least-progress first exploration order of the sweep-line method while current algorithms for explicit-state CTL model checking are based on backwards traversal. Some initial algorithms for a fragment of CTL were investigated in [13], but they do not cover full CTL model checking.

References

1. Baier, C., Katoen, J.: Principles of model checking. MIT Press (2008)
2. Brim, L., Černá, I., Moravec, P., Šimša, J.: Accepting Predecessors Are Better than Back Edges in Distributed LTL Model-Checking. In: Hu, A.J., Martin, A.K. (eds.) FMCAD 2004. LNCS, vol. 3312, pp. 352–366. Springer, Heidelberg (2004). https://doi.org/10.1007/978-3-540-30494-4_25
3. Christensen, S., Kristensen, L.M., Mailund, T.: A sweep-line method for state space exploration. In: Tools and Algorithms for the Construction and Analysis of Systems, 7th International Conference, TACAS. Lecture Notes in Computer Science, vol. 2031, pp. 450–464. Springer, Cham (2001). https://doi.org/10.1007/3-540-45319-9_31
4. Duret-Lutz, A., Kordon, F., Poitrenaud, D., Renault, E.: Heuristics for checking liveness properties with partial order reductions. In: Artho, C., Legay, A., Peled, D. (eds.) Automated Technology for Verification and Analysis - 14th International Symposium, ATVA 2016, Chiba, Japan, October 17–20, 2016, Proceedings. Lecture Notes in Computer Science, vol. 9938, pp. 340–356 (2016). https://doi.org/10.1007/978-3-319-46520-3_22
5. Evangelista, S., Kristensen, L.M.: A sweep-line method for büchi automata-based model checking. Fundam. Informaticae **131**(1), 27–53 (2014). https://doi.org/10.3233/FI-2014-1003
6. Evangelista, S.: High Level Petri Nets Analysis with Helena. In: Ciardo, G., Darondeau, P. (eds.) ICATPN 2005. LNCS, vol. 3536, pp. 455–464. Springer, Heidelberg (2005). https://doi.org/10.1007/11494744_26
7. Evangelista, S.: Experimenting with stubborn sets on petri nets. In: Gomes, L., Lorenz, R. (eds.) Application and Theory of Petri Nets and Concurrency - 44th International Conference, PETRI NETS 2023, Lisbon, Portugal, June 25-30, 2023, Proceedings. Lecture Notes in Computer Science, vol. 13929, pp. 346–365. Springer, Cham (2023). https://doi.org/10.1007/978-3-031-33620-1_19
8. Holzmann, G.J., Peled, D.A., Yannakakis, M.: On nested depth first search. In: The Spin Verification System, Proceedings of a DIMACS Workshop. DIMACS Series in Discrete Mathematics and Theoretical Computer Science, vol. 32, pp. 23–31. DIMACS/AMS (1996). https://doi.org/10.1090/DIMACS/032/03
9. Kordon, F., et al.: Complete Results for the 2025 Edition of the Model Checking Contest (2025). https://mcc.lip6.fr/2025/results.php
10. Kristensen, L.M., Mailund, T.: A generalised sweep-line method for safety properties. In: FME 2002: Formal Methods - Getting IT Right, International Symposium of Formal Methods Europe. Lecture Notes in Computer Science, vol. 2391, pp. 549–567. Springer, Cham (2002). https://doi.org/10.1007/3-540-45614-7_31
11. Peled, D.A.: Ten years of partial order reduction. In: Computer Aided Verification, 10th International Conference, CAV. Lecture Notes in Computer Science, vol. 1427, pp. 17–28. Springer, Cham (1998). https://doi.org/10.1007/BFB0028727
12. Peled, D.: All from one, one for all: on model checking using representatives. In: Courcoubetis, C. (ed.) CAV 1993. LNCS, vol. 697, pp. 409–423. Springer, Heidelberg (1993). https://doi.org/10.1007/3-540-56922-7_34
13. Rodríguez, A., Kristensen, L.M., Rutle, A.: Verification of the MQTT iot protocol using property-specific CTL sweep-line algorithms. Trans. Petri Nets Other Model. Concurr. **15**, 165–183 (2021). https://doi.org/10.1007/978-3-662-63079-2_8
14. Schmidt, K.: Automated generation of a progress measure for the sweep-line method. Int. J. Softw. Tools Technol. Transf. **8**(3), 195–203 (2006). https://doi.org/10.1007/S10009-005-0201-1

15. Valmari, A.: Stubborn sets for reduced state space generation. In: Rozenberg, G. (ed.) ICATPN 1989. LNCS, vol. 483, pp. 491–515. Springer, Heidelberg (1991). https://doi.org/10.1007/3-540-53863-1_36
16. Valmari, A.: The state explosion problem. In: Reisig, W., Rozenberg, G. (eds.) ACPN 1996. LNCS, vol. 1491, pp. 429–528. Springer, Heidelberg (1998). https://doi.org/10.1007/3-540-65306-6_21
17. Wolf, K.: Petri Net Model Checking with LoLA 2. In: Khomenko, V., Roux, O.H. (eds.) PETRI NETS 2018. LNCS, vol. 10877, pp. 351–362. Springer, Cham (2018). https://doi.org/10.1007/978-3-319-91268-4_18

Conformance Checking for Partially Ordered Event Logs Using Token-Based Replay

Sabine Folz-Weinstein[1]([✉]) [iD], Michael Gößwein[1] [iD], Christian Beecks[1] [iD], and Robin Bergenthum[2] [iD]

[1] Chair of Data Science, FernUniversität in Hagen, Hagen, Germany
`sabine.folz-weinstein@fernuni-hagen.de`
[2] Faculty of Mathematics and Computer Science, FernUniversität in Hagen, Hagen, Germany

Abstract. Conformance checking aims to quantify process compliance by comparing an event log and a reference model. While event logs have traditionally been represented as sequential traces, real-life processes often require partially ordered traces to accurately reflect causal dependencies, concurrency, durations, overlapping, or uncertainty in the observed behavior. Consequently, there is a growing need for algorithms that can directly handle partially ordered input. In conformance checking for partially ordered input, two different paradigms exist. The goal is either to evaluate whether at least *one* sequential execution order of a partially ordered trace conforms to the model (*uncertain* semantics). This is useful, e.g., in case of quality issues of the event log, where the main goal is to evaluate whether the observed behavior can be explained by the model. Alternatively, the goal is to evaluate whether *all possible* execution orders of a partially ordered trace conform to the model, including concurrent executions of unordered events (*certain* semantics). This is important in use cases where some execution orders may be harmful and guarantees are required, or where concurrency of events is relevant.

So far, research has primarily focused on *uncertain* semantics. To close this gap, in this paper, we introduce a token-based replay conformance checking approach for *certain* semantics, which evaluates the full behavior defined by a partially ordered trace or event log, and calculates a fitness value in polynomial time. We additionally propose a fast variant that calculates lower and upper fitness bounds in linear time.

Keywords: Business Process Modeling · Conformance Checking · Event Data · Partial Orders · Workflow Nets

1 Introduction

The goal of conformance checking is to evaluate and quantify process compliance by comparing an event log and a reference process model [2,3,18]. To illustrate

© The Author(s), under exclusive license to Springer Nature Switzerland AG 2026
J. Desel and A. Kalenkova (Eds.): PETRI NETS 2026, LNCS 16567, pp. 200–222, 2026.
https://doi.org/10.1007/978-3-032-27879-1_10

this, consider the following (simplified) example of an emergency care process in a hospital (Fig. 1). In the reference model of the process, represented by a Petri net, an *Initial Examination* is conducted, then an arbitrary number of *Blood* samples are taken and *Medication* given, which may be done in any order but not in parallel. Independently of these, the *Temperature* of the patient is measured, which can be done in parallel or in any order to *Medication* and *Blood* samples. Ultimately, a *Final Examination* is conducted.

Figure 2 shows an example event log, where each line represents an event. If we aggregate by patient ID, we identify sequences of events which represent the treatment of one patient, so-called traces. Traditionally, the events of a trace are totally ordered by their timestamps. However, in our example, the available timestamp information has different granularity. Data quality issues like these are very common in real-life event logs [8, 17, 25]. Here, we are not sure where to order *Medication*, which only has a date, with respect to *Initial Examination*, *Blood* and *Temperature*, as any sequential order would be arbitrary. To leave such events unordered, we need a partially ordered representation (right).

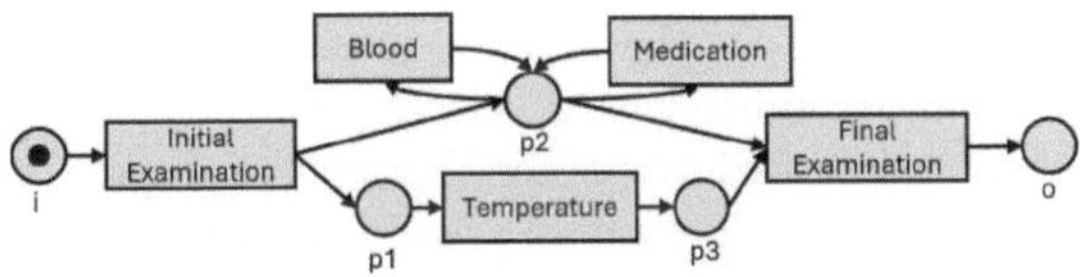

Fig. 1. A Petri net as reference model of an emergency care process in a hospital.

Let us consider that in another hospital, the personnel is equipped with mobile devices to get reliable log information including possible durations, i.e. a start and completion time (Fig. 3) where applicable. Our example shows that the *Initial Examination* was started and completed, then a *Medication* was started, in parallel, a *Blood* sample was taken and *Temperature* was measured, the *Medication* was completed, and the *Final Examination* was started and completed. Very frequently, such valuable additional information on the actual causal relations or concurrencies of events is available in real-life use cases (e.g., data flows, domain knowledge, lifecycle information) [8, 32]. Also in this case, we need a partially ordered representation (Fig. 3, right) to express that *Medication* happened concurrently to *Blood* sample and *Temperature* but before *Final Examination*, because a sequential order would add arbitrary order relations (e.g., *Medication* before *Initial Examination*), for which there is no evidence in the log.

Both examples illustrate that the use of sequential traces may unintentionally introduce order relations between events that are purely coincidental, while possibly losing valuable process information that is available but cannot be represented. To directly express concurrency, duration, overlapping or uncertainty [16, 27], and depict the underlying real-life process more precisely, an increasing amount of work therefore represents traces as partial orders of events [4, 11, 13, 20].

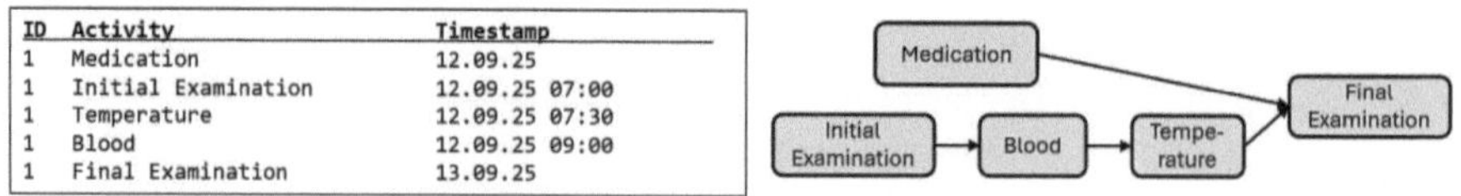

Fig. 2. Extract of an event log with data quality issues (left) and the partially ordered trace depicting the treatment of patient ID 1 (right).

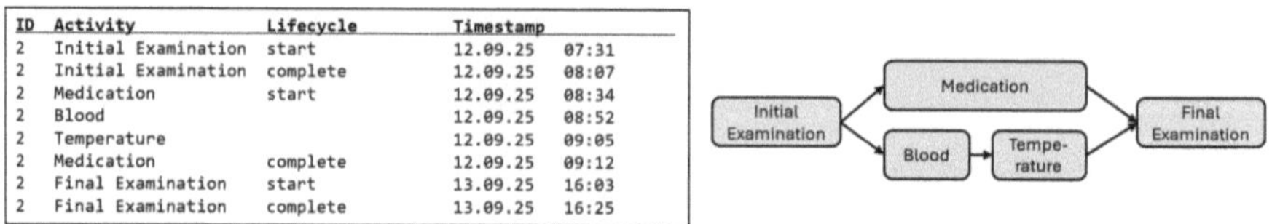

Fig. 3. Extract of an event log with lifecycle information (left) and the partially ordered trace depicting the treatment of patient ID 2 (right).

Our goal in conformance checking is to evaluate for every trace of the event log whether it is executable in the model, or to quantify the deviation. For sequential traces, this is straightforward, as every trace represents one sequential execution order. A partially ordered trace, however, formally represents several possible execution orders. Considering all possible permutations of the unordered events, it represents several sequential execution orders (so-called sequentializations or interleavings). Considering, in addition, all possible combinations where unordered events are executed simultaneously, there are even more execution orders. As we have seen in our examples, a partially ordered trace as such does not show the reason why some of the events are not ordered. This reason depends on the use case. Likewise, the question that we want to answer in conformance checking, i.e., how to evaluate whether a partially ordered trace conforms to the model, depends on the use case. Typically, two questions are addressed in conformance checking for partially ordered traces.

The first question is to evaluate whether the model allows at least *one of the sequential execution orders* represented by a partially ordered trace. For the partially ordered traces in both examples in Fig. 2 and 3, the sequential execution order *Initial Examination - Medication - Blood - Temperature - Final Examination* is executable in the model. Thus, both partially ordered traces would be classified as conforming. This interpretation of a partially ordered trace is called *uncertain* semantics [20]. It is, e.g., useful in use cases with event log data quality issues, and the main goal is to evaluate whether an "imprecise" observation can be explained by the reference model or is clearly deviating. Practically all existing conformance checking algorithms for partially ordered input are based on this semantics. Typically, the (best fitting) sequential execution order is calculated using alignments, i.e., by finding a shortest path in the reachability graph of the so-called synchronous product, which combines the process model and a net-based representation of the partially ordered trace [23,28,30]. Alternative approaches focus on specific representations, e.g., scheduled processes [29] or planning algorithms [22], consider probabilities of the execution orders, or compute bounds for the conformance values [21,26,31].

However, to determine and quantify an accordance or deviation with respect to the model based on one single sequential execution order of a partially ordered trace is a best-case assumption and ignores potential deviations of all other possible execution orders. In our example in Fig. 2, it is possible that *Initial Examination* occurred before *Medication*, which is conforming, but it is also possible that a *Medication* was given before or in parallel to *Initial Examination* (or in parallel to *Blood* sample) which would be deviating from the reference model. Thus, even in use cases where we are not sure ("uncertain") about the actual execution order of events due to log quality issues, it may not be enough to identify one executable sequential order and just ignore all others. In addition, *uncertain* semantics still inherently impose sequential structures on the partially ordered input, and cannot detect issues related to concurrent executions of events. This is especially relevant in use cases where the events of a process have a duration. In our example in Fig. 3, we have reliable information that the *Blood* sample was taken during *Medication*, and this concurrent execution is not allowed in the model, while both sequential execution orders (*Blood* sample is taken before or after *Medication*) are executable. In both our hospital use cases, all these deviations are important to be detected and taken into account. Giving medication before the initial examination may mask symptoms. Similarly, a blood sample taken during a medication, e.g. an infusion, may lead to wrong analysis results. Both can lead to a wrong diagnosis and treatment, and a thorough review on the process is required to ensure future process compliance.

Thus, the second question addressed by conformance checking for partially ordered input is to evaluate whether the model allows *all execution orders* represented by the partially ordered trace (and quantify any deviations). This interpretation of a partially ordered trace is called *certain* semantics in the process mining community and conforms to the formal interpretation of behavior represented by a partial order in concurrency theory (*interleaving* referring to all sequential execution orders, *true concurrent* to all execution orders including simultaneous executions of unordered events, where both can infer a potentially exponential number of different execution variants [20]). This approach is important in use cases where, e.g., some execution orders are harmful, where we must guarantee to assess the full behavior represented by the partially ordered trace, or where concurrent executions of events are relevant. Only *certain* semantics leverage the full potential of a partially ordered representation.

So far, no conformance checking algorithm for *certain* semantics exists. The partial-order alignments recently proposed by Geurtjens and Lu [14] aim to take concurrency into account using directed Petri net unfoldings, and use heuristics to cope with state-space explosion. However, this approach does not evaluate the full behavior defined by a partially ordered trace (or log), it is computationally very expensive, and it does not quantify deviations in terms of a fitness value.

To close this gap, we propose fitness metrics and a conformance checking algorithm for *certain* semantics. The proposed fitness metrics are based on the concept of token-based replay, a well-established fitness measure. We adapt an existing verification algorithm [6] from Petri net theory to compute the fitness of a workflow net with respect to a partially ordered event log in polynomial time. Additionally, we introduce a fast algorithm variant which has linear runtime and computes both a lower and an upper bound for the fitness. We implemented and evaluated both algorithm variants using public event logs.

The remainder of this paper is structured as follows. After introducing the necessary preliminaries, we define fitness for partially ordered traces. We then present the algorithm core, formulated as a maximum flow problem. Based on this core, we introduce our efficient token-based replay approach and a fast variant computing upper and lower fitness bounds. Finally, we evaluate the proposed algorithms and conclude the paper.

2 Preliminaries

We model observed behavior as partially ordered traces and event logs [3].

Definition 1 (Partially Ordered Trace). *A labeled partial order (lpo) is a triple $(V, \prec, l)$ where V is a finite set of events, $\prec \subseteq V \times V$ is a transitive and irreflexive relation, and l is a labeling function. Let A be a set of activities and let $<^*$ denote the transitive closure of a relation $<$. A triple $(V, <, l)$ is a partially ordered trace if $(V, <^*, l)$ is a labeled partial order and the labeling function $l\colon V \to A$ assigns an activity label to every event.*

A partially ordered event log is a multiset of partially ordered traces.

If each partially ordered trace of an event log is a total order, then the above definition coincides with the classical definitions of traces and event logs.

Distributed systems are usually represented by marked Petri nets [10].

Definition 2 (Marked Petri Net). *A Petri net is a tuple (P, T, W) where P is a finite set of places, T is a finite set of transitions such that $P \cap T = \emptyset$ holds, and $W\colon (P \times T) \cup (T \times P) \to \mathbb{N}$ is a multiset of arcs. Let $t \in T$ be a transition. We call $\bullet t = \sum_{p \in P} W(p, t) \cdot p$ the preset and $t \bullet = \sum_{p \in P} W(t, p) \cdot p$ the postset of t. A marking of (P, T, W) is a multiset $m\colon P \to \mathbb{N}$. Let m_0 be a marking, then we call $N = (P, T, W, m_0)$ a marked Petri net and m_0 the initial marking of N.*

Petri nets have a simple firing rule. A transition t is enabled (can fire) in marking m if $m \geq \bullet t$ holds. Firing a transition changes the marking from m to $m' = m - \bullet t + t \bullet$. Repeatedly processing the firing rule produces so-called firing sequences.

Concurrent behavior cannot be expressed using firing sequences. Therefore step semantics [16], process net semantics [15], or tokenflows [7] are used in Petri net theory to define whether a partially ordered trace or lpo is "executable" (enabled) in a Petri net. Currently, compact tokenflows offer the most efficient algorithmic approaches and also use the concept of initial and final markings. A tokenflow is a distribution of tokens on the arcs and events of a partially ordered trace. Such a distribution is valid for a marked Petri net if it satisfies certain conditions [7]: (i) every event receives enough tokens, (ii) no event must pass on too many tokens, and (iii) the initial marking is not exceeded.

Definition 3 (Enabled Partially Ordered Trace). *Let (P, T, W, m_0) be a marked Petri net and $(V, <, l)$ be a partially ordered trace such that $l(V) \subseteq T$ holds. A tokenflow is a function $x \colon (V \cup <) \to \mathbb{N}$. x is valid for $p \in P$, if and only if, the following conditions hold:*

i. $\forall_{v \in V} \colon x(v) + \sum_{v' < v} x(v', v) \geq W(p, l(v))$
ii. $\forall_{v \in V} \colon \sum_{v < v'} x(v, v') \leq x(v) + \sum_{v' < v} x(v', v) - W(p, l(v)) + W(l(v), p)$
iii. $\sum_{v \in V} x(v) \leq m_0(p)$

A partially ordered trace is enabled in a marked Petri net if and only if there exists a valid tokenflow for every place.

Within process mining, the above definition of an enabled partially ordered trace corresponds to the interpretation of the behavior represented by a partially ordered trace in *certain* semantics.

In *uncertain* semantics, which is very common in the process mining community, an alternative interpretation of the behavior represented by a partially ordered trace is used, and thus, an alternative definition whether a partially ordered trace is "executable" in a Petri net is required. In this paper, we call such partially ordered traces explainable.

Definition 4 (Explainable Partially Ordered Trace). *Let $M = (P, T, W, m_0)$ be a marked Petri net and $(V, <, l)$ be a partially ordered trace such that $l(V) \subseteq T$ holds. $(V, <, l)$ is explainable in M if and only if a total order $\ll$ exists, such that $< \subseteq \ll$, and $(V, \ll, l)$ is a firing sequence of M.*

Reference models of business processes are usually represented by workflow nets [1], which are a restricted subclass of marked Petri nets.

Definition 5 (Workflow Net). *A workflow net N is a tuple (P, T, W, i, o) where (P, T, W, i) is a marked Petri net, $W \colon (P \times T) \cup (T \times P) \to \{0, 1\}$ is a set of arcs, $i, o \in P$ are two places such that i has no incoming arc and o has no outgoing arc, where every node $n \in (P \cup T)$ is on a directed path from i to o. We call o the final marking of N.*

3 Fitness of a Partially Ordered Trace

The goal of this work is to introduce a fitness measure and a conformance checking algorithm for partially ordered traces in *certain* semantics. The key notion for this measure is the definition of an enabled partially ordered trace. Here, we build on the core concept of tokenflows, understood as token distributions over a partially ordered trace, and adapt it to workflow nets. This adaptation allows us to characterize the enabledness of a partially ordered trace and, at the same time, to quantify deviations for partially ordered traces that are not enabled.

As outlined in the introduction, alignment-based techniques are not suitable for conformance checking in *certain* semantics, as they focus on evaluating whether a partially ordered trace is explainable in a given workflow net by identifying one (best possible) sequentialization of the partial order that is a firing sequence. In *certain* semantics, by contrast, we evaluate the full behavior represented by a partially ordered trace. Therefore, we follow a token-based replay approach. As a consequence, our approach inherits strengths and limitations of token-based replay, which have been extensively discussed in the literature [3,9].

In classical token-based replay, all sequential traces of an event log are replayed in the workflow net. If an event cannot be executed because a place in the preset of the related transition does not carry a token, we count a so-called missing token, add the token and fire the transition. Tokens that are not consumed after the trace has been fully replayed (except for the expected final token on place o) are called remaining tokens. The token replay fitness [2] relates the number of missing and consumed tokens to the number of remaining and produced tokens. If a trace is a firing sequence of the workflow net, there are no missing and remaining tokens and the fitness equals 1.

Definition 6 (Token Replay Fitness). *Let s be a trace and let N be a workflow net. Then we define $fitness(s, N) = 1 - \frac{1}{2}\left(\frac{missing(s,N)}{consumed(s,N)} + \frac{remaining(s,N)}{produced(s,N)}\right)$.*

The fitness value $fitness(L, N)$ of an event log L is defined analogously, taking into account the number of produced, consumed, missing, and remaining tokens across all traces, weighted by the respective trace multiplicities.

Token-based replay for sequential traces proceeds step by step, transitions are fired one after another, which makes checking token availability and identifying remaining tokens at the end straightforward. This is no longer possible for a partially ordered trace in *certain* semantics, as unordered events may occur in any order or concurrently, and multiple events can produce and consume tokens. Consequently, enabledness must be evaluated with respect to all orderings (and including concurrent executions) consistent with the partial order, i.e., we must evaluate whether in all possible cases, all required tokens are available before they are needed. To generalize token-based replay to partially ordered traces, we identify consuming and producing events for each place of the workflow net and determine how tokens can be distributed between them. We identify matching pairs of events, such that each pair consists of a producing event and a consuming event, and the producing event must be ordered before the consuming event in the partial order. Since multiple producing and consuming events may exist, several possible distributions of tokens can exist for every place of the net. We formalize the notion of token distributions for a workflow net as follows.

Definition 7 (Token Distribution). *Let $N = (P, T, W, i, o)$ be a workflow net, $p \in P \setminus \{i, o\}$ a place, and $tr = (V, <, l)$ a partially ordered trace. We denote*

$In(p) := \{v \in V | (l(v), p) \in W\}$ the set of events producing a token in p and $Out(p) := \{v \in V | (p, l(v)) \in W\}$ the set of events consuming a token from p. Replaying the partially ordered trace in any order will produce $|In(p)|$ tokens in p and consume $|Out(p)|$ tokens from p. Let $<^*$ be the transitive closure of $<$. We define the set of possible matches $M(p) := \{(u, v) | u \in In(p), v \in Out(p), u <^* v\}$. We define a distribution of tokens $D(p) \subseteq M(p)$ as a set of matches such that for all $(u, v), (u', v') \in D(p)$: $u \neq u'$ and $v \neq v'$ holds.

Note that the sets of producing and consuming events are not necessarily disjoint since there may be short loops (i.e., transitions that are connected to a place by both an ingoing and outgoing arc) in a workflow net. However, since a partial order is irreflexive, no event can consume a token that it produced itself. As workflow nets have no arc weights, each event can produce and/or consume at most one token.

Token distributions are a well-known concept in Petri net verification. Token-flows define token distributions on the arcs (and events) of a partially ordered trace. A partially ordered trace is enabled in a workflow net if the induced token distributions constitute a valid tokenflow. Since workflow nets are a restriction of marked Petri nets, all conditions of Def. 3 must be satisfied for a valid tokenflow: All events that require a token to fire must receive a token (i), only produced tokens may be passed on (ii), and in a workflow net, there is no initial marking available on inner places of the workflow net to be consumed (iii). In addition, all tokens that are produced must be consumed, as no final marking is allowed on inner places either. Consequently, for workflow nets, a valid tokenflow is equivalent to a perfect matching of consuming and producing events [7,13].

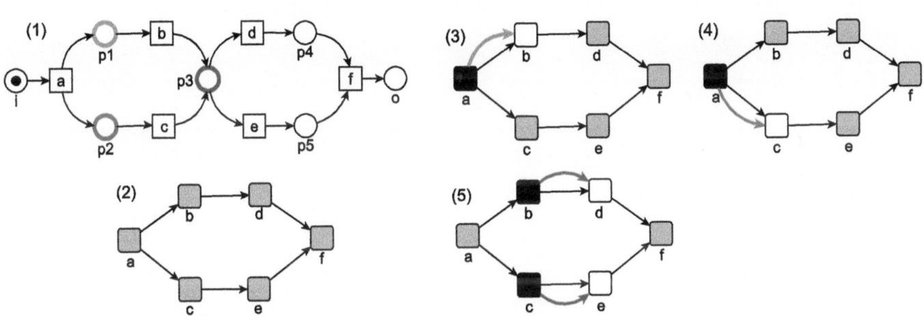

Fig. 4. A workflow net (1), a partially ordered trace (2), and token distributions on the partially ordered trace with respect to *p1* (3), *p2* (4) and *p3* (5).

Example 1. Consider the workflow net (1) and the partially ordered trace (2) in Fig. 4. To highlight producing and consuming events, producing events are marked black and consuming events marked white, and a match by a colored directed transitive arc in the same color as the considered place. (3) shows the token distribution for *p1*, where *a* produces a token and *b* consumes a token, which is the only possible match. (4) shows the token distribution for place $p2$, where the only match is from a to c. In the token distribution of (5) for place $p3$, two matches (b, d) and (c, e) exist, which is also the only possible token distribution. All distributions represent valid tokenflows, since all required tokens are produced, and all produced tokens can be consumed. Not depicted in the illustration are $p4$ and $p5$, for both of which token distributions exist where all produced tokens are consumed. Since also the token from i is consumed by a and the token on o produced by f, perfect matchings exist for all places of the net and the partially ordered trace is enabled in the workflow net.

For some places, several token distributions may exist on a partially ordered trace, with different numbers of matched and unmatched events, and it may be the case that none of them is a perfect matching. In such cases, we aim to quantify the deviation by counting missing and remaining tokens.

Verification algorithms calculate the best possible token distribution and evaluate whether it is a valid tokenflow. Likewise, classical token-based replay counts missing tokens only if they are truly missing, i.e., every available token will be consumed and not simply ignored. Thus, for partially ordered traces, a fitness measure must be based on a token distribution that maximizes the number of matches between producing and consuming events. We call such a distribution the maximal token distribution. We call a perfect matching a valid token distribution. A valid token distribution directly coincides with a valid tokenflow.

Definition 8 (Maximal Token Distribution). *Let $N = (P, T, W, i, o)$ be a workflow net, $p \in P \setminus \{i, o\}$ be a place, tr be a partially ordered trace, and $\Delta(p)$ be the set of all possible token distributions for p. We call a token distribution $D(p) \in \Delta(p)$ maximal if for all $D'(p) \in \Delta(p) : |D(p)| \geq |D'(p)|$ holds. We call a token distribution valid, if $|D(p)| = |Out(p)| = |In(p)|$ holds.*

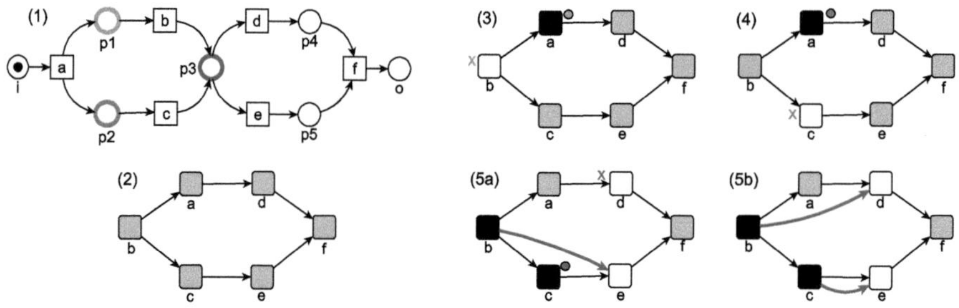

Fig. 5. A workflow net (1), a partially ordered trace (2), and token distributions for this partially ordered trace with respect to $p1$ (3), $p2$ (4) and $p3$ (5a and 5b) of the workflow net.

Example 2. Consider the partially ordered trace and workflow net in Fig. 5, which are the same as in Fig. 4, except that b and a are switched in the partially ordered trace. Producing events are again marked black and consuming events white, a match by a colored directed transitive arc, a remaining token by a dot and a missing token by an x (color referring to the evaluated place). For this partially ordered trace, no matches exist for $p1$ (3) nor for $p2$ (4), as a produces a token but occurs later than b, and a is not ordered to c. For $p3$, the set of possible matches is $\{(b,e),\ (b,d),\ (c,e)\}$, such that two different token distributions exist: either $\{(b,e)\}$, which leaves c and d unmatched as they are not in an order relation (5a), or $\{(b,d),(c,e)\}$ with two matches and no unmatched events (5b), which is consequently the maximal distribution for $p3$.

Based on the notations of Defs. 7 and 8, we define the numbers of produced, consumed, missing, and remaining tokens for a partially ordered trace with respect to a place based on the maximal token distribution. Although the maximal distribution itself may not be unique, the maximal number of matches and the numbers of remaining and missing tokens are uniquely determined.

Definition 9 (Consumed, Produced, Missing and Remaining tokens with respect to a Place). *Let $N = (P,T,W,i,o)$ be a workflow net, tr be a partially ordered trace, $p \in P \setminus \{i,o\}$ be a place, and $D(p)$ be a maximal token distribution. We define the number of consumed, produced, missing and remaining tokens as*

$$consumed(p,tr,N) := |Out(p)|, \qquad missing(p,tr,N) := |Out(p)| - |D(p)|,$$
$$produced(p,tr,N) := |In(p)|,\ and \qquad remaining(p,tr,N) := |In(p)| - |D(p)|.$$

The initial and final places i and o of a workflow net can be evaluated directly based on purely structural properties of the workflow net and the partially ordered trace without using token distributions, since only the initial marking is available on i and only the final marking on o. If there are more than one event in the partially ordered trace that consume tokens from i, there are missing tokens on i; if there is no consuming event, the initial token is remaining. If there are more than one event which produce tokens on o, there are remaining tokens on o; if there is no producing event, the final token is missing.

To define the fitness of a workflow net with respect to a partially ordered trace, we aggregate the numbers of produced, consumed, missing and remaining tokens over all places of N, including i and o.

Definition 10 (Consumed, Produced, Missing and Remaining tokens with respect to a Workflow Net). *Let $N = (P,T,W,i,o)$ be a workflow net, tr be a partially ordered trace, and for every $p \in P \setminus \{i,o\}$: $D(p)$ be a maximal token distribution. We define the number of consumed, produced, missing and remaining tokens as*

$$consumed(tr,N) \quad := 1 + \sum_{p \in P} |Out(p)|$$
$$produced(tr,N) \quad := 1 + \sum_{p \in P} |In(p)|$$
$$missing(tr,N) \quad := \sum_{p \in P \setminus \{i,o\}} missing(p,tr,N) + max(0, |Out(i)| - 1)$$
$$+ max(0, 1 - |In(o)|)$$

$$remaining(tr, N) \ := \ \sum_{p \in P \setminus \{i,o\}} remaining(p, tr, N) + max(0, 1 - |Out(i)|)$$
$$+ \ max(0, |In(o)| - 1)$$

Using the values of Def. 10, Def. 6 can now be applied on partially ordered traces and partially ordered event logs accordingly.

Resuming Example 2, we identify a missing and a remaining token for $p1$, a missing and a remaining token for $p2$, and no missing or remaining tokens on other places. Over all places, eight tokens are produced and eight tokens are consumed, resulting in a fitness value of 0.75.

Finally, since every total order is a special case of a partial order, it is worth noting that for sequential traces, both token-replay fitness measures coincide, because for sequential traces, all token distributions are maximal.

4 Token-Based Replay for Partially Ordered Traces as Maximum Flow Problem

To calculate the fitness value for a partially ordered trace, we must determine a maximal token distribution for each place of the reference workflow net, as introduced in Sect. 4. By definition, this corresponds to matching producing and consuming events such that the number of matches is maximized. This problem can be reduced to a bipartite matching problem, which is well-studied in the literature. In Petri net verification, similar matching problems are commonly addressed by translating them into maximum flow problems [6,7,19]. In the following, we show how maximal token distributions for partially ordered traces can be calculated using a maximum flow-based approach.

Flow Network Construction. To transform a partially ordered trace into its associated flow network, we add a unique source node and a unique sink node. Since it is possible that events of the partially ordered trace are both consuming and producing with respect to the same place of the workflow net (i.e., short loops), we split each event of the partially ordered trace into a consumption node and a production node, such that the consumption precedes production. Consumption and production node of an event are connected by a directed arc. We then connect the source node to the production nodes of all producing events, and the consumption nodes of all consuming events to the sink node by directed arcs. All arcs to and from source and sink have capacity 1, as transitions in a workflow net can produce and/or consume one token. Arcs between events in the partially ordered trace have a capacity equal to the total number of producing events ($|In(p)|$), because produced tokens may accumulate on arcs while flowing through the network.

Example 2 (continued). Resuming our Example 2 and Fig. 5, Fig. 6 (1) depicts the flow network of the given partially ordered trace and place $p3$ of the workflow net. All events of the partially ordered trace are split into a consumption and a production part (adding label-suffixes "_cons" and "_prod"). Again, producing events are marked black and consuming events are marked white. We connect producing events to the source and consuming events to sink node by directed arcs. The arcs from the source and to the sink have a capacity of 1, while all other arcs have a capacity of 2 due to two producing events.

Maximum Flow and Token Distributions. The so-called *flow* is a function assigning non-negative integers to the arcs of the flow network, respecting capacity and flow conservation constraints. Flow conservation requires that the sum of flow reaching a node is equal to the sum of flow leaving a node (except for the source and sink nodes). Flow is only generated at the source and propagated along directed paths until it reaches the sink. In our setting, flow directly corresponds to tokens. The *value* of a flow function equals the total flow reaching the sink and thus represents the number of matched producing and consuming events. Maximizing the flow value therefore corresponds to maximizing the number of matches, i.e., to computing a maximal token distribution as defined in Sect. 4. Since all arcs from source and to sink have capacity 1, the maximal flow value equals the number of saturated sink arcs.

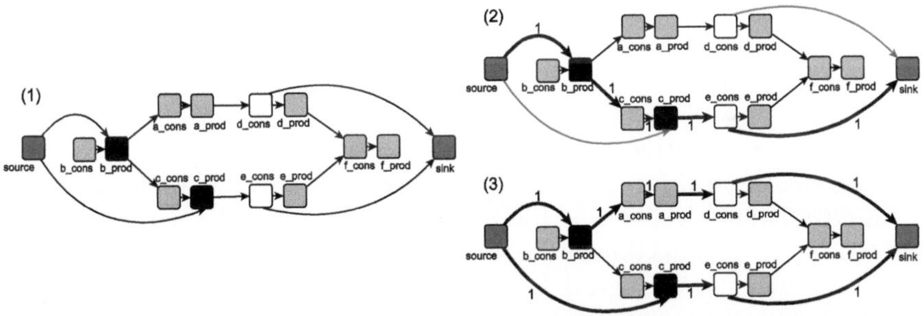

Fig. 6. The associated flow network (1) for the partially ordered trace and *p3* in Fig. 5, a flow (2) and a maximal flow (3), corresponding to the non-maximal and maximal token distributions in Fig. 5 (5a and 5b).

Example 2 (continued). In Fig. 5, (2) and (3) depict the two different possible flows in this flow network. The numbers of tokens, i.e., flow that is passed on from node to node, is noted on the arcs, and the respective arcs are marked in bold (flow of 0 is not noted). The flow in (2) has a value of 1 and directly corresponds to the token distribution in Fig. 5 (5a). Node e_cons can consume one token, i.e., pass on one token to the sink. Due to the flow conservation condition, it can therefore only accept one token on its ingoing arc from c_prod. Since c_cons has already an incoming flow (token) from b_prod, which it cannot consume but must pass on to c_prod, it cannot accept more flow from the source. This means, it cannot produce another token, and the arc from the source must remain unsaturated. Similarly, the arc from d_cons to the sink remains unsaturated, since no flow reaches this node which could be consumed and passed on to the sink. For illustration purposes, we mark unsaturated arcs from the source in blue and unsaturated arcs to the sink in red. The flow in (3) has a value of 2, which is the maximum flow value, as it is higher than the flow in (2), and no other flow is possible. This flow saturates all arcs from the source and to the sink, and corresponds to the maximal token distribution in Fig. 5 (5b). The example illustrates that each unit of flow from source to sink represents a match between a producing and a consuming event, and that maximal flows coincide with maximal token distributions.

Computing Missing and Remaining Tokens. Missing and remaining tokens can be identified directly using the maximal flow value. Missing tokens correspond to unsaturated arcs to the sink and are obtained by subtracting the maximal flow value from the number of consuming events in the partially ordered trace $|D(p)|$, as defined in Def. 9. Analogously, remaining tokens correspond to unsaturated arcs from the source and are obtained by subtracting the maximal flow value from the number of producing events. In our example, the non-maximal flow in Fig. 5 (2) yields one missing and one remaining token. The maximal flow in Fig. 5 (3) yields no missing and no remaining token, and these values are then used for fitness calculation.

Complexity and Runtime. The maximum flow approach guarantees to calculate an exact fitness value for a partially ordered trace and a workflow net. Its runtime is directly determined by the complexity of the maximum flow algorithm. Different algorithms exist to solve maximum flow problems, whereof the Preflow-Push-algorithm is the most prominent. All maximum flow algorithms run in $O(n^3)$ time, where n is the number of events of the partially ordered trace. Although this maximum flow-based approach provides a sound algorithmic core, a polynomial runtime may not be efficient enough for large real-life event logs. Therefore, in the following, we improve the efficiency of our token-based replay algorithm by extending this core approach using concepts of the so-called Firing LPO verification algorithm [6].

5   Efficient Token Replay for Partially Ordered Traces

The main idea of the Firing LPO verification algorithm [6] is to exploit the observation that, for many places of a workflow net, it is easy to determine whether a valid compact tokenflow, i.e., a perfect matching of consuming and producing events, exists for a given lpo. To identify such places, it applies two heuristics steps first, the *forward* and *backward strategies*, which run in linear time. It applies an exact maximum flow algorithm, which explores all possible distributions of tokens, only for those places that cannot be resolved by the heuristics.

In this section, we adapt these heuristics for token-based replay on workflow nets. In contrast to verification, we do not only decide enabledness, but identify maximal token distributions and derive missing and remaining tokens required for fitness calculation. For a fixed place p and a partially ordered trace, the heuristics construct concrete token distributions by simulating one possible totally ordered execution of the partially ordered trace. While these distributions are not guaranteed to be maximal, they provide immediate information about producing and consuming events, as well as about missing and remaining tokens. Based on this information, it is often possible to conclude that no better distribution exists. Only if this cannot be decided by the heuristics steps, a maximum-flow computation is required.

Forward and Backward Strategy. In the forward strategy, we start with an empty marking and traverse the partially ordered trace with respect to the considered place by simply firing all events in one total order which respects the partial order. If an event is consuming, we check whether it receives a token from one of its predecessors in the partially ordered trace. If no token is available, a missing token is counted. If an event produces a token, or if tokens are not needed by the current event, all available tokens are pushed forward by the current event along the arcs of the partially ordered trace to successor events. In case there are tokens available and there are several successors to an event (i.e., a branching), all available tokens are simply pushed to one of the successors. In this case, alternative distributions would be possible, but are not explored by the heuristic. Remaining tokens accumulate on the final event(s).

The backward strategy is symmetric to the forward strategy, but all arcs of the partially ordered trace are inverted. The events are processed in the reverse total order used in the forward strategy, starting with the last event, firing inversely, and also inverting the count of missing and remaining tokens.

Decision Logic. After executing the forward strategy heuristics, we have brute-force constructed one possible token distribution. This distribution may or may not be maximal, but we obtained information on producing and consuming events, as well as on missing and remaining tokens. From the definitions in Sect. 4, it directly follows that the number of matches in a token distribution is bounded by $|D(p)| \leq min(|In(p)|, |Out(p)|)$, since each match requires a producing and a consuming event. Based on the constructed distribution, the following cases can be decided immediately:

1) If no tokens are missing, all consuming events received a token, and no better distribution is possible even if there are remaining tokens. Consequently, the constructed token distribution is maximal.
2) Analogously, if there were no remaining tokens, all produced tokens are consumed, and no better distribution is possible, even if there are missing tokens. Again, the constructed token distribution is maximal.

In these cases, we already found a maximal token distribution and can directly calculate the fitness value. If neither case applies, a better distribution of tokens yielding less missing and remaining tokens may still exist. If so, the backward strategy and eventually an exact maximum flow algorithm is applied.

Example 2 (continued). Consider again the partially ordered trace and place $p3$ of the workflow net in our running example. Figure 7 (top) illustrates the execution of the forward strategy heuristics. Again, we highlight producing events

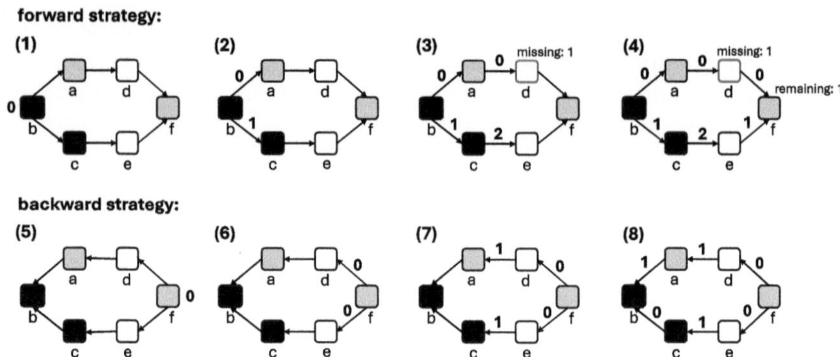

Fig. 7. Forward and backward strategy for the partially ordered place with respect to place *p3* of the workflow net in Fig. 5.

in black and consuming events in white. We use the sequential order *b c a d e f*, which respects the partial order of the events, to traverse the partially ordered trace. In (1), we start with an empty marking, i.e. no available tokens at event *b*. *b* is a producing event, such that now one token is available. After b, there is a branching and two successors to b. All available tokens are always pushed to one of the successors; here, we assume that the available token is pushed to *c* and none to *a* (2). *c* is again producing, adds one token to the token that it received and pushes two tokens to *e*. *a* is neither producing nor consuming on *p3*. It received no token and pushes no token to *d*. *d* is consuming but no token is available, such that we note one missing token (3). *d* can push no tokens to *f*. *e* consumes one of the two tokens that it received from *c*, and pushes one token to *f*. *f* is neither consuming nor producing. It receives no token from *d* and one token from *e*, such that one token is remaining (4). After the forward strategy, we count two consuming events, two producing events, one match, one missing and one remaining token. Due to two producing and consuming events, two matches are possible, which we have not yet met in the forward strategy. Therefore, we execute the backward strategy (bottom). We switch the direction of all arcs, and process the partially ordered trace in the inverse sequential order used in the forward strategy, i.e., *f e d a c b*. We start with no tokens on *f* (5). *f* is neither consuming nor producing and pushes no tokens to either of its successors *e* and *d* (6). *e* produces a token and pushes it to *c*. *d* produces a token and pushes it to *a* (7). *c* consumes the token that it received and pushes no token to *b*. *a* is neither consuming nor producing and pushes the token that it received to *b*. Finally, *b* receives one token from *c* and no token from *a*. It consumes the received token, and no token is remaining (8). In the end, there are no remaining and no missing tokens and two matches. Thus, we found the maximal distribution.

Complexity and Runtime. Algorithm 1 summarizes the efficient, exact token-based replay algorithm. For each place of the workflow net, the start and end

---

Algorithm 1: Exact Token Replay Fitness for a Partially Ordered Trace

---

Data: tr, $N = (P, T, F, i, o)$

1   $con \leftarrow |Out(i)| + 1$
2   $pro \leftarrow |In(o)| + 1$
3   $mis, rem \leftarrow 0$
4   **if** $Out(i) = 0$ **then** $rem \leftarrow rem + 1$
5   **if** $Out(i) > 1$ **then** $mis \leftarrow mis + Out(i) - 1$
6   **if** $In(o) = 0$ **then** $mis \leftarrow mis + 1$
7   **if** $In(o) > 1$ **then** $rem \leftarrow rem + In(o) - 1$
8   **foreach** $p \in P \setminus \{i, o\}$ **do**
9      $con \leftarrow con + |Out(p)|$
10     $pro \leftarrow pro + |In(p)|$
11     $D(p) \leftarrow forwardstrategy(p, tr)$
12   **if** $|D(p)| = min(|In(p)|, |Out(p)|)$ **then** goto 17
13     $D(p) \leftarrow backwardstrategy(p, tr)$
14   **if** $|D(p)| = min(|In(p)|, |Out(p)|)$ **then** goto 17
15     $G(p) \leftarrow bipartiteMatchingProblem(p)$
16     $D(p) \leftarrow maxFlowAlgorithm(G(p))$
17     $mis \leftarrow mis + |Out(p)| - |D(p)|$
18     $rem \leftarrow rem + |In(p)| - |D(p)|$
19 **end**
20 **return** $1 - \frac{1}{2}\left(\frac{mis}{con} + \frac{rem}{pro}\right)$

---

places are evaluated structurally. For all other places, the forward and backward strategies are applied first. Both run in linear time in the length of the sequence used to traverse the partially ordered trace (i.e., number of events of the trace), since each event is processed once like in classical token-based replay. After executing both strategies, we apply our decision logic. In case the current distribution is maximal, no further computation is required, and the runtime per place remains linear. Only if no decision can be made, a maximum flow algorithm with cubic runtime is applied. Based on empirical evaluation results of the original Firing LPO algorithm, which show quadratic or even close-to linear runtime for most use cases, we also expect a close-to linear runtime for our token-based replay algorithm, which would be similar to classical token-based replay. However, since some places may still require an exact maximum flow algorithm, a close-to linear runtime cannot be guaranteed.

6   Fast Variant for Upper and Lower Fitness Bounds

To obtain a fast token-based replay algorithm with guaranteed linear runtime, we do not apply a maximal flow algorithm at all, but only use the information obtained in the two heuristic steps. As a consequence, the resulting algorithm cannot calculate an exact fitness value, but instead derives upper and lower bounds for the number of missing and remaining tokens and, thus, for the fitness

---

Algorithm 2: Token Replay Fitness Interval

Data: tr, $N = (P, T, F, i, o)$

7 $\ldots$

8 **foreach** $p \in P \setminus \{i, o\}$ **do**

9 $\quad$ $con \leftarrow con + |Out(p)|$

10 $\quad$ $pro \leftarrow pro + |In(p)|$

11 $\quad$ **if** $|Out(p)| > |In(p)|$ **then** $mis_{up} \leftarrow mis_{up} + |Out(p)| - |In(p)|$

12 $\quad$ **if** $|Out(p)| < |In(p)|$ **then** $rem_{up} \leftarrow rem_{up} + |In(p)| - |Out(p)|$

13 $\quad$ $D_f(p) \leftarrow forwardstrategy(p, tr)$

14 $\quad$ $D_b(p) \leftarrow backwardstrategy(p, tr)$

15 $\quad$ $mis_{lo} \leftarrow mis_{lo} + |Out(p)| - max(|D_f(p)|, |D_b(p)|)$

16 $\quad$ $rem_{lo} \leftarrow rem_{lo} + |In(p)| - max(|D_f(p)|, |D_b(p)|)$

17 **end**

18 **return** $1 - \frac{1}{2}\left(\frac{mis_{up}}{con} + \frac{rem_{up}}{pro}\right), 1 - \frac{1}{2}\left(\frac{mis_{lo}}{con} + \frac{rem_{lo}}{pro}\right)$

---

value. In scenarios where we are interested in a quick estimate to compare the fitness of several alternative nets, or analyzing very large real-life event logs, this algorithm may be a useful alternative.

We obtain the upper bound of the fitness value by determining the theoretically achievable maximal, i.e. best possible token distribution. We know that the numbers of produced and consumed tokens are independent of a token distribution. We also know that the maximal amount of matches in a token distribution is bound by $min(|Out(p)|, |In(p)|)$. Thus, if there are more consuming events than producing events, we can count the difference as missing tokens in a (theoretically achievable) maximal token distribution, and if there are more producing events than consuming events, we can count the difference as remaining tokens in a (theoretically) maximal token distribution. Aggregated with the values for place i and o, we obtain the numbers of produced, consumed, missing and remaining tokens for calculating the upper fitness bound.

Concerning the lower bound of the fitness value, we apply both heuristics steps and compare the amount of matches in both constructed token distributions, which may both be non-maximal. We use the better one of the constructed token distributions to calculate the numbers of missing and remaining tokens for the lower fitness bound. Note that lower and upper fitness bound are equal in case a maximal distribution is identified by the heuristics.

Algorithm 2 replaces the foreach-loop of Algorithm 1 and calculates the upper and lower fitness bounds of a partially ordered trace with respect to a workflow net by applying only the two heuristics steps. Note that in this algorithm, we use two variables for missing and remaining tokens, one for the upper (mis_{up}, rem_{up}) and one for the lower bound calculation (mis_{lo}, rem_{lo}), and we return two fitness values ($fitness_{up}, fitness_{lo}$).

7   Evaluation

To this date, no public "standard" partially ordered event logs are available yet, and no other conformance checking approaches for *certain* semantics exist. Partial-order alignments aim to take concurrency of events into account, but they do not evaluate the full behavior of a partially ordered event log, nor calculate a fitness value, such that the approaches are not directly comparable. For an overall runtime comparison, note that identifying one optimal partial-order alignment for a partially ordered trace takes seconds to minutes [14].

Therefore, to evaluate our approach in a reproducible way and on a broader basis, we use several well-known public totally ordered event logs for our experiments: the BPI Challenge logs 2012 (a), 2012 (o), 2019 (filtered for the "consignment" trace attribute), road traffic fine management (RTFM), reviewing, and teleclaims [2]. The BPI and RTFM logs are accessible via https://data.4tu.nl.

We transform these event logs into partially ordered event logs using a concurrency oracle [5,20], the so-called Alpha concurrency oracle implemented in the CCO tool[1] [12], which exploits directly-follows relations to identify concurrency. The partially ordered event logs still represent all traces of the original event log (sequential executions of the partial orders), but in addition, they may contain some "concurrent" behavior which cannot be expressed using sequential traces. Note that the partially ordered log versions will also be compressed in size compared to the original versions. For our experiments, we consider the partially ordered version as the reference log.

The first goal of our experiments is to evaluate whether, and how many, deviations can be detected assuming different semantics in conformance checking. The second goal is to assess the general efficiency of our algorithms.

To clearly identify semantic-based differences and to exclude other effects, we assume perfect fitness for one variant in the first two of our experiments.

In our first experiment, we use a workflow net for every test log which, by construction, allows the execution of (at least) one sequentialization per partially ordered trace in the event log, such that any conformance checking approach based on *uncertain* semantics will calculate perfect fitness for this net ("Net pf-unc"). This experiment allows a comparison of *uncertain* and *certain* semantics.

In our second experiment, we use a workflow net for every test event log, which has perfect fitness with respect to the original sequential test event logs ("Net pf-seq"). This experiment evaluates deviations which are only detectable when taking concurrency into account.

In a third experiment, which is independent of semantics, we use a workflow net for every test event log with a lower overall fitness ("Net lf") to evaluate whether the fitness intervals calculated by the fast algorithm variant get broader and whether the runtimes increase in case of a lower fitness value of the test net.

We generated our test workflow nets using the eST Miner process discovery algorithm [24], for the primary reason that this discovery algorithm allows us to balance the fitness of the generated test nets by selecting the fraction of traces of

---

[1] https://github.com/sabinefw/ConfigurableConcurrencyOracleTool.

the event log which must be (at least) executable. To generate the test nets for the first experiment ("Net pf-unc"), we derived an event log from the partially ordered version which contains one random sequentialization of each partially ordered trace. All test data used in the experiments can be found on Github[5]. Note that due to the specific characteristics of the eST Miner, unique start and end events are added to every (partially ordered) trace, also by the algorithm implementation for this evaluation.

Implementation and Experimental Setup. We evaluated both algorithms, ("Exact") and ("Fast"), presented in Sect. 5 and 6 of this paper. Both are implemented in Python, accessible on GitHub[2]. The evaluation was performed single-threaded on a 6-core AMD Ryzen 5 3600 3.60 GHz, 16 GB main memory.

Table 1. Fitness.

Log	Exact			Fast		
	Net pf-unc	Net pf-seq	Net lf	Net pf-unc	Net pf-seq	Net lf
Teleclaims	0.963	1	0.935	[0.963, 0.963]	[0.985, 1]	[0.935, 0.935]
Reviewing	1	1	0.880	[1, 1]	[1, 1]	[0.880, 0.880]
RTFM	0.973	0.973	0.892	[0.972, 1]	[0.972, 1]	[0.892, 0.893]
BPI12(a)	0.955	1	0.915	[0.955, 0.955]	[0.971, 1]	[0.915, 0.915]
BPI12(o)	1	1	0.951	[0.995, 1]	[0.995, 1]	[0.951, 0.951]
BPI19(c)	0.970	0.975	0.969	[0.970, 1]	[0.975, 1]	[0.967, 0.969]

Table 2. Runtime in milliseconds.

Log		Exact			Fast		
Name	Size	Net pf-unc	Net pf-seq	Net lf	Net pf-unc	Net pf-seq	Net lf
Teleclaims	8	0.52	0.97	0.71	0.48	0.78	0.70
Reviewing	93	28.63	23.21	19.70	24.17	23.15	19.43
RTFM	85	35.18	67.31	100.11	3.76	5.35	10.36
BPI12(a)	12	0.39	1.02	0.48	0.36	0.69	0.42
BPI12(o)	75	15.31	22.17	7.60	5.24	6.82	7.40
BPI19(c)	206	43.03	76.87	18.78	4.21	9.91	11.16

Fitness Values. Table 1 depicts the fitness values calculated by the "Exact" algorithm, and the intervals calculated by the "Fast" algorithm for the partially ordered event log version with respect to all test net versions. All values are

---

[2] https://github.com/sabinefw/TokenBasedReplayForPartialOrders.

rounded to three decimals, 1 is precise. For "Net pf-unc" with perfect fitness under *uncertain* semantics, the fitness calculated by the "Exact" algorithm for *certain* semantics is in most cases not 1, which means that the algorithm detected deviations of model and partially ordered event log which approaches based on *uncertain* semantics cannot detect. Also for "Net pf-seq" with perfect fitness for the sequential event log version, we detected cases where the calculated fitness value is not 1, which are deviations only detectable when taking concurrency into account. Depending on the use case, these deviations of model and log may be important to identify and quantify. The fitness values of "Net lf" are only interesting with respect to the fitness intervals calculated by the "Fast" algorithm, which are generally very narrow, and do not broaden with decreasing overall fitness.

Runtime. Table 2 depicts the runtime for the "Exact" and "Fast" algorithm. We include the respective log sizes (amount of traces) as a reference. As the primary goal of our approach was not a speed-up compared to other approaches, but the ability to calculate the fitness of a workflow net and a partially ordered event log in *certain* semantics, this evaluation is mainly done to assess the general efficiency of the new approaches, which is excellent. Note that the runtime of the "Fast" algorithm corresponds to classical token-based replay on sequential traces and event logs. For most event logs, the runtime of the exact approach is not significantly higher than for the fast approach, which shows the evaluation can be done by the heuristics steps for most of the places, and the maximum flow algorithm step must be applied on very few places only. In our experiments, the runtime of the exact token-based replay approach for partially ordered event logs is in fact still linear with a factor or quadratic, although the problem is now cubic, such that both approaches are applicable on real-life event logs. Although the runtime difference of "Exact" and "Fast" is not significant in our experiments, this may be different in other practical use cases.

8   Conclusion

Only partially ordered event logs allow to directly model complex relations of real-life processes like concurrency, duration, and overlap of events, or uncertainty resulting from data quality problems frequently observed in event logs. Consequently, there is a growing need for process mining algorithms which can directly handle partially ordered input. Rethinking existing procedures and overcoming the sequential order assumption is especially relevant in fields such as healthcare, education, and logistics, which exhibit highly concurrent behavior.

In conformance checking for partially ordered event logs, two paradigms exist. Conformance checking under *uncertain* semantics focuses on identifying a sequential execution order of a partially ordered trace which can be executed in (or is "closest" to) the reference model, and by doing so, still inherently imposes sequential interpretations on partially ordered behavior. This interpretation is useful to detect clear deviations, but for the above named reasons it tends to overestimate compliance and it is not able to detect deviations related to concurrent executions. Therefore, for many practical use cases and to fully exploit the benefits of a partially ordered representation, an evaluation of *all* possible behavior represented by a partially ordered trace in *certain* semantics is required. While many different approaches exist for conformance checking under *uncertain* semantics, very little research has been conducted yet on conformance checking under *certain* semantics. In addition to handling concurrency, a potentially vast number of execution variants must be evaluated, such that an efficient runtime is an important additional factor to be coped with in order to make an approach applicable on real-life event data.

To bridge this gap, we introduce a token-based replay algorithm which calculates the fitness of an partially ordered trace or a partially ordered event log under *certain* semantics with respect to a workflow net in polynomial time, and a fast algorithm variant which calculates upper and lower fitness bounds in guaranteed linear time. We conducted several experiments with well-known public event logs that we transformed to partially ordered logs, and different test nets. We were able to identify deviations which can only be detected when evaluating the full behavior represented by a partially ordered trace, and considering concurrent executions. Unless using *certain* semantics, such deviating behavior will be classified as fully conforming. Concerning efficiency, runtimes of milliseconds for both proposed algorithms allow to apply them on large real-life event logs.

In future work, we plan to compare our approach to other new conformance checking algorithms for *certain* semantics, as well as to broaden our evaluation by using (non-public) generically partially ordered event logs as input. Furthermore, since our approach "inherited" all advantages and limitations of classical token based replay, we also plan to address specific issues like handling labeled workflow nets and silent transitions by extending the approach, e.g., using heuristics.

References

1. van der Aalst, W.M.P.: Verification of workflow nets. In: ICATPN. LNCS, vol. 1248, pp. 407–426. Springer (1997)
2. van der Aalst, W.M.P.: Process Mining - Data Science in Action, 2nd edn. Springer (2016)
3. van der Aalst, W.M.P.: Process mining: a 360 degree overview. In: Process Mining Handbook. LNBIP, vol. 448, pp. 3–34. Springer (2022)
4. van der Aalst, W.M.P., Santos, L.F.R.: May I take your order? - on the interplay between time and order in process mining. In: Business Process Management Workshops. LNBIP, vol. 436, pp. 99–110. Springer (2021)
5. Armas-Cervantes, A., Dumas, M., Rosa, M.L., Maaradji, A.: Local concurrency detection in business process event logs. ACM Trans. Internet Techn. **19**(1), 16:1–16:23 (2019)
6. Bergenthum, R.: Firing partial orders in a petri net. In: Petri Nets. LNCS, vol. 12734, pp. 399–419. Springer (2021)
7. Bergenthum, R., Lorenz, R.: Verification of scenarios in petri nets using compact tokenflows. Fundam. Informaticae **137**(1), 117–142 (2015)

8. Bose, J.C.J.C., Mans, R.S., van der Aalst, W.M.P.: Wanna improve process mining results? In: CIDM, pp. 127–134. IEEE (2013)

9. Carmona, J., van Dongen, B., Solti, A., Weidlich, M.: Conformance Checking. Springer (2018)

10. Desel, J., Reisig, W.: Place/transition petri nets. In: Petri Nets. LNCS, vol. 1491, pp. 122–173. Springer (1996)

11. Dumas, M., García-Bañuelos, L.: Process mining reloaded: Event structures as a unified representation of process models and event logs. In: Petri Nets. LNCS, vol. 9115, pp. 33–48. Springer (2015)

12. Folz-Weinstein, S., Bergenthum, R., Beecks, C.: Partially ordered event logs and concurrency oracles. In: AWPN 2024 Workshop Proceedings. Gesellschaft für Informatik eV (2024)

13. Folz-Weinstein, S., Rennert, C., Mannel, L.L., Bergenthum, R., van der Aalst, W.: est2 miner - process discovery based on firing partial orders. In: CAiSE 2025, Proceedings, Part II, pp. 59–75. Springer (2025)

14. Geurtjens, D., Lu, X.: Folda: computing partial-order alignments using directed net unfoldings. In: Senderovich, A., Cabanillas, C., Vanderfeesten, I., A. Reijers, H. (eds.) Business Process Management, pp. 126–143. Springer (2026)

15. Goltz, U., Reisig, W.: Processes of place/transition-nets. In: Diaz, J. (ed.) Automata, Languages and Programming, pp. 264–277. Springer (1983)

16. Grabowski, J.: On partial languages. Fundam. Informaticae 4(2), 427 (1981)

17. ter Hofstede, A.H.M., et al.: Process-data quality: the true frontier of process mining. ACM J. Data Inf. Qual. 15(3), 29:1–29:21 (2023)

18. IEEE Task Force on Process Mining: Process mining manifesto. In: Business Process Management Workshops (1). LNBIP, vol. 99, pp. 169–194. Springer (2011)

19. Juhás, G., Lorenz, R., Desel, J.: Can I execute my scenario in your net? In: ICATPN. LNCS, vol. 3536, pp. 289–308. Springer (2005)

20. Leemans, S.J.J., van Zelst, S.J., Lu, X.: Partial-order-based process mining: a survey and outlook. Knowl. Inf. Syst. 65(1), 1–29 (2023)

21. Leemans, S., Brockhoff, T., van der Aalst, W., Polyvyanyy, A.: Partially ordered stochastic conformance checking. Knowl. Inf. Syst. 67, 2291–2319 (2025)

22. de Leoni, M., Lanciano, G., Marrella, A.: Aligning partially-ordered process-execution traces and models using automated planning. In: Proceedings of the International Conference on Automated Planning and Scheduling, vol. 28, no. 1, pp. 321–329 (2018)

23. Lu, X., Fahland, D., van der Aalst, W.M.P.: Conformance checking based on partially ordered event data. In: Fournier, F., Mendling, J. (eds.) BPM 2014. LNBIP, vol. 202, pp. 75–88. Springer, Cham (2015). https://doi.org/10.1007/978-3-319-15895-2_7

24. Mannel, L.L., van der Aalst, W.M.P.: Finding complex process-structures by exploiting the token-game. In: Donatelli, S., Haar, S. (eds.) PETRI NETS 2019. LNCS, vol. 11522, pp. 258–278. Springer, Cham (2019). https://doi.org/10.1007/978-3-030-21571-2_15

25. Mans, R., van der Aalst, W.M.P., Vanwersch, R.J.B.: Process Mining in Healthcare - Evaluating and Exploiting Operational Healthcare Processes. Springer Briefs in Business Process Management. Springer (2015)

26. Pegoraro, M., Uysal, M.S., van der Aalst, W.M.: Conformance checking over uncertain event data. Inf. Syst. 102, 101810 (2021)

27. Pratt, V.: Modelling concurrency with partial orders. Int. J. Parallel Program. 15(1) (1986)

28. Raun, K., Tommasini, R., Awad, A.: Back to the order: partial orders in streaming conformance checking. Inf. Syst. **133**, 102566 (2025)
29. Senderovich, A., et al.: Conformance checking and performance improvement in scheduled processes: a queueing-network perspective. Inf. Syst. **62**, 185–206 (2016)
30. Sommers, D., Sidorova, N., van Dongen, B.: Exact and approximated log alignments for processes with inter-case dependencies. In: Petri Nets. Springer (2023)
31. van der Aa, H., Leopold, H., Weidlich, M.: Partial order resolution of event logs for process conformance checking. Decis. Support Syst. **136**, 113347 (2020)
32. Wynn, M.T., van der Aalst, W.M.P., Verbeek, E., Stefano, B.N.D.: The IEEE XES standard for process mining: experiences, adoption, and revision. IEEE Comput. Intell. Mag. **19**(1), 20–23 (2024)

Maximal Firing Semantics for Continuous and Ordinary Petri Nets

Serge Haddad[1]($\boxtimes$) and Amber Agarwal[2]

[1] Université Paris-Saclay, LMF, ENS Paris-Saclay, CNRS, Gif-sur-Yvette, France
`shaddad@lmf.cnrs.fr`
[2] Department of Computer Science and Engineering, I.I.T. Delhi, Hauz Khas, New Delhi 110016, India

Abstract. Continuous Petri nets (CPNs) is a formalism for dynamic systems that has been successfully explored for modelling and theoretical purposes. Since its firing rule entails an uncountably infinite set of reachable markings, it is interesting to introduce firing rules that combine the relaxation of the enabling condition of CPNs and the countability of the set of reachable markings ensured by Petri nets (PNs). The maximal firing semantics which consists in firing the maximal amount possible of a transition is such a rule that in addition corresponds to some modelling needs. Moreover it can also be applied to PNs. Here we study the theoretical implications of such a choice. First we show that contrary to CPNs (resp. PNs) where all the standard properties like reachability, liveness, ... can be decided with low (resp. high) complexity, in CPNs and PNs with maximal firing semantics these properties become undecidable. Then we focus on the subclass of free-choice nets establishing that the Commoner's condition for liveness and the rank theorem for well-formedness are still valid with the maximal firing semantics. More precisely, we define two families of firing rules that include the standard and maximal firing semantics, and ensure the validity of the characterisation of liveness and well-formedness in free-choice nets.

Keywords: Continuous Petri nets · Free-Choice Petri nets · Liveness · Boundedness

1 Introduction

Continuous Petri Nets. At the end of the eighties, continuous Petri nets (CPNs) were introduced in [3] for: (1) alleviating the combinatory explosion triggered by discrete Petri nets (i.e. usual Petri nets) and, (2) modelling the behaviour of physical systems whose state is composed of continuous variables. While the structure of a CPN is identical to a Petri net (PN), its states are specified by a nonnegative real number of tokens in places. The dynamics of

The work of S. Haddad has been supported by project BISOUS (ANR-22-CE48-0012) and the work of A. Agarwal has been supported by UMI ReLaX.

the system is triggered either by discrete events or by a continuous evolution ruled by speed of firings. In the former case such nets are simply called CPNs while in the latter they are called timed CPNs. In both cases, the evolution is due to a fractional transition firing (infinitesimal and simultaneous for timed CPNs).

Theoretical Developments on CPN. In [19], it was shown that reachability is undecidable in timed CPNs even for bounded ones. Thus the theoretical research has focused on CPNs. In [6] among other results, the authors proved that the reachability, coverability, and boundedness problems are PTIME-complete while the deadlock-freeness problem is NP-complete. This line of research has been pursued in [1], where it was established that the reachability set inclusion is coNP-complete and that the existential home state problem belongs to the class Σ_2^{P1}. In addition, still in [1] the authors design an algorithm for coverability in PNs that combines the standard methods with the algorithms for CPNs and whose experiments have shown that it outperforms previous tools. One defines the *mode* of a marking as the set of transitions still fireable in the future starting from this marking and given a firing sequence its *trajectory* is the sequence of decreasing modes along this firing sequence. Motivated by biological applications, the trajectory problem consists to check whether a trajectory is feasible. In [8], it was shown that the trajectory problem is NP-complete and the authors design an exponential time algorithm to build the set of feasible trajectories.

Modelling with CPNs. CPNs have been used in several significant application fields. The modelling of manufacturing systems with CPNs has been proposed in [22]. In [2], a method based on CPNs is proposed for the fault diagnosis of manufacturing systems in order to circumvent the combinatory explosion when dealing with PNs. A bottom-up modelling methodology based on CPNs to represent cell metabolism and solve in this framework the regulation control problem has been described in [20]. Combining PNs and CPNs yields hybrid Petri nets with applications to modelling and simulation of water distribution systems [7] and to the analysis of traffic in urban networks [21]. Similarly [16] extends CPNs to fuzzy CPNs for the analysis of biological systems.

Our Contributions. One key point of the semantics of CPN, is the ability to consume/produce a fraction of tokens proportional to the fraction of a transition firing. This ability naturally leads to (1) a non deterministic firing of a transition corresponding to the choice of the fraction inside a real interval and (2) an uncountable reachability set. These two features may be inappropriate for some applications. However it is possible to avoid these issues while preserving the flexibility allowed by the fractional firing: choosing to fire the maximal possible amount of firing. In fact, this rule, called the *maximal firing rule*, can be chosen both for CPNs and for PNs where in the latter case it will correspond to the maximal integer amount.

---

[1] A complexity class which lies at a low level of the polynomial hierarchy, see https://en.wikipedia.org/wiki/Polynomial_hierarchy.

- So we introduce the maximal firing rule both for CPN (denoted MF-CPN) and PN (denoted MF-PN) and exhibit relevant differences related to usual properties of nets between the standard semantics and this semantics;
- Then we show that deciding whether standard properties (reachability, coverability, boundedness, etc.) hold for a MF-CPN or a MF-PN is undecidable;
- So we focus on the subclass of free-choice nets and the two famous theorems about them, the Commoner theorem that characterizes liveness [9] and the rank theorem that characterizes well-formedness [4]. For each theorem, we introduce a family of semantics both including the ordinary semantics and the maximal firing semantics and we establish that these theorems hold for the corresponding family of semantics.

Organization. In Sect. 2, we define the maximal firing rule and illustrate its difference with the usual semantics w.r.t. usual properties and prove that checking whether most of these properties hold is undecidable. Section 3 introduces the family of appropriate semantics and show that the Commoner theorem holds for this family. Section 4 introduces the family of regular semantics and show that the rank theorem holds for this family. Finally in Sect. 5, we conclude and give some perspectives to this work.

2   Maximal Firing Semantics for Petri Nets

2.1   Definitions and Illustration

Notations. $\mathbb{N}$ (resp. $\mathbb{Z}$, $\mathbb{R}$, $\mathbb{R}_+$) denote the set of naturals (resp. integers, reals, non negative reals). Let $\mathbf{v} \in \mathbb{R}^X$ where X is a finite set. $[[\mathbf{v}]] \triangleq \{x \in X \mid \mathbf{v}[x] \neq 0\}$ and will be called the *support* of $\mathbf{v}$. $\|\mathbf{v}\|_\infty \triangleq \max_{x \in X}(|\mathbf{v}[x]|)$. $\mathbf{v}$ is positive (resp. non negative) if for all $x \in X$, $\mathbf{v}[x] > 0$ (resp. $\mathbf{v}[x] \geq 0$). $\mathbf{v}$ is semi-positive if it is non negative and there exists $x \in X$ such that $\mathbf{v}[x] > 0$. $\mathbf{0}$ denotes the null vector. Let $Y \subseteq X$, then $\mathbf{1}_Y$ is defined by $[[\mathbf{1}_Y]] = Y$ and $\mathbf{1}_Y[y] = 1$ for all $y \in Y$. We use the following intuitive notation: $\mathbf{v} = \sum_{x \mid \mathbf{v}[x] > 0} \mathbf{v}[x]x$, where coefficients equal to 1 and null terms are omitted. Let f be a partial function from E to F, $domain(f) \triangleq \{x \in E \mid f(x) \text{ is defined}\}$ and $range(f) \triangleq \{y \in F \mid \exists x \in E, y = f(x)\}$.

All kinds of nets that we study here have the same structure. We add the slight restriction that every transition should have a least one input place. Indeed transitions without inputs and their output places are more or less irrelevant for the behaviour of the net.

Definition 1 (net). *A net is a tuple* $\mathcal{N} = (P, T, \mathsf{Pre}, \mathsf{Post})$, *where:*

- *P a finite nonempty set of places;*
- *T is a finite nonempty set of transitions with $P \cap T = \emptyset$;*
- *Pre (resp. Post) is the backward (resp. forward) $P \times T$ incidence matrix, whose entries belong to $\mathbb{N}$ such that $\forall t \in T, \exists p \in P$ with $\mathsf{Pre}[p, t] > 0$.*

Notations. The *incidence matrix* of $\mathcal{N}$ is the matrix $\mathbf{C} \triangleq \mathsf{Post} - \mathsf{Pre}$. For $p \in P$, set ${}^{\bullet}p \triangleq \{t \in T : \mathsf{Post}[p,t] > 0\}$ and $p^{\bullet} \triangleq \{t \in T : \mathsf{Pre}[p,t] > 0\}$. Dually, for $t \in T$, set ${}^{\bullet}t \triangleq \{p \in P : \mathsf{Pre}[p,t] > 0\}$ and $t^{\bullet} \triangleq \{p \in P : \mathsf{Post}[p,t] > 0\}$. If $x \in P \cup T$, write ${}^{\bullet}x^{\bullet} \triangleq {}^{\bullet}x \cup x^{\bullet}$. These notations are extended to sets of items: ${}^{\bullet}X \triangleq \bigcup_{x \in X} {}^{\bullet}x$, $X^{\bullet} \triangleq \bigcup_{x \in X} x^{\bullet}$ and ${}^{\bullet}X^{\bullet} \triangleq \bigcup_{x \in X} {}^{\bullet}x^{\bullet}$. For $t \in T$, $\mathsf{Pre}[t]$ (resp. $\mathsf{Post}[t]$, $\mathbf{C}[t]$) denotes the column vector of Pre (resp. Post, $\mathbf{C}$) indexed by t. For $T' \subseteq T$, $\mathsf{Pre}_{T'}$ (resp. $\mathsf{Post}_{T'}$, $\mathbf{C}_{T'}$) denotes the matrix of Pre (resp. Post, $\mathbf{C}$) restricted to columns of T'.

We will consider *integer markings* $\mathbf{m} \in \mathbb{N}^P$ and *real markings* $\mathbf{m} \in \mathbb{R}_+^P$. When the context entails the kind of markings or when it applies to both kinds of markings, we will simply say markings. A pair $(\mathcal{N}, \mathbf{m})$ will be called a *marked net*. Let $Q \subseteq P$, $\mathbf{m}(Q) \triangleq \sum_{p \in Q} \mathbf{m}(p)$.

Definition 2 (Enabling degree). *Let* $\mathcal{N} = (P, T, \mathsf{Pre}, \mathsf{Post})$ *be a net,* $t \in T$ *and* $\mathbf{m}$ *be a marking. Then* $\mathsf{enab}(t, \mathbf{m}) \in \mathbb{R}_+$, *the* enabling degree *of* t *in* $\mathbf{m}$*, is defined by* $\min_{p \in {}^{\bullet}t} \frac{\mathbf{m}(p)}{\mathsf{Pre}[p,t]}$.

The next definition introduces four firing rules corresponding to different kinds of net. In the sequel the context will make clear which firing rule is used.

Definition 3 (Firing rules). *Let* $(\mathcal{N}, \mathbf{m})$ *be a marked net and* $t \in T$*. Then:*

1. *When* $\mathcal{N}$ *is a Petri net (PN),* $\mathbf{m}$ *is an integer marking,* t *is enabled in* $\mathbf{m}$ *if* $\mathsf{enab}(t, \mathbf{m}) \geq 1$ *and* $\mathbf{m} \xrightarrow{1t}_{\mathcal{N}} \mathbf{m} + \mathbf{C}[t]$*;*
2. *When* $\mathcal{N}$ *is a maximal firing Petri net (MF-PN),* $\mathbf{m}$ *is an integer marking,* t *is enabled in* $\mathbf{m}$ *if* $\mathsf{enab}(t, \mathbf{m}) \geq 1$ *and* $\mathbf{m} \xrightarrow{\lfloor \mathsf{enab}(t, \mathbf{m}) \rfloor t}_{\mathcal{N}} \mathbf{m} + \lfloor \mathsf{enab}(t, \mathbf{m}) \rfloor \mathbf{C}[t]$*;*
3. *When* $\mathcal{N}$ *is a continuous Petri net (CPN),* $\mathbf{m}$ *is a real marking,* t *is enabled in* $\mathbf{m}$ *if* $\mathsf{enab}(t, \mathbf{m}) > 0$ *and for all* $0 < \alpha \leq \mathsf{enab}(t, \mathbf{m})$*,* $\mathbf{m} \xrightarrow{\alpha t}_{\mathcal{N}} \mathbf{m} + \alpha \mathbf{C}[t]$*;*
4. *When* $\mathcal{N}$ *is a maximal firing continuous Petri net (MF-CPN),* $\mathbf{m}$ *is a real marking,* t *is enabled in* $\mathbf{m}$ *if* $\mathsf{enab}(t, \mathbf{m}) > 0$ *and* $\mathbf{m} \xrightarrow{\mathsf{enab}(t, \mathbf{m}) t}_{\mathcal{N}} \mathbf{m} + \mathsf{enab}(t, \mathbf{m})\, \mathbf{C}[t]$*.*

Observation. In Petri nets, there are alternative semantics known as step semantics (see [11] for instance) where simultaneous firings of different transitions may occur while here whatever the firing rule of the previous definition, a single transition can be fired.

Notation. By analogy with Petri nets, we sometimes rewrite $\mathbf{m} \xrightarrow{1t}_{\mathcal{N}} \mathbf{m}'$ as $\mathbf{m} \xrightarrow{t}_{\mathcal{N}} \mathbf{m}'$. We also rewrite $\rightarrow_{\mathcal{N}}$ as $\rightarrow$ when there is no possible confusion about $\mathcal{N}$ or as $\rightarrow_{PN}$, $\rightarrow_{MF-PN}$, $\rightarrow_{CPN}$ or $\rightarrow_{MF-CPN}$ to indicate which firing rule is used. We also omit the new marking as in $\mathbf{m} \xrightarrow{\alpha t}$ when it does not matter. $\mathcal{Z}_{\mathbb{R}} \triangleq \mathbb{R}_+ \times T$ denotes the set of *real firing steps* and $\mathcal{Z}_{\mathbb{N}} \triangleq \mathbb{N} \times T$ denotes the set of *integer firing steps*.

Illustration. Consider the net $\mathcal{N}$ and $\mathbf{m}$ the marking depicted below. Then $\mathsf{enab}(t, \mathbf{m}) = \frac{x}{2}$.

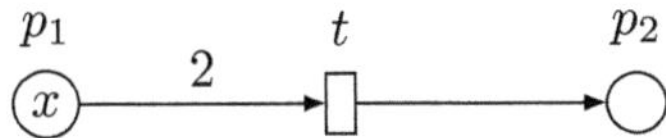

- Let $x = 1$. If $\mathcal{N}$ is a PN or a MF-PN then t is disabled. If $\mathcal{N}$ is a CPN or a MF-CPN then t is enabled. If $\mathcal{N}$ is a CPN then from $\mathbf{m}$, αt can be fired leading to $\mathbf{m}' = (1 - 2\alpha)p_1 + \alpha p_2$ with $\alpha \in \,]0, \frac{1}{2}]$. If $\mathcal{N}$ is a MF-CPN $\frac{1}{2}t$ can be fired leading to $\mathbf{m}' = \frac{1}{2}p_2$;
- Let $x = 5$. Then t is enabled whatever the kind of net. From $\mathbf{m}$, αt can be fired leading to $\mathbf{m}' = (5 - 2\alpha)p_1 + \alpha p_2$ with:
  - $\alpha = 1$ when $\mathcal{N}$ is a PN;
  - $\alpha = 2$ when $\mathcal{N}$ is a MF-PN;
  - $\alpha \in \,]0, \frac{5}{2}]$ when $\mathcal{N}$ is a CPN;
  - $\alpha = \frac{5}{2}$ when $\mathcal{N}$ is a MF-CPN.

Let $(\mathcal{N}, \mathbf{m}_0)$ be marked net (whatever its kind). Let $\sigma = \alpha_1 t_1 \ldots \alpha_n t_n$ be a sequence of steps, its *Parikh image* $\boldsymbol{\sigma} \in \mathbb{R}_+^T$ is defined by $\sum_{t \in T}(\sum_{t_i = t} \alpha_i)t$. $\sigma = \alpha_1 t_1 \ldots \alpha_n t_n$ is a *firing sequence* that leads to $\mathbf{m}_n$ if there exists a sequence $(\mathbf{m}_i)_{0 < i < n}$ such that for all $0 < i \leq n$, $\mathbf{m}_{i-1} \xrightarrow{\alpha_i t_i} \mathbf{m}_i$. $\sigma = (\alpha_i t_i)_{i \geq 1}$ is an infinite firing sequence if for all $n \geq 1$, $\alpha_1 t_1 \ldots \alpha_n t_n$ is a firing sequence. A marking $\mathbf{m}$ is *reachable* if there exists a firing sequence that leads to $\mathbf{m}$. A marking $\mathbf{m}$ is *coverable* if there exists a firing sequence that leads to some $\mathbf{m}' \geq \mathbf{m}$.

Observation. Let $(\mathcal{N}, \mathbf{m}_0)$ be a marked MF-CPN such that $\mathbf{m}_0$ is an integer marking and for all $p \in P$ and $t \in T$, $\mathsf{Pre}[p, t] \leq 1$. Then using a straightforward induction, it can be shown that every reachable marking $\mathbf{m}$ is an integer marking and for all $t \in T$, $\mathsf{enab}(t, \mathbf{m})$ is an integer. Thus the behaviour of the MF-CPN $(\mathcal{N}, \mathbf{m}_0)$ is identical to the behaviour of the MF-PN $(\mathcal{N}, \mathbf{m}_0)$.

2.2 Properties

Here we recall most of the standard properties of Petri nets in our more general framework.

Definition 4. *Let $(\mathcal{N}, \mathbf{m}_0)$ be a marked net (whatever its kind). Then:*

- *$(\mathcal{N}, \mathbf{m}_0)$ is* deadlock-free *if for all reachable marking $\mathbf{m}$, there exists αt such that $\mathbf{m} \xrightarrow{\alpha t}$;*
- *$(\mathcal{N}, \mathbf{m}_0)$ is* live *if for all reachable marking $\mathbf{m}$ and all $t \in T$, there exists σ a sequence of steps and α such that $\mathbf{m} \xrightarrow{\sigma \alpha t}$;*
- *$(\mathcal{N}, \mathbf{m}_0)$ is* bounded *if there exists $B \in \mathbb{N}$ such that for all reachable marking $\mathbf{m}$, $\|\mathbf{m}\|_\infty \leq B$;*
- *$\mathcal{N}$ is* well-formed *if there exists a marking $\mathbf{m}_0$ such that $(\mathcal{N}, \mathbf{m}_0)$ is live and bounded;*
- *$(\mathcal{N}, \mathbf{m}_0)$ admits $\mathbf{m}_h$ as a* home state *if for all reachable marking $\mathbf{m}$, $\mathbf{m}_h$ is reachable in $(\mathcal{N}, \mathbf{m})$.*
- *$(\mathcal{N}, \mathbf{m}_0)$ is* reversible *if $\mathbf{m}_0$ is a home state.*

As usual, the reachability set is the set of reachable markings. The following proposition is a straightforward consequence of the specification of firing rules.

Proposition 1. *The reachability set of the marked MF-PN $(\mathcal{N}, \mathbf{m}_0)$ is included in the reachability set of the marked PN $(\mathcal{N}, \mathbf{m}_0)$. Consequently if the marked PN $(\mathcal{N}, \mathbf{m}_0)$ is bounded then the marked MF-PN $(\mathcal{N}, \mathbf{m}_0)$ is bounded.*

Proof. Every firing sequence $\mathbf{m}_0 \xrightarrow{\sigma} \mathbf{m}$ with $\sigma = \alpha_1 t_1 \ldots \alpha_n t_n$ of the marked MF-PN $(\mathcal{N}, \mathbf{m}_0)$ can be mimicked in the marked PN $(\mathcal{N}, \mathbf{m}_0)$ by the firing sequence $\mathbf{m}_0 \xrightarrow{t_1^{\alpha_1} \ldots t_n^{\alpha_n}} \mathbf{m}$. $\qquad\square$

While the maximal firing semantics seems to be a slight variant of the ordinary semantics in the case of Petri nets, several standard properties of PNs are not satisfied by MF-PNs.

Let $\mathcal{N}$ be a PN, then for all $\mathbf{v} \in \mathbb{N}^T$ there exists a marking $\mathbf{m}$ and a sequence $\sigma \in T^*$ such that $\mathbf{m} \xrightarrow{\sigma}$ and $\boldsymbol{\sigma} = \mathbf{v}$.

Proposition 2. *There exist a MF-PN (viewed also as a MF-CPN) $\mathcal{N}$ and a vector $\mathbf{v} \in \mathbb{N}^T$ such that for all marking $\mathbf{m}$ and all $\sigma \in \mathcal{Z}_{\mathbb{N}}^*$ with $\boldsymbol{\sigma} = \mathbf{v}$, σ is not a firing sequence starting from $\mathbf{m}$.*

Proof. Consider the MF-PN $\mathcal{N}$ depicted below and $\mathbf{v} = t_1 + t_2$.

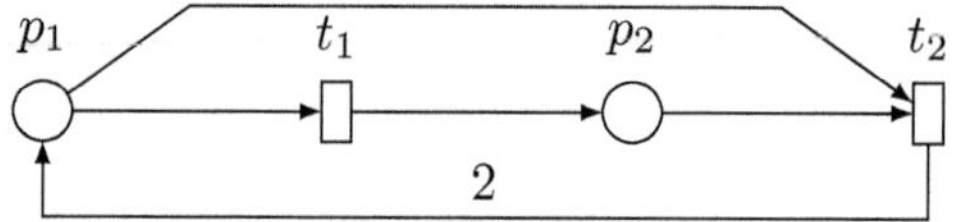

• Let $\sigma = 1t_1 1t_2$. In order to fire t_1 from some marking $\mathbf{m}$, one must have $\mathbf{m}(p_1) = 1$. So marking $\mathbf{m}'$ reached by this firing fulfills $\mathbf{m}'(p_1) = 0$ and t_2 is disabled in $\mathbf{m}'$.

• Let $\sigma = 1t_2 1t_1$. Once t_2 is fired from some marking $\mathbf{m}$ reaching $\mathbf{m}'$, one has $\mathbf{m}'(p) \geq 2$ and thus $\mathbf{m}' \xrightarrow{\alpha t_1}$ with $\alpha > 1$.

$\qquad\square$

Let $\mathcal{N}$ be a PN, then for all markings $\mathbf{m} \leq \mathbf{m}'$ and sequence $\sigma \in T^*$, if $\mathbf{m} \xrightarrow{\sigma}$ then $\mathbf{m}' \xrightarrow{\sigma}$ which is called the monotonicity property. In order to take into account the maximal firing semantics, we relax this property. A MF-PN $\mathcal{N}$ is *weakly monotonic* if for all markings $\mathbf{m} \leq \mathbf{m}'$ and sequence $\sigma = (\alpha_i t_i)_{i \leq n} \in \mathcal{Z}_{\mathbb{N}}^*$ such that $\mathbf{m} \xrightarrow{\sigma}$, there exists $\sigma' = (\beta_i t_i)_{i \leq n} \in \mathcal{Z}_{\mathbb{N}}^*$ with for all $i \leq n$, $\alpha_i \leq \beta_i$ and $\mathbf{m}' \xrightarrow{\sigma'}$.

Proposition 3. *There exists a MF-PN (viewed also as a MF-CPN) which is not weakly monotonic.*

Proof. Consider the marked MF-PN $(\mathcal{N}, \mathbf{m})$ depicted below and $\mathbf{m}' = 2p_1 + 2p_2 \geq \mathbf{m}$. $\mathbf{m} \xrightarrow{1t_1 1t_2}$ while from $\mathbf{m}'$, there are only firing sequences $2t_1$ and $2t_2$.

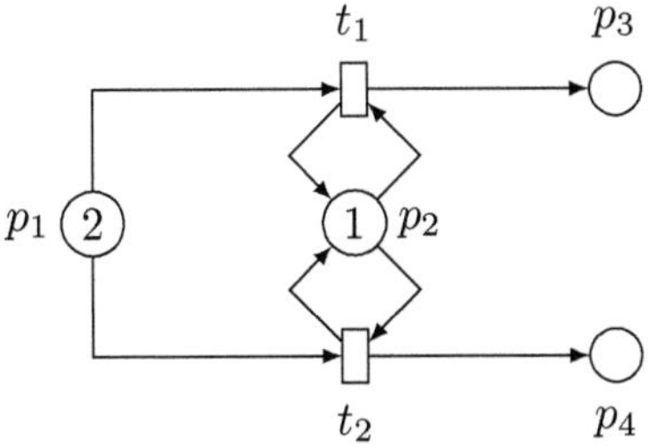

By the way, $p_3 + p_4$ is reachable in the PN $(\mathcal{N}, \mathbf{m})$ while not even coverable in the MF-PN $(\mathcal{N}, \mathbf{m}')$.

$\square$

A net can be viewed as a directed graph. So standard graph terminology will be used: connected, strongly connected, strongly connected component (SCC), bottom strongly connected component (BSCC), top strongly connected component (TSCC), etc.

In connected Petri nets, strong connectivity is a necessary condition to be well-formed. The next proposition studies whether this holds for some other kinds of nets.

Proposition 4. *There exist a connected well-formed CPN and a connected well-formed MF-CPN which are not strongly connected.*

Proof
Let us examine the marked CPN (or MF-CPN) on the right. It is bounded since for all reachable marking $\mathbf{m}$, $\mathbf{m}(p_1) + \mathbf{m}(p_2) = 1$. It is live since every firing of t marks p_1 and it is not strongly connected.

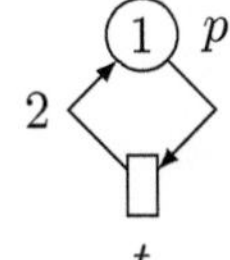

$\square$

A linear set of markings L is defined by a base $\mathbf{v}_0$ and a finite family of generators $(\mathbf{v}_i)_{i \in I}$ such that $L \triangleq \{\mathbf{v} \mid \exists \mathbf{x} \in \mathbb{N}^I, \mathbf{v} = \mathbf{v}_0 + \sum_{i \in I} \mathbf{x}[i]\mathbf{v}_i\}$. A semi-linear set is a finite union of linear sets. A net is *persistent* if once a transition is enabled it cannot be disabled until it is fired. The reachability set of a persistent PN is a semi-linear set [13].

Proposition 5. *There exists a persistent MF-PN (viewed also as a MF-CPN) whose reachability set is not semi-linear.*

Proof
Consider the marked persistent MF-PN $(\mathcal{N}, \mathbf{m})$ depicted on the right. Its reachability set is $\{2^n p \mid n \in \mathbb{N}\}$ which is not semi-linear.

$\square$

For the next result, we introduce *counter programs*. A *d-counter program* $\mathcal{P}$ is defined by a set of d counters $\{c_1, \ldots, c_d\}$ and a set of $n+1$ instructions labelled by $\{0, \ldots, n\}$, where for all $i < n$, the instruction i is of type

- either incrementation (1) $c_j \leftarrow c_j + 1; \mathbf{goto}\ i'$ with $1 \leq j \leq d$ and $0 \leq i' \leq n$
- or test (2) $\mathbf{if}\ c_j > 0\ \mathbf{then}\ c_j \leftarrow c_j - 1; \mathbf{goto}\ i'\ \mathbf{else\ goto}\ i''$ with $1 \leq j \leq d$ and $0 \leq i', i'' \leq n$

and the instruction n is **halt**. The program starts at instruction 0 and null values of counters and halts if it reaches the instruction n.

The halting problem for two-counter programs asks, given a two-counter program $\mathcal{P}$ whether $\mathcal{P}$ eventually halts. It is undecidable [17]. W.l.o.g., we may assume that on termination the program resets its counters before halting and that the instruction identifier n occurs at least in another instruction.

Theorem 1. *The reachability, coverability and boundedness, deadlock-freeness and liveness problems are undecidable for marked MF-PNs and MF-CPNs even when persistent.*

Proof. Let $\mathcal{P}$ be a two-counter program with $I \subseteq \{0, \ldots, n-1\}$, the set of test instructions. The marked MF-PN $(\mathcal{N}, \mathbf{m}_0)$ is defined as follows (see also the figure below):

- $P = \{q_i\}_{0 \leq i \leq n} \cup \{c_1, c_2\} \cup \{r_i\}_{i \in I}$;
- $T = \{t_{i,1}, t_{i,2}, t_{i,3}\}_{i \in I} \cup \{t_i\}_{i \in \{0, \ldots, n-1\} \setminus I}$;
- for all instruction i, $c_j \leftarrow c_j + 1; \mathbf{goto}\ i'$, $\mathsf{Pre}[t_i] = q_i$ and $\mathsf{Post}[t_i] = q_{i'} + c_j$;
- for all instruction i, if $c_j > 0$ then $c_j \leftarrow c_j - 1; \mathbf{goto}\ i'$, **else goto** i'',
  $\mathsf{Pre}[t_{i,1}] = q_i + c_j$ and $\mathsf{Post}[t_{i,1}] = r_i$
  $\mathsf{Pre}[t_{i,2}] = 2r_i + c_j$ and $\mathsf{Post}[t_{i,2}] = 2q_{i'} + c_j$
  $\mathsf{Pre}[t_{i,3}] = q_i + r_i$ and $\mathsf{Post}[t_{i,3}] = 2q_{i''} + c_j$.

The initial marking is $\mathbf{m}_0 = 2q_0 + c_1 + c_2$.

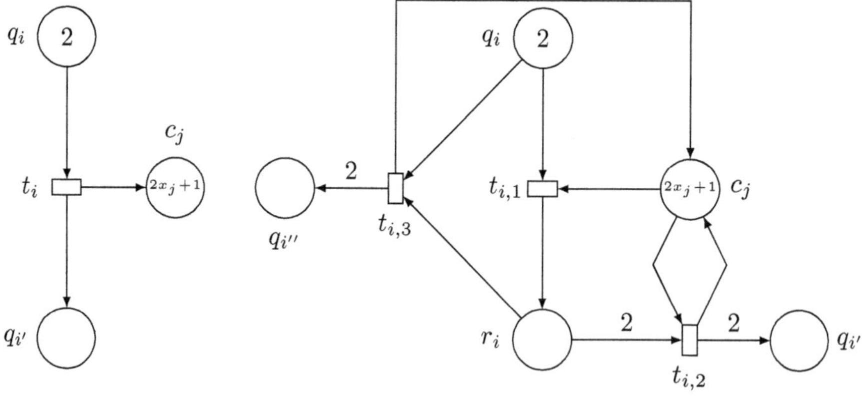

Let us define the function φ from configurations of $\mathcal{P}$, $s = (i, x_1, x_2)$ where i is the current instruction and x_j is the current value of counter c_j for $j \in \{1, 2\}$, to markings of $\mathcal{N}$ by: $\varphi(s) = 2q_i + (2x_1 + 1)c_1 + (2x_2 + 1)c_2$. Observe that $\mathbf{m}_0 = \varphi(s_0)$ where s_0 is the initial configuration of $\mathcal{P}$.

We will establish by induction that in $(\mathcal{N}, \mathbf{m}_0)$, there is a single maximal firing sequence $\rho = \varphi(s_0) \xrightarrow{\sigma_0} \varphi(s_1) \xrightarrow{\sigma_1} \varphi(s_2) \cdots$ where s_k is the configuration after k instructions of $\mathcal{P}$ (if it has not halted before). Thus ρ is finite if and only if $\mathcal{P}$ halts reaching $\mathbf{m}_f = 2q_n + c_1 + c_2$. We will also show that otherwise the reached markings never cover $\mathbf{m}_f$.

So let us assume that the current marking is $\mathbf{m}_k = 2q_i + (2x_1+1)c_1 + (2x_2+1)c_2$ where $i \neq n$.

• **Case** i is the instruction $c_j \leftarrow c_j + 1; \mathbf{goto}\ i'$, say $j = 1$.

Thus t_i is the single enabled transition with $\mathsf{enab}(t_i, \mathbf{m}_k) = 2$

and $\mathbf{m}_k \xrightarrow{2t_i} 2q_{i'} + (2x_1 + 3)c_1 + (2x_2 + 1)c_2 = \varphi(s_{k+1})$.

• **Case** i is the instruction $\mathbf{if}\ c_j > 0\ \mathbf{then}\ c_j \leftarrow c_j - 1; \mathbf{goto}\ i'$, $\mathbf{else\ goto}\ i''$, say $j = 1$. Thus $t_{i,1}$ is the single enabled transition.

   – **Subcase** $x_1 = 0$ implying $\mathsf{enab}(t_{i,1}, \mathbf{m}_k) = 1$.

      $2q_i + c_1 + (2x_2 + 1)c_2 \xrightarrow{t_{i,1}} q_i + r_i + (2x_2 + 1)c_2$.
      Now only $t_{i,3}$ is enabled with enabling degree 1.

      $q_i + r_i + (2x_2 + 1)c_2 \xrightarrow{t_{i,3}} 2q_{i''} + c_1 + (2x_2 + 1)c_2 = \varphi(s_{k+1})$.

   – **Subcase** $x_1 > 0$ implying $\mathsf{enab}(t_{i,1}, \mathbf{m}_k) = 2$.

      $2q_i + (2x_1 + 1)c_1 + (2x_2 + 1)c_2 \xrightarrow{2t_{i,1}} 2r_i + (2x_1 - 1)c_1 + (2x_2 + 1)c_2$.
      Now only $t_{i,2}$ is enabled with enabling degree 1.

      $2r_i + (2x_1 - 1)c_1 + (2x_2 + 1)c_2 \xrightarrow{t_{i,2}} 2q_{i'} + (2x_1 - 1)c_1 + (2x_2 + 1)c_2 = \varphi(s_{k+1})$.

Thus $\mathbf{m}_f$ is reachable or coverable if and only if $\mathcal{P}$ halts. Moreover $(\mathcal{N}, \mathbf{m}_0)$ is deadlock-free if and only if $\mathcal{P}$ does not halt.

For the liveness problem it suffices to add a transition end with $\mathsf{Pre}[end] = q_n$ and $\mathsf{Post}[end] = \sum_{p \in P} p$. If $\mathcal{P}$ halts then once q_n is marked, end is always fireable enabling any other transition and thus ensuring liveness. Otherwise every $t \in {}^\bullet q_n$ is never fireable (and there is at least one).

For the boundedness problem it suffices to add a place $count$ which is output of every transition. If $\mathcal{P}$ halts then the reachability set is finite and so $(\mathcal{N}, \mathbf{m}_0)$ is bounded. Otherwise due to the infinite firing sequence, $count$ is unbounded. Observe that along the maximal firing sequence, the enabling degrees of the enabled transitions are integers. So this result also applies to MF-CPNs.

$\square$

The following table summarizes the decidability and complexity status of three main decision problems for nets: reachability, coverability and boundedness. While for the ordinary semantics these problems exhibit a huge gap between

CPNs and PNs, for the maximal firing semantics whatever the kind of nets, these problems become undecidable.

	PN	CPN
Ordinary semantics		
reachability	Ackermanian-complete [14,15]	in PTIME [6]
coverability	EXPSPACE-complete [18]	in PTIME [6]
boundedness	EXPSPACE-complete [18]	in PTIME [6]
Maximal firing semantics	reachability, coverability, boundedness, undecidable	

3   Revisiting Commoner Theorem

While for general PNs, the ordinary and maximal firing semantics lead to largely different behaviours, it is interesting to study whether this gap still holds for subclasses of nets. Among the potential subclasses, free-choice PNs have a particular status: several structural characterizations of properties, low complexity bounds for some decision problems, complete synthesis rules, etc. In the sequel of this paper, we focus on the similarities and differences of free-choice PNs under ordinary and maximal firing semantics.

Definition 5. *A net $\mathcal{N}$ is a free-choice net if:*

- *for all $p \in P$ and $t \in T$, $\mathsf{Pre}(p,t) \leq 1$ and $\mathsf{Post}(p,t) \leq 1$;*
- *for all $t, t' \in T$ such that ${}^\bullet t \cap {}^\bullet t' \neq \emptyset$, ${}^\bullet t = {}^\bullet t'$.*

Observe that in free-choice nets, the preset and postset of a transition t determine $\mathsf{Pre}[t]$ and $\mathsf{Post}[t]$. So one could substitute in the definition of a free-choice net the backward and forward incidence matrices by the presets and postsets of transitions. Since $\mathsf{Pre}(p,t) \leq 1$ for all $p \in P$ and $t \in T$, free-choice MF-PNs and MF-CPNs with integer initial markings have identical behaviours. Let us first exhibit some straightforward differences between the semantics when applied to free-choice nets.

Proposition 6. *There exists a marked free-choice net which is live and bounded both as a PN and a MF-PN but with different reachability sets.*

Proof. Consider the free-choice-net below. The reachability graphs are presented on the right with thin edges for the PN and thick edges for the MF-PN. They are both strongly connected with t_1 and t_2 occurring in the two graphs. However $p_1 + p_2$ is not reachable in the MF-PN.

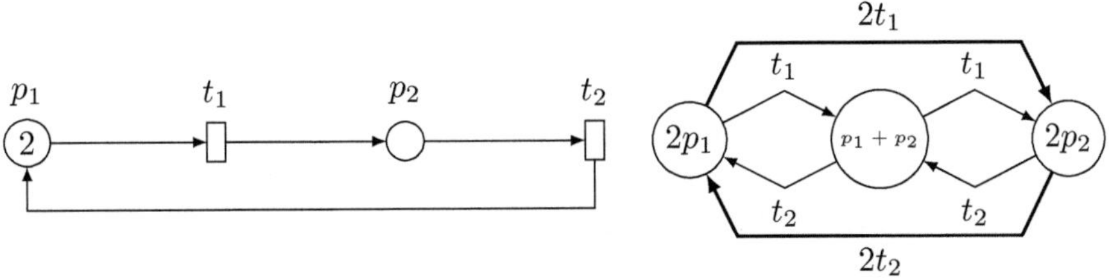

Even when live and bounded free-choice PNs and MF-PNs have the same reachability set, they do not have necessarily the same home states.

Proposition 7. *There exists a marked free-choice net which is live and bounded both as a PN and a MF-PN with same reachability sets such that the PN is reversible and the MF-PN is not reversible.*

Proof. Consider the free-choice-net below. The reachability graphs are presented on the right with thin edges for the PN and thick edges for the MF-PN. Here the initial marking $p_1 + p_2$ is a home state in the PN but not in the MF-PN.

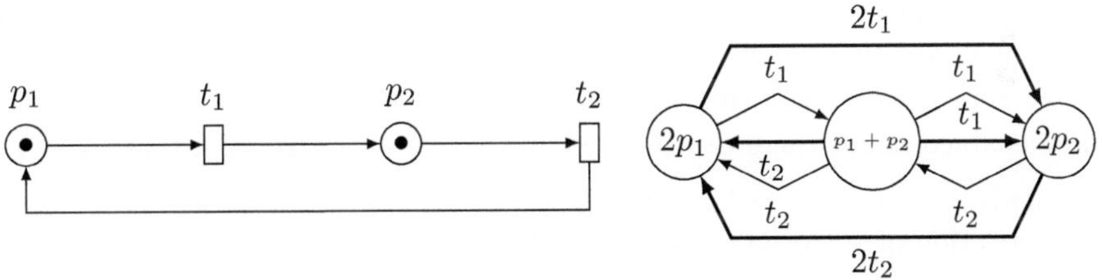

Introducing *clusters*, *traps* and *siphons* of free-choice nets is useful for their analysis.

Definition 6 (cluster). *Let $\mathcal{N}$ be free-choice net and $t \in T$, the cluster gener-ated by t, $c(t) \subseteq P \cup T$, is defined by:*

- $c(t) \cap P = {}^{\bullet}t$;
- $c(t) \cap T = \{t' \mid {}^{\bullet}t' = {}^{\bullet}t\}.$

The set of clusters of a free-choice net denoted $\mathcal{C\ell}(\mathcal{N})$ constitutes a partition of $P \cup T$.

Definition 7 (trap and siphon). *Let $\mathcal{N}$ be free-choice net and $Q \subseteq P$. Then:*

- Q *is a trap if* $Q^{\bullet} \subseteq {}^{\bullet}Q$;
- Q *is a siphon if* ${}^{\bullet}Q \subseteq Q^{\bullet}$.

Observation. Whatever the semantics, when a trap is initially marked it always remains marked and when a siphon is initially unmarked it always remains unmarked. Let $P' \subseteq P$, since a union of traps (resp. siphons) is a trap (resp. a siphon), there exists a unique maximal trap (resp. maximal siphon) included in P'. Let us recall the Commoner characterisation of liveness for free-choice PNs. A place p is *isolated* if $\bullet p^\bullet = \emptyset$.

Proposition 8 ([5,9]). *A marked free-choice PN with no isolated place is live if and only if every siphon includes a trap initially marked.*

From a complexity point of view, this characterisation implies that the liveness problem belongs to NP (in fact it is NP-complete). In [6], it is shown that liveness problem for general CPNs is also NP-complete. So in the sequel, we focus on free-choice MF-PNs.

We want to show that the Commoner characterisation holds for MF-PNs. Instead of proceeding directly, we choose an indirect path that we find more fruitful. We introduce the notion of *appropriate* firing rule for free-choice nets. By definition the firing rule of MF-PNs (and the one of PNs) is appropriate for free-choice nets. But as illustrated by some examples, appropriate firing rules are much more general. Afterwards we establish that Commoner characterisation holds for marked free-choice net under any appropriate firing rule.

An appropriate firing rule is not necessarily deterministic: for all marking $\mathbf{m}$ and t enabled in $\mathbf{m}$, there can be even an infinite set of $\mathbf{m}'$ such that $\mathbf{m} \xrightarrow{t} \mathbf{m}'$.

Notation. Let $\mathbf{v} \in \mathbb{R}^X$ where X is a finite set. Let $Y \subseteq X$ then $\mathbf{v}^{\downarrow Y} \in \mathbb{R}^Y$ is the restriction of $\mathbf{v}$ to the components of Y.

Definition 8 (appropriate firing rule). *Let $\mathcal{N}$ be a free-choice net. An* appropriate firing rule *is defined by:*

1. *for all $t \in T$ and $\mathbf{m} \in \mathbb{N}^{\bullet t^\bullet}$, a set $next(\mathbf{m}, t) \subseteq \mathbb{N}^{t^\bullet}$ non empty if and only if for all $p \in {}^\bullet t$, $\mathbf{m}(p) > 0$;*
2. *for all $t \in T$ and $\mathbf{m} \in \mathbb{N}^{\bullet t^\bullet}$, all $\mathbf{m}' \in next(\mathbf{m}, t)$, and all $p \in t^\bullet$, $\mathbf{m}'(p) > 0$;*
3. *for all $t \in T$ and $\mathbf{m} \in \mathbb{N}^{\bullet t^\bullet}$, all $\mathbf{m}' \in next(\mathbf{m}, t)$, and all $p \in {}^\bullet t \setminus t^\bullet$, $\mathbf{m}'(p) < \mathbf{m}(p);$*

Let $\mathbf{m} \in \mathbb{N}^P$. Then t is enabled in $\mathbf{m}$ if $\mathbf{m}^{\downarrow \bullet t}$ is positive and $\mathbf{m} \xrightarrow{t} \mathbf{m}'$ if $\mathbf{m}'^{\downarrow t^\bullet} \in next(\mathbf{m}^{\downarrow \bullet t^\bullet}, t)$ and for all $p \notin t^\bullet$, $\mathbf{m}'(p) = \mathbf{m}(p)$.

Discussion. First an appropriate firing rule for t is local: it only depends on the marking of input and output places of t. Then Condition 1 states that the enabling condition is the one of PNs and MF-PNs. The set $next(\mathbf{m}, t)$ represents the different *modes* for firing a transition. For instance, when t is enabled in PNs and MF-PNs, this set is a singleton. Condition 2 states that every output place of t is marked after any firing of t. Condition 3 states that the marking of every input place of t which is not an output place of t decreases after any firing of t.

One could imagine that an appropriate rule also ensures the Commoner characterisation with real markings instead of integer markings. However this does not work. Consider the 'middle' firing rule for CPN: there is a single mode which consists when firing an enabled transition t at $\mathbf{m}$, to fire $\frac{\mathsf{enab}(t,\mathbf{m})}{2}t$. This firing rule satisfies Conditions 1 and 2 and 3. Let us look at the net below: p is a siphon but not a trap. So it does not fulfill the Commoner condition. However there is a single infinite firing sequence $(2^{-n}t)_{n>0}$ implying that the net is live.

The next lemma shows that the property of siphons and traps in PNs also holds for nets with an appropriate firing rule.

Lemma 1. *Let $(\mathcal{N}, \mathbf{m}_0)$ be a free-choice net with an appropriate firing rule.*

- *Let S be a siphon such that $\mathbf{m}_0(S) = 0$. Then for all reachable marking $\mathbf{m}$, $\mathbf{m}(S) = 0$;*
- *Let Q be a trap such that $\mathbf{m}_0(Q) > 0$. Then for all reachable marking $\mathbf{m}$, $\mathbf{m}(Q) > 0$.*

Proof. The first assertion is obtained by induction using Condition 1 of Definition 8. The second assertion is obtained by induction using Condition 2 of Definition 8. $\square$

Example of Appropriate Firing Rules. Let us look at the subnet below. As for PNs t is enabled in $\mathbf{m}$ if $\mathbf{m}(p_1) > 0$ and $\mathbf{m}(p_2) > 0$. Let $\mathbf{m}'$ be a marking reached after the firing of t. Then:

- $\mathbf{m}'(p_1) = \lfloor \frac{\mathbf{m}(p_1)}{2} \rfloor < \mathbf{m}(p_1)$;
- $\mathbf{m}'(p_2) = \max(\mathbf{m}(p_2) - \mathbf{m}(p_1), 0) < \mathbf{m}(p_2)$
  (this is consistent with Condition 1 which only requires $\mathbf{m}(p_2)$ to be marked);
- $\mathbf{m}'(q_1) \in \mathbb{N}_{>0}$;
  (this is consistent with Condition 3 which only requires $\mathbf{m}'(q_1)$ to be marked);
- $\mathbf{m}'(q_2) = \mathbf{m}(q_2) + \mathbf{m}(p_2) > 0$;
- $\mathbf{m}'(q_3) = \max(\mathbf{m}(q_3) + 2, 3) > 0$.

This example shows that appropriate firing rules can express numerous behaviours depending on the modelling needs.

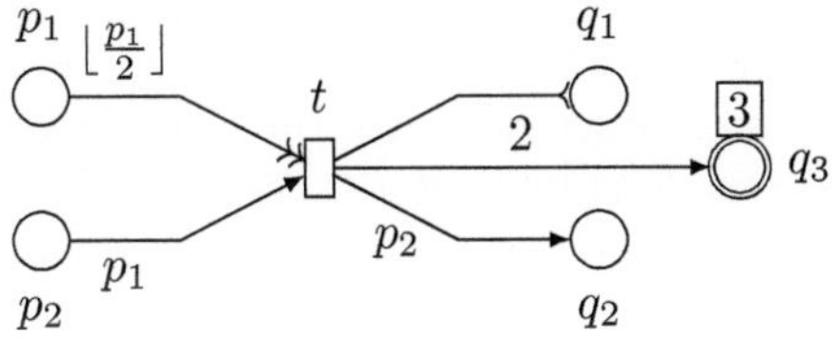

The next definition and lemma do not depend on the firing rule and apply to general nets.

Definition 9. *Let $\mathcal{N}$ be a net, $t \in T$ and $\mathbf{m}$ be a marking.*

- *t is* live *at $\mathbf{m}$ if for all $\mathbf{m}'$ reachable in $(\mathcal{N}, \mathbf{m})$, there exists a firing sequence $\mathbf{m}' \xrightarrow{\sigma t}$;*
- *t is* dead *at $\mathbf{m}$ if for all $\mathbf{m}'$ reachable in $(\mathcal{N}, \mathbf{m})$, t is disabled at $\mathbf{m}'$.*

The set of live (resp. dead) transitions at $\mathbf{m}$ is denoted $L_\mathbf{m}$ (resp. $D_\mathbf{m}$).

Observation. By definition, $L_\mathbf{m} \cap D_\mathbf{m} = \emptyset$. On the other hand, $L_\mathbf{m} \cup D_\mathbf{m}$ is not necessary equal to T.

The next lemma holds more generally for transition systems with a finite set of transitions whatever the firing rule may be.

Lemma 2. *Let $(\mathcal{N}, \mathbf{m}_0)$ be a marked net. There exists a reachable marking $\mathbf{m}$ such that $L_\mathbf{m} \cup D_\mathbf{m} = T$. Furthermore if $L_{\mathbf{m}_0} \neq T$ then $D_\mathbf{m} \neq \emptyset$.*

Proof. Observe that along a firing sequence, the set of live transitions and the set of dead transitions are non decreasing.
If $L_{\mathbf{m}_0} \cup D_{\mathbf{m}_0} = T$ we are done otherwise pick some $t \in T \setminus (L_{\mathbf{m}_0} \cup D_{\mathbf{m}_0})$. Since t is not live, there is a marking $\mathbf{m}_1$ reachable from $\mathbf{m}_0$ such that $L_{\mathbf{m}_0} \subseteq L_{\mathbf{m}_1}$ and $D_{\mathbf{m}_0} \cup \{t\} \subseteq D_{\mathbf{m}_1}$. Since T is finite, this process must stop and one reaches a marking $\mathbf{m}$ such that $L_\mathbf{m} \cup D_\mathbf{m} = T$.
Assume that $L_{\mathbf{m}_0} \neq T$. If $D_{\mathbf{m}_0} \neq \emptyset$ we are done. Otherwise the transition t above belongs to $D_{\mathbf{m}_1}$ and we are also done.

$\square$

Let $(\mathcal{N}, \mathbf{m})$ be a free-choice marked net with an appropriate firing rule. By definition of free-choice nets, $D_\mathbf{m}$ are the transitions of a set of clusters.

Notation. Let $\mathbf{m}$ be a marking and $p \in P$. We say that p is a *dead place* of $\mathbf{m}$ if for every reachable marking from $\mathbf{m}$, p is unmarked.

Lemma 3. *Let $\mathcal{N}$ be a free-choice net with an appropriate firing rule and $\mathbf{m}$ be a marking such that $D_\mathbf{m} \neq \emptyset$. There exists a marking $\mathbf{m}'$ reachable from $\mathbf{m}$ such that for all cluster c with $c \cap T \subseteq D_\mathbf{m}$, there is a place $p_c \in c \cap P$ such that p_c is a dead place of $\mathbf{m}'$.*

Proof. Let c be a cluster such that $c \cap T \subseteq D_\mathbf{m}$ and assume that some $p \in c \cap P$ can be marked in a reachable marking $\mathbf{m}_p$ of $(\mathcal{N}, \mathbf{m})$. Observe that in any reachable marking of $(\mathcal{N}, \mathbf{m}_p)$, p remains marked since $D_\mathbf{m} \subseteq D_{\mathbf{m}_p}$. So from $\mathbf{m}_p$ we can iterate this process reaching a marking $\mathbf{m}_c$ such that no more $p \in c \cap P$ can be marked in a reachable marking of $(\mathcal{N}, \mathbf{m}_c)$. On the other hand, $D_\mathbf{m} \subseteq D_{\mathbf{m}_c}$ implying that there is at least one dead place $p_c \in c \cap P$ at $\mathbf{m}_c$. Starting from $\mathbf{m}_c$ and iterating this process for the other clusters c' such that $c' \cap T \subseteq D_\mathbf{m}$ concludes the proof.

$\square$

The next proposition shows that the Commoner characterisation of liveness for marked free-choice PNs is also valid for marked free-choice nets with an appropriate firing rule. Its proof is based on the one of Petr Jančar [10].

Proposition 9. *A marked free-choice net with an appropriate firing rule and no isolated place is live if and only if every siphon includes an initially marked trap.*

Proof. Let $(\mathcal{N}, \mathbf{m}_0)$ be a marked free-choice net with an appropriate firing rule and no isolated place.

• Assume that $(\mathcal{N}, \mathbf{m}_0)$ is not live and let $\mathbf{m}$ be the reachable marking of Lemma 2 with $L_\mathbf{m} \cup D_\mathbf{m} = T$ and $D_\mathbf{m} \neq \emptyset$. Apply Lemma 3 and let $\mathbf{m}'$ the marking reachable from $\mathbf{m}$ such that for all cluster c fulfilling $c \cap T \subseteq D_\mathbf{m}$, there is a dead place p_c. Observe that since $L_\mathbf{m} \cup D_\mathbf{m} = T$, $L_{\mathbf{m}'} = L_\mathbf{m}$ and $D_{\mathbf{m}'} = D_\mathbf{m}$. Let R be the set of places $\{p_c \mid c \cap T \subseteq D_\mathbf{m}\}$, by definition $R^\bullet = D_\mathbf{m}$. Consider $t \in {}^\bullet p_c$ for some p_c. Since p_c is a dead place, $t \notin L_\mathbf{m}$, otherwise there would be a sequence $\mathbf{m}' \xrightarrow{\sigma t} \mathbf{m}''$ with $\mathbf{m}''(p_c) > 0$ (Condition 2 of Definition 8). Thus $t \in D_\mathbf{m} = R^\bullet$ and so R is a siphon unmarked by $\mathbf{m}'$. Moreover it cannot contain a trap initially marked since this trap would be still marked in $\mathbf{m}'$ (Lemma 1).

• Assume that $\mathcal{N}$ has a non empty siphon S such the maximal trap $Q \subseteq S$ fulfills $\mathbf{m}_0(Q) = 0$. $S^\bullet \neq \emptyset$ since there is no isolated place and S is a siphon. We will show that $S^\bullet \cap L_{\mathbf{m}_0} = \emptyset$ implying non liveness of $(\mathcal{N}, \mathbf{m}_0)$.

If $Q = S$ then $S^\bullet \subseteq D_{\mathbf{m}_0}$ by application of Lemma 1 and we are done.

Otherwise $Q \subsetneq S$ and since S is not a trap there is some $t_1 \in S^\bullet$ such that $t_1{}^\bullet \cap S = \emptyset$. Furthermore since Q is a trap ${}^\bullet t_1 \cap Q = \emptyset$. Define $R_1 = S \setminus {}^\bullet t_1 \subsetneq S$. If $R_1 \neq Q$, we iterate the process: since R_1 is not a trap there is some $t_2 \in R_1{}^\bullet$ such that $t_2{}^\bullet \cap R_1 = \emptyset$. Furthermore since Q is a trap ${}^\bullet t_2 \cap Q = \emptyset$. Define $R_2 = R_1 \setminus {}^\bullet t_2 \subsetneq R_1$. This process must stop and one gets $Q = R_k \subsetneq R_{k-1} \subsetneq \cdots R_1 \subsetneq R_0 = S$ with for all $1 \leq i \leq k$, $t_i{}^\bullet \cap R_{i-1} = \emptyset$, and $R_i = R_{i-1} \setminus {}^\bullet t_i$ (see the figure below). Let c_i be the cluster of t_i and $T_i = c_i \cap T$. By construction, $S^\bullet = Q^\bullet \cup \bigcup_{1 \leq i \leq k} T_i$.

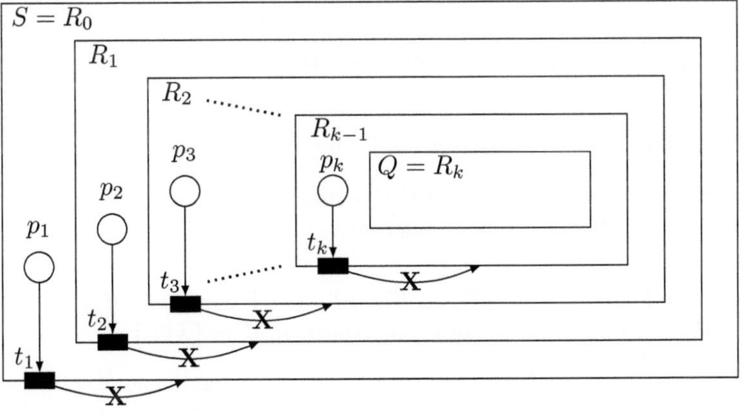

Let $U = \{t_1, \ldots, t_k\}$. We build a firing sequence σ as follows. If there exists a sequence $\mathbf{m}_0 \xrightarrow{\sigma_0} \mathbf{m}_1$ with $\sigma_0 \in (T \setminus S^\bullet)^* \cdot U$ then σ starts by σ_0 otherwise $\sigma = \varepsilon$. We repeat the process in $\mathbf{m}_1$: if there exists a sequence $\mathbf{m}_1 \xrightarrow{\sigma_1} \mathbf{m}_2$ with $\sigma_1 \in (T \setminus S^\bullet)^* \cdot U$ then σ starts by $\sigma_0 \sigma_1$ otherwise $\sigma = \sigma_0$.

We claim that this process produces a finite sequence σ. For all $1 \leq i \leq k$, we arbitrarily choose some $p_i \in {}^\bullet t_i \cap R_{i-1}$. First firing transitions of $T \setminus S^\bullet = T \setminus {}^\bullet S^\bullet$ does not modify the marking of S. There is a finite number of occurrences of t_k since $p_k \notin \bigcup_{j<k} t_j{}^\bullet$ and as long as the marking of p_k remains positive the firing of t_k decreases this marking (Condition 3 of Definition 8). There is a finite number of occurrences of t_{k-1} before the first occurrence of t_k, between two consecutive occurrences of t_k and after the last occurrence of t_k since $p_{k-1} \notin \bigcup_{j<k-1} t_j{}^\bullet$ and as long as the marking of p_{k-1} remains positive the firing of t_{k-1} decreases this marking (Condition 3 of Definition 8). Iterating this reasoning, from $k-2$ to 1, establishes that σ is a finite sequence.[2]

Let $\mathbf{m}$ be the marking reached by σ. We claim that $S^\bullet \subseteq D_{\mathbf{m}}$ which will achieve the proof. Observe that $\mathbf{m}(Q) = 0$ since the transitions of $T \setminus S^\bullet \cup U$ do not produce tokens in Q. Assume by contradiction that $\mathbf{m} \xrightarrow{\sigma'} \mathbf{m}' \xrightarrow{t}$ with $\sigma' \in (T \setminus S^\bullet)^*$ and $t \in S^\bullet$. $\mathbf{m}'(Q)$ is still null and thus $t \in T_i$ for some i. So t_i is also fireable at $\mathbf{m}'$ contradicting the maximality of σ.

$\square$

4   Revisiting the Rank Theorem

This section is devoted to the analysis of well-formedness in free-choice MF-PNs. Since this analysis is closely related to the properties of the incidence matrix $\mathbf{C}$, we first recall the concept of invariants.

Let $\mathbf{C}$ be the incidence matrix of a net. A vector $\mathbf{v} \in \mathbb{Z}^T$ (resp. $\mathbf{v} \in \mathbb{Z}^P$) is a T-invariant (resp. P-invariant) if $\mathbf{C} \cdot \mathbf{v} = \mathbf{0}$ (resp. $\mathbf{v} \cdot \mathbf{C} = \mathbf{0}$).

Observation. Assume there exists a positive P-invariant $\mathbf{v}$. Then there cannot exist $\mathbf{w} \in \mathbb{N}^T$ such that $\mathbf{C} \cdot \mathbf{w} \gneq \mathbf{0}$. Otherwise $\mathbf{v} \cdot \mathbf{C} \cdot \mathbf{w} = \mathbf{0} \cdot \mathbf{w} = 0$ and $\mathbf{v} \cdot \mathbf{C} \cdot \mathbf{w} = \mathbf{v} \cdot (\mathbf{C} \cdot \mathbf{w}) > 0$.

P-nets are a very simple class of free-choice nets which fulfills several interesting properties.

Definition 10. *A P-net $\mathcal{N}$ is a net fufilling the following properties:*

- *for all $p \in P$ and $t \in T$, $\mathsf{Pre}[p, t] \leq 1$ and $\mathsf{Post}[p, t] \leq 1$;*
- *for all $t \in T$, ${}^\bullet t$ and $t^\bullet$ are singletons.*

Proposition 10. *Let $\mathcal{N}$ be a P-net. Then $\mathbf{1}_P$ is a P-invariant of $\mathcal{N}$. Furthermore if $\mathcal{N}$ is connected then the set of P-invariants of $\mathcal{N}$ is $\{\lambda \mathbf{1}_P\}_{\lambda \in \mathbb{Z}}$.*

---

[2] Contrary to the proof in [10] there is no possible bound on the number of occurrences of transitions of U due to the infinite branching allowed by the appropriate firing rules.

Proof. For all $t \in T$, $\mathbf{1}_P \cdot \mathbf{C}[t] = \mathbf{1}_P \cdot (t^\bullet - {}^\bullet t) = 1 - 1 = 0$.

Let $\mathbf{v}$ be a P-invariant. $\mathbf{v} \cdot \mathbf{C}[t] = \mathbf{v} \cdot (t^\bullet - {}^\bullet t) = \mathbf{v}[t^\bullet] - \mathbf{v}[{}^\bullet t] = 0$. By induction, this implies that for all p, p' connected $\mathbf{v}[p] = \mathbf{v}[p']$ which concludes the proof. $\square$

Below is the most standard definition of subnets as it operates by restriction to subsets of places and transitions and elimination of the corresponding lines and columns of the backward and forward incidence matrices.

Definition 11. *Let $\mathcal{N}$ be a net and $X \subseteq P \cup T$. Then $\mathcal{N}[X]$ the net generated by X has $P \cap X$ for set of places, $T \cap X$ for set of transitions with the backward and forward incidences being restrictions of* Pre *and* Post *to $P \cap X \times T \cap X$.*

Definition 12. *Let $\mathcal{N}$ be a net and $\emptyset \neq Q \subseteq P$. Then $\mathcal{N}[Q \cup {}^\bullet Q^\bullet]$ is a P-component of $\mathcal{N}$ if it is a strongly connected P-net.*

Observation. Let $\mathcal{N}[Q \cup {}^\bullet Q^\bullet]$ be a P-component. Then Q is both a minimal trap and a minimal siphon of $\mathcal{N}$.

The two following results are fundamental in the study of free-choice PNs. First the associated characterisations are purely structural and second they entail polynomial time algorithms which is quite rare in the field of PNs (contrary to the theory of CPNs).

Theorem 2 (rank theorem [4]). *Let $\mathcal{N}$ be a connected free-choice PN without isolated place. Then $\mathcal{N}$ is well-formed if and only if:*

- *$\mathcal{N}$ has a positive P-invariant;*
- *and $\mathcal{N}$ has a positive T-invariant;*
- *and $rank(\mathbf{C}) = |\mathcal{C}\ell(\mathcal{N})| - 1$.*

Theorem 3 ([4]). *Let $(\mathcal{N}, \mathbf{m}_0)$ be a connected marked free-choice PN without isolated place. Then $(\mathcal{N}, \mathbf{m}_0)$ is live and bounded if and only if:*
$\mathcal{N}$ is well-formed and for every siphon $\emptyset \neq Q \subseteq P$, $\mathbf{m}_0(Q) > 0$.

Based on these two theorems, in [12] efficient polynomial time algorithms have been designed to check whether a free-choice PN is well-formed and whether a marked free-choice PN is live and bounded. Here we prove that these characterisations also hold for MF-PNs. We will proceed as in the previous section by introducing a family of firing rules for which these characterisations are valid.

A regular firing rule is not necessarily deterministic but must preserve a close connection with the incidence matrix.

Definition 13 (regular firing rule). *Let $\mathcal{N}$ be a free-choice net. A regular firing rule is defined for all $t \in T$ and $\mathbf{m} \in \mathbb{N}^{t^\bullet}$ by:*
$$a \ set \ next(\mathbf{m}, t) \subseteq \{\mathbf{m} + \alpha \mathbf{C}[t]^{\downarrow t^\bullet} \mid \alpha \in \mathbb{N} \wedge 0 < \alpha \leq \mathsf{enab}(t, \mathbf{m})\}$$
$$where \ next(\mathbf{m}, t) \neq \emptyset \ iff \ for \ all \ p \in {}^\bullet t, \ \mathbf{m}(p) > 0.$$

Let $\mathbf{m} \in \mathbb{N}^P$. Then t is enabled in $\mathbf{m}$ if $\mathbf{m}^{\downarrow {}^\bullet t}$ is positive and $\mathbf{m} \xrightarrow{\alpha t} \mathbf{m}'$ if $\mathbf{m}'^{\downarrow t^\bullet} = (\mathbf{m} + \alpha \mathbf{C}[t])^{\downarrow t^\bullet} \in next(\mathbf{m}^{\downarrow {}^\bullet t}, t)$ and for all $p \notin {}^\bullet t^\bullet$, $\mathbf{m}'(p) = \mathbf{m}(p)$.

Observations. A regular firing rule is an appropriate firing rule. Assume that $\mathcal{N}$ has a positive P-invariant $\mathbf{v}$. For any regular firing rule if $\mathbf{m} \xrightarrow{\alpha t} \mathbf{m}'$ then $\mathbf{m}' = \mathbf{m} + \alpha \mathbf{C}[t]$. Thus for every reachable $\mathbf{m}$ in $(\mathcal{N}, \mathbf{m}_0)$, $\mathbf{v} \cdot \mathbf{m} = \mathbf{v} \cdot \mathbf{m}_0$, implying that for all $p \in P$, $\mathbf{m}(p) \leq \mathbf{v} \cdot \mathbf{m}_0$ and so the boundedness of $(\mathcal{N}, \mathbf{m}_0)$.

Most of the proofs in the remaining part of this section are similar to those stated for PNs in [5]. In particular, the two following propositions do not depend on the firing rule.

Proposition 11. *Let $\mathcal{N}$ be a free-choice net and $\emptyset \neq Q \subseteq P$. Then Q is a minimal siphon iff (a) every cluster contains at most one place of Q and (b) $\mathcal{N}[Q \cup {}^\bullet Q]$ is strongly connected.*

Proof.

• Assume Q is a minimal siphon. If a cluster contains two places p and p', define $Q' = Q \setminus \{p'\}$. Then $Q'^\bullet = Q^\bullet$ and ${}^\bullet Q' \subseteq {}^\bullet Q$ implying that Q' is a siphon contradicting the minimality of Q.

Let X be a TSCC of $\mathcal{N}[Q \cup {}^\bullet Q]$. It cannot consist of a single transition t. Otherwise t would have no input in Q and at least one output in Q and Q would not be a siphon. Let Q' be the set of places of X. We claim that Q' is a siphon. Let $p \in Q'$ and $t \in {}^\bullet p$. By definition of X, $t \in X$ and there is some $p' \in X$ such that $t \in p'^\bullet$. So Q' is a siphon and by minimality equal to Q. Thus $\mathcal{N}[Q \cup {}^\bullet Q]$ is strongly connected.

• Assume that Conditions (a) and (b) are satisfied. Since $\mathcal{N}[Q \cup {}^\bullet Q]$ is strongly connected, for all $t \in {}^\bullet Q$ there is some $p \in Q$ such that $t \in p^\bullet$. So Q is a siphon. Let $\emptyset \neq Q' \subsetneq Q$. Due to the strong connectivity there is $p' \in Q'$, $p \in Q \setminus Q'$ $t \in T$ such that $t \in p^\bullet \cap {}^\bullet p'$. Due to Condition (a), ${}^\bullet t \cap Q = \{p\}$. Thus Q' is not a siphon.

$\square$

Proposition 12. *Let $\mathcal{N}$ be a strongly connected free-choice net. Then every place is contained in a minimal siphon.*

Proof.

Let $p \in P$. One builds $\{p\} = X_0 \subsetneq X_1 \subsetneq \cdots$ a finite increasing sequence of subsets of $P \cup T$ such that for all n,

- $\mathcal{N}[X_n]$ is strongly connected;
- for all cluster $c \in \mathcal{Cl}(\mathcal{N})$, $|c \cap X_n \cap P| \leq 1$.

These properties are satisfied by X_0. Assume that X_n fulfills these properties and that there exists $t \in T \cap ({}^\bullet X_n \setminus X_n)$.

Since $\mathcal{N}$ is strongly connected, there is a shortest path from X_n to t: $\pi = x_1 \ldots x_k$ with $x_k = t$ implying that $x_1 \in X_n$ and for all $i > 1$, $x_i \notin X_n$.

Let us define $X_{n+1} = X_n \cup \{x_i\}_{i \leq k}$. Observe that the path π can be extended by a place in X_n. Since X_n is strongly connected and this extended path starts from an item in X_n and ends in an item in X_n, X_{n+1} is strongly connected.

Assume by contradiction that two places of $\mathcal{N}[X_{n+1}]$ belong to the same cluster. One of them must be some x_i for $i > 1$. The second one is either some x_j (say with $i < j < k$) or some $p' \in X_n$. If it is x_j then the path $x_1 \ldots x_i x_{j+1} \ldots x_k$ is a shorter path, a contradiction. If it is p' then the path $p' x_{i+1} \ldots x_k$ is a shorter path, another contradiction.

Consider X_N the last item of the sequence and $Q = X_N \cap P$. Then for all $t \in {}^\bullet Q$, $t \in X_N$. Since $\mathcal{N}[X_N]$ is strongly connected, for all $t \in X_N$, $t \in {}^\bullet Q$. Thus $X_N = Q \cup {}^\bullet Q$ which entails that Q satisfies the characterization of minimal siphon stated by Proposition 11.

$\square$

Observe that the next proposition holds for every connected net and not only for free-choice nets.

Proposition 13. *A well-formed connected net $\mathcal{N}$ with a regular firing rule is strongly connected.*

Proof. Consider a live connected marked net $(\mathcal{N}, \mathbf{m}_0)$ with a regular firing rule which is not strongly connected and let $\mathcal{N}'$ be a subnet corresponding to a TSCC of $\mathcal{N}$: $\mathcal{N}'$ is also live. There are two cases to consider.

• There is some $t \in \mathcal{N}'$ and $p \notin \mathcal{N}'$ such that $p \in t^\bullet$. Since $\mathcal{N}'$ is live there is an infinite firing sequence σ of $\mathcal{N}'$ with an infinite number of occurrences of t. σ is also a firing sequence of $\mathcal{N}$ with the marking of p going to ∞. Thus $(\mathcal{N}, \mathbf{m}_0)$ is unbounded.

• There is some $p \in \mathcal{N}'$ and $t \notin \mathcal{N}'$ such that $t \in p^\bullet$. Since $\mathcal{N}$ is live there is an infinite firing sequence σ of $\mathcal{N}$ with an infinite number of occurrences of t. Thus the marking of p is infinitely often increased by transitions of $\mathcal{N}'$ (and may also be decreased by these transitions). Let σ' be the projection of σ over the transitions of $\mathcal{N}'$. σ' is also an infinite firing sequence of $\mathcal{N}$. Since the occurrences of t have disappeared, the marking of p goes to ∞ along σ'. Thus $(\mathcal{N}, \mathbf{m}_0)$ is unbounded.

$\square$

Corollary 1. *Let $\mathcal{N}$ be a well-formed connected free-choice net $\mathcal{N}$ with a regular firing rule. Then every place is contained in a minimal siphon.*

Proof. Due to Proposition 13, $\mathcal{N}$ with a regular firing rule is strongly connected and thus the statement of the corollary is established by Proposition 12.

$\square$

Proposition 14. *Let $\mathcal{N}$ be a well-formed free-choice net $\mathcal{N}$ with a regular firing rule and R be a minimal siphon. Then $\mathcal{N}[R \cup {}^\bullet R^\bullet]$ is a P-component.*

Proof. Let Q be the maximal trap included in R. Since $\mathcal{N}$ is well-formed there is a marking $\mathbf{m}_0$ such that $(\mathcal{N}, \mathbf{m}_0)$ is live and bounded. By Commoner characterization, R must contain a marked trap. So $Q \neq \emptyset$.

Let $t \in Q^\bullet$.

- $1 \leq |{}^{\bullet}t \cap Q| \leq |{}^{\bullet}t \cap R| \leq 1$ by item (a)
  of the characterization of minimal siphons. Thus $|{}^{\bullet}t \cap Q| = 1$;
- Since Q is a trap $Q^{\bullet} \subseteq {}^{\bullet}Q$ and so $|t^{\bullet} \cap Q| \geq 1$.

As a consequence, $\mathbf{m}(Q)$ is non decreasing along a firing sequence. Assume by contradiction that there is some $t \in Q^{\bullet} = {}^{\bullet}Q$ such that $|t^{\bullet} \cap Q| > |{}^{\bullet}t \cap Q|$. Since $(\mathcal{N}, \mathbf{m}_0)$ is live there is an infinite firing sequence with t occurring infinitely often and so $\mathbf{m}(Q)$ will go to ∞ contradicting the boundedness. Thus $t \in {}^{\bullet}Q \Leftrightarrow |t^{\bullet} \cap Q| > 0 \Rightarrow |{}^{\bullet}t \cap Q| \geq |t^{\bullet} \cap Q| > 0$. Thus $Q^{\bullet} = {}^{\bullet}Q$. In particular Q is a siphon and so $Q = R$. Moreover for all $t \in Q^{\bullet}$, $|{}^{\bullet}t \cap Q| = |t^{\bullet} \cap Q| = 1$.

Now $\mathcal{N}[R \cup {}^{\bullet}R^{\bullet}] = \mathcal{N}[R \cup {}^{\bullet}R]$ is strongly connected by item (b) of the characterisation of minimal siphons. So $\mathcal{N}[R \cup {}^{\bullet}R^{\bullet}]$ is a P-component.

□

Corollary 2. *Let $\mathcal{N}$ be a well-formed connected free-choice net $\mathcal{N}$ with a regular firing rule. Then $\mathcal{N}$ has a positive P-invariant.*

Proof. By Corollary 1 every place p is included in a minimal siphon R_p which is a P-component by Proposition 14. So $\sum_{p \in P} \mathbf{1}_{R_p}$ is a positive invariant.

□

Theorem 4. *Let $(\mathcal{N}, \mathbf{m}_0)$ be a free-choice marked net. If there exists some regular firing rule such that $(\mathcal{N}, \mathbf{m}_0)$ is live and bounded then for every regular firing rule $(\mathcal{N}, \mathbf{m}_0)$ is live and bounded*

Proof. Since isolated places are irrelevant for liveness and boundedness, w.l.o.g, we suppose that $\mathcal{N}$ has no isolated place.

Assume there exists some regular firing rule such that $(\mathcal{N}, \mathbf{m}_0)$ is live and bounded. Since a regular firing rule is an appropriate firing rule, applying Proposition 9 one gets that for every regular firing rule $(\mathcal{N}, \mathbf{m}_0)$ is live.

On the other hand, applying Corollary 2, $\mathcal{N}$ has a positive P-invariant implying that for every regular firing rule $(\mathcal{N}, \mathbf{m}_0)$ is bounded.

□

The following corollary is an immediate consequence of the previous theorem and extends the results about well-formed PNs.

Corollary 3. *Let $\mathcal{N}$ be a connected free-choice net with some regular firing rule. Then $\mathcal{N}$ is well-formed if and only if:*

- *$\mathcal{N}$ has a positive P-invariant;*
- *and $\mathcal{N}$ has a positive T-invariant;*
- *and $rank(\mathbf{C}) = |\mathcal{Cl}(\mathbf{C})| - 1$.*

Moreover $(\mathcal{N}, \mathbf{m}_0)$ is live and bounded if and only if:
$\mathcal{N}$ is well-formed and for every siphon $\emptyset \neq Q \subseteq P$, $\mathbf{m}_0(Q) > 0$.

5 Conclusion

We have introduced a new semantics both for CPNs and PNs: the maximal firing semantics. Whilst the semantics seems to combine the advantages of the two formalisms, we have shown that all standard decision problems for nets (reachability, coverability, boundedness, liveness) are undecidable.

Thus we have focused our analysis on the well-known subclass of free-choice nets. In order to obtain generic results, we have introduced two families of semantics; the appropriate firing semantics and the (more restricted) regular firing semantics. Both families include the ordinary and the maximal firing semantics of nets. Then we have proved that the Commoner characterisation of liveness holds for nets with any appropriate firing semantics and that the rank theorem about well-formedness holds for nets with any regular firing semantics.

There are some open questions that we want to address in the future:

- The decidability status of properties for bounded CPNs with maximal firing semantics;
- The complexity of well-formedness for free-choice CPNs with standard semantics;
- The characterisation of home states in live and bounded free-choice PNs with maximal firing semantics.

We are also investigating other subclasses of Petri nets with low-complexity decision algorithms to analyze the impact of the maximal firing semantics. At last we want to revisit the applications developed for CPNs (biology, control theory, etc.) to study the relevance of the maximal firing semantics in this context.

References

1. Blondin, M., Finkel, A., Haase, C., Haddad, S.: The logical view on continuous Petri nets. ACM Trans. Comput. Logic **18**(3), 24:1–24:28 (2017)
2. Cabasino, M.P., Seatzu, C., Mahulea, C., Suárez, M.S.: Fault diagnosis of manufacturing systems using continuous Petri nets. In: Proceedings of the IEEE International Conference on Systems, Man and Cybernetics, Istanbul, Turkey, 10–13 October 2010, pp. 534–539. IEEE (2010)
3. David, R., Alla, H.: Continuous Petri nets. In: Proceedings of the 8th European Workshop on Application and Theory of Petri Nets, Zaragoza, Spain (1987)
4. Desel, J.: A proof of the Rank Theorem for extended free choice nets. In: Jensen, K. (ed.) ICATPN 1992. LNCS, vol. 616, pp. 134–153. Springer, Heidelberg (1992). https://doi.org/10.1007/3-540-55676-1_8
5. Desel, J., Esparza, J.: Free Choice Petri Nets. Cambridge University Press (1995)
6. Fraca, E., Haddad, S.: Complexity analysis of continuous Petri nets. Fund. Inform. **137**(1), 1–28 (2015)
7. Gudiño-Mendoza, B., López-Mellado, E., Alla, H.: Modeling and simulation of water distribution systems using timed hybrid Petri nets. SIMULATION **88**(3), 329–347 (2012)

8. Haar, S., Haddad, S.: On the expressive power of transfinite sequences for continuous Petri nets. In: Kristensen, L.M., van der Werf, J.M.E.M. (eds.) Application and Theory of Petri Nets and Concurrency - 45th International Conference, PETRI NETS 2024, Geneva, Switzerland, 26–28 June 2024, Proceedings. LNCS, vol. 14628, pp. 109–131. Springer, Cham (2024)

9. Hack, M.H.T.: Analysis of production schemata by Petri nets. Technical report, MIT, Dept. Electrical Engineering, M.S. thesis (1972)

10. Jancar, P.: A concise proof of Commoner's theorem. Petri Net Newsl. **49**, 43 (1995)

11. Juhás, G., Lehocki, F., Lorenz, R.: Semantics of Petri nets: a comparison. In: Proceedings of the Winter Simulation Conference, WSC 2007, Washington, DC, USA, 9–12 December 2007, pp. 617–628. WSC (2007)

12. Kemper, P., Bause, F.: An efficient polynomial-time algorithm to decide liveness and boundedness of free-choice nets. In: Jensen, K. (ed.) ICATPN 1992. LNCS, vol. 616, pp. 263–278. Springer, Heidelberg (1992). https://doi.org/10.1007/3-540-55676-1_15

13. Landweber, L.H., Robertson, E.L.: Properties of conflict-free and persistent Petri nets. J. ACM **25–3**, 352–364 (1978)

14. Leroux, J.: The reachability problem for Petri nets is not primitive recursive. In: 62nd IEEE Annual Symposium on Foundations of Computer Science, FOCS 2021, Denver, CO, USA, 7–10 February 2022, pp. 1241–1252. IEEE (2021)

15. Leroux, J., Schmitz, S.: Reachability in vector addition systems is primitive-recursive in fixed dimension. In: 34th Annual ACM/IEEE Symposium on Logic in Computer Science, LICS 2019, Vancouver, BC, Canada, 24–27 June 2019, pp. 1–13. IEEE (2019)

16. Liu, F., Sun, W., Heiner, M., Gilbert, D.R.: Hybrid modelling of biological systems using fuzzy continuous Petri nets. Briefings Bioinform. **22**(1), 438–450 (2021)

17. Minsky, M.L.: Computation: Finite and Infinite Machines. Prentice-Hall Inc. (1967)

18. Rackoff, C.: The covering and boundedness problems for vector addition systems. Theor. Comput. Sci. **6**, 223–231 (1978)

19. Recalde, L., Haddad, S., Suárez, M.S.: Continuous Petri nets: expressive power and decidability issues. Int. J. Found. Comput. Sci. **21**(2), 235–256 (2010)

20. Ross-Leon, R., Ramirez-Trevino, A., Morales, J.A., Ruiz-Leon, J.: Control of metabolic systems modeled with timed continuous Petri nets. In: ACSD/Petri Nets Workshops. CEUR Workshop Proceedings, vol. 827, pp. 87–102 (2010)

21. Renato Vázquez, C., Sutarto, H.Y., Boel, R., Silva, M.: Hybrid Petri net model of a traffic intersection in an urban network. In: Proceedings of the IEEE International Conference on Control Applications, CCA 2010, Yokohama, Japan, pp. 658–664 (2010)

22. Zerhouni, N., Alla, H.: Dynamic analysis of manufacturing systems using continuous Petri nets. In: Proceedings of the IEEE International Conference on Robotics and Automation, Cincinnati, OH, USA, vol. 2, pp. 1070–1075 (1990)

Towards General Trace Theory

Ryszard Janicki[1], Maciej Koutny[1,2], Łukasz Mikulski[3(✉)], and Rajiv Ranjan[2]

[1] Department of Computing and Software, McMaster University, Hamilton, ON L8S 4K1,
Canada
`janicki@mcmaster.ca`

[2] School of Computing, Newcastle University, 1 Science Square, Newcastle upon Tyne NE4
5TG, UK
`{maciej.koutny,raj.ranjan}@ncl.ac.uk`

[3] Faculty of Mathematics and Computer Science and Institute of Advanced Studies, Nicolaus
Copernicus University in Toruń, Chopina 12/18, Toruń, Poland
`lukasz.mikulski@mat.umk.pl`

Abstract. This paper investigates abstract semantical foundations for modelling and verifying concurrent systems by analysing the fundamental relationships between events that shape individual executions. We unify and generalise the established trace-based approaches by developing two label-free semantical frameworks: a step-based model grounded in stratified orders, and an interval-based model capturing observable durations through interval orders. Building on the view of traces as classes of equivalent executions, we introduce novel rewriting mechanisms (including a new rule for sequences of maximal antichains) that characterise these classes independently of concrete syntactic representations. Our approach employs closed relational structures as invariants of concurrent processes and studies their maximal extensions to derive canonical representatives of behaviours. The resulting trace-like equivalences provide, for the first time, an abstract treatment of unlabelled partial orders that isolates the essence of concurrent behaviour and has the potential to support more principled verification techniques.

Keywords: concurrency · total order · partial order · stratified order · interval order · relational structure · trace theory · causality · acyclicity

1 Introduction

Faithful behaviour capture and effective verification techniques of concurrent systems usually require analyses of relationships, such as causality and independence, between the events (occurrences of actions) involved in a system run. Such an approach greatly improves the practical design, construction, and verification of complex software and hardware concurrent systems [25].

Discrete formal semantics of concurrent systems can be defined and investigated at different (consistent) levels of abstraction: from representations of non-branching individual runs, to structures specifying sets of related individual runs, to system models such as Petri nets. Moreover, the choice of representations of individual runs or executions leads to different semantical frameworks across these levels.

© The Author(s), under exclusive license to Springer Nature Switzerland AG 2026
J. Desel and A. Kalenkova (Eds.): PETRI NETS 2026, LNCS 16567, pp. 245–266, 2026.
https://doi.org/10.1007/978-3-032-27879-1_12

Considering non-branching individual runs often amounts to choosing a class of partial orders (e.g., total, stratified, or interval) according to which event occurrences are arranged. Then, suitable relational structures encompassing sets of individual runs may be able to capture intrinsic relationships between events, e.g., causality.

The original trace models captured simple precedence relationships between events [3,26]. Subsequent generalisations introduced the notion of weak precedence and step-wise concurrency in order to express simultaneous or partially-ordered events more explicitly [11,13,17].

This paper aims to push those generalisations further by identifying and focusing on an abstract, label-free essence of traces. We discuss instantaneous (step-based) model based on stratified orders that treat concurrent steps as basic computational atoms as well as a more pragmatic model based on interval orders where such an atomicity is relaxed to accommodate observable durations and interval-like behaviours (according to Wiener [35] executions that can be observed by a single observer are interval orders).

Beyond formal modelling, traces are most useful when interpreted as classes of equivalent executions, a viewpoint that underpins practical verification techniques such as partial-order reduction. In a nutshell, by collapsing all interleavings that represent the same partial-order behaviour into a single representative, one can greatly reduce the state space explored during verification [4,22,28].

In the proposed approach, we use the invariants of concurrent processes (closed relational structures) as specifications of sets of individual executions under consideration, and examine the structure of their maximal extensions (which represent such executions). So far traces (and trace equivalences) were constructed on the basis of rewriting rules applied to language-theoretic representations of concrete observations, viz. sequences in the case of total orders, step sequences in the case of stratified orders, and sequences of beginnings and endings in the case of interval orders. We add to this list a novel rewriting rule for sequences of maximal antichains, which furnish an alternative representation of interval orders and, moreover, a unique sequential one, analogous to chains of atomic elements and chains of unordered sets for total and stratified orders, respectively.

For the two discussed cases, we propose one-step transformations of appropriate partial orders that originate from the inclusion relation. In this way, we obtain to the best of our knowledge for the first time, trace-like equivalences on abstract, unlabelled objects, capturing the essence of traces.

1.1   Introductory Example

Consider the following design scenario. It has been specified that one needs to schedule actions for execution in the set $\Delta = \{x, y, z, w\}$ under the following restrictions:

- x should precede y (precedence denoted $x \prec y$)
- x should precede or be simultaneous with w (weak precedence denoted $x \sqsubset w$)
- y should precede or be simultaneous with z.   (weak precedence denoted $y \sqsubset z$)

We represent the above restrictions as a relational structure:

$$rs_{intro} = \langle \Delta, \prec, \sqsubset \rangle = \langle \{x,y,z,w\}, \{\langle x,y \rangle\}, \{\langle x,w \rangle, \langle y,z \rangle\} \rangle . \tag{1}$$

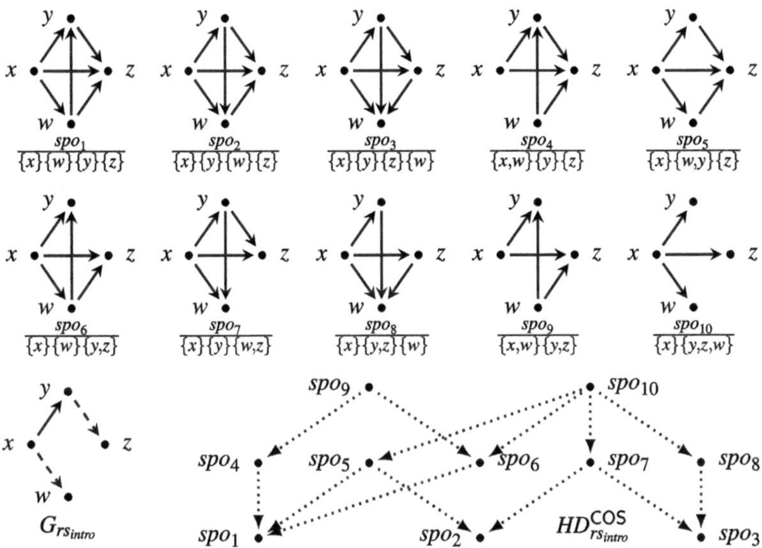

Fig. 1. Deriving and structuring stratified order schedules for rs_{intro}.

Figure 1 depicts a directed graph $G_{rs_{intro}}$ with two kinds of arcs representing rs_{intro} (solid arcs represent $\prec$ and dashed arcs represent $\sqsubset$).

It has further been specified that all schedules should be in the form of stratified partial orders, where a partial order $po = \langle \Delta, \prec' \rangle$ is stratified if its domain Δ can be split into a sequence $\Delta_1 \ldots \Delta_k$ of non-empty sets of unordered elements, and $\Delta_i \times \Delta_{i+1} \subseteq \prec'$, for every $1 \leq i < k$. We then interpret po as a schedule where first all the actions in Δ_1 are executed simultaneously, then all the actions in Δ_2 are executed simultaneously, and so on. Sequences like $\Delta_1 \ldots \Delta_k$ provide language theoretic representations of stratified orders, and are called 'sequences of maximal antichains (strata)'.

There is a simple check verifying that a partial order $spo = \langle \Delta, \prec' \rangle$ conforms to the restrictions captured by rs_{intro}, namely, it must be the case that, for all $t \neq u \in \Delta$:

$$ t \prec u \implies t \prec' u \quad \text{and} \quad t \sqsubset u \implies u \not\prec' t . \tag{2} $$

Now, we might want to find out whether there is at least one stratified order schedule conforming to the restrictions in Eq. (2). Rather than searching the whole set of 75 stratified orders with domain Δ, this can be done by checking that in the graph $G_{rs_{intro}}$ there is no cycle which traverses a solid arc. As this holds for $G_{rs_{intro}}$ in Fig. 1 (in fact, there is no cycle at all in this case), we know that at least one stratified order schedule can be found. More precisely, there are 10 schedules, $spo_1, \ldots, spo_{10}$, depicted in Fig. 1 together with the corresponding sequences of maximal antichains. For example, spo_5 is a stratified order with the corresponding sequence of maximal antichains $\{x\}\{w,y\}\{z\}$ (i.e., first execute x, then simultaneously execute w and y, and finally execute z).

The ten schedules in Fig. 1 are all equivalent as far as the specification rs_{intro} is concerned. However, one might wish to make further checks and select for the actual deployment schedule(s) satisfying some additional criteria. It is possible, of course, to list them all and apply suitable procedures. However, in general, the number of potential schedules may be too big to handle in this way.[1] A practical alternative is to structure the space into an abstract connected graph which is then explored by a suitable (distributed) search technique (see, e.g., [21]). In such an approach the graph is not supplied to the search algorithm upfront, but it is generated on the fly during search. This is done by a neighbourhood function that returns the neighbours of a node. The neighbourhood function takes a node as an input and computes, and should return the neighbours without looking further into the graph.

In this paper, we propose to structure the set of all schedules into a search graph using the partial order inclusion. In addition, we would like to keep the number of the neighbours of a given node low, which is a desirable property by the exploration strategies mentioned above. We will therefore restrict partial order inclusion to its transitive reduction represented by Hasse diagrams.

Figure 1 shows Hasse diagram $HD^{COS}_{rs_{intro}}$ of partial order inclusion for the set of stratified orders $\{spo_1, \ldots, spo_{10}\}$. Since $HD^{COS}_{rs_{intro}}$ is connected (though not strongly connected), after starting at any spo_i we can reach any other spo_j by traversing the arcs of $HD^{COS}_{rs_{intro}}$ in either direction. Therefore, an exploration algorithm starting from a single node is in principle capable of eventually visiting the entire set of schedules.

What we have described above is very close to the idea of a concurrent trace which comprises equivalent executions and is equipped with a similarity allowing one to move between any pair of equivalent executions. And, in our view, graphs like $HD^{COS}_{rs_{intro}}$ can be regarded as a pre-requisite to a trace model for stratified order schedules conforming to rs_{intro}. For such a scheme to work, we need to provide a neighbourhood function for the links between the nodes of $HD^{COS}_{rs_{intro}}$, and a proof that $HD^{COS}_{rs_{intro}}$ is indeed connected.

In this paper, we will provide Hasse diagrams also for schedules represented by interval orders [6]. We would also like to stress that although the introductory example was presented in terms of specifications of schedules, the framework we are aiming at has much wider potential applicability, including verification of concurrent systems.

1.2   Contribution of This Paper

The paper proposes a generic approach to defining trace-like structures for concurrent system executions represented by unlabelled partial orders, and generated by relational structures capable of expressing precedence and weak precedence. In particular, it proposes to structure sets of related executions into abstract graphs on the basis of partial order inclusion. The resulting Hasse diagrams can be seen as precursors of trace models based, e.g., , on language-theoretic representations and labelled partial orders. In the case of interval orders, the paper introduces a novel axiomatisation of partial order inclusion used in the corresponding Hasse diagrams.

The paper provides an instantiation of the proposed approach for two concrete classes of partial orders, namely stratified orders and interval orders. In each case, the equivalence between different partial orders results from the satisfaction of specifications built from two relationships, viz. precedence and weak precedence.

---

[1] For example, if $\prec = \sqsubset = \varnothing$, then the number of different total order schedules is $|\Delta|!$, a huge number even for medium size Δ.

1.3   Related Research

Formal models of operational semantics of concurrent systems are often based on models for representing individual (non-branching) system runs, where two events (executed actions) can be observed as either happening one after another, or as being simultaneous. Such behaviours can be represented as partial orders of events where execution precedence is transitive, and simultaneity or overlapping is captured by the lack of ordering. In this paper, we are interested in two kinds of partial orders modelling concurrent system executions: stratified orders modelling sequential executions of sets of simultaneous (instantaneous) actions, and interval orders modelling behaviours where events can take time and the corresponding time intervals can overlap.

The use of interval orders in the area of concurrency theory can be traced back to [16,18,20,23,24,33,34]. It is important to emphasize that Petri net theories in, e.g., [1,2,31,32], consider executions to be general partial orders, where ordering represents causality, and unorderness represents event independence. In the approach followed in this paper, partial orders of this kind are considered as specifications of sets of individual executions (see below and Sect. 8).

In order to use specifications encompassing sets of related executions one needs to be able to derive them directly from single executions using the relevant structural properties of the concurrent system. In the case of total order executions, this brings into focus relational structures with acyclic relations on events (e.g., dependence graphs introduced in [27] and analysed in detail in [8]), which yield invariant structures (partial orders) after applying a suitable closure operation (e.g., the transitive closure for acyclic relations).

A generalisation of the above approach has been initiated in the late 1980s [7,15, 16,18,24]. More recently, the approach has been substantially revised and generalised first in [10] and then in [12–14].

1.4   Structure of This Paper

In the next section, we recall basic facts involved in the modelling of concurrent behaviours using partial orders, while Sect. 3 recalls the definitions concerning the abstract specifications of behaviours represented by the stratified and interval orders. The following section re-visits the two motivating examples. Section 5 introduces a generic approach to defining trace models for partial orders. The next section instantiates the proposed approach for stratified orders, and Sect. 7 does the same for interval orders. Section 8 briefly discusses additional cases of building trace structures, and outlines potential future work.

2   Partial Orders

Let Δ be a finite non-empty set and $\prec$ be an irreflexive binary relation over Δ; moreover, let $x \frown y$ if $x \neq y$ and $x \not\prec y \not\prec x$. Then $po = \langle \Delta, \prec \rangle$ is called (below $x, y, z, w \in \Delta$):

- *(strict) partial order* if $\qquad\qquad\qquad\qquad\qquad\qquad x \prec z \prec y \implies x \prec y.$
- *total order* if $\qquad\qquad\qquad\qquad x \neq y \not\prec x \lor x \prec z \prec y \implies x \prec y.$

- *stratified order* if $\qquad\qquad x \prec z \prec y \;\vee\; x \frown z \prec y \;\vee\; x \prec z \frown y \implies x \prec y.$
- *interval order* if $\qquad\qquad\qquad\quad\; x \prec y \wedge z \prec w \implies x \prec w \;\vee\; z \prec y.$

The partial, total, stratified, and interval orders are denoted by PO, TPO, SPO, and IPO, respectively. We have $\mathsf{TPO} \subset \mathsf{SPO} \subset \mathsf{IPO} \subset \mathsf{PO}$. We can denote $\prec$ by $\prec_{po}$, etc.

The inclusion between partial orders with the same domain is denoted by $\trianglelefteq$, and the strict inclusion by $\triangleleft$ (i.e., $(\Delta, \prec) \trianglelefteq (\Delta, \prec')$ if $\prec \,\subseteq\, \prec'$, and $(\Delta, \prec) \trianglelefteq (\Delta, \prec')$ if $\prec \,\subset\, \prec'$).

Let $\mathscr{P}$ be a set of partial orders. Then $\triangleleft_{\mathscr{P}} \,\subseteq\, \triangleleft$ is a binary relation on $\mathscr{P}$ such that $po \triangleleft_{\mathscr{P}} po'$ if there is no $po'' \in \mathscr{P}$ such that $po \triangleleft po'' \triangleleft po'$. Intuitively, $\triangleleft_{\mathscr{P}}$ is the transitive reduction (Hasse reduction) of the strict partial order inclusion restricted to $\mathscr{P}$. And, by the assumed finiteness of partial orders,[2] we have $\triangleleft \cap\, (\mathscr{P} \times \mathscr{P}) = \triangleleft_{\mathscr{P}}^{+}$.

The idea of using stratified and interval orders as executions (schedules) comes from the following two facts (cf. [5,6]):

- $\langle \Delta, \prec \rangle$ is a stratified order iff there is a mapping $\mathbf{t} : \Delta \to \mathbb{R}^{+}$ such that $x \prec y \iff \mathbf{t}(x) < \mathbf{t}(y)$, for all $x \neq y \in \Delta$.
- $\langle \Delta, \prec \rangle$ is an interval order iff there are mappings $\mathbf{b}, \mathbf{e} : \Delta \to \mathbb{R}^{+}$ such that $\mathbf{b}(x) \leq \mathbf{e}(x)$ and $x \prec y \iff \mathbf{e}(x) < \mathbf{b}(y)$, for all $x \neq y \in \Delta$.
  Intuitively, each x can be assigned a real 'execution interval' $[\mathbf{b}(x), \mathbf{e}(x)]$.

3   Relational Structures

We employ relational structures of the form $rs = \langle \Delta, \prec, \sqsubset \rangle$, where $\prec$ and $\sqsubset$ are irreflexive binary relations over a finite non-empty domain Δ. We can denote $\prec$ by $\prec_{rs}$, etc.

A relational structure $rs' = \langle \Delta', \prec', \sqsubset' \rangle$ is an *extension* of rs if $\Delta = \Delta'$, $\prec \,\subseteq\, \prec'$ and $\sqsubset \,\subseteq\, \sqsubset'$. We denote this by $rs \trianglelefteq rs'$, and also denote $rs \triangleleft rs'$ if $rs \neq rs'$. Moreover, for all $x \neq y \in \Delta$:

$$rs[x \longrightarrow y] = \langle \Delta, \prec \cup \{\langle x,y \rangle\}, \sqsubset \rangle \quad \text{and} \quad rs[x \,\text{-}\!\!\longrightarrow y] = \langle \Delta, \prec, \sqsubset \cup \{\langle x,y \rangle\} \rangle .$$

3.1   Maximal and Closed Structures

Let $\mathscr{R}$ be a set of relational structures which will provide specifications. There are two subsets of $\mathscr{R}$ playing a central role in our discussion.

A relational structure $rs \in \mathscr{R}$ is *maximal* in $\mathscr{R}$ if there is no other structure in $\mathscr{R}$ extending it. We denote this by $rs \in \mathscr{R}^{max}$. Also, for every relational structure $rs \in \mathscr{R}$, $\max_{\mathscr{R}}(rs')$ are all the structures in $\mathscr{R}^{max}$ extending it. Note that $\max_{\mathscr{R}}(rs)$ is a non-empty set as rs is finite.

A relational structure $rs \in \mathscr{R}$ is *closed* in $\mathscr{R}$ if $\max_{\mathscr{R}}(rs') \subset \max_{\mathscr{R}}(rs)$, for every relational structure $rs' \in \mathscr{R}$ which strictly extends rs. We denote this by $rs \in \mathscr{R}^{clo}$. Also, for every relational structure $rs \in \mathscr{R}$, there is a unique closed relational structure $\mathrm{close}_{\mathscr{R}}(rs)$ which extends rs.

---

[2] Hence there cannot be infinitely many partial orders 'in-between' any two given partial orders.

Referring to the scheduling example, maximal structures are generators of schedules, and closed structures are the most informative specifications (as adding any extra restriction invalidates some of the current schedules). To express this formally, we 'convert' partial orders into relational structures. More precisely, for every partial order $po = \langle \Delta, \prec \rangle$, we denote $\partial(po) = \langle \Delta, \prec, \prec \cup \frown \rangle$ and $\partial^{-1}(\langle \Delta, \prec, \prec \cup \frown \rangle) = po$. Then, for all $x \neq y \in \Delta$, we have:

$$
\begin{array}{lllll}
\text{either} & x \prec y \not\prec x \not\succ y & \wedge & x \prec_{\partial(po)} y \not\prec_{\partial(po)} x \sqsubset_{\partial(po)} y \not\sqsubset_{\partial(po)} x \\
\text{or} & x \not\prec y \prec x \not\succ y & \wedge & x \not\prec_{\partial(po)} y \prec_{\partial(po)} x \not\sqsubset_{\partial(po)} y \sqsubset_{\partial(po)} x & (3) \\
\text{or} & x \not\prec y \not\prec x \frown y & \wedge & x \not\prec_{\partial(po)} y \not\prec_{\partial(po)} x \sqsubset_{\partial(po)} y \sqsubset_{\partial(po)} x \, .
\end{array}
$$

3.2   CO-Structures and CI-Structures

Each relational structure $rs = \langle \Delta, \prec, \sqsubset \rangle$ can be represented by G_{rs} which is a graph with vertices Δ and two kinds of arcs: $x \longrightarrow y$ (representing $x \prec y$) and $x \dashrightarrow y$ (representing $x \sqsubset y$). Paths and cycles in G_{rs} are defined similarly as in the standard directed graphs. For example, $(x \longrightarrow y \dashrightarrow z)$ is a path in $G_{rs_{intro}}$ of Fig. 1. The path resulting from joining a path π ending in a vertex which is the start of path π' is denoted by $\pi \circ \pi'$, e.g., $(x \longrightarrow y) \circ (y \dashrightarrow z) = (x \longrightarrow y \dashrightarrow z)$.

We distinguish six specific kinds of paths in G_{rs} leading from vertex x to vertex y:

- $\mathrm{paths}_{rs}^{w}(x,y)$ are paths without any $\longrightarrow$ arcs.
- $\mathrm{paths}_{rs}^{s}(x,y)$ are paths traversing at least one $\longrightarrow$ arc.
- $\mathrm{paths}_{rs}^{*}(x,y)$ are paths not traversing two successive $\dashrightarrow$ arcs.
- $\mathrm{paths}_{rs}^{s*}(x,y)$ are paths in $\mathrm{paths}_{rs}^{*}(x,y)$ starting with an $\longrightarrow$ arc.
- $\mathrm{paths}_{rs}^{*s}(x,y)$ are paths in $\mathrm{paths}_{rs}^{*}(x,y)$ ending with an $\longrightarrow$ arc.
- $\mathrm{paths}_{rs}^{s*s}(x,y)$ are paths in $\mathrm{paths}_{rs}^{*}(x,y)$ starting and ending with an $\longrightarrow$ arc.

In this paper, we consider two specific sets of relational structures, namely the CO-*structures* COS, and the CI-*structures* CIS:

- $rs \in$ COS if no cycle in G_{rs} traverses an $\longrightarrow$ arc.
- $rs \in$ CIS if every cycle in G_{rs} contains a pair of consecutive $\dashrightarrow$ arcs.

There is a strong connection between the above sets of relational structures and the stratified and interval orders as we have $\mathrm{COS}^{max} = \partial(\mathrm{SPO})$ and $\mathrm{CIS}^{max} = \partial(\mathrm{IPO})$.

Suppose now that $\mathscr{R} = \mathrm{COS}$ and $\mathscr{P} = \mathrm{SPO}$ (or $\mathscr{R} = \mathrm{CIS}$ and $\mathscr{P} = \mathrm{IPO}$).

We can now state formally that $\mathrm{exec}_{\mathscr{P}}(rs) = \partial^{-1}(\max_{\mathscr{R}}(rs))$ are the *executions* in $\mathscr{P}$—called 'schedules' in the introductory example—conforming to $rs \in \mathscr{R}$. It is then immediate that, for all $rs \in \mathscr{R}$ and $po \in \mathscr{P}$, we have (cf. Eq. (2)):

$$
po \in \mathrm{exec}_{\mathscr{P}}(rs) \iff \Delta_{rs} = \Delta_{po} \wedge \prec_{rs} \subseteq \prec_{po} \wedge \prec_{po} \cap \sqsubset_{rs}^{-1} = \varnothing \, . \tag{4}
$$

Proposition 1. *Let $rs \in \mathscr{R}$.*

1. $\max_{\mathscr{R}}(rs) = \max_{\mathscr{R}}(\mathrm{close}_{\mathscr{R}}(rs)) = \partial(\mathrm{exec}_{\mathscr{P}}(rs)) \neq \varnothing$.
2. $\mathrm{exec}_{\mathscr{P}}(rs) = \mathrm{exec}_{\mathscr{P}}(\mathrm{close}_{\mathscr{R}}(rs)) = \partial^{-1}(\max_{\mathscr{R}}(rs)) \neq \varnothing$.
3. $po_2 \lhd po_1 \lhd po_3 \implies po_1 \in \mathrm{exec}_{\mathscr{P}}(rs)$, *for all $po_1 \in \mathscr{P}$ and $po_2, po_3 \in \mathrm{exec}_{\mathscr{P}}(rs)$.*

Note that the last part of the above result makes it explicit that specifications in COS and CIS generate 'convex' sets of partial order executions in SPO and IPO, respectively.

3.3   Closed CO-Structures and CI-Structures

The closed structures in COS—called SO-*structures*—comprise relational structures $\langle \Delta, \prec, \sqsubset \rangle$ such that, for all $x, y, z \in \Delta$:

$$x \prec y \implies x \sqsubset y \ (\text{SO:1}) \qquad x \sqsubset y \prec z \lor x \prec y \sqsubset z \implies x \prec z \ (\text{SO:3})$$
$$x \sqsubset y \sqsubset z \neq x \implies x \sqsubset z \ (\text{SO:2})$$

One can construct the closure of a CO-structure *cos* by exhaustingly applying the axioms SO:1–SO:3. More precisely, for all $x \neq y$:

$$\begin{aligned}
x \prec_{\text{close}_{\text{COS}}(cos)} y \ &\Longleftrightarrow \ \text{paths}^{s}_{cos}(x,y) \neq \varnothing \\
x \sqsubset_{\text{close}_{\text{COS}}(cos)} y \ &\Longleftrightarrow \ \text{paths}^{s}_{cos}(x,y) \cup \text{paths}^{w}_{cos}(x,y) \neq \varnothing \ .
\end{aligned} \tag{5}$$

The closed structures in CIS—called IC-*structures*—comprise relational structures $\langle \Delta, \prec, \sqsubset \rangle$ such that, for all $x, y, z, w \in \Delta$:

$$\begin{aligned}
x \prec y &\implies x \sqsubset y \not\sqsubset x \ (\text{IC:1}) & x \sqsubset y \prec z \sqsubset w \neq x &\implies x \sqsubset w \ (\text{IC:4}) \\
x \prec y \prec z &\implies x \prec z \quad (\text{IC:2}) & x \sqsubset y \prec z \lor x \prec y \sqsubset z &\implies x \sqsubset z \ (\text{IC:5}) \\
x \prec y \sqsubset z \prec w &\implies x \prec w \quad (\text{IC:3})
\end{aligned}$$

One can construct the closure of an IC-structure *cis* by exhaustingly applying the axioms IC:1–IC:5. More precisely, for all $x \neq y$:

$$\begin{aligned}
x \prec_{\text{close}_{\text{CIS}}(cis)} y \ &\Longleftrightarrow \ \text{paths}^{s*s}_{cis}(x,y) \neq \varnothing \\
x \sqsubset_{\text{close}_{\text{CIS}}(cis)} y \ &\Longleftrightarrow \ \text{paths}^{s*}_{cis}(x,y) \cup \text{paths}^{*s}_{cis}(x,y) \neq \varnothing \ .
\end{aligned} \tag{6}$$

Note that, by the definition of CIS (including the irreflexivity of weak precedence) and Eq. (6), we immediately obtain that:

$$(\Delta, \prec, \sqsubset) \in \text{CIS} \iff \forall x \in \Delta : \ \text{paths}^{s*}_{(\Delta, \prec, \sqsubset)}(x,x) = \text{paths}^{*s}_{(\Delta, \prec, \sqsubset)}(x,x) = \varnothing. \tag{7}$$

It is sometimes possible to add new relationships to a closed structure without violating the acyclicity constraints (cf. [13, 14]). More precisely, if $\mathscr{R} \in \{\text{COS}, \text{CIS}\}$ and $rs = \langle \Delta, \prec, \sqsubset \rangle \in \mathscr{R}^{clo}$, then, for $x \neq y \in \Delta$:

$$\begin{aligned}
x \not\prec y & \implies rs[y \dashrightarrow x] \in \mathscr{R} \\
x \not\sqsubset y & \implies rs[y \longrightarrow x] \in \mathscr{R} \\
x \not\sqsubset y \not\sqsubset x & \implies rs[x \dashrightarrow y][y \dashrightarrow x] \in \mathscr{R} \ .
\end{aligned} \tag{8}$$

The second line of Eq. (8) can be generalised in the following way.

Proposition 2. *Let* $ccis = \langle \Delta, \prec, \sqsubset \rangle \in \text{CIS}^{clo}$ *and* $\Delta', \Delta'' \subseteq \Delta$ *be non-empty and disjoint sets such that* $(\Delta' \times \Delta'') \cap \sqsubset = \varnothing$. *Then* $cis = \langle \Delta, \prec \cup (\Delta'' \times \Delta'), \sqsubset \rangle \in \text{CIS}$.

We end this section with an axiomatisation of the maximal CI-structures: $\langle \Delta, \prec, \sqsubset \rangle$ belongs to CIS^{max} if for all $x, y, z, w \in \Delta$:

$$\begin{aligned}
x \prec y &\implies x \sqsubset y & (\text{MCI:1}) & \qquad x \prec y \land z \prec w \implies x \prec w \lor z \prec y \ (\text{MCI:3}) \\
x \prec y &\Longleftrightarrow y \not\sqsubset x \neq y & (\text{MCI:2})
\end{aligned}$$

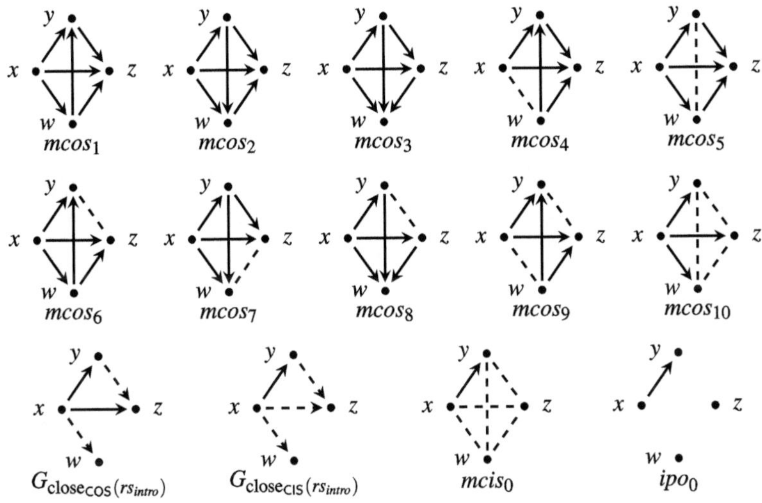

Fig. 2. Example from Fig. 1 continued. A solid arc implies a dashed arc between the same pair of nodes, and a dashed edge implies two dashed arcs in opposite directions.

4   Examples

Consider again the example in Fig. 1, where rs_{intro} there is a CO-structure though it is not a closed CO-structure. The closure of rs_{intro} in COS, $\text{close}_{COS}(rs_{intro})$, is shown in Fig. 2. Compared to rs_{intro}, there are three new relationships: $x \sqsubset y$ because we already have $x \prec y$ and SO:1 can be applied; $x \prec z$ because we already have $x \prec y$ and $y \sqsubset z$ and SO:3 can be applied; and $x \sqsubset z$ because we just derived $x \prec z$ and SO:1 can be applied. Hence,

$$\text{close}_{COS}(rs_{intro}) = rs_{intro}[x \longrightarrow z][x - \longrightarrow z][x - \longrightarrow y].$$

The maximal extensions of rs_{intro} and $\text{close}_{COS}(rs_{intro})$ in COS are the relational structures $mcos_1, \ldots, mcos_{10}$ depicted in Fig. 2. Note that $mcos_i = \partial(spo_i)$, where spo_i is as in Fig. 1 (for $i = 1, \ldots, 10$).

We then observe that rs_{intro} is also a CI-structure as COS $\subset$ CIS. One could therefore extend the search for acceptable schedules to interval order schedules. Figure 2 depicts the closure of rs_{intro} in CIS:

$$\text{close}_{CIS}(rs_{intro}) = rs_{intro}[x - \longrightarrow z][x - \longrightarrow y].$$ The set of schedules now includes interval orders which are not stratified, e.g., ipo_0 depicted in Fig. 2. This schedule can be implemented by: $x \mapsto [1,2]$, $y \mapsto [3,4]$, and $z, w \mapsto [1,4]$.

As another example, consider $cis_{example}$ depicted in Fig. 3. It contains a cycle traversing a solid arc which means that in this case there is no chance of finding a stratified order schedule. However, such a cycle does not rule out schedules given in the form of interval partial orders satisfying Eq. (2). The required property in this case is that every cycle contains at least one pair of consecutive dashed arcs. As this holds for $cis_{example}$, at least one interval order schedule can be found.[3] In fact, there are five interval order

---

[3] Note also that $cis_{example}$ is not a closed CI-structure, and its closure is shown in Fig. 3.

schedules, $ipo_1,\ldots,ipo_5$, shown in Fig. 3. For example, ipo_5 can be implemented by the following assignment of intervals: $z \mapsto [1,2]$, $w \mapsto [3,4]$, $x \mapsto [5,6]$, and $y \mapsto [1,6]$.

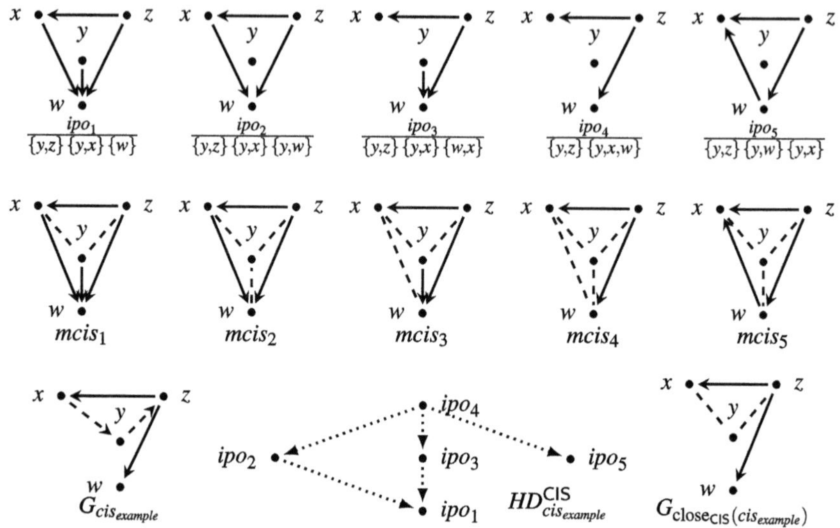

Fig. 3. Deriving and structuring interval order schedules for $cis_{example}$. Note that, for each ipo_i, its maximal antichain representation achrep(ipo_i) is also shown.

The maximal extensions of $cis_{example}$ in CIS are $mcis_1,\ldots,mcis_5$ satisfying $mcis_i = \partial(ipo_i)$ (for $i = 1,\ldots,5$) are also depicted in Fig. 3.

Finally, Fig. 3 shows Hasse diagram $HD^{CIS}_{cis_{example}}$ which is connected, and so after starting at one ipo_i we can reach any other ipo_j by traversing the arcs in either direction.

5   General Idea for Traces of Partial Orders

In this section, we consider any set of partial orders $\mathscr{P}$ together with a set of relational structures $\mathscr{R}$ representing precedence and weak precedence such that $\partial(\mathscr{P}) = \mathscr{R}^{max}$. In particular, we could have $\mathscr{P} = $ SPO and $\mathscr{R} = $ COS, or $\mathscr{P} = $ IPO and $\mathscr{R} = $ CIS.

Each $rs \in \mathscr{R}$ generates a set of partial orders $\mathscr{P}_{rs} = \text{exec}_{\mathscr{P}}(rs)$ included in $\mathscr{P}$, and we regard all partial orders in $\mathscr{P}_{rs}$ as 'equivalent' due to being generated by the same abstract specification rs. We then aim at structuring $\text{exec}_{\mathscr{P}}(rs)$—the space of all partial orders conforming to rs—to allow for effective manipulation and processing, as well as capturing salient semantical features of behaviours represented by relational structures in $\mathscr{R}$ and partial orders in $\mathscr{P}$.

As already indicated, we propose to structure $\text{exec}_{\mathscr{P}}(rs)$ into a *Hasse diagram*

$$HD^{\mathscr{R}}_{rs} = \langle \text{exec}_{\mathscr{P}}(rs), \lhd_{\mathscr{R}} \cap (\text{exec}_{\mathscr{P}}(rs) \times \text{exec}_{\mathscr{P}}(rs)) \rangle$$

which is a directed graph, provided that it is connected. Note also that since $\lhd_{\mathscr{R}}$ is a partial order, $HD^{\mathscr{R}}_{rs}$ is strongly connected iff $|\text{exec}_{\mathscr{P}}(rs)| = 1$.

For the proposed structuring of the 'solution space' $\mathrm{exec}_{\mathscr{P}}(rs)$ to work in practice, we need to provide a proof that $HD_{rs}^{\mathscr{R}}$ is a connected directed graph, and also have means to calculate the neighbourhood of $po \in \mathrm{exec}_{\mathscr{P}}(rs)$ without referring to other partial orders in $\mathrm{exec}_{\mathscr{P}}(rs)$.

The first issue above needs to be treated on a case-by-case basis. The latter issue is addressed in the next immediate result (see Eq. (4)).

Proposition 3. *Let $rs \in \mathscr{R}$ and $po, po' \in \mathscr{P}$ be such that $po \lhd_{\mathscr{P}} po'$.*

1. *If $po \in \mathscr{P}_{rs}$, then $po' \in \mathscr{P}_{rs} \iff (\prec_{po'} \setminus \prec_{po}) \cap \sqsubset_{rs}^{-1} = \varnothing$.*
2. *If $po' \in \mathscr{P}_{rs}$, then $po \in \mathscr{P}_{rs} \iff \prec_{rs} \subseteq (\prec_{po'} \setminus \prec_{po})$.*

Hence, to calculate the neighbourhood of $po \in \mathrm{exec}_{\mathscr{P}}(rs)$ without referring to other partial orders in $\mathrm{exec}_{\mathscr{P}}(rs)$, all one requires is an effective way of calculating the relation $\lhd_{\mathscr{P}}$. Again, this needs to be treated on the case-by-case basis.

In the next two sections, we consider in turn stratified orders with CO-structures, and interval orders with CI-structures. In each case, we prove that $HD_{rs}^{\mathscr{R}}$ is connected, and provide a straightforward way of calculating $\lhd_{\mathscr{P}}$.

6   The Case of Stratified Orders and CO-structures

Several properties of stratified and interval orders can be expressed using maximal sets on unordered elements. This is also the case in this and the next sections, where we use such sets to formulate and prove our results.

A *maximal antichain* (cf. [6, 16]) of a partial order $po = \langle \Delta, \prec \rangle$ is a non-empty set $A \subseteq \Delta$ such that: (i) $x \frown y$, for all $x \neq y \in A$; and (ii) for every $w \in \Delta \setminus A$, there is $z \in A$ satisfying $z \prec w$ or $w \prec z$.

The set of all maximal antichains is denoted by $\mathrm{maxach}(po)$, and we denote $A \ll B$ if $(A \setminus B) \times (B \setminus A) \subseteq \prec$, for all $A \neq B \in \mathrm{maxach}(po)$. It turns out that $\mathrm{principal}(po) = \langle \mathrm{maxach}(po), \ll \rangle$ is a partial order, called the *principal order* of po.

Principal orders uniquely identify the partial orders from which they are derived as $x \prec y \iff A \ll B$, for all $A, B \in \mathrm{maxach}(po)$ such that $x \in A$ and $y \in B$. Moreover, they provide a simple characterisation of the stratified and interval orders: (i) a partial order is interval iff its principal order is total; and (ii) a partial order is stratified iff its principal order is total, and its maximal antichains are mutually disjoint. Thus, each interval (and so also stratified) order po is unambiguously identified by the sequence $\mathrm{achrep}(po) = A_1 \ldots A_n$ of its maximal antichains listed in the order given by $\ll$, called the *maximal antichain representation* (or *maxach-representation*) of po (see Fig. 3).

In general, a finite sequence $\sigma = A_1 \ldots A_n$ of finite non-empty sets is an *antichain sequence* (or *ach-sequence*) (cf. [19]) if, for all $1 \leq i, j, k \leq n$:

$$i < n \implies A_i \not\subseteq A_{i+1} \not\subseteq A_i \tag{9a}$$

$$i < j < k \implies A_i \cap A_k \subseteq A_j . \tag{9b}$$

It can be shown that all *maxach*-representations of interval orders are *ach*-sequences, and all *ach*-sequences are *maxach*-representations of interval orders. Then, given an *ach*-sequence σ, $\mathrm{achrep}^{-1}(\sigma)$ is the interval order such that $\sigma = \mathrm{achrep}(\mathrm{achrep}^{-1}(\sigma))$.

We begin this section with a characterisation of Hasse inclusion for stratified orders.

Theorem 1. *Let* $spo, spo' \in \mathsf{SPO}$. *Then* $spo \lhd_{\mathsf{SPO}} spo'$ *if and only if* $\mathrm{achrep}(spo) = \sigma A \sigma'$ *and* $\mathrm{achrep}(spo') = \sigma BC \sigma'$, *where* $A = B \uplus C$ *(A is the disjoint union of B and C).*

It is worth observing that the above result justifies the axiom

$$\frac{A = B \uplus C}{A = BC} \tag{10}$$

used in the definition of traces of step sequences recently investigated, e.g., in [13].

As an indication of possible exponential reduction of the solution space resulting from using $\lhd_{\mathsf{SPO}}$ rather than $\unlhd$, we observe that if $spo \in \mathsf{SPO}$ is such that $\mathrm{achrep}(spo) = A_1 \ldots A_n$ and $|A_1| = \cdots = |A_n| = 2$, then

$$\begin{aligned}
|spo' \in \mathsf{SPO} \mid spo' \lhd_{\mathsf{SPO}} spo \ \vee \ spo \lhd_{\mathsf{SPO}} spo'\}| &= 3 \cdot n - 1 \\
|spo' \in \mathsf{SPO} \mid spo' \unlhd spo \ \vee \ spo \unlhd spo'\}| &= 3^n + 2^{n-1} - 3 \, .
\end{aligned} \tag{11}$$

The connectedness of Hasse diagrams for CO-structures is shown next, first for the class of closed CO-structures.

Proposition 4. HD^{COS}_{ccos} *is a connected graph, for every* $ccos \in \mathsf{COS}^{clo}$.

Proof. We proceed by reverse induction on the 'extends' relation $\lhd$. The second case in the inductive step is illustrated in Fig. 4.

In the base case, $ccos \in \mathsf{COS}^{max}$. Then HD^{COS}_{ccos} has one node, and so the result holds.

In the inductive case, $ccos = \langle \Delta, \prec, \sqsubset \rangle \notin \mathsf{COS}^{max}$ and we assume that the result holds for every $ccos' \in \mathsf{COS}^{clo}$ such that $ccos \lhd ccos'$. We then consider two cases.

Case 1: There are $x \neq y \in \Delta$ such that $x \not\sqsubset y \not\sqsubset x$. Then, by SO:1, we have $x \not\prec y \not\prec x$. Hence, by Eq. (8), $cos_1 = ccos[x \dashrightarrow y]$, $cos_2 = ccos[y \dashrightarrow x]$, and $cos_3 = ccos[x \dashrightarrow y][y \dashrightarrow x]$ all belong to COS.

Let $ccos_i = \mathrm{close}_{\mathsf{COS}}(cos_i) \in \mathsf{COS}^{clo}$ (for $i = 1, 2, 3$). Then

$$\mathrm{max}_{\mathsf{COS}}(ccos) = \bigcup_{i=1,2,3} \mathrm{max}_{\mathsf{COS}}(ccos_i) \ \text{ implying } \ \mathrm{exec}_{\mathsf{SPO}}(ccos) = \bigcup_{i=1,2,3} \mathrm{exec}_{\mathsf{SPO}}(ccos_i) \, .$$

Moreover, for $i = 1, 2, 3$, we have $ccos \lhd ccos_i$, and so $HD^{\mathsf{COS}}_{ccos_i}$ is connected by the induction hypothesis. Also, we have (see Proposition 1(1,2)):

$$\mathrm{max}_{\mathsf{COS}}(ccos_3) \subseteq \bigcap_{i=1,2} \mathrm{max}_{\mathsf{COS}}(ccos_i) \ \text{ implying } \ \mathrm{exec}_{\mathsf{SPO}}(ccos_3) \subseteq \bigcap_{i=1,2} \mathrm{exec}_{\mathsf{SPO}}(ccos_i),$$

as well as $\varnothing \neq \mathrm{max}_{\mathsf{COS}}(ccos_3)$ implying $\varnothing \neq \mathrm{exec}_{\mathsf{SPO}}(ccos_3)$. As a result, HD^{COS}_{ccos} is connected.

Case 2: For all $x \neq y \in \Delta$, $x \sqsubset y$ or $y \sqsubset x$. Then, by SO:2 and $ccos \notin \mathsf{COS}^{max}$, there are non-empty mutually disjoint sets $\Delta_1, \ldots, \Delta_k$ $(k \geq 2)^4$ such that

$$\Delta = \Delta_1 \cup \cdots \cup \Delta_k \ \text{ and } \ \sqsubset = \bigcup_{1 \leq i \leq j \leq k} \Delta_i \times \Delta_j \setminus id_\Delta \, . \tag{12}$$

---

[4] Note that the Δ_i's are the strongly connected components of the directed graph $\langle \Delta, \sqsubset \rangle$.

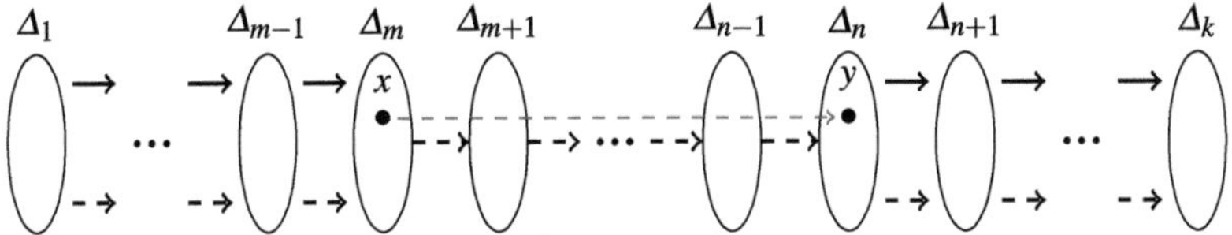

Fig. 4. Relations graph G_{ccos} in Case 2 of the proof of Proposition 4.

Moreover, by SO:1,3 and the irreflexivity of $\prec$, we have

$$\prec \; \subseteq \; \bigcup_{1 \le i < j \le k} \Delta_i \times \Delta_j \, . \tag{13}$$

Now, if the last inclusion was equality, then $ccos \in \mathsf{COS}^{max}$, a contradiction. Hence, there are $1 \le m < n \le k$ as well as $x \in \Delta_m$ and $y \in \Delta_n$ such that $x \not\prec y$. Moreover,

$$\prec \; \cap \; \bigcup_{m \le i < j \le n} \Delta_i \times \Delta_j \; = \; \varnothing \tag{14}$$

as otherwise $\mathrm{paths}^s_{ccos}(x,y) \ne \varnothing$, contradicting $ccos \in \mathsf{COS}^{clo}$ and $x \not\prec y$.

As a result, $cos_1 = ccos[y - \!\!\longrightarrow x]$ and $cos_2 = ccos[x \longrightarrow y]$ both belong to COS, by Eq. (8) and Eq. (12).

Let $ccos_i = \mathrm{close}_{\mathsf{COS}}(cos_i) \in \mathsf{COS}^{clo}$ (for $i = 1,2$). Then (see Proposition 1(1,2)):

$$\mathrm{max}_{\mathsf{COS}}(ccos) = \bigcup_{i=1,2} \mathrm{max}_{\mathsf{COS}}(ccos_i) \ \text{ implying } \ \mathrm{exec}_{\mathsf{SPO}}(ccos) = \bigcup_{i=1,2} \mathrm{exec}_{\mathsf{SPO}}(ccos_i) \, .$$

Moreover, for $i = 1,2$, we have that $ccos \lhd ccos_i$, and so $HD^{\mathsf{COS}}_{ccos_i}$ is connected by the induction hypothesis. We then observe that, by Eq. (8) and Eq. (12) and Eq. (14), we have:

$$mcos_1 = \partial(\mathrm{achrep}^{-1}(\Delta_1 \ldots \Delta_{i-1}(\Delta_i \cup \cdots \cup \Delta_j)\Delta_{j+1} \ldots \Delta_k)) \in \mathrm{max}_{\mathsf{COS}}(cos_1)$$
$$mcos_2 = \partial(\mathrm{achrep}^{-1}(\Delta_1 \ldots \Delta_k)) \in \mathrm{max}_{\mathsf{COS}}(cos_2) \, .$$

Let $spo_i = \partial^{-1}(mcos_i)$ (for $i = 1,2$). We have $spo_1 \lhd spo_2$, and so, by Proposition 1(3) and $\lhd \cap \, (\mathsf{COS} \times \mathsf{COS}) \; = \lhd^+_{\mathsf{SPO}}$, there is a directed path between spo_1 and spo_2 in HD^{COS}_{ccos}. The latter is therefore connected. $\qquad\square$

Then, directly from Propositions 4 and 1(2), we obtain

Theorem 2. HD^{COS}_{cos} *is a connected graph, for every* $cos \in \mathsf{COS}$.

7   The Case of Interval Orders and CI-structures

We begin this section with a characterisation of Hasse inclusion for interval orders. To ease the presentation, we will now consider amended *ach*-sequences with empty sets added at the beginning and end. That is, we consider *guarded ach-sequences* of the form $\gamma = \varnothing \sigma \varnothing$, where σ is an *ach*-sequence. Moreover, we denote $\mathrm{achrep}^{-1}(\gamma) = \mathrm{achrep}^{-1}(\sigma)$.

We need a single axiom to capture Hasse inclusion for the interval orders:

$$BAC =_{x,y} \alpha\beta \tag{15}$$

where A, B, C are sets, $x \in A \setminus C$, $y \in A \setminus B$, and $x \neq y$ as well as

$$\alpha = \begin{cases} B & \text{if } A \setminus \{y\} \subseteq B \\ B(A \setminus \{y\}) & \text{otherwise} \end{cases} \qquad \beta = \begin{cases} C & \text{if } A \setminus \{x\} \subseteq C \\ (A \setminus \{x\})C & \text{otherwise}. \end{cases}$$

There are 9 instantiations of the above, where $x \neq y$, $A_x = A \setminus \{x\}$, and $A_y = A \setminus \{y\}$:

$$\frac{x,y \in A}{\varnothing A \varnothing \to \varnothing A_y A_x \varnothing} \qquad \frac{x \in A \setminus C \quad y \in A \quad A_x \not\subseteq C}{\varnothing AC =_{x,y} \varnothing A_y A_x C} \qquad \frac{x \in A \quad y \in A \setminus B \quad A_y \not\subseteq B}{BA\varnothing =_{x,y} BA_y A_x \varnothing}$$

$$\frac{x \in A \setminus C \quad y \in A \quad A_x \subseteq C}{\varnothing AC =_{x,y} \varnothing A_y C} \qquad \frac{x \in A \setminus C \quad y \in A \setminus B \quad A_x \not\subseteq C \quad A_y \not\subseteq B}{BAC =_{x,y} BA_y A_x C} \qquad \frac{x \in A \setminus C \quad y \in A \setminus B \quad A_x \subseteq C \quad A_y \subseteq B}{BAC =_{x,y} BC} \tag{16}$$

$$\frac{x \in A \quad y \in A \setminus B \quad A_y \subseteq B}{BA\varnothing =_{x,y} BA_x \varnothing} \qquad \frac{x \in A \setminus C \quad y \in A \setminus B \quad A_x \subseteq C \quad A_y \not\subseteq B}{BAC =_{x,y} BA_y C} \qquad \frac{x \in A \setminus C \quad y \in A \setminus B \quad A_x \not\subseteq C \quad A_y \subseteq B}{BAC =_{x,y} BA_x C}$$

Here we can observe more clearly what is the meaning of the transformation we have captured. Basically, the axioms are concerned with a maximal antichain A where event x ends and event y begins. Intuitively, this means that x overlaps y in the weakest possible way. In such a case, one can break this overlapping by splitting A into two consecutive maximal antichains, $A \setminus \{y\}$ and $A \setminus \{x\}$, making x precede y in the tightest possible way. No other relationships between the events are affected.

As an example illustrating an axiom given in Eq. (15), we use it to 'justify' the axiom given in Eq. (10) (recall that stratified orders are also interval orders). Note that it does not matter what is before and after maximal antichains in *maxach*-representations of stratified orders as the component sets are mutually disjoint. Below we show that starting from $\{x, y, z, w\}$ we can derive $\{x, y\}\{z, w\}$:

$$\{x, y, z, w\} =_{x,z} \{x, y, w\}\{y, z, w\} =_{x,w} \{x, y\}\{y, z, w\}$$
$$=_{y,w} \{x, y\}\{y, z\}\{z, w\} =_{y,z} \{x, y\}\{z, w\}.$$

We then obtain a characterisation of Hasse inclusion for the interval orders.

Theorem 3. *Let* $ipo, ipo' \in \mathsf{IPO}$. *Then* $ipo \lhd_{\mathsf{IPO}} ipo'$ *if and only if there are guarded ach-sequences* $\gamma = \tau BAC \omega$ *and* $\gamma' = \tau \alpha \beta \omega$, *where* $BAC =_{x,y} \alpha\beta$ *is as in Eq. (15), such that* $\gamma = \mathrm{achrep}(ipo)$ *and* $\gamma' = \mathrm{achrep}(ipo')$.

Hence Eq. (15) provides an axiomatisation of interval order inclusion. The connectedness of Hasse diagrams is shown next, first for the closed CI-structures, after proving an auxiliary result.

Let $rs = \langle \Delta, \prec, \sqsubset \rangle$ be a relational structure. Then $x \in \Delta$ *irreducibly precedes* $y \in \Delta$ in rs if $x \prec y$ and there are no $z, w \in \Delta$ such that $x \prec z \prec y$ or $x \prec z \sqsubset w \prec y$. We denote this by $x \prec_{rs}^{irred} y$.

Proposition 5. *Let* $mcis = \langle \Delta, \prec, \sqsubset \rangle \in \mathsf{CIS}^{max}$ *and* $x \prec_{mcis}^{irred} y$. *Then* $rs = \langle \Delta, \prec', \sqsubset' \rangle \in \mathsf{CIS}^{max}$, *where* $\prec'$ *and* $\sqsubset'$ *are given by* $\prec \setminus \{\langle x, y \rangle\}$ *and* $\sqsubset \cup \{\langle y, x \rangle\}$, *respectively.*

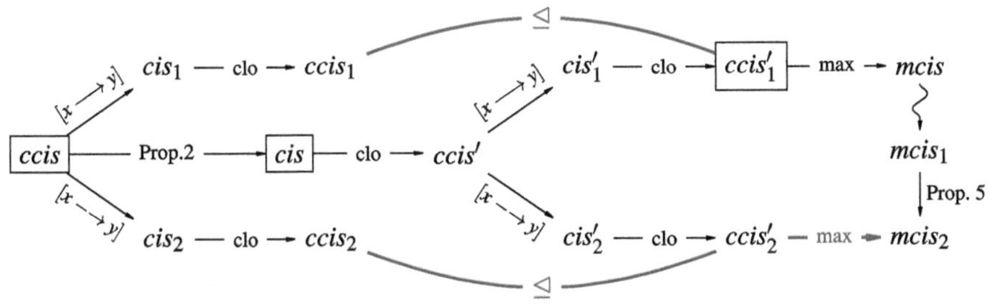

Fig. 5. Flow diagram containing all relational structures used in the proof of Proposition 6. Boxes refer to the vertices from Fig. 6, black arcs are elements of the construction, while red thick arcs are proven properties.

Proof. We note that $x \neq y$ since $\prec$ and $\sqsubset$ are irreflexive. Hence, $\sqsubset'$ is irreflexive. Clearly, MCI:1 and MCI:2 hold for rs, as we only added $\langle y,x \rangle$ to $\sqsubset$, and erased $\langle x,y \rangle$ from $\prec$.

Suppose that MCI:3 does not hold for rs. Then there exist $x',y',z',w' \in \Delta$ such that: $x' \prec' w'$, $z' \prec' y'$, $x' \not\prec' y'$, and $z' \not\prec' w'$. Hence, since $\prec' \subseteq \prec$ and $mcis \in \mathsf{CIS}^{max}$ satisfies MCI:3, we have: $x' \prec w'$, $z' \prec y'$, and $x' \prec y' \vee z' \prec w'$. It therefore follows that $\langle x',y' \rangle$ or $\langle z',w' \rangle$ is the only relationship $\langle x,y \rangle$ erased from $\prec$. Without loss of the generality, we can assume that $\langle x,y \rangle = \langle x',y' \rangle$.

As $\prec$ and $\sqsubset$ are irreflexive, $x = x' \neq w'$ and $y = y' \neq z'$. We consider two cases.

Case 1: $w' = z'$. Then we have $x \prec z' \prec y$, producing a contradiction with $x \prec_{mcis}^{irred} y$.

Case 2: $w' \neq z'$. Then $z' \not\prec w'$. Indeed, suppose that $z' \prec w'$. Then, as $z' \not\prec' w'$, we have $\langle x,y \rangle = \langle z',w' \rangle$, and so $y' = w'$. This, however, contradicts $x' \prec' w'$ and $x' \not\prec' y'$. Then, from $z' \not\prec w' \neq z'$, and MCI:1,2, we obtain $w' \sqsubset z'$. As a consequence, we have $x \prec w' \sqsubset z' \prec y$, producing a contradiction with $x \prec_{mcis}^{irred} y$. $\qquad\square$

The 'operational' intuition behind Proposition 5 is as follows. Having an interval order in the form of a relational structure $mcis$, one can replace strong precedence $x \prec y$ by a self-loop of weak precedences $x \sqsubset y \sqsubset x$ provided that the strong precedence $x \prec y$ cannot be derived by applying MCI:3. More precisely, if in a concrete representation of the interval order (i.e., interval sequence of beginnings and ends) we have at some place the end of x followed by a sequence of other ends, followed by a sequence of other begins, and finally followed by a beginning of y, then we can construct (by proper swaps): (i) an equivalent representation of this interval order, where the end of x directly precedes the beginning of y; and (ii) then extend a bit the the intervals representing x and y, so that they begin to overlap. In this way, we obtain another interval order, where the only change is the erasing of the strong precedence $x \prec y$.

Proposition 6. HD_{ccis}^{CIS} *is a connected graph, for every* $ccis \in \mathsf{CIS}^{clo}$.

Proof. We proceed by reverse induction on the 'extends' relation $\lhd$.

In the base case, $ccis \in \mathsf{CIS}^{max}$. Then HD_{ccis}^{CIS} has one node, and so the result holds.

In the inductive case, $ccis = \langle \Delta, \prec, \sqsubset \rangle \notin \mathsf{CIS}^{max}$, and we assume that the result holds for every $ccis' \in \mathsf{CIS}^{clo}$ such that $ccis \lhd ccis'$. We then consider two cases.

Case 1: There are $x \neq y \in \Delta$ such that $x \not\sqsubset y \not\sqsubset x$.

Then, by IC:1, we also have $x \not\prec y \not\prec x$. Hence, by Eq. (8), $cis_1 = ccis[x \: \text{--}\!\!\longrightarrow y]$, $cis_2 = ccis[y \: \text{--}\!\!\longrightarrow x]$, and $cis_3 = ccis[y \: \text{--}\!\!\longrightarrow x][x \: \text{--}\!\!\longrightarrow y]$ all belong to CIS.

Let $ccis_i = \mathrm{close}_{\mathsf{CIS}}(cis_i) \in \mathsf{CIS}^{clo}$ (for $i = 1, 2, 3$). Then (see Proposition 1(1,2)):

$$\max_{\mathsf{CIS}}(ccis) = \bigcup_{i=1,2,3} \max_{\mathsf{CIS}}(ccis_i) \quad \text{implying} \quad \mathrm{exec}_{\mathsf{IPO}}(ccis) = \bigcup_{i=1,2,3} \mathrm{exec}_{\mathsf{IPO}}(ccis_i) .$$

Moreover, for $i = 1, 2, 3$, we have $ccis \lhd ccis_i$, and so $HD^{\mathsf{CIS}}_{ccis_i}$ is connected by the induction hypothesis. Also, we have (see Proposition 1(1,2)):

$$\max_{\mathsf{CIS}}(ccis_3) \subseteq \bigcap_{i=1,2} \max_{\mathsf{CIS}}(ccis_i) \quad \text{implying} \quad \mathrm{exec}_{\mathsf{IPO}}(ccis_3) \subseteq \bigcap_{i=1,2} \mathrm{exec}_{\mathsf{IPO}}(ccis_i),$$

as well as $\varnothing \neq \max_{\mathsf{CIS}}(ccis_3)$ implying $\varnothing \neq \mathrm{exec}_{\mathsf{IPO}}(ccis_3)$. As a result, HD^{CIS}_{ccis} is connected.

Case 2: For all $x \neq y \in \Delta$, $x \sqsubset y$ or $y \sqsubset x$. To help the reader navigate through this case, Figs. 5 and 6 visualise the idea behind the proof.

In this case we have $x \neq y \in \Delta$ such that $x \sqsubset y \not\sqsubset x \not\prec y$. Hence, by Eq. (8), $cis_1 = ccis[x \longrightarrow y]$ and $cis_2 = ccis[y \: \text{--}\!\!\longrightarrow x]$ both belong to CIS.

Let $ccis_i = \mathrm{close}_{\mathsf{CIS}}(cis_i) \in \mathsf{CIS}^{clo}$ (for $i = 1, 2$). Then

$$\max_{\mathsf{CIS}}(ccis) = \bigcup_{i=1,2} \max_{\mathsf{CIS}}(ccis_i) \quad \text{implying} \quad \mathrm{exec}_{\mathsf{IPO}}(ccis) = \bigcup_{i=1,2} \mathrm{exec}_{\mathsf{IPO}}(ccis_i) .$$

Moreover, for $i = 1, 2$, we have that $ccis \lhd ccis_i$, and so $HD^{\mathsf{CIS}}_{ccis_i}$ is connected by the induction hypothesis. To prove that HD^{CIS}_{ccis} is connected, we will now show that there are $ipo_i \in \mathrm{exec}_{\mathsf{IPO}}(ccis_i)$ (for $i = 1, 2$) such that $ipo_2 \lhd_{\mathsf{IPO}} ipo_1$.

First, we extend $ccis$ as follows. Let $\Delta' = \{z \mid x \prec z\}$ and $\Delta'' = \{w \mid w \prec y\}$. Note that $x, y \notin \Delta' \cup \Delta''$ and $\Delta' \cap \Delta'' = \varnothing$. We observe that $z \in \Delta'$ and $w \in \Delta''$ implies $z \not\sqsubset w$, due to $x \not\prec y$ and IC:3. Hence $(\Delta' \times \Delta'') \cap \sqsubset = \varnothing$. Therefore, by Proposition 2, $cis = \langle \Delta, \prec \cup (\Delta'' \times \Delta'), \sqsubset \rangle$ belongs to CIS.

Let $ccis' = \mathrm{close}_{\mathsf{CIS}}(cis) \in \mathsf{CIS}^{clo}$. Then $x \not\prec_{ccis'} y \not\sqsubset_{ccis'} x$, as shown below.

- Suppose that $x \prec_{ccis'} y$. Then, by Eq. (6) and $x \not\prec_{cis} y$, there is $\pi \in \mathrm{paths}^{s*s}_{cis}(x, y)$ (but $\pi \notin \mathrm{paths}^{s*s}_{ccis}(x, y)$). Let π' be the initial part of π leading to $w \in \Delta''$ for the first time. Then $\pi' \circ (w \longrightarrow y) \in \mathrm{paths}^{s*s}_{ccis}(x, y)$. Hence, by Eq. (6), we obtain $x \prec y$, yielding a contradiction.
- Suppose next that $y \sqsubset_{ccis'} x$. Then, by Eq. (6) and $y \not\sqsubset_{cis} x$, there is π in $\mathrm{paths}^{s*}_{cis}(y, x) \cup \mathrm{paths}^{*s}_{cis}(y, x)$ (but $\pi \notin \mathrm{paths}^{s*}_{ccis}(y, x) \cup \mathrm{paths}^{*s}_{ccis}(y, x)$). Let π' be the initial part of π leading to $w \in \Delta''$ for the first time. Then $\pi' \circ (w \longrightarrow y) \in \mathrm{paths}^{*s}_{cis}(y, y)$. Hence, by Eq. (7), we obtain $cis \notin$ CIS, yielding a contradiction.

Moreover, $\{z \mid x \prec_{ccis'} z\} = \Delta'$ and $\{w \mid w \prec_{ccis'} y\} = \Delta''$. Indeed, if $x \prec_{ccis'} z \notin \Delta'$ then there is $\pi \in \mathrm{paths}^{s*s}_{cis}(x, z)$ (but $\pi \notin \mathrm{paths}^{s*s}_{ccis}(x, z)$). Hence π must contain $\longrightarrow$ arcs from Δ'' to Δ'. Let d be the last element of Δ' that appears in π, and so $\pi = \pi' \circ \pi''$, where d is the last element of π' and the first element of π''. Then $\pi'' \in \mathrm{paths}^{*s}_{ccis}(d, z)$,

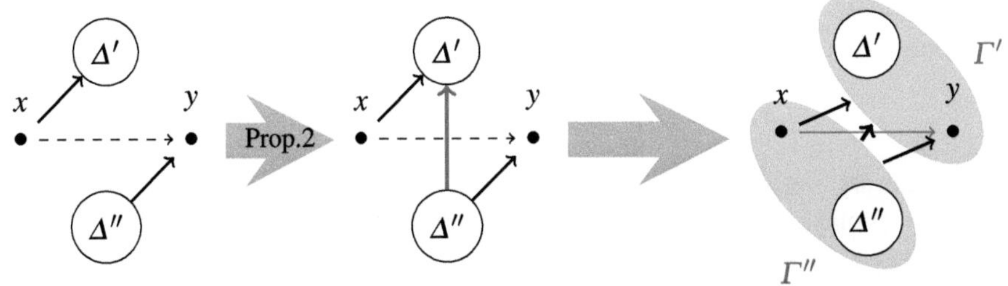

Fig. 6. Constructing *cis* and $ccis'_1$ from *ccis* in the proof of Proposition 6.

and so $(x \longrightarrow d) \circ \pi'' \in \mathrm{paths}^{s*s}_{ccis}(x,z)$. Hence, by Eq. (6), we obtain $x \prec z$. As a result, we have $z \in \Delta'$, yielding a contradiction. The proof for Δ'' is similar.

Since $x \not\prec_{ccis'} y \not\sqsubset_{ccis'} x$, by Eq. (8), we have that $cis'_1 = ccis'[x \longrightarrow y]$ and $cis'_2 = ccis'[y -\!-\!\rightarrow x]$ both belong to CIS.

Let $ccis'_i = \mathrm{close}_{\mathsf{CIS}}(cis'_i) \in \mathsf{CIS}^{clo}$ (for $i = 1,2$). Then we have

$$\max_{\mathsf{CIS}}(ccis') = \bigcup_{i=1,2} \max_{\mathsf{CIS}}(ccis'_i) \quad \text{implying} \quad \mathrm{exec}_{\mathsf{IPO}}(ccis') = \bigcup_{i=1,2} \mathrm{exec}_{\mathsf{IPO}}(ccis'_i) .$$

Moreover, $ccis_i \trianglelefteq ccis'_i$ (for $i = 1,2$).

Let $\Gamma' = \{z \mid x \prec_{ccis'_1} z\}$ and $\Gamma'' = \{w \mid w \prec_{ccis'_1} y\}$. We then observe the following:

- $\Gamma'' \times \Gamma' \subseteq \prec_{ccis'_1}$. Indeed, $\Delta'' \times \Delta' \subseteq \prec_{ccis'_1}$ holds by the construction of *cis*. Recall that extending *ccis* to *cis* and closing it to *ccis'* does not change Δ' and Δ''. Hence, each arc $x \longrightarrow z'$ with $z' \in \Gamma' \setminus \Delta'$, and each arc $w' \longrightarrow y$ with $w' \in \Gamma'' \setminus \Delta''$, was generated by suitable path during the closure of cis'_1. These suitable paths had to traverse $x \longrightarrow y$ and so, w.l.o.g. we have $\pi \in \mathrm{paths}^{s*}_{ccis'}(w',x) \neq \varnothing$ and $\pi' \in \mathrm{paths}^{*s}_{ccis'}(y,z') \neq \varnothing$. As a consequence, we obtain the following:
  - $\pi \circ (x \longrightarrow y) \circ \pi' \in \mathrm{paths}^{s*s}_{ccis'}(w',z')$, and so $w' \prec_{ccis'_1} z'$.
    Hence $(\Gamma'' \setminus \Delta'') \times (\Gamma' \setminus \Delta') \subseteq \prec_{ccis'_1}$.
  - For every $z \in \Delta'$, $\pi \circ (x \longrightarrow z) \in \mathrm{paths}^{s*s}_{ccis'}(w',z)$, and so $w' \prec_{ccis'_1} z$.
    Hence $(\Gamma'' \setminus \Delta'') \times \Delta' \subseteq \prec_{ccis'_1}$.
  - For every $w \in \Delta''$, $(w \longrightarrow y) \circ \pi' \in \mathrm{paths}^{s*s}_{ccis'}(w,z')$, and so $w \prec_{ccis'_1} z'$.
    Hence $\Delta'' \times (\Gamma' \setminus \Delta') \subseteq \prec_{ccis'_1}$.
- $y \in \Gamma'$ and $x \in \Gamma''$.

For each maximal CI-structure $mcis \in \max_{\mathsf{CIS}}(ccis'_1)$, let

$$\psi(mcis) = |\mathrm{succ}^{irred}_{mcis}(x) \setminus \Gamma'| + |\mathrm{pred}^{irred}_{mcis}(y) \setminus \Gamma''| .$$

where $\mathrm{succ}^{irred}_{mcis}(x) = \{z \mid x \prec^{irred}_{mcis} z\}$ and $\mathrm{pred}^{irred}_{mcis}(y) = \{w \mid w \prec^{irred}_{mcis} y\}$.

We then observe that there is $mcis_1 \in \max_{\mathsf{CIS}}(ccis'_1)$ satisfying:

$$y \in \mathrm{succ}^{irred}_{mcis_1}(x) \subseteq \Gamma' \quad \text{and} \quad x \in \mathrm{pred}^{irred}_{mcis_1}(y) \subseteq \Gamma'' . \tag{17}$$

Indeed, let $mcis_1$ be any CI-structure in $\max_{\mathsf{CIS}}(ccis'_1)$ with the lowest value of $\psi()$.

Suppose first that $\psi(mcis_1) > 0$. Then, w.l.o.g. $\mathrm{succ}^{irred}_{mcis}(x) \nsubseteq \Gamma'$. Let

$$mcis' = \langle \Delta, \prec_{mcis} \setminus \{\langle x,z \rangle\}, \sqsubset_{mcis} \cup \{\langle z,x \rangle\} \rangle \,,$$

for an arbitrary $z \in \mathrm{succ}^{irred}_{mcis}(x) \setminus \Gamma' \neq \varnothing$. By Proposition 5, $mcis' \in \mathsf{CIS}^{max}$. Moreover, $mcis' \in \max_{\mathsf{CIS}}(ccis'_1) = \max_{\mathsf{CIS}}(cis'_1)$, as otherwise $\langle x,z \rangle \in \prec_{ccis'_1}$ and so $z \in \Gamma'$. As $\psi(mcis') < \psi(mcis_1)$, we have a contradiction with the choice of $mcis_1$. Hence, $\psi(mcis_1) = 0$, and so $\mathrm{succ}^{irred}_{mcis_1}(x) \subseteq \Gamma'$ and $\mathrm{pred}^{irred}_{mcis_1}(y) \subseteq \Gamma''$.

Suppose now that $y \notin \mathrm{succ}^{irred}_{mcis_1}(x)$. This means that there are paths in $\mathrm{paths}^{s*s}_{mcis_1}(x,y)$ with at least two arcs. Let $\pi = (x \longrightarrow z) \circ \pi' \circ (w \longrightarrow y)$ be one of the longest simple paths (i.e., without repeated nodes) in $\mathrm{paths}^{s*s}_{mcis_1}(x,y)$.

We then observe that $z \in \Gamma'$ and $w \in \Gamma''$. Indeed, suppose w.l.o.g. that $z \notin \Gamma'$. Then, by $\mathrm{succ}^{irred}_{mcis_1}(x) \subseteq \Gamma'$, $z \notin \mathrm{succ}^{irred}_{mcis_1}(x)$. Hence, there is a path $\pi'' \in \mathrm{paths}^{s*s}_{mcis_1}(x,z)$ with at least two arcs. Suppose that π'' shares a node $h \neq z$ with $\pi' \circ (w \longrightarrow y)$. Then we can find an initial fragment π''' of $\pi' \circ (w \longrightarrow y)$ ending at h, and a final fragment π'''' of π'' ending at h, so that $\pi''' \circ \pi'''' \in \mathrm{paths}^{*s}_{mcis_1}(z,z)$, contradicting $mcis_1 \in \mathsf{CIS}$. Hence $\pi'' \circ \pi \in \mathrm{paths}^{s*s}_{mcis_1}(x,y)$ has no repeated nodes, contradicting the choice of π.

We therefore have $\pi' \in \mathrm{paths}^*_{mcis_1}(z,w)$ and, by $z \in \Gamma'$ and $w \in \Gamma''$, $w \prec_{mcis_1} z$. Hence, $\mathrm{paths}^{*s}_{mcis_1}(w,w) \neq \varnothing$, contradicting Eq. (7) and $mcis_1 \in \mathsf{CIS}$. Therefore, $y \in \mathrm{succ}^{irred}_{mcis_1}(x)$ and $x \in \mathrm{pred}^{irred}_{mcis_1}(y)$ can be shown in a similar way.

As a result, the chosen $mcis_1$ satisfies Eq. (17).

Let $mcis_2 = \langle \Delta, \prec_{mcis_1} \setminus \{\langle x,y \rangle\}, \sqsubset_{mcis_1} \cup \{\langle y,x \rangle\} \rangle$. We then observe that $mcis_2$ belongs to $\max_{\mathsf{CIS}}(cis'_2)$. Indeed, by Proposition 5, $mcis_2 \in \mathsf{CIS}^{max}$. Moreover, $cis'_2 \trianglelefteq mcis_2$ since, by $cis'_1 \trianglelefteq mcis_1$, we have:

$$\prec_{cis'_2} = \prec_{cis'_1} \setminus \{\langle x,y \rangle\} \subseteq \prec_{mcis_1} \setminus \{\langle x,y \rangle\} = \prec_{mcis_2}$$
$$\sqsubset_{cis'_2} = \sqsubset_{cis'_1} \cup \{\langle y,x \rangle\} \subseteq \sqsubset_{mcis_1} \cup \{\langle y,x \rangle\} = \sqsubset_{mcis_2} \,.$$

Hence, $ipo_2 \triangleleft_{\mathsf{IPO}} ipo_1$, where $ipo_i = \partial^{-1}(mcis_i)$ (for $i = 1,2$). It then suffices to observe that $ipo_i \in \mathrm{exec}_{\mathsf{IPO}}(ccis'_i) \subseteq \mathrm{exec}_{\mathsf{IPO}}(ccis_i)$ (for $i = 1,2$).   □

Then, directly from Propositions 6 and 1(2), we obtain the following

Theorem 4. HD^{CIS}_{cis} *is a connected graph for every* $cis \in \mathsf{CIS}$.

8   Concluding Remarks

In the final section of the paper we first briefly discuss two examples; in particular, we explain how the classical Mazurkiewicz traces [26] could fit with the approach proposed here. We end with remarks concerning future work.

8.1   Strong Simultaneity

Consider a variation of the design scenario in Fig. 1, assuming the following restrictions: x should precede y and z ($x \prec y$ and $x \prec z$), w should not precede x ($x \sqsubset w$), and z and y should be scheduled together or 'be strongly simultaneous' ($y \sqsubset z \sqsubset y$). We represent such restrictions using $rs_{strongsim}$ with $G_{rs_{strongsim}}$ depicted in Fig. 7. In fact, $rs_{strongsim} = \mathsf{close}_{\mathsf{COS}}(rs_{intro})[z \longrightarrow y]$, where rs_{intro} is as in Eq. (1) and $G_{\mathsf{close}_{\mathsf{COS}}(rs_{intro})}$ is shown in Fig. 2. Figure 7 depicts $HD^{\mathsf{COS}}_{rs_{intro}}$, where spo_6, spo_8, spo_9, and spo_{10} are the only four schedules from Fig. 1 conforming to the current specification.

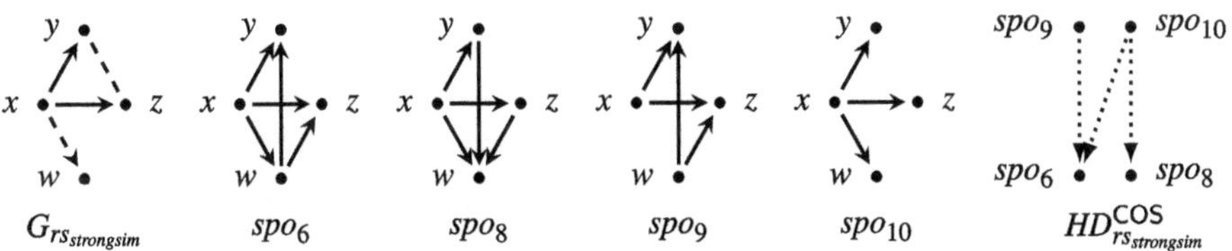

Fig. 7. Deriving and structuring stratified order schedules for $rs_{strongsim}$.

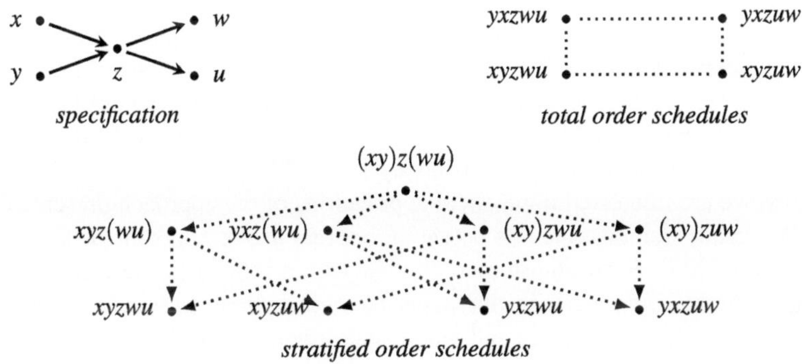

Fig. 8. Deriving stratified order schedules (represented by sequences of maximal antichains, where the curly brackets around singleton sets are omitted, and otherwise replaced by the round brackets), and total order schedules (represented by sequences).

This example demonstrates that we can use relationships like 'strong simultaneity' (cf. [13]) in the specification. It also confirms that extending closed specifications necessarily reduces solution spaces.

8.2   Backward Compatibility (simple Traces)

Consider a design scenario with five actions, x, y, z, w, u, and two restrictions: both x and y should precede z ($x \prec z$ and $y \prec z$), and z should precede both w and u ($z \prec w$ and $z \prec u$). In terms of relational structures, the specification is an acyclic graph depicted in Fig. 8. The task of deriving all stratified order schedules results in Hasse diagram also shown there.

The situation changes when we start considering total order schedules, as the inclusion order for total orders is empty, and trying to connect the schedules using the partial order inclusion is bound to fail. However, there is a remarkably effective and time-honoured solution to this problem, viz. 'relate total orders by swapping neighbouring elements'. This idea has indeed been the basis of Mazurkiewicz traces [26] and Fig. 8 shows how the total order schedules for our example are accordingly related.

> This example suggests that the presented solution may be easily extended to cover a situation where maximal structures are incomparable by inclusion.

8.3   Future Work

Following the idea of backward compatibility examples, it is worth to investigate combining the two types of relationship for structuring solution spaces presented there. Another interesting research question refers to more general subclasses of partial orders.

It is natural to involve the labels of individual events and discuss both classical, static relationships known from Mazurkiewicz traces [26], and more involved local traces, where the dependencies between labels are dynamic [9].

Finally, we are interested in practical applications of the approach presented in this paper. We already demonstrated its potential, but real use requires the design of effective algorithms [29]. Such algorithmic challenges are related to the graph properties of Hasse diagrams introduced in this paper as well as their undirected versions. For example, one might ask whether and when such graphs are Eulerian or Hamiltonian, provide total orders (Hamiltonian paths) that make it easier to search a set of equivalent computations, or allows one to go through all equivalence classes with a given property, generating each of them only once [30].

Acknowledgement. We are grateful to the anonymous referees, whose comments contributed to the revised version of this paper. Support by the Leverhulme Trust grant RPG-2022-025, EPSRC grant EP/Y028813/1, Discovery NSERC of Canada grant No. 2025-07015, and the Minister of Science and Higher Education programme "Excellence Initiative - Research University" is acknowledged.

References

1. Best, E., Fernández, C.: Nonsequential Processes: A Petri Net View. EATCS Monographs on Theoretical Computer Science. Springer, Cham (1988)
2. Best, E., Devillers, R.R.: Sequential and concurrent behaviour in Petri net theory. Theoret. Comput. Sci. **55**(1), 87–136 (1987)
3. Diekert, V., Métivier, Y.: Partial commutation and traces. In: Rozenberg, G., Salomaa, A. (eds.) Handbook of Formal Languages, Volume 3: Beyond Words, pp. 457–533. Springer, Cham (1997)
4. Esparza, J., Heljanko, K.: Unfoldings - A Partial-Order Approach to Model Checking. Monographs in Theoretical Computer Science. An EATCS Series. Springer, Cham (2008)
5. Fishburn, P.C.: Intransitive indifference with unequal indifference intervals. J. Math. Psychol. **7**, 144–149 (1970)
6. Fishburn, P.C.: Interval Orders and Interval Graphs. Wiley, Hoboken (1985)
7. Gaifman, H., Pratt, V.R.: Partial order models of concurrency and the computation of functions. In: Proceedings of the Symposium on Logic in Computer Science (LICS 1987), Ithaca, New York, USA, 22–25 June 1987, pp. 72–85. IEEE Computer Society (1987)
8. Hoogeboom, H.J., Rozenberg, G.: Dependence graphs. In: Diekert, V., Rozenberg, G. (eds.) The Book of Traces, pp. 43–67. World Scientific (1995)
9. Hoogers, P., Kleijn, H., Thiagarajan, P.: A trace semantics for petri nets. Inf. Comput. **117**(1), 98–114 (1995)
10. Janicki, R., Kleijn, J., Koutny, M., Mikulski, Ł: Characterising concurrent histories. Fund. Inform. **139**(1), 21–42 (2015)
11. Janicki, R., Kleijn, J., Koutny, M., Mikulski, Ł: Step traces. Acta Informatica **53**(1), 35–65 (2016)
12. Janicki, R., Kleijn, J., Koutny, M., Mikulski, Ł: Relational structures for concurrent behaviours. Theor. Comput. Sci. **862**, 174–192 (2021)
13. Janicki, R., Kleijn, J., Koutny, M., Mikulski, Ł.: Paradigms of Concurrency - Observations, Behaviours, and Systems - a Petri Net View. Studies in Computational Intelligence, vol. 1020. Springer, Cham (2022)
14. Janicki, R., Kleijn, J., Koutny, M., Mikulski, Ł.: Relational structures for interval order semantics of concurrent systems. In: Kristensen, L.M., van der Werf, J.M.E.M. (eds.) Application and Theory of Petri Nets and Concurrency - 45th International Conference, PETRI NETS 2024, Geneva, Switzerland, 26–28 June 2024. LNCS, vol. 14628, pp. 153–174. Springer, Cham (2024)
15. Janicki, R., Koutny, M.: Invariants and paradigms of concurrency theory. In: Aarts, E.H.L., van Leeuwen, J., Rem, M. (eds.) PARLE 1991. LNCS, vol. 506, pp. 59–74. Springer, Heidelberg (1991). https://doi.org/10.1007/3-540-54152-7_58
16. Janicki, R., Koutny, M.: Structure of concurrency. Theor. Comput. Sci. **112**(1), 5–52 (1993)
17. Janicki, R., Koutny, M.: Semantics of inhibitor nets. Inf. Comput. **123**(1), 1–16 (1995)
18. Janicki, R., Koutny, M.: Fundamentals of modelling concurrency using discrete relational structures. Acta Informatica **34**(5), 367–388 (1997)
19. Janicki, R., Koutny, M.: Operational semantics, interval orders and sequences of antichains. Fund. Inform. **169**(1–2), 31–55 (2019)
20. Janicki, R., Yin, X.: Modeling concurrency with interval traces. Inf. Comput. **253**, 78–108 (2017)
21. Khemani, D.: Search Methods in Artificial Intelligence. Cambridge University Press (2024)
22. Khomenko, V., Koutny, M., Vogler, W.: Canonical prefixes of Petri net unfoldings. Acta Infortmatica **40**(2), 95–118 (2003)

23. Lamport, L.: Time, clocks, and the ordering of events in a distributed system. Commun. ACM **21**(7), 558–565 (1978)
24. Lamport, L.: The mutual exclusion problem: part I - a theory of interprocess communication. J. ACM **33**(2), 313–326 (1986)
25. Ponce de León, H., Mokhov, A.: Compact and efficiently verifiable models for concurrent systems. Formal Methods Syst. Des. **53**(3), 407–431 (2018). https://doi.org/10.1007/s10703-018-0316-0
26. Mazurkiewicz, A.: Concurrent program schemes and their interpretations. DAIMI report. PB 78, Aarhus University (1977)
27. Mazurkiewicz, A.W.: Trace theory. In: Brauer, W., Reisig, W., Rozenberg, G. (eds.) Petri Nets: Central Models and Their Properties, Advances in Petri Nets 1986, Part II, Proceedings of an Advanced Course, Bad Honnef, Germany, 8–19 September 1986. LNCS, vol. 255, pp. 279–324. Springer, Cham (1986)
28. McMillan, K.L.: Using unfoldings to avoid the state explosion problem in the verification of asynchronous circuits. In: von Bochmann, G., Probst, D.K. (eds.) Computer Aided Verification, Fourth International Workshop, CAV 1992, Montreal, Canada, 29 June–1 July 1992. LNCS, vol. 663, pp. 164–177. Springer, Cham (1992)
29. Mikulski, Ł, Niewiadomski, A., Piątkowski, M., Smyczyński, S.: On generation of context-abstract plans. In: Canal, C., Idani, A. (eds.) SEFM 2014. LNCS, vol. 8938, pp. 376–388. Springer, Cham (2015). https://doi.org/10.1007/978-3-319-15201-1_25
30. Mikulski, Ł., Piątkowski, M., Smyczyński, S.: Lexicographical generations of combined traces. In: Proceedings of ACSD 2013, pp. 196–205. IEEE Computer Society (2013)
31. Petri, C.A.: Fundamentals of a theory of asynchronous information flow. In: Information Processing, Proceedings of the 2nd IFIP Congress 1962, Munich, Germany, 27 August–1 September 1962, pp. 386–390. North-Holland (1962)
32. Petri, C.A.: Concepts of net theory. In: Mathematical Foundations of Computer Science: Proceedings of Symposium and Summer School, Strbské Pleso, High Tatras, Czechoslovakia, 3–8 September 1973, pp. 137–146. Mathematical Institute of the Slovak Academy of Sciences (1973)
33. Pratt, V.R.: Modeling concurrency with partial orders. Int. J. Parallel Prog. **15**(1), 33–71 (1986)
34. Vogler, W.: Partial order semantics and read arcs. Theoret. Comput. Sci. **286**(1), 33–63 (2002)
35. Wiener, N.: A contribution to the theory of relative position. Proc. Cambridge Philos. Soc. **33**(2), 313–326 (1914)

Safety Analysis in Broadcast Networks Defined by Graph Grammars

Christoffer Lind Andersen[1(✉)], Radu Iosif[1], and Arnaud Sangnier[2]

[1] Univ. Grenoble Alpes, CNRS, Grenoble INP, VERIMAG, Grenoble, France
`{christoffer-lind.andersen,radu.iosif}@univ-grenoble-alpes.fr`
[2] DIBRIS, Università di Genova, Genoa, Italy
`arnaud.sangnier@unige.it`

Abstract. We consider families of networks where processes communicate by synchronous broadcast (i.e., a sent message is received by all the neighbours of the emitter) and we study the following safety problem: is there a network in the given family, such that some process can reach an error location? Specifically, we focus on families of network topologies defined by graph grammars, where each node of the produced graph can be either a clique or a cloud (anti-clique) of processes of unbounded sizes. We show that, in general, the considered safety problem is undecidable, when the communication is reliable, and becomes decidable with unreliable communication (i.e., broadcast messages can be lost) or whenever the protocols executed by the different processes cannot send and receive messages from the same control state (also known as the wait-only syntactic restriction).

1   Introduction

Present-day computer systems are increasingly distributed, with myriads of devices that communicate over a network. The design of such networks often requires structuring the network into subnets that have specific communication topologies, which determines their functionalities. The physical layers may also vary inside the same network, from wired, i.e., obeying a point-to-point communication policy, to wireless, i.e., relying on a broadcast communication model. The huge scale of networks, such as datacenter networks with $\geq 10^4$ routers for a regional hub and continuously growing, requires *parametric* verification techniques, that work no matter how many servers in a rack, switches in a cluster, clusters in a layer, etc. Despite decades of theoretical research and an impressive body of results (see [5] for a nice survey), these techniques consider a small number of fixed communication topologies, such as cliques that model broadcast [15], or rings that model point-to-point communication [7].

One of the reasons for focusing on particular hard-coded architectures (i.e., cliques and rings) is that the parameterized verification of safety properties (i.e., reachability of a given error state by some process in the network) is undecidable

Partly Supported by ANR PaVeDyS (ANR-23-CE48-0005).

J. Desel and A. Kalenkova (Eds.): PETRI NETS 2026, LNCS 16567, pp. 267–288, 2026.
https://doi.org/10.1007/978-3-032-27879-1_13

even for ring topologies that circulate sufficiently many message types [7]. For clique (or equivalently complete) topologies instead, the problem has been shown to be decidable when the communication is performed by synchronized pairwise rendez-vous [15] (with a polynomial time algorithm) or by broadcast [13].

Some of the limitations imposed by such hard-coded architectures have been overcome in recent years, with the development of parametric verification techniques for systems described using logic [3] or graph grammars [6,17,19]. Logic-based specification languages are particularly suitable for describing graph properties, such as planarity, k-colorability or connectivity. On the other hand, graph grammars are appealing for their constructive aspect, i.e., the ability of specifying how large graphs are built from smaller subgraphs. Both logics and graph grammars are at the core of the theory of formal languages, due to seminal result of Courcelle and Engelfriet, which shows the decidability of Monadic Second Order (MSO) logic over the classes of context-free languages produced by hyperedge-replacement (HR) and vertex-replacement (VR) grammars (see [8] for a survey).

Contributions. In this paper, we consider the parameterized safety problem for families of topologies described by VR-grammars, with a broadcast communication model, where each process sends messages to all its neighbors in the graph. Since the parametric safety problem is known to be undecidable even for line topologies [10], we consider families of underlying *clique-and-cloud* topologies, in which each vertex represents either a clique (i.e., a complete graph) or a cloud (i.e., a completely disconnected graph). The vertices of the expanded cliques/clouds are pairwise linked as indicated by the underlying graph. In other words, our topology specification language does not allow to describe a line, ring or tree of processes, but instead lines, rings or trees of cliques/clouds of arbitrary sizes, respectively. This style of specification is inspired by the layered architectures of modern datacenters, such as Azure [18], where each layer consists of a cloud of pairwise disconnected switches, that are connected with all the switches from the upper and lower layers.

When considering such families of structured cliques/clouds, the parametric safety problem becomes undecidable under the reliable communication model, even for the simplest clique-and-cloud topologies, such as a line. This is not surprising, because, for instance, reliable broadcast in a clique can be used to determine a leader (i.e., send all non-leader processes into an inactive sink state), hence a line of cliques can become a line of leaders, able to simulate a 2-counter Minsky machine [10].

To recover decidability, we consider two natural restrictions: *unreliable communication*, i.e., where messages can be lost, and *reliable communication* with the wait-only restriction on the local control of a process, i.e., no states from which messages can be both sent and received [16]. All decidability results are obtained by an encoding of the safety verification problem in an MSO formula, that is checked against the VR-grammar that describes the family of clique-and-cloud topologies. The latter problem (i.e., the existence of a model of an MSO formula in the language of a VR-grammar) is decidable, by a classical result of Courcelle and Engelfriet [8].

Related Works. Parametric systems with clique topologies, that communicate using reliable broadcast have been among the first to be studied in the literature [13]. The technique for establishing decidability of the parametric safety problem relies on an algorithm based on well-structured transition systems [1]. These results strongly depend on the hypothesis that the topology of each system is a single clique. In contrast, our work goes beyond single cliques and allows to reason about parametric systems whose topologies consist of arbitrary graph-like structures, with vertices representing cliques and/or clouds.

In [9,10], Delzanno et al. study the parametric safety problem for reliable and unreliable broadcast seeking for a topology which breaks the safety property. They show that this problem is undecidable for reliable broadcast [10] and decidable in polynomial time when the communication is unreliable [9]. Another way to regain decidability with reliable broadcast is to restrict the set of considered topologies [11]. In these works, it is not possible to describe precise architectures, there are either unrestricted or the restriction regards only the length of paths in the graphs.

Graph grammars and MSO have been used to specify families of topologies using the point-to-point (rendez-vous) communication model. In this model, each network of processes is simulated by an equivalent 1-safe Petri net [14], meaning that the parametric verification problem amounts to reasoning about infinite families of 1-safe Petri nets, that share a certain structural invariant. On one hand, parametric systems specified using HR-grammars can be translated into a finite set of Petri nets, that over approximate their behavior [6]. On the other hand, parametric systems specified by VR-grammars can be converted into HR-systems that preserve the safety properties [17]. For token-passing systems, where the rendez-vous communication is restricted to the passing of a single token among processes, the decidability of the safety verification problems is established using the same decidability result of Courcelle and Engelfriet [8] used in this paper [2]. Unfortunately, such decidability results for rendez-vous systems cannot be easily transferred to broadcast systems, mainly because, with broadcast, the number of participants in an interaction is unbounded.

Notation. The set of natural numbers is denoted by $\mathbb{N}$. Given numbers $i, j \in \mathbb{N}$, we write $[i,j] \stackrel{\text{def}}{=} \{i, i+1, \ldots, j\}$, assumed to be empty if $i > j$. The cardinality of a finite set A is denoted by $\|A\|$. The disjoint union $A \uplus B$ is defined as the union of A and B, if $A \cap B = \emptyset$, and undefined, otherwise. For a relation $R \subseteq A \times A$, we denote by R^* its reflexive and transitive closure. We write $[x \to y]$ for the function $f : A \to A$ that sends x into y and behaves as the identity over A everywhere else.

Due to lack of space, omitted proofs can be found in [4].

2   Network Model

This section defines the network model. Intuitively, a network is a binary graph, whose vertices run identical copies of one or more finite-state machines, called process types. Instances of process types, also called processes, placed on adjacent vertices in the network graph may engage in broadcast communication along the edges: a process sends a message and several or all of its neighbours receive it. As usual, sending or receiving a message is modelled by a local state change within a sender or receiver process, respectively.

2.1 Process Types and Network Topologies

Let Σ be a finite alphabet of messages, considered fixed in the rest of this paper.

Definition 1 (Process Type). *A process type P is a tuple (Q, q_{in}, δ) where:*

- *Q is a finite set of control states,*
- *$q_{in} \in Q$ is the initial state,*
- *$\delta \subseteq Q \times \{?m, !m \mid m \in \Sigma\} \times Q$ is the action relation; as usual, we write $q \xrightarrow{\alpha}_\delta q'$ instead of $(q, \alpha, q') \in \delta$.*

In the rest of this paper, we fix a set of process types $\mathcal{P} = \{P_1, \ldots, P_\ell\}$, for which we assume that $P_i \stackrel{\text{def}}{=} (Q_i, q_{in,i}, \delta_i)$ for all $i \in [1, \ell]$ and that $Q_i \cap Q_j = \emptyset$ for all $i \neq j \in [1, \ell]$, i.e., each process type has a set of local states disjoint from the other. Let $Q \stackrel{\text{def}}{=} \biguplus_{1 \leq i \leq \ell} Q_i$ and $\delta \stackrel{\text{def}}{=} \biguplus_{1 \leq i \leq \ell} \delta_i$ denote the overall set of states and transition relation, respectively.

We represent a network by a simple undirected binary graph, with no self-loops, such that each vertex of the network graph is labelled by a process type.

Definition 2 (Network Topology). *A network topology is a $\mathcal{P}$-labeled graph $G = (V, E, lab)$ such that:*

- *V is a finite set of vertices,*
- *$E \subseteq (V \times V) \setminus \{(v, v) \mid v \in V\}$ is an irreflexive and symmetric edge relation, i.e., $(u, v) \in E \iff (v, u) \in E$ for all $u, v \in V$,*
- *$lab : V \mapsto \mathcal{P}$ labels each vertex with a process type.*

2.2 Execution Semantics

The execution of a network is formally described by a transition relation over configurations. A configuration is a snapshot of the network, that gives the local state of each process. Formally, a configuration of a network topology $G = (V, E, lab)$ is a function $c : V \mapsto Q$, where Q is the set of states, such that $c(v) \in Q_i$ if $lab(v) = P_i$, for each vertex $v \in V$. The *initial* configuration c_{in}^G is given by $c_{in}^G(v) = q_{in,i}$, for each vertex $v \in V$ such that $lab(v) = P_i$. For a given configuration c, we denote by $\#_c(q) \stackrel{\text{def}}{=} \|\{v \in V \mid c(v) = q\}\|$, whenever the network topology in question is understood. We denote by $\mathcal{C}_G$ the set of configurations of the network topology G.

For a given network topology G, we define a transition relation $\Rightarrow_G \subseteq \mathcal{C}_G \times \mathcal{C}_G$. For a family $\mathcal{G}$ of topologies, we define the relation $\Rightarrow_\mathcal{G} \stackrel{\text{def}}{=} \bigcup_{G \in \mathcal{G}} \Rightarrow_G$. We omit mentioning the network topology of the family thereof, whenever this is understood or not important. In the definition of $\Rightarrow_G$, hence of $\Rightarrow_{\mathcal{G}}$, we distinguish the cases of *reliable* and *unreliable* communication. The communication is reliable if each message that has been sent, via a $!m$ action, will be received, via a matching $?m$ action, by each process able to take that action. Instead, in unreliable communication, some messages may be lost. Let $G = (V, E, lab)$ be a network graph, in the following.

Reliable Communication. The execution semantics for reliable communication is the relation $\Rightarrow_r \subseteq \mathcal{C}_G \times \mathcal{C}_G$, such that $c \Rightarrow_r c'$ if and only if, (i) there exists a vertex $u \in V$ and a message $m \in \Sigma$ such that $c(u) \xrightarrow{!m}_\delta c'(u)$ and (ii) for each vertex $v \in V \setminus \{u\}$, exactly one of the following holds:

- $(u, v) \in E$ and $c(v) \xrightarrow{?m}_\delta c'(v)$, i.e., v is a neighbour of u that can receive m,
- $(u, v) \in E$, $c(v) \xrightarrow{?m}_\delta q'$ for no $q' \in Q$ and $c'(v) = c(v)$, i.e., v is a neighbour of u, but it cannot receive m, or
- $(u, v) \notin E$ and $c'(v) = c(v)$, i.e., v is not a neighbour of u.

In this case, we also say that u sends the message m. Note that a send action is always non-blocking, meaning that it will be executed even if no neighbouring process can execute a matching reception.

Unreliable Communication. The execution semantics for unreliable communication is the relation $\Rightarrow_u \subseteq \mathcal{C}_G \times \mathcal{C}_G$, such that $c \Rightarrow_u c'$ if and only if, (i) there exists a vertex $u \in V$ and a message $m \in \Sigma$ such that $c(u) \xrightarrow{!m}_\delta c'(u)$ and (ii) for each vertex $v \in V \setminus \{u\}$ one of the following holds:

- $(u, v) \in E$ and $c(v) \xrightarrow{?m}_\delta c'(v)$, i.e., v is a neighbour of u that receives m,
- $c'(v) = c(v)$, i.e., v does not receive the message.

In this case, even if a neighbour of the sender can receive a message m, via a $q \xrightarrow{?m} q'$ action, it may not take that action.

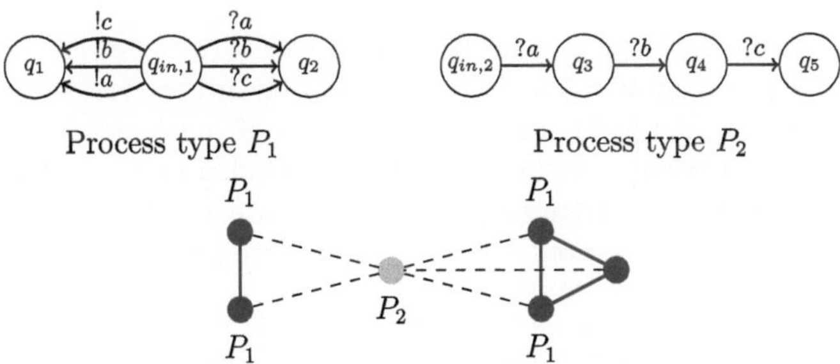

Fig. 1. Two process types and a network topology G

Example 1. We draw an example of two process types P_1 and P_2 on Fig. 1 with respective initial states $q_{in,1}$ and $q_{in,2}$ together with a possible network topology G (all the lines are considered to be edges, even if they are dashed). We wonder whether in this latter topology, there is an execution from the initial configuration which leads to a configuration where the central vertex is in state q_5. Note that to reach such a configuration, the central vertex needs to receive sequentially the messages a, b and c. These messages can be produced by the vertices with process type P_1. However, if we consider the reliable semantics, it is not possible to reach this configuration, because in the two cliques of vertices with process type P_1 (on the left and on the right), as soon as a process broadcasts a message, it sends its neighbours in state q_2 from which they cannot broadcast messages any more. It is hence impossible that the three messages are sent. As a matter of fact, it is possible to put the central node in state q_4 but not in q_5. If we consider the unreliable semantics, then it is possible to reach a configuration with the central node in q_5, the reason being that when a process in one of the clique of process types P_1 has a neighbour which broadcasts a message, it does not have to receive it. We can hence have three vertices, where the first one broadcasts a, then the second one broadcasts b and finally the last one broadcasts c.

3   Families of Topologies

We introduce a specification method for describing infinite families of network topologies. The definition is given in two stages. First, we consider graphs whose vertices can be replaced either by a clique (i.e., complete graphs) or a cloud (i.e., a set of disconnected vertices), depending on a special label. These graphs are called *clique-and-cloud* topologies. Second, we introduce *graph grammars* that describe infinite sets of clique-and-cloud topologies. A family of topologies described in this way consists of the set of concrete expansions of the clique-and-cloud topologies from the language of the graph grammar.

3.1   Clique-and-Cloud Topologies

A *clique-and-cloud topology* $\mathcal{G}$ is a tuple $(\mathcal{V}, \mathcal{E}, Lab, \mathcal{V}_\Delta, \mathcal{V}_{\cdot\cdot})$ where $(\mathcal{V}, \mathcal{E}, Lab)$ is a network topology and $\mathcal{V} = \mathcal{V}_\Delta \uplus \mathcal{V}_{\cdot\cdot}$ is a partition of the vertices. Each vertex in $\mathcal{V}_\Delta$ represents a clique (i.e., a complete graph of unbounded size), whereas each vertex in $\mathcal{V}_{\cdot\cdot}$ represents a cloud (i.e., an unbounded number of disconnected vertices). Each undirected edge $(u, v) \in \mathcal{E}$ represents a set of undirected edges, between each vertex represented by u and each vertex represented by v.

Definition 3 (Expansion). *A network topology $G = (V, E, lab)$ is an expansion of the clique-and-cloud topology $\mathcal{G} = (\mathcal{V}, \mathcal{E}, Lab, \mathcal{V}_\Delta, \mathcal{V}_{\cdot\cdot})$ via a surjective function $f : V \mapsto \mathcal{V}$ if and only if the following hold:*

- *$lab(v) = Lab(f(v))$ for all $v \in V$, i.e., all vertices in the expansion inherit the process type of the originating vertex from the clique-and-cloud topology,*

- $(u, v) \in E$ *and (i)* $(f(u), f(v)) \in \mathcal{E}$ *or (ii)* $f(u) = f(v)$ *and* $f(u) \in \mathcal{V}_\Delta$, *i.e., an edge in the expansion corresponds to an edge from the clique-and-cloud topology or from one of the expanded cliques.*

We denote by $\llbracket \mathcal{G} \rrbracket$ *the set of pairs* $\{(G, f) \mid G$ *expands* $\mathcal{G}$ *via* $f\}$.

We sometimes write $G \in \llbracket \mathcal{G} \rrbracket$, meaning that there exists a surjective function f, such that $(G, f) \in \llbracket \mathcal{G} \rrbracket$. Note that each clique-and-cloud topology represents an infinite number of network topologies.

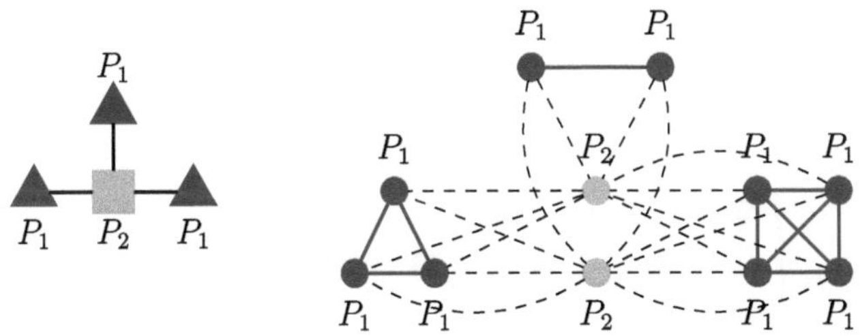

Fig. 2. A clique-and-cloud topology and one of its expansion

Example 2. On Fig. 2, we depicted on the left a clique-and-cloud topology consisting of three cliques ▲ labeled with P_1 and one cloud node ▮ labeled with P_2. The right side of the figure shows an expansion of this topology (we have drawn in dashed lines the edges which connect the cliques and clouds together, and in plain lines the edges of the cliques).

3.2 Graph Grammars

We consider graph grammars written using the vertex-replacement (VR) algebra, see [8,12] for surveys. This algebra manipulates graphs having a set of designated vertices, called *ports*. In our setting, ports are labelled (in addition to process types and clique/cloud labels) with elements from a countably infinite set Π of *port labels*. An *open* clique-and-cloud topology is a tuple $(\mathcal{V}, \mathcal{E}, Lab, \mathcal{V}_\Delta, \mathcal{V}_{\therefore}, Port)$, where $(\mathcal{V}, \mathcal{E}, Lab, \mathcal{V}_\Delta, \mathcal{V}_{\therefore})$ is a clique-and-cloud topology as before and $Port : \mathcal{V} \rightharpoonup \Pi$ is a partial mapping associating several vertices with port labels. Note that the same port label can be used to label zero or more vertices. In the following, we do not distinguish isomorphic graphs, i.e., that differ only by a renaming of vertices.

We consider the VR algebra, whose domain is the set of open clique-and-cloud topologies, and whose operations are described below:

- *Unit graph*: for a port label $\pi \in \Pi$, $P \in \mathcal{P}$ and $\kappa \in \{\Delta, \therefore\}$, the constant (i.e., nullary operation) $\bullet\pi(P, \kappa)$ denotes the open topology with a single vertex, labelled with the process type P and port π, that belongs to $\mathcal{V}_\kappa$, and no edges.

- *Edge creation*: $\mathrm{edg}_{\pi_1,\pi_2}(\mathcal{G})$ adds an undirected edge between each vertex labelled π_1 and each vertex labelled π_2 from $\mathcal{G}$, see Fig. 3 (a).
- *Port redefinition*: $\mathrm{relab}_\alpha(\mathcal{G})$ renames each port label π of $\mathcal{G}$ as $\alpha(\pi)$, for a mapping $\alpha : \Pi \mapsto \Pi$, see Fig. 3 (c).
- *Disjoint union*: $\mathcal{G}_1 \oplus \mathcal{G}_2$ takes the component-wise disjoint union of $\mathcal{G}_1$ and $\mathcal{G}_2$ (if the graphs are not disjoint, we take an isomorphic copy of one of them), see Fig. 3 (b).

Given a set of variables $\mathcal{X}$ (of arity 0), a term of the VR-algebra over Π and $\mathcal{X}$ is built as usual from the operations of VR and the variables from $\mathcal{X}$. A *ground term* is a term with no variable. We write θ^{VR} for the clique-and-cloud topology obtained by interpreting the function symbols in the ground term θ as the corresponding VR-operations and by removing the port labels.

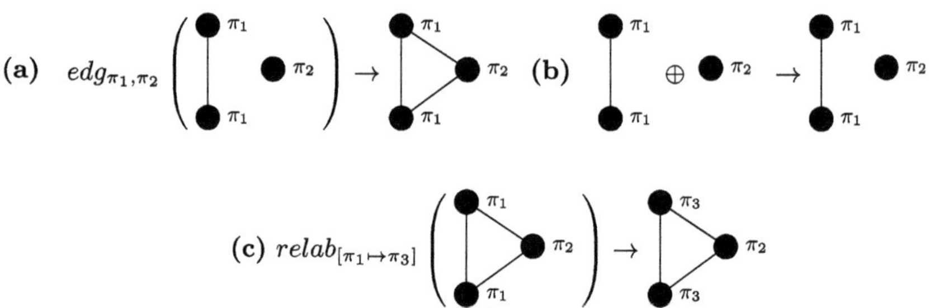

Fig. 3. The operations of the vertex replacement algebra; **(a)** edge creation, **(b)** disjoint union and **(c)** port redefinition

We use grammars (i.e., finite sets of recursive rules) over the signature of the VR-algebra to describe infinite sets of clique-and-cloud topologies:

Definition 4 (VR-graph grammar). *A VR-graph grammar Γ is a tuple $(\mathcal{X}, \Pi, \mathcal{R}, X_0)$ where $\mathcal{X}$ is a set of* non-terminals *(i.e., variables), Π is a set of vertex ports, X_0 is an initial non-terminal and $\mathcal{R}$ is a set of rules of the form $X \hookrightarrow \theta$ where $X \in \mathcal{X}$ and θ is a VR-term with variables from $\mathcal{X}$.*

Given two terms ξ and η, a derivation step $\xi \leadsto_\Gamma \eta$ means that η is obtained from ξ by replacing an occurrence of a non-terminal X with the term θ for some rule $X \hookrightarrow \theta$ from Γ. A X-derivation is then a sequence of steps starting with the non-terminal X. The derivation is complete if it ends with a ground term (a term is ground if it has no variable occurrences). This allows us to define the language of a grammar $\mathcal{L}(\Gamma)$ which is the set of clique-and-cloud topologies defined by $\{\Theta^{\mathrm{VR}} \mid X_0 \leadsto_\Gamma^* \Theta$ is a complete X_0-derivation$\}$.

Example 3. We present a grammar which generates clique-and-cloud topologies in form of a star with a central cloud vertex of process type P_2 and adjacent clique vertices around it of process type P_1. The grammar is defined as $\Gamma_{Star} = (\{X_0, S, C, O\}, \{\pi_1, \pi_2\}, \mathcal{R}, X_0)$ and the rules of the grammar are given by:

$$\begin{aligned} X_0 &\hookrightarrow edg_{\pi_1, \pi_2}(S) & C &\hookrightarrow \bullet\pi_2(P_2, \because) \\ S &\hookrightarrow (S \oplus O) & O &\hookrightarrow \bullet\pi_1(P_1, \Delta) \\ S &\hookrightarrow (C \oplus O) \end{aligned}$$

The clique-and-cloud topology depicted on the right of Fig. 2 belongs to $\mathcal{L}(\Gamma_{Star})$ and it corresponds to the term: $edg_{\pi_1, \pi_2}(((\bullet\pi_2(P_2, \because) \oplus \bullet\pi_1(P_1, \Delta)) \oplus \bullet\pi_1(P_1, \Delta)) \oplus \bullet\pi_1(P_1, \Delta))$

3.3 Safety Problem for Topologies Defined by Grammars

We can now present the safety verification problem we are interested in. It consists in checking, given a VR-graph grammar and a control state q, whether there is a network topology G belonging the expansion of a clique-and-cloud topology in the language of the grammar such that there exists an execution in G where a process leads the control state q. This problems is connected to a safety question, because if q is an error state and the answer to the problem is negative, it means that in all the clique-and-cloud topologies obtained from the grammar, the error state can never be reached. We move to the formal definition of the control state reachability problem which is parameterized by a network transition relation $\Rightarrow$ for the broadcast (reliable vs unreliable):

$\underline{\texttt{CSReach}(\Rightarrow)}$

Input: A VR-graph grammar Γ and a control state q

Question: Does there exist $\mathcal{G} \in L(\Gamma)$ and $G \in [\![\mathcal{G}]\!]$ and $c \in \mathcal{C}_G$ such that: $\#_c(q) > 0$ and $c_{in}^G \Rightarrow^* c$?

4 Undecidability for Reliable Communication

We prove that the parametric safety problem is undecidable for clique-and-cloud systems with reliable communication. In [10], this problem was proven to be undecidable, without any restriction on the considered topologies, i.e. given a single process type and a state, does there exist a network topology where a process can reach the given state? However, if one considers the networks to be cliques, the problem is known to be decidable [13]. This is what motivates us to look at network topologies generated by clique-and-cloud topologies. The undecidability proof presented in [10] relies on two ingredients, first the authors show it is possible to 'extract' a line from a network topology (here extracting means that the other vertices are sent in an error state from which they cannot perform any actions any more) and then they explain how to simulate the behavior of a Minsky machine with two counters (programs that manipulate two natural variables, which can be only incremented, decremented and compared to 0) on lines of unbounded size. Our undecidability proof uses similar techniques.

4.1   Leader Election in Clique-and-Cloud Topologies

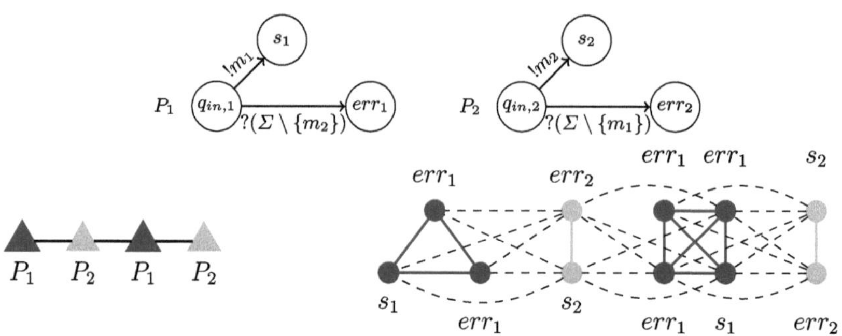

Fig. 4. Two process-types, a cloud-and-clique topology in $\mathcal{L}(\Gamma_L)$ and one of its expansion

We first give the intuition on how one can extract a line of single vertices with different process types, from a line of cliques. We consider the two process types given in Fig. 4 and a grammar which produces lines of clique vertices, where the process associated to each vertex alternates between P_1 and P_2. The grammar in question is $\Gamma_L = (\{L_0\}, \{\pi_1, \pi_2, \pi_3, \pi_4, \pi_5\}, \mathcal{R}, L_0)$, having the following rules:

$$L_0 \hookrightarrow edg_{\pi_1,\pi_2}(\bullet\pi_1(P_1, \Delta) \oplus \bullet\pi_2(P_2, \Delta))$$
$$L_0 \hookrightarrow relab_{[\pi_3 \mapsto \pi_5, \pi_4 \mapsto \pi_5]}(edg_{\pi_3,\pi_4}(relab_{[\pi_2 \mapsto \pi_3]}(L_0) \oplus relab_{[\pi_1 \mapsto \pi_4]}(L_0)))$$

One can see that the language $\mathcal{L}(\Gamma_L)$ is composed by clique-and-cloud topologies of the following form for some $m \geq 1$: $\mathcal{G} = (\{v_i\}_{1 \leq i \leq 2m}, \mathcal{E}, Lab, \mathcal{V}_\Delta, \mathcal{V}_{..})$ with $\mathcal{E} = \{(v_i, v_{i+1}), (v_{i+1}, v_i) \mid 1 \leq i \leq 2m - 1\}$, $Lab(v_i)$ is equal to P_1 if i is odd and to P_2 otherise, and $\mathcal{V}_\Delta = \{v_i\}_{1 \leq i \leq 2m}$. The left-hand side of Fig. 4 provides an example of such a topology with $m = 2$ and on the right is depicted one of its expansion. On this latter network topology, we have indicated control states to represent a possible configuration reachable from the corresponding initial configuration with reliable broadcast. One specificity of the reachable configurations (with the reliable semantics) is that for each clique component in the clique-and-cloud topology, there is at any moment at most one vertex in state s_1 or s_2 and if such a vertex is present, the other vertices of the same clique are in the state err_1 or err_2. Indeed, to reach such a state the corresponding process needs to broadcast the message m_1 or m_2 sending all its neighbors with the same process type into the error state err_1 or err_2. Note that if we consider the unreliable semantics, this property will not hold anymore, and we could find two vertices in the same clique labeled for instance with s_1. If we consider that the vertices in states err_1 and err_2 will not participate in any further communication, and we extend the process types starting from s_1 or s_2, this means that we are able, in a certain sense, to extract a single line from any clique-and-cloud topology where process types of type P_1 and P_2 alternate. While this method of leader selection is only viable for clique topologies, we present a second method in [4] which deals with clique-and-cloud topologies.

4.2 Simulating a Minsky Machine

As said before, our undecidability result comes from the possibility to simulate the behavior of a (two-counter) Minsky machine with line topologies of unbounded length. A (two-counter) Minsky machine is a program which manipulates two natural variables (or counters) c_1 and c_2, and which is composed of a finite set of instructions. Each of the instruction is either of the form (1) $L : c_i := c_i + 1$; goto L' or (2) $L :$ if $c_i = 0$ then goto L' else $c_i := c_i - 1$; goto L'' where $i \in \{1, 2\}$ and L, L', L'' are labels preceding each instruction. Furthermore, there is a special label L_F from which nothing can be done. The behavior of each instruction is the expected one. The *halting problem* of deterministic Minsky machines, which consists in deciding whether the unique execution that starts from L_0 with counters equal to 0 reaches L_F, is undecidable [20].

Fig. 5. Network topology to simulate a run of a 2-counter Minsky machine

Our proof follows the same line as the one presented in [10]. We use a grammar which builds lines of unbounded length as the one presented in Fig. 5. Using a technique similar to the one presented before we extract a single leader process in each clique vertex, and then we simulate the counter machine. The process type Ctl, called the controller, simulates the instructions of the Minsky machine and the processes of type K_i encode the value of the counter c_i thanks to specific states Z and NZ. Intuitively, the number of processes in states NZ represent the value of the counter and all the processes in the 'extracted line' from the controller to the first node in state NZ are in state Z. To increase a counter, a message is transmitted to the last vertex in Z who changes its state to NZ and sends back an acknowledgment to the controller. To decrement the counter, we proceed the same way but this time, it is the last node to be in state NZ (which has a neighbor in state Z) which moves to Z. Finally, to test if the counter c_i is 0, the controller checks (thanks to an exchange of messages) whether its first neighbor of process type K_i is in state Z. One key idea of this reduction is that if there are not enough processes in the line to simulate the counter value, the simulation is blocked. However we have that the Minsky machine halts iff there exists a line (big enough) where the controller ends in the state L_f. The detailed construction is provided in [4].

Theorem 1. CSReach($\Rightarrow_r$) *is undecidable.*

5   Describing Reachable States in MSO

The previous negative result motivates the search for sufficient conditions that guarantee decidability of the safety problem, without jeopardizing the expressivity of the model we consider. In this section, we show that, under certain assumptions, we are able to build a formula of the Monadic Second Order logic (MSO) that characterizes the local control states which can be reached by an execution of some network topology obtained by the expansion of a given clique-and-cloud topology.

5.1   Monadic Second Order Logic for Clique-and-Cloud Topologies

We begin by giving the definition of MSO, adapted to our context. Monadic Second Order (MSO) logic is the set of formulae built inductively, from a set *var* of first-order variables and a set *Var* of set variables, according to the syntax:

$$\varphi ::= \varphi \wedge \varphi \mid \neg\varphi \mid \exists x.\varphi \mid \exists X.\varphi \mid X(y) \mid E(x,y) \mid P_i(x) \mid \Delta(x) \mid \therefore (x)$$

where $x, y \in var$, $X \in Var$, $\wedge$ and $\neg$ are the classical boolean operations for conjunction and complement, $X(y)$ is the unary predicate for membership $E(x,y)$ is the binary predicate for edges and, $P_i(x)$ (with $P_i \in \mathcal{P}$) is the labeling predicate for process type and $\Delta(x)$ and $\therefore (x)$ are the labeling predicates for expansion types. Additionally, we assume classical shorthand notations for $\vee$ (disjunction), $\implies$ (implication), $\forall v.\varphi$ and $\forall V.\varphi$ (universal quantification). As usual, we call *sentence* a formula with no free occurrence of first order or second order variables (all variables are in the scope of some quantifiers).

The MSO formulae we consider in this paper are then interpreted over clique-and-cloud topologies. A variable store for a clique-and-cloud topology $\mathcal{G}$ is a mapping ν from first order (resp. set) variables into vertices (resp. sets of vertices) of $\mathcal{G}$. Let $\mathcal{G} = (\mathcal{V}, \mathcal{E}, Lab, \mathcal{V}_\Delta, \mathcal{V}_{\therefore})$ be a clique-and-cloud topology, we define the satisfaction relation $\models$ parameterized by a store inductively as follows (boolean cases are omitted):

$$
\begin{aligned}
&\mathcal{G} \models_\nu E(x,y) \iff (\nu(x), \nu(y)) \in \mathcal{E} \qquad && \mathcal{G} \models_\nu X(x) \iff \nu(x) \in \nu(X) \\
&\mathcal{G} \models_\nu \Delta(x) \iff \nu(x) \in \mathcal{V}_\Delta && \mathcal{G} \models_\nu P_i(x) \iff Lab(\nu(x)) = P_i \text{ for } P_i \in \mathcal{P} \\
&\mathcal{G} \models_\nu \therefore (x) \iff \nu(x) \in \mathcal{V}_{\therefore} && \mathcal{G} \models_\nu \exists x.\varphi \iff \mathcal{G} \models_{\nu[x \mapsto u]} \varphi \text{ for some } u \in \mathcal{V} \\
& && \mathcal{G} \models_\nu \exists X.\varphi \iff \mathcal{G} \models_{\nu[X \mapsto \mathcal{U}]} \varphi \text{ for some } \mathcal{U} \subseteq \mathcal{V}
\end{aligned}
$$

where $\nu[x \mapsto u]$ (resp. $\nu[X \mapsto \mathcal{U}]$) represents the store that agrees with ν everywhere except for the variable x (resp. X) which is mapped to the vertex u (resp. the set of vertices $\mathcal{U}$). If the MSO formula φ is a sentence, the store is not important, hence we omit it. In this case, we write $[\![\varphi]\!] \overset{\text{def}}{=} \{\mathcal{G} \mid \mathcal{G} \models \varphi\}$ for the set of models of a sentence φ. We point out that considering MSO with vertex quantifiers[1] is crucial for the following decidability theorem (Theorem 2).

In our context, MSO formulae appear as a very useful tool because we are able to decide whether there exists a clique-and-cloud topology in the language of

---

[1] Usually also denoted MSO_1 in the literature [8].

a grammar which satisfies a given formula. This is thanks to a result of Courcelle and Engelfriet, which states that given a VR-graph grammar Γ and an MSO sentence φ, it is possible to build a "refined" grammar Γ_φ that produces the set of graphs from the language of Γ that, moreover, satisfy φ, see, e.g., [8, Theorem 1.22]. Since emptiness of a grammar is decidable in polynomial time (in the size of the input grammar), we obtain the following theorem:

Theorem 2. *The following problem is decidable:* given a graph grammar Γ written in the VR-algebra and an MSO sentence φ, does $\mathcal{L}(\Gamma) \cap [\![\varphi]\!] = \emptyset$?

5.2   Reachable States Labeling of Clique-and-Cloud Topologies

In the rest of the section, let $\mathcal{G} = (\mathcal{V}, \mathcal{E}, Lab, \mathcal{V}_\Delta, \mathcal{V}_\cdot)$ be a fixed clique-and-cloud topology and $\Rightarrow$ a network transition relation (i.e., a relation between network configurations, defined by either the reliable or unreliable semantics). We recall that Lab maps each vertex in $\mathcal{V}$ to a process type in $\mathcal{P}$, where $\mathcal{P} = \{P_1, \ldots, P_\ell\}$, $P_i = (Q_i, q_{in,i}, \delta_i)$, for all $i \in [1, \ell]$ and $Q = \biguplus_{1 \leq i \leq \ell} Q_i$ and $\delta = \biguplus_{1 \leq i \leq \ell} \delta_i$ denote the overall set of states and transition relation, respectively.

A *states-labeling* of $\mathcal{G}$ is a function $\gamma : \mathcal{V} \mapsto 2^Q$, that associates each vertex to a subset of Q. In particular, we are interested in a specific states labeling, namely the *reachable states-labeling for $\mathcal{G}$* $\gamma_{reach}^{\Rightarrow} : \mathcal{V} \mapsto 2^Q$, defined as follows:

Definition 5 (Reachable states-labeling).   *For all $v \in \mathcal{V}$, we have $q \in \gamma_{reach}^{\Rightarrow}(v)$ if and only if there exist $(G, f) \in [\![\mathcal{G}]\!]$, $u \in f^{-1}(v)$ and $c \in \mathcal{C}_G$ verifying the following conditions: (1) $c(u) = q$ and (2) $c_{in}^G \Rightarrow^* c$.*

The next lemma, which can be directly proven by applying the definition, shows the connection between the reachable states-labeling and the control state reachability problem for the clique-and-cloud topology $\mathcal{G}$.

Lemma 1. *Let $q \in Q$ be a state. Then, there exists $G \in [\![\mathcal{G}]\!]$ and $c \in \mathcal{C}_G$ such that $\#_c(q) > 0$ and $c_{in}^G \Rightarrow^* c$ iff there exists $v \in \mathcal{V}$ such that $q \in \gamma_{reach}^{\Rightarrow}(v)$*

In the sequel, we will show that, under certain conditions, we can build an MSO formula which characterizes the clique-and-cloud topology whose reachable states labeling exhibit a given control state q. By Theorem 2, this provides a general argument for the decidability of the $\mathtt{CSReach}(\Rightarrow)$ problem.

5.3   Two Sufficient Criteria

We state here the conditions we rely on to establish our decidability results. We begin by introducing some new notations. Given a clique-and-cloud topology $\mathcal{G} = (\mathcal{V}, \mathcal{E}, Lab, \mathcal{V}_\Delta, \mathcal{V}_\cdot)$ and two disjoint network topologies $G_1 = (V_1, E_1, lab_1)$ and $G_2 = (V_2, E_2, lab_2)$ such that $(G_1, f_1), (G_2, f_2) \in [\![\mathcal{G}]\!]$, we denote by $G_1 \uplus G_2$ the network topology (V, E, lab) where: $V \stackrel{\text{def}}{=} V_1 \uplus V_2$ and $E \stackrel{\text{def}}{=} E_1 \uplus E_2 \uplus \{(v, v') \in V_i \times V_{3-i} \mid f_i(v) = f_{3-i}(v'), f_i(v) \in \mathcal{V}_\Delta, i = 1, 2\}$ and $lab(v) \stackrel{\text{def}}{=} lab_i(v)$ for all $v \in V_i$ and all $i \in \{1, 2\}$. Note that, by construction, we have the following:

Lemma 2. *If $f : V \mapsto V$ is the surjective function such that $f(v) = f_1(v)$ for all $v \in V_1$ and $f(v) = f_2(v)$ for all $v \in V_2$ then $(G_1 \uplus G_2, f) \in [\![\mathcal{G}]\!]$.*

Finally, given two configurations $c_1 \in \mathcal{C}_{G_1}$ and $c_2 \in \mathcal{C}_{G_2}$, we denote by $c_1 \uplus c_2$ the configuration in $\mathcal{C}_{G_1 \uplus G_2}$ such that $c_1 \uplus c_2(v) = c_1(v)$ for all $v \in V_1$ and $c_1 \uplus c_2(v) = c_2(v)$ for all $v \in V_2$. We say that a state $q \in Q$ is an *emitter* state, whenever there exists $m \in \Sigma$ and $q' \in Q$ such that $(q, !m, q') \in \delta$ and denote by $Q_E \subseteq Q$ the set of emitter states. The first criterion is stated below. We say that the network transition relation $\Rightarrow$ respects the *composition property for emitters* when it satisfies the following statement, for all clique-and-cloud topologies $\mathcal{G}$:

Definition 6 (Composition property for emitters). *For all $G_1, G_2 \in [\![\mathcal{G}]\!]$, if $c_i \in \mathcal{C}_{G_i}$ are configurations, where $c_{in}^{G_i} \Rightarrow^* c_i$, $i = 1, 2$, then there exists a configuration $c_1' \in \mathcal{C}_{G_1}$ verifying $c_{in}^{G_1 \uplus G_2} \Rightarrow^* (c_1' \uplus c_2)$ and $c_1'(v) = c_1(v)$, for all vertices v of G_1, such that $c_1(v) \in Q_E$.*

Intuitively this property expresses that if two configurations are reachable then we can reach another configuration where all the emitters states (resp. all the states) of the first (resp. second) configuration are present. This allows us to compose configurations on demand to compute reachable states. Even if not stated this way, this property is the key ingredient of the polynomial time algorithm for safety under the unreliable semantics proposed in [9] and a similar property has been established for wait-only protocols under the reliable semantics in [16].

The second criterion is a condition on when control state of a vertex can change when taking one step of the network transition relation $\Rightarrow$. We say that the network transition relation $\Rightarrow$ respects the *local successor property* whenever if satisfies the following statements, for all clique-and-cloud topologies $\mathcal{G}$:

Definition 7 (Local successor property). *For all $G = (V, E, lab) \in [\![\mathcal{G}]\!]$, $c \in \mathcal{C}_G$ and $v \in V$, there exists $c' \in \mathcal{C}_G$ such that $c \Rightarrow c'$ and $c(v) \neq c'(v)$ if and only if there exists $m \in \Sigma$, such that one of the following properties holds:*

(1) $c(v) \xrightarrow{!m}_\delta c'(v)$, or,

(2) $c(v) \xrightarrow{?m}_\delta c'(v)$ and there exists $u \in V$ and $q \in Q$ such that $(u, v) \in E$ and $c(u) \xrightarrow{!m}_\delta q$.

Relying on the definitions of the reliable and unreliable semantics, we can prove the following lemma:

Lemma 3. *The network transition relations $\Rightarrow_r$ (reliable communication) and $\Rightarrow_u$ (unreliable communication) both meet the local successor property.*

Note that whereas we prove that the unreliable semantics respects the composition property for emitters, this does not hold for the reliable semantics, and we will see that we need another restriction in this context.

5.4   Characterization of the Reachable States-Labeling

We now provide a characterization of the reachable states labeling of clique-and-cloud topologies for the network transitions relations respecting both the composition property for emitters (Definition 6) and the local successor property (Definition 7). Let $\mathcal{G} = (\mathcal{V}, \mathcal{E}, Lab, \mathcal{V}_\Delta, \mathcal{V}_{..})$ be a clique-and-cloud topology and $\Rightarrow$ be a transition relation. We define the following property of states-labelings:

Definition 8 (Inductive reachability property). *A states-labeling $\gamma : \mathcal{V} \mapsto 2^Q$ meets the inductive reachability property for $\Rightarrow$ if and only if the following hold, for all $v, v' \in \mathcal{V}$ and $q, q', q'', q''' \in Q$:*

1. *if $Lab(v) = P_i$ then $q_{in,i} \in \gamma(v)$,*
2. *if $q \in \gamma(v)$ and $q \xrightarrow{!m}_\delta q'$ then $q' \in \gamma(v)$,*
3. *if $v \in \mathcal{V}_\Delta$ and $q, q' \in \gamma(v)$, $q \xrightarrow{!m}_\delta q''$ and $q' \xrightarrow{?m}_\delta q'''$ then $q''' \in \gamma(v)$,*
4. *if $q \in \gamma(v)$, $q' \in \gamma(v')$, $(v, v') \in \mathcal{E}$, $q \xrightarrow{!m}_\delta q''$ and $q' \xrightarrow{?m}_\delta q'''$ then $q''' \in \gamma(v')$.*

Furthermore, given states-labelings γ and γ', we write $\gamma \sqsubseteq \gamma'$ for the pointwise inclusion of γ into γ'. We state below an important lemma relating the reachable states-labeling γ_{reach} and the inductive reachability property:

Lemma 4. *If a network transition relation $\Rightarrow$ respects both the composition property for emitters and the local successor property, then $\gamma_{reach}^{\Rightarrow}$ is the least states-labeling (in the sense of $\sqsubseteq$) having the inductive reachability property.*

Proof. Let $\Rightarrow$ be a network transition relation respecting both the composition property for emitters and the local successor property and let $\mathcal{G} = (\mathcal{V}, \mathcal{E}, Lab, \mathcal{V}_\Delta, \mathcal{V}_{..})$ be a clique-and-cloud topology. We recall that we assumed $\mathcal{P} = \{P_1, \ldots, P_\ell\}$ where $P_i = (Q_i, q_{in,i}, \delta_i)$ for all $i \in [1, \ell]$ and that we denote $Q \stackrel{\text{def}}{=} \bigcup_{1 \leq i \leq \ell} Q_i$ and $\delta = \biguplus_{1 \leq i \leq \ell} \delta_i$. We fist show that $\gamma_{reach}^{\Rightarrow}$ verifies each case of the inductive reachability property.

1. Let $v \in \mathcal{V}$ and assume $Lab(v) = P_i$. Consider the network topology $G = (\mathcal{V}, \mathcal{E}, Lab)$. Note that $(G, id) \in [\![\mathcal{G}]\!]$ where id is the identity function (which is surjective). By definition of c_{in}^G, we have $c_{in}^G(v) = q_{in,i}$. Since $id^{-1}(v) = v$ and since $c_{in} \Rightarrow^* c_{in}$, applying the definition of $\gamma_{reach}^{\Rightarrow}$, we deduce $q_{in,i} \in \gamma_{reach}^{\Rightarrow}(v)$.
2. Let $v \in \mathcal{V}$ and assume that $q \in \gamma_{reach}^{\Rightarrow}(v)$ and that there exists $q \xrightarrow{!m}_\delta q'$. We assume that $q \neq q'$, otherwise the property is obviously satisfied. By definition of $\gamma_{reach}^{\Rightarrow}$, there exist $(G, f) \in [\![\mathcal{G}]\!]$ and a vertex $v' \in f^{-1}(v)$ and a configuration $c \in \mathcal{C}_G$ such that $c(v') = q$ and $c_{in}^G \Rightarrow c$. Since we have $(q \xrightarrow{!m}_\delta q')$, using the fact that $\Rightarrow$ respects the local successor property, we deduce there exists a configuration c' such that $c \Rightarrow c'$ and $c'(v') = q'$. Hence, we have as well $c_{in}^G \Rightarrow c'$. We conclude that $q' \in \gamma_{reach}^{\Rightarrow}(v)$.
3. Let $v \in \mathcal{V}$ and assume that there exist $q, q' \in \gamma_{reach}^{\Rightarrow}(v)$ and $v \in \mathcal{V}_\Delta$ and $q \xrightarrow{!m}_\delta q''$ and $q' \xrightarrow{?m}_\delta q'''$. By definition of $\gamma_{reach}^{\Rightarrow}$, there exist $(G_1, f_1) \in [\![\mathcal{G}]\!]$ and a vertex $v_1 \in f_1^{-1}(v)$ and a configuration $c_1 \in \mathcal{C}_{G_1}$ such that $c_1(v_1) = q$

and $c_{in}^{G_1} \Rightarrow^* c_1$. Similarly, there exist $(G_2, f_2) \in [\![\mathcal{G}]\!]$ and a vertex $v_2 \in f_2^{-1}(v)$ and a configuration $c_2 \in \mathcal{C}_{G_2}$ such that $c_2(v_2) = q'$ and $c_{in}^{G_2} \Rightarrow^* c_2$. We now show that $q''' \in \gamma_{reach}^{\Rightarrow}(v)$ (assuming $q''' \neq q'$ otherwise, the result is obvious). We consider the network topology $G = G_1 \uplus G_2$ with $G = (V, E, lab)$. Thanks to Lemma 2, we have $(G, f) \in [\![\mathcal{G}]\!]$ (where f is the function given in the lemma). Since $\Rightarrow$ respects the composition property for emitters and since q is an emitter state, we deduce that there exists a configuration $c_1' \in \mathcal{C}_{G_1}$ verifying $c_{in}^G \Rightarrow^* (c_1' \uplus c_2)$ and $c_1'(v_1) = c_1(v_1) = q$. We denote by c the $(G_1 \uplus G_2)$-configuration $c_1' \uplus c_2$. By definition of $\uplus$, we have hence $c(v_1) = q$ and $c(v_2) = q'$. Furthermore, by definition of (G, f), since $f(v_1) = f_1(v_1) = f_2(v_2) = f(v_2) = v$ and $v \in V_\Delta$, we have $(v_1, v_2) \in E$. Using the fact that $\Rightarrow$ respects the local successor property, we deduce there exists a configuration c' such that $c \Rightarrow c'$ and $c'(v_2) = q'''$. Hence, we have as well $c_{in}^G \Rightarrow^* c'$. We conclude that $q''' \in \gamma_{reach}^{\Rightarrow}(v)$.

4. Let $v, v' \in V$ such that $(v, v') \in \mathcal{E}$ and assume that there exist $q \in \gamma_{reach}^{\Rightarrow}(v)$ and $q' \in \gamma_{reach}^{\Rightarrow}(v')$ and $q \xrightarrow{!m}_\delta q''$ and $q' \xrightarrow{?m}_\delta q'''$. By definition of $\gamma_{reach}^{\Rightarrow}$, there exist $(G_1, f_1) \in [\![\mathcal{G}]\!]$ and a vertex $v_1 \in f_1^{-1}(v)$ and a configuration $c_1 \in \mathcal{C}_{G_1}$ such that $c_1(v_1) = q$ and $c_{in}^{G_1} \Rightarrow^* c_1$. Similarly, there exist $(G_2, f_2) \in [\![\mathcal{G}]\!]$ and a vertex $v_2 \in f_2^{-1}(v)$ and a configuration $c_2 \in \mathcal{C}_{G_2}$ such that $c_2(v_2) = q'$ and $c_{in}^{G_2} \Rightarrow^* c_2$. To prove that $q''' \in \gamma_{reach}^{\Rightarrow}(v')$, we apply the same reasoning as the previous case, the main difference being that we have here $(v_1, v_2) \in E$ because $(f(v_1), f(v_2)) = (v, v') \in \mathcal{E}$.

Consequently, $\gamma_{reach}^{\Rightarrow}$ respects the inductive reachability property.

We consider now a states labeling γ respecting the inductive reachability property, and we show that $\gamma_{reach}^{\Rightarrow} \sqsubseteq \gamma$. For this matter, we define the family $(\gamma_n)_{n \in \mathbb{N}}$ of states labelings as follows: for all $n \in \mathbb{N}$, for all $v \in V$, we have $q \in \gamma_n(v)$ iff there exists $G = (V, E, lab)$ such that $(G, f) \in [\![\mathcal{G}]\!]$ and there exists $v' \in f^{-1}(v)$ and $c \in \mathcal{C}_G$ verifying the following conditions:

1. $c(v') = q$, and,
2. $c_{in} \Rightarrow^m c$ for $0 \leq m \leq n$ (where $\Rightarrow^0$ is the identity relation and $\Rightarrow^k$ is the relation $\Rightarrow \cdot \Rightarrow^{k-1}$ for all $k > 1$)

We show by induction that $\gamma_n \sqsubseteq \gamma$.

- We first show that $\gamma_0 \sqsubseteq \gamma$. Let $v \in V$ and q be a control state in $\gamma_0(v)$. We prove that $q \in \gamma(v)$. By definition of γ_0, there exists $(G, f) \in [\![\mathcal{G}]\!]$ with $G = (V, E, lab)$ and a vertex $v' \in f^{-1}(v)$ such that $c_{in}^G(v') = q$. Assume now that $lab(v') = P_i$. In that case we have $lab(v') = P_i$ and $q = q_{in,i}$. By definition of $[\![\mathcal{G}]\!]$, we have $Lab(v) = Lab(f(v')) = lab(v') = P_i$. Since γ respects the inductive reachability property (in particular statement 1), we deduce that $q = q_{in,i}$ belongs to $\gamma(v)$. Hence, we have $\gamma_0 \sqsubseteq \gamma$.
- We now move to the inductive step. Let $n \in \mathbb{N}$ and assume that $\gamma_n \sqsubseteq \gamma$. We show that $\gamma_{n+1} \sqsubseteq \gamma$. Let $v \in V$ and q be a control state in $\gamma_{n+1}(v)$. First note that if $q \in \gamma_n(v)$, then using the induction hypothesis we obtain directly

that $q \in \gamma(v)$, we assume hence that $q \notin \gamma_n(v)$. By definition of γ_{n+1}, there exists $(G, f) \in [\![\mathcal{G}]\!]$ with $G = (V, E, lab)$ and a vertex $v' \in f^{-1}(v)$ and $c \in \mathcal{C}$ and $m \leq (n + 1)$ such that $c_m(v') = q$ and $c_{in}^G \Rightarrow^m c_m$. Note that since $q \notin \gamma_n(v)$, we know that $m = n + 1$ and there must exist a configuration $c' \in \mathcal{C}$ such that $c_{in}^G \Rightarrow^n c' \Rightarrow c$ with $c'(v') = q'$ (where $q \neq q'$). By definition we have $q' \in \gamma_n(v)$ and thanks to induction hypothesis, we deduce $q' \in \gamma(v)$. Now since $c' \Rightarrow c$, and since $\Rightarrow$ respects the local successor property, we can perform the following case analysis:

1. Either, $q' \xrightarrow{!m}_\delta q$, in that case since $q' \in \gamma(v)$ and since γ respects the inductive reachability property (in particular statement 2.), we deduce that $q \in \gamma(v)$.

2. Or $q' \xrightarrow{?m}_\delta q$ and there exists $u \in V$ and $q'' \in Q$ such that $f(u) = f(v')$ and $(u, v') \in E$ and $c(u) \xrightarrow{!m}_\delta q''$. First note that since $f(u) = f(v') = v$ and $(u, v') \in E$, by definition of $(G, f) \in [\![\mathcal{G}]\!]$, we have $v \in \mathcal{V}_\Delta$. Then the same way, we show that $q' = c'(v') \in \gamma(v)$, we can show that $c'(u) \in \gamma^n(v) \subseteq \gamma(v)$. Since γ respects the inductive reachability property (in particular statement 3.), we deduce that $q \in \gamma(v)$.

3. Or $q' \xrightarrow{?m}_\delta q$ and there exists $u \in V$ and $q'' \in Q$ such that $f(u) \neq f(v')$ and $(u, v') \in E$ and $c(u) \xrightarrow{!m}_\delta q''$. First note that since $f(u) \neq f(v')$ and $(u, v') \in E$, by definition of $(G, f) \in [\![\mathcal{G}]\!]$, we have $(f(u), v) \in \mathcal{E}$. Then the same way we show that $q' = c'(v') \in \gamma(v)$, we can show that $c'(u) \in \gamma^n(f(u)) \subseteq \gamma(f(u))$. Since γ respects the inductive reachability property (in particular statement 4.), we deduce that $q \in \gamma(v)$.

Hence, we obtain $\gamma_{n+1} \sqsubseteq \gamma$.

We have then proved that $\gamma_n \sqsubseteq \gamma$ for all $n \in \mathbb{N}$. By applying the definition, we have that $\gamma_{reach}^{\Rightarrow}(v) = \bigcup_{n \in \mathbb{N}} \gamma_n(v)$ for all $v \in \mathcal{V}$. Hence, we have $\gamma_{reach}(v) \subseteq \gamma(v)$ for all $v \in \mathcal{V}$, and consequently $\gamma_{reach} \sqsubseteq \gamma$. $\qquad\square$

5.5   Defining the Reachable States-Labeling in MSO

In this section, we consider a network transition relation $\Rightarrow$ meeting both the composition property for emitters (Definition 6) and the local successor property (Definition 7). We define the equivalent property of the reachable states-labeling stated in Lemma 4 using an MSO formula. More specifically, assuming w.l.o.g. an enumeration $Q = \{q_1, \ldots, q_n\}$ of the set of states, we build an MSO formula $\Phi_{reach}(X_1, \ldots, X_n)$ where each X_i is a free set variable, such that $\mathcal{G} \models_\nu \Phi_{reach}(X_1, \ldots, X_n)$ if and only if $\nu(X_i)$ is the set of vertices v verifying $q_i \in \gamma_{reach}^{\Rightarrow}(v)$ for all $i \in [1, n]$.

First, we build an MSO formula $\Phi_{ind}(X_1, \ldots, X_n)$ to characterize the states-labelings which respect the inductive reachability property (Definition 8). This formula uses four families of formulae: $\varphi_{1,i}(x, X_1, \ldots, X_n)$, $\varphi_{2,i}(x, X_1, \ldots, X_n)$, $\varphi_{3,i}(x, X_1, \ldots, X_n)$ and $\varphi_{4,i}(x, X_1, \ldots, X_n)$, for $i \in [1, n]$, where x is a free first-order variable. Each of this family is related to one of the conditions of Definition

8 and expresses the respective condition regarding the state q_i and the vertex associated to x. Formally, for all $i \in [1, n]$, we define:

$$\varphi_{1,i}(x, X_1, \ldots, X_n) \stackrel{\text{def}}{=} \begin{cases} \text{false if } \nexists j \in [1, \ell].\ q_i = q_{in,j}, \\ P_j(x) \text{ if } q_i = q_{in,j} \end{cases}$$

$$\varphi_{2,i}(x, X_1, \ldots, X_n) \stackrel{\text{def}}{=} \bigvee_{j \in \mathcal{J}} X_j(x), \text{ where}$$
$$\mathcal{J} = \{ j \in [1, n] \mid q_j \xrightarrow{!m}_\delta q_i,\ m \in \Sigma \}$$

$$\varphi_{3,i}(x, X_1, \ldots, X_n) \stackrel{\text{def}}{=} \bigvee_{(k,k') \in \mathcal{K}} X_k(x) \wedge X_{k'}(x) \wedge \Delta(x), \text{ where}$$
$$\mathcal{K} = \left\{ (k, k') \in [1, n] \times [1, n] \;\middle|\; \begin{array}{l} (q_k \xrightarrow{!m}_\delta q) \\ (q_{k'} \xrightarrow{?m}_\delta q_i) \\ m \in \Sigma \\ q \in Q \end{array} \right\}$$

$$\varphi_{4,i}(x, X_1, \ldots, X_n) \stackrel{\text{def}}{=} \bigvee_{(k,k') \in \mathcal{K}} \exists x'.\, X_k(x') \wedge X_{k'}(x) \wedge E(x', x)$$

$$\Phi_{ind}(X_1, \ldots, X_n) \stackrel{\text{def}}{=} \forall x.\Big(\bigwedge_{i=1}^{n} \big(\bigvee_{s=1}^{4} \varphi_{s,i}(x, X_1, \ldots, X_n) \big) \implies X_i(x) \Big)$$

In order to state and prove the property of this formula below, for a store ν, we define the states-labeling $\gamma_\nu(v) \stackrel{\text{def}}{=} \{ q_i \in Q \mid v \in \nu(X_i) \}$, for all $v \in \mathcal{V}$. Thanks to the definition of the inductive reachability property and relying on the semantics of MSO, we deduce the following lemma.

Lemma 5. *For each clique-and-cloud topology $\mathcal{G}$ and store ν, we have $\mathcal{G} \models_\nu \Phi_{ind}(X_1, \ldots, X_n)$ if and only if γ_ν satisfies the inductive reachability property.*

From Lemma 4, we know that since the considered network transition relation respects the composition property for emitters and the local successor property, then the reachable states labeling is the smallest states labeling (with respect to $\sqsubseteq$) having the inductive reachability property. In MSO, we can state that a state-labeling is the smallest, by the following formula: $\Phi_{reach}(X_1, \ldots, X_n) = \Phi_{ind}(X_1, \ldots, X_n) \wedge \forall X_1'.\ldots.\forall X_n'.\forall x.\Phi_{ind}(X_1', \ldots, X_n') \wedge \bigwedge_{i=1}^{n}(X_i(x) \implies X_i'(x))$. Thanks to Lemmas 4 and 5, we obtain the following:

Lemma 6. *For each clique-and-cloud topology $\mathcal{G}$ and each store ν, we have $\mathcal{G} \models_\nu \Phi_{reach}(X_1, \ldots, X_n)$ if and only if γ_ν is the reachable states-labeling.*

5.6   General Decidability Result

Lemma 6 gives a characterization of the reachable states-labeling in MSO, when the transition relation of the network meets the criteria of Definitions 6 and 7. By Lemma 1, knowing the reachable states-labeling is key to solving the control state reachability problem. In the following, assume that the network transition relation $\Rightarrow$ respects the two criteria of Subsect. 5.3, Γ be a VR-graph grammar and a q be a control state (w.l.o.g. we assume $q = q_1$). We consider then the MSO formula $\Phi = \exists x.\exists X_1.\ldots.\exists X_n.\Phi_{reach}(X_1, \ldots, X_n) \wedge X_1(x)$. If we interpret this formula, it means that in the reachable states-labeling described by $\Phi_{reach}(X_1, \ldots, X_n)$, there is at least one vertex such that q_1 is associated to this vertex. Thanks to Lemma 1, this answers the control state reachability question. Furthermore, Theorem 2 states that we can decide whether there exists a network topology $\mathcal{G}$ in the language of the grammar Γ and which satisfies Φ. This gives us a general decidability framework, which can stated as follows:

Theorem 3. *If the network transition relation $\Rightarrow$ respects the composition property for emitters (Definition 6) and the local successor property (Definition 7), then* CSReach($\Rightarrow$) *is decidable.*

6   Specific Decidability Results

6.1   The Case of Unreliable Broadcast

We show that the control state reachability for the unreliable network transition relation $\Rightarrow_u$ is decidable. Thanks to Theorem 3, it is sufficient to prove that $\Rightarrow_u$ respects the composition property for emitters and the local successor property. Note that we have already proved with Lemma 3 that this relation respects the local successor property. It remains to show the other property.

Lemma 7. *The networks transition relations $\Rightarrow_u$ respects the composition property for emitters.*

Proof. Let $\mathcal{G}$ be a clique-and-cloud topology and let $G_1, G_2 \in [\![\mathcal{G}]\!]$. Assume we have $c_1 \in \mathcal{C}_{G_1}$ such that $c_{in}^{G_1} \Rightarrow_u^* c_1$ and $c_2 \in \mathcal{C}_{G_2}$ such that $c_{in}^{G_2} \Rightarrow_u^* c_2$. We show that $c_{in}^{G_1 \uplus G_2} \Rightarrow^* (c_1 \uplus c_2)$ (this is actually a bit stronger than the composition property for emitters and this is possible thanks to the unreliable broadcast). Since we have unreliable broadcast, nodes are free to ignore incoming messages and not change state, from this we can let $G_1 \uplus G_2$ perform the sequence $c_{in}^{G_1 \uplus G_2} \Rightarrow_u^* c_1 \uplus c_{in}^{G_2}$ and have all nodes in G_2 do not receive the broadcast messages so that they stay in their initial state. Similarly, for the G_2 transition sequence, we will have all nodes in G_1 not react and then have that $c_1 \uplus c_{in}^{G_2} \Rightarrow_u^* c_1 \uplus c_2$ which concludes the proof, showing $c_{in}^{G_1 \uplus G_2} \Rightarrow^* (c_1 \uplus c_2)$. $\square$

Consequently, we can deduce the result of Theorem 4 by Lemma 3 and Lemma 7 using Theorem 3:

Theorem 4. CSReach($\Rightarrow_u$) *is decidable.*

6.2   Reliable Broadcast with the Wait-Only Restriction

As we have seen in Sect. 4, the control state reachability is undecidable for the reliable network transition relation $\Rightarrow_r$. This implies that this relation does not satisfy the composition property for emitters (we know by Lemma 3, that it respects the local successor property). However, the wait-only restriction on process types introduced in [16], allows us to regain decidability in our context. This restriction states that in the considered process types, there should never be a state from which messages can be both sent and received. Formally a process type $P = (Q, q_{in}, \delta)$ is said to be *wait-only* iff for all states $q \in Q$, there does not exist $m, m' \in \Sigma$ and $q', q'' \in Q$ such that $q \xrightarrow{!m}_\delta q'$ and $q \xrightarrow{?m'}_\delta q''$ and there does not exist $q''' \in Q$ and $m'' \in \Sigma$ such that $q_{in} \xrightarrow{?m''}_\delta q'''$. We call q an *action state* if there exists $m \in \Sigma$ and $q' \in Q$ verifying $q \xrightarrow{!m}_\delta q'$ or from which there is no outgoing transitions. Due to the restriction on q_{in}, it is always an *action state*. The other states are called *waiting states*. Finally, we say that the set of process types $\mathcal{P} = \{P_1, \dots, P_\ell\}$ is wait-only iff P_i is wait-only for all $i \in [1, \ell]$. We now show that this restriction allows us to regain decidability.

Lemma 8. *When dealing with wait-only process types, the network transition relation $\Rightarrow_r$ respects the composition property for emitters.*

Proof. Let $\mathcal{G}$ be a clique-and-cloud topology and let $G_1, G_2 \in [\![\mathcal{G}]\!]$. Assume we have $c_1 \in \mathcal{C}_{G_1}$ such that $c_{in}^{G_1} \Rightarrow_r^* c_1$ and $c_2 \in \mathcal{C}_{G_2}$ such that $c_{in}^{G_2} \Rightarrow_r^* c_2$. We let $G_1 \uplus G_2$ perform the sequence of transitions $c_{in}^{G_1 \uplus G_2} \Rightarrow_r^* c_1 \uplus c_{in}^{G_2}$. This is sequence of transitions is possible as all vertices in $c_{in}^{G_2}$ are in an initial state, which are by definition action states from which no message can be received. Next, we let $G_1 \uplus G_2$ perform the same sequence of transitions as in $c_{in}^{G_2} \Rightarrow_r^* c_2$, this gives rise to $c_1 \uplus c_{in}^{G_2} \Rightarrow_r^* c_1' \uplus c_2$. Note that the taken transitions can have an impact only on the vertices of c_1 which are in a waiting state, but for all vertices $v \in V_1$ where $c_1(v) \in Q_E$ we know that these cannot react to a broadcast and will not move. Hence, for $v \in V_1$ if $c_1(v) \in Q_E$ we have $c_1'(v) = c_1(v)$. $\square$

Finally, by Lemma 3 and Lemma 8 we fulfil the hypothesis of Theorem 3 and obtain the result of Theorem 5.

Theorem 5. `CSReach(`$\Rightarrow_r$`)` *restricted to wait-only process types is decidable.*

7   Conclusion

We have shown in this paper, that the parametric safety problem for broadcast networks where the set of accepted topologies is given thanks to a graph grammar recognizing clique-and-cloud topologies is decidable for unreliable communication, undecidable for reliable communication and that in the latter case, decidability can be regained by imposing the wait-only restriction on the processes running in each node of the networks. Our technique to obtain decidability relies on a reduction to the model-checking problem for MSO formulae over graph grammars. First, this technique does not allow us to have precise complexity bounds and in the future we plan to find to which complexity classes belong the considered problems. Second, we also think that this general decidability procedure can be applied in different context or for other properties.

References

1. Abdulla, P.A., Cerans, K., Jonsson, B., Tsay, Y.-K.: General decidability theorems for infinite-state systems. In: LICS'96, LICS '96, page 313, USA. IEEE Computer Society (1996)
2. Aminof, B., Kotek, T., Rubin, S., Spegni, F., Veith, H.: Parameterized model checking of rendezvous systems. Distrib. Comput. **31**(3), 187–222 (2018)
3. Aminof, B., Kotek, T., Rubin, S., Spegni, F., Veith, H.: Parameterized model checking of rendezvous systems. Distributed Comput. **31**(3), 187–222 (2018)
4. Andersen, C.L., Iosif, R., Sangnier, A.: Safety analysis in broadcast networks defined by graph grammars. Technical report, VERIMAG (2026). https://hal.science/hal-05528808
5. Bloem, R., et al.: Decidability of parameterized verification. In: Synthesis Lectures on Distributed Computing Theory. Morgan & Claypool Publishers (2015)
6. Bozga, M., Iosif, R., Sangnier, A., Villani, N.: Counting abstraction and decidability for the verification of structured parameterized networks. In: Piskac, R., Rakamarić, Z. (eds.) CAV'25. LNCS, vol. 15933, pp. 238–262. Springer, Cham (2025). https://doi.org/10.1007/978-3-031-98682-6_13
7. Browne, M.C., Clarke, E.M., Grumberg, O.: Reasoning about networks with many identical finite state processes. Inf. Comput. **81**(1), 13–31 (1989)
8. Courcelle, B., Engelfriet, J.: Graph Structure and Monadic Second-Order Logic: A Language-Theoretic Approach. Encyclopedia of Mathematics and its Applications. Cambridge University Press (2012)
9. Delzanno, G., Sangnier, A., Traverso, R., Zavattaro, G.: On the complexity of parameterized reachability in reconfigurable broadcast networks. In: FSTTCS'12, volume 18 of LIPIcs, pp. 289–300. Schloss Dagstuhl - Leibniz-Zentrum für Informatik (2012)
10. Delzanno, G., Sangnier, A., Zavattaro, G.: Parameterized verification of ad hoc networks. In: Gastin, P., Laroussinie, F. (eds.) CONCUR 2010. LNCS, vol. 6269, pp. 313–327. Springer, Heidelberg (2010). https://doi.org/10.1007/978-3-642-15375-4_22
11. Delzanno, G., Sangnier, A., Zavattaro, G.: On the power of cliques in the parameterized verification of ad hoc networks. In: Hofmann, M. (ed.) FoSSaCS 2011. LNCS, vol. 6604, pp. 441–455. Springer, Heidelberg (2011). https://doi.org/10.1007/978-3-642-19805-2_30
12. Engelfriet, J.: Context-Free Graph Grammars, pp. 125–213. Springer, Heidelberg (1997)
13. Esparza, J., Finkel, A., Mayr, R.: On the verification of broadcast protocols. In: LICS'99, p. 352, USA. IEEE Computer Society (1999)
14. J. Esparza and M. Nielsen. Decidability issues for Petri nets - a survey. Elektronische Informationsverarbeitung und Kybernetik **30**, 143–160 (1994)
15. German, S.M., Sistla, A.P.: Reasoning about systems with many processes. J. ACM **39**(3), 675–735 (1992)

16. Guillou, L., Sangnier, A., Sznajder, N.: Safety verification of wait-only non-blocking broadcast protocols. In: Kristensen, L.M., van der Werf, J.M. (eds.) PETRI NETS 2024, pp. 291–311. LNCS, vol. 14628. Springer, Cham (2024). https://doi.org/10.1007/978-3-031-61433-0_14
17. Iosif, R., Sangnier, A., Villani, N.: Verifying parameterized networks specified by vertex-replacement graph grammars. In: Lahlou, S., Mukund, M. (eds.) NETYS 2025. LNCS, vol. 15736, pp. 57–80. Springer, Cham (2025)
18. Jayaraman, K., et al.: Validating datacenters at scale. In: Proceedings of the ACM Special Interest Group on Data Communication, SIGCOMM '19, New York, NY, USA, pp. 200–213. Association for Computing Machinery (2019)
19. Le Metayer, D.: Describing software architecture styles using graph grammars. IEEE Trans. Software Eng. **24**(7), 521–533 (1998)
20. Minsky, M.: Computation, Finite and Infinite Machines. Prentice Hall (1967)

Coming Home for Blocking Transitions Fast

Christopher T. Schwanen[1]([envelope]) [ID], Wied Pakusa[2] [ID],
and Wil M. P. van der Aalst[1] [ID]

[1] Chair of Process and Data Science (PADS), RWTH Aachen University,
Aachen, Germany
{schwanen,wvdaalst}@pads.rwth-aachen.de
[2] Faculty of Mathematics, Informatics and Technology, Koblenz University of
Applied Sciences, Koblenz, Germany
pakusa@hs-koblenz.de

Abstract. Gaujal, Haar, and Mairesse proved in 2003 that every cluster in a live, bounded, free-choice Petri net system (i.e., transitions that share the same set of input places) has a blocking marking, i.e., a marking which enables only the transitions of the cluster. Moreover, for each cluster, this blocking marking is unique and reachable from every state of the system. In this paper, we study live, bounded, free-choice systems that, in addition, possess a home cluster, i.e., a cluster whose blocking marking only marks the places of the cluster. Prototypical examples are sound free-choice workflow nets, a commonly used subset of the standard model class in process mining. Our main result shows that in such systems, shortest firing sequences to blocking markings have linear length. Moreover, we obtain a quadratic upper bound on the length of shortest firing sequences between arbitrary markings once a home cluster is present. Both results strongly improve known bounds, as the best currently known bound is the cubic one of Desel and Esparza for general live, bounded, free-choice systems. Furthermore, we show that each transition can be enabled via an elementary firing sequence, i.e., a firing sequence which contains at most one transition per cluster. These insights might be of independent interest for future research on free-choice systems.

Keywords: Petri Nets · Free-Choice Petri Nets · Home Cluster · Blocking Marking · Reachability · Process Mining

1 Introduction

We study live, bounded, free-choice Petri net systems (LBFC-systems). Recall that a Petri net system is *live* if at each state, every transition can be enabled eventually; it is *bounded* if there is a global constant $b \in \mathbb{N}$ such that no place contains more than b tokens in any reachable marking; and it is *free-choice* if all pairs of transitions either share all or none of their input places.

© The Author(s), under exclusive license to Springer Nature Switzerland AG 2026
J. Desel and A. Kalenkova (Eds.): PETRI NETS 2026, LNCS 16567, pp. 289–311, 2026.
https://doi.org/10.1007/978-3-032-27879-1_14

LBFC-systems enjoy several useful algorithmic and structural properties. For instance, they allow for a combinatorial, graph-theoretic characterization of their recurrent states. To be precise, Best, Desel, and Esparza [7] proved that *home markings* (i.e., markings to which the system can return from every reachable state) in LBFC-systems can be characterized by so-called *traps*, which are sets of places that, once marked, cannot become unmarked in the future. Since maximal unmarked traps can be computed via a polynomial-time fixed-point process, this gives rise to efficient (polynomial-time) algorithms to analyze LBFC-systems. As one application, it becomes tractable to decide if an LBFC-system is cyclic (i.e., whether the initial marking is a home marking). Note that such behavioral properties of systems are usually very hard to compute, so it is beneficial to have efficient algorithms at hand (see, e.g., [14]). Another remarkable property of LBFC-systems is that the diameter of their reachability graph is polynomially bounded in the size of the system and the bound b. In fact, as Desel and Esparza [11,12] established in their *Shortest Sequence Theorem* whenever two markings are reachable from each other, then there exists a connecting firing sequence of length $\mathcal{O}(b \cdot n^3)$ where n is the number of transitions and b is the bound of the LBFC-system. A third notable result on LBFC-systems is due to Gaujal, Haar, and Mairesse [17] and is known as the *Blocking Theorem*. It states that for every *cluster* in an LBFC-system (i.e., sets of transitions with the same input places), there exists a unique *blocking marking* that enables only the transitions of the cluster. Moreover, this blocking marking is reachable from every reachable marking of the system *and* without firing any transition of the cluster itself. Uniqueness implies that whenever a marking enables only a single cluster, it coincides with the blocking marking associated to that cluster already. This means that the state space of an LBFC-system, when restricted to blocking markings, is in a one-to-one correspondence with the clusters of the system. In particular, it is linearly bounded in the number of transitions. Of course this raises many questions, for example, of how this "simple" part of the state space is connected to the remainder of the system or how we can reach this part as fast as possible. We study these questions in the present paper.

Given the outstanding properties of LBFC-systems, it is not surprising that they found numerous applications also in related areas of computer science, such as process mining [2,9]. Here, algorithmic problems involving large amounts of event data have to be solved efficiently. Process models, which are often based on Petri nets, play a central role in this context. In particular, *sound workflow nets*, a prominent subclass of *live* and *bounded* Petri net systems, tailored for business process modeling, are nowadays the de-facto standard for models in the process mining community [1,2]. Since sound workflow nets are live and bounded already, by adding the free-choice restriction, we obtain a subclass of LBFC-systems to which we can apply the host of structure-theoretical and algorithmic results such as the ones outlined above. This route has been investigated by several researchers in the process mining community, see, e.g., [1,16,18–23], and has lead to important algorithmic improvements.

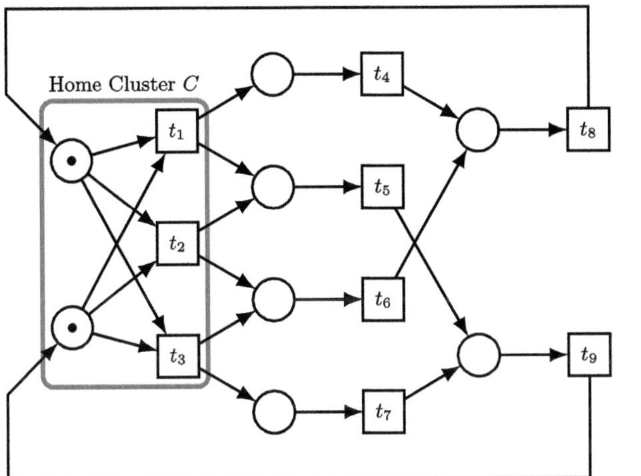

Fig. 1. Example LBFC-system with a home cluster C. The current marking of the system is the marking of the home cluster C: all places of C are marked and there are no tokens outside of C. It is therefore also the blocking marking associated to the home cluster C.

While this shows how Petri net theory helps to solve problems in different application domains, the converse is true as well: notions arising, e.g., in the area of process mining, lead to new insights for the underlying Petri net models. A prominent example is the concept of *home clusters*, which first arose in the context of process mining and was later investigated from the viewpoint of Petri net theory. An example of an LBFC-system with a home cluster is shown in Fig. 1. A home cluster in a free-choice Petri net is a cluster whose (unique) blocking marking only marks the places of the cluster itself. This means that the transitions in a home cluster bind *all* tokens such that no tokens are left outside the cluster. This particular blocking marking is also the current marking of the system in Fig. 1. The consequence is that a home cluster captures the idea of having a "regeneration point" of the system which can serve as a "terminal/initial state", i.e., a state where the system completes its computation or restarts afresh. While in general, as also explained in [17], each blocking marking of each cluster can serve as a regeneration point of the system, a home cluster ensures that, in addition, all tokens are concentrated in the cluster when it becomes enabled. This assumption fits very well for process models, for example, where all flows must simultaneously reach a final state in order to complete the process.

In fact, in *sound* workflow nets, one requirement is that the model has the *option to complete* which corresponds to the existence of a home cluster (where the net reaches its final marking). While home clusters clearly have practical motivation, they also have very interesting structure-theoretical properties. As the third author proved in a series of articles [3–6], the effect of having a home cluster alone in combination with the free-choice assumption is sufficient in order to guarantee *lucency* of the system. Lucency is a generalization of the blocking marking property saying that each (reachable) marking is uniquely determined by the *set of enabled transitions*. In other words, it is impossible to have two different reachable markings that enable the same set of transitions. An immediate consequence is that the state space of a free-choice system with a home cluster is finite, i.e., the system is bounded.

Motivated by these results, we continue the study of LBFC-systems with a home cluster (which includes sound free-choice workflow nets). We focus on the question of how efficiently blocking markings can be reached in such systems, i.e., we ask for the *length of minimal firing sequences to blocking markings*—that is, sequences which cannot be shortened by permuting and/or removing transitions. Our motivation comes twofold: first, blocking markings are a fundamental concept in the theory of free-choice systems, and understanding how efficiently they can be reached is of theoretical interest. In fact, establishing tight bounds on the length of minimal firing sequences to blocking markings sheds light on the expressive power of the underlying Petri net systems. More specifically, our results imply that each blocking sequence of super-linear length must contain redundant computation (formally, it must contain a T-invariant which does not change the marking of a system). Hence, super-linear (non-recurrent) computations cannot be simulated by free-choice systems with a home cluster. This gives clear indication of their limited expressive power.

Second, in algorithmic contexts and application domains such as process mining, it is often necessary to construct firing sequences to blocking markings explicitly (e.g., when replaying event logs on process models). In such cases, of course, having good upper bounds on the length of firing sequences is important to derive performance guarantees for the algorithms. When we consider, for instance, a naive algorithm which (non-deterministically) guesses firing sequences (e.g., in form of an integer linear programming formulation), then bounds on the length of such sequences are of utmost importance to obtain a succinct encoding and to ensure that the solvers can run efficiently.

We also improve upon the best previously known bounds for general LBFC-systems. Note that, in general, there can be minimal firing sequences of different lengths, so another question is: what is the length of the *shortest* of all minimal firing sequences to a blocking marking? We already have a cubic upper bound due to Desel and Esparza's Shortest Sequence Theorem [11,12] (see above), but one could hope that this bound can be improved by the presence of a home cluster. Indeed, our results imply such improvements for blocking and for general markings (to linear and quadratic length, respectively).

2   Preliminaries

In this paper, $\mathbb{N} := \{0, 1, 2, \ldots\}$ denotes the natural numbers (including 0). For a set of sets $\mathcal{A}$, we use $\cup \mathcal{A}$ as a shortcut for $\bigcup_{A \in \mathcal{A}} A$. Let us fix our notation for the basic Petri net concepts used throughout the paper.

Definition 1 (Indicator Function). Let A be a set and let $B \subseteq A$. The function $\chi_B \colon A \to \{0,1\}$, defined by

$$\chi_B(x) := \begin{cases} 1 & x \in B, \\ 0 & x \notin B, \end{cases}$$

is the *indicator function* (or *characteristic function*) of set B. As a special case, we have $\chi_\emptyset =: \mathbf{0}$ where $\mathbf{0}$ is the zero function mapping every element of A to 0.

Definition 2 (Multiset). A *multiset* M over a set A is a function $M \colon A \to \mathbb{N}$; thus, for any $a \in A$, $M(a)$ indicates the multiplicity of a in the multiset M. The set of all multisets over A is given by $\mathbb{N}^A$. We also use the notation $\left[a^{M(a)} \mid a \in A \right]$ for a multiset $M \in \mathbb{N}^A$. Any set $B \subseteq A$ can be considered a multiset using its indicator function, i.e., $\chi_B = \left[b^1 \mid b \in B \right] = [b \in B] \in \mathbb{N}^A$. For multisets $M, M' \in \mathbb{N}^A$, we use the standard notation for functions, e.g., $M + M'$, $M \leq M'$, etc. The *support* (or *domain*) of a multiset $M \in \mathbb{N}^A$, denoted by $\langle M \rangle$, is the set of elements contained in M, i.e., $\langle M \rangle := \{a \in A \mid M(a) > 0\}$. The restriction of a multiset $M \in \mathbb{N}^A$ to a set $B \subseteq A$ is the multiset $M|_B := \left[a^{M(a)} \mid a \in B \right] \in \mathbb{N}^B$ containing only the elements of B with their respective multiplicity.

Whenever we compare or operate with multisets $M \in \mathbb{N}^A$ and $M' \in \mathbb{N}^B$ over different domains A and B, we implicitly assume that they are first extended with 0 for each item of $A \cup B$ not in their domain.

Definition 3 (Sequence, Parikh Vector, Permutation). *Sequences* with index set I over a set A are denoted by $\sigma = \langle a_i \rangle_{i \in I} \in A^I$. The *length* of a sequence σ is written as $|\sigma|$ and the set of all finite sequences over A is denoted by A^*. Given two sequences σ and σ', $\sigma \cdot \sigma'$ (or $\sigma\sigma'$ in short) denotes the concatenation of the two sequences. The restriction of a sequence $\sigma \in A^*$ to a set $B \subseteq A$ is the subsequence $\sigma|_B$ of σ consisting of all elements in B. The *Parikh vector* of a sequence $\sigma \in A^*$, denoted by $\boldsymbol{\sigma}$, is its multiset representation defined by $\boldsymbol{\sigma}(a) := \left| \sigma|_{[a]} \right|$ for every $a \in A$, and $\langle \boldsymbol{\sigma} \rangle$ provides the *support* of σ. A sequence $\sigma' \in A^*$ is a *permutation* of σ if and only if $\boldsymbol{\sigma'} = \boldsymbol{\sigma}$.

Definition 4 (Petri Net). A *Petri net* N is a bipartite directed graph $N = (P, T, F)$ where P and T, $P \cap T = \emptyset$, are disjoint finite sets of vertices and $F \subseteq (P \times T) \cup (T \times P)$ is the set of arcs. In a Petri net, P is called the set of *places*, T the set of *transitions*, and F the *flow relation*. Given a vertex $v \in P \cup T$, its *pre-set* $\bullet v$ and *post-set* $v \bullet$ are defined by $\bullet v := \{u \in P \cup T \mid (u, v) \in F\}$ and $v \bullet := \{u \in P \cup T \mid (v, u) \in F\}$. With regard to a place (transition), its pre- and post-set are also called *input* and *output transitions* (*places*). This $\bullet$-notation is extended to a set of vertices $V \subseteq P \cup T$ by $\bullet V := \bigcup_{v \in V} \bullet v$ and $V \bullet := \bigcup_{v \in V} v \bullet$.

Definition 5 (Subnet). Let $N = (P, T, F)$ be a Petri net. $N' = (P', T', F')$ is a *subnet* of N if $P' \subseteq P$, $T' \subseteq T$, and $F' = F \cap ((P' \times T') \cup (T' \times P'))$. Note that a subnet is always an *induced* subnet (since F' contains all arcs of F between the nodes in $P' \cup T'$). For a non-empty set of nodes $V \subseteq P \cup T$, we denote this *induced subnet* as $N[V]$ given as $N[V] := (P \cap V, T \cap V, F \cap (V \times V))$.

Definition 6 (Free-Choice Petri Net). A Petri net $N = (P, T, F)$ is *free-choice* if any two transitions either share all or none of their input places, i.e., $\forall t, t' \in T \colon {\bullet}t = {\bullet}t' \vee {\bullet}t \cap {\bullet}t' = \emptyset$.

Definition 7 (Marking, System, Firing Rule). Given a Petri net $N = (P, T, F)$, a *marking* $M \in \mathbb{N}^P$ is a multiset where $M(p)$ is the number of *tokens* at place $p \in P$. A place $p \in P$ is *marked* at M if $M(p) > 0$. A place that is not marked is called *empty*. The pair (N, M) of a weakly-connected Petri net $N = (P, T, F)$ with at least one place and at least one transition, i.e., $P \neq \emptyset$ and $T \neq \emptyset$, and a marking $M \in \mathbb{N}^P$ is a *system*. A transition $t \in T$ is *enabled* in M, denoted by $(N, M)[t\rangle$, if and only if each of its input places $p \in {\bullet}t$ is marked, i.e., $M \geq \chi_{{\bullet}t}$. An enabled transition may *fire*, denoted by $(N, M)[t\rangle(N, M')$, and firing results in a new marking $M' := M - \chi_{{\bullet}t} + \chi_{t{\bullet}}$.

A sequence of transitions $\sigma := \langle t_i \rangle_{i=1}^{|\sigma|} \in T^*$ is called a *firing sequence* of (N, M) if for every transition t_i of the sequence holds that $(N, M_{i-1})[t_i\rangle$ and $(N, M_{i-1})[t_i\rangle(N, M_i)$ where $M_0 := M$ and $M_{|\sigma|} =: M'$. Firing such a sequence is denoted by $(N, M)[\sigma\rangle(N, M')$. The empty sequence $\langle\rangle$ is always enabled and firing the empty sequence leaves the marking unchanged, i.e., $(N, M)[\langle\rangle\rangle(N, M)$. A marking M' is *reachable* if a firing sequence $\sigma \in T^*$ exists such that M' is the resulting marking, i.e., $(N, M)[\sigma\rangle(N, M')$. The set of all reachable markings of (N, M) is defined by $[N, M\rangle := \{M' \in \mathbb{N}^P \mid \exists \sigma \in T^* \colon (N, M)[\sigma\rangle(N, M')\}$.

In this paper, we restrict systems to consist of at least one place and one transition and being weakly connected. This allows us to disregard edge cases that are of little relevance anyway and thereby simplify our proofs.

Definition 8 (Boundedness, Safeness). Given some $k \in \mathbb{N}$, a system (N, M_0) with $N = (P, T, F)$ is *k-bounded* if k is a bound for any reachable marking, i.e., $\forall M \in [N, M_0\rangle \colon \forall p \in P \colon M(p) \leq k$. (N, M_0) is *safe* if it is 1-bounded.

Definition 9 (Liveness). A system (N, M_0) with $N = (P, T, F)$ is *live* if for every transition $t \in T$ and reachable marking $M \in [N, M_0\rangle$, there exists a reachable marking $M' \in [N, M\rangle$ which enables t, i.e., $\forall t \in T \colon \forall M \in [N, M_0\rangle \colon \exists M' \in [N, M\rangle \colon (N, M')[t\rangle$.

Definition 10 (Well-Formedness). A Petri net N is *well-formed* if there exists a marking M_0 of N such that (N, M_0) is a live and bounded system.

Well-formed Petri nets are strongly connected, see also [11, Theorem 2.25].

Definition 11 (LBFC-System). We denote a system which is live, (b-) bounded, and free-choice as *LBFC-system*. Note that the place bound b may vary across different LBFC-systems, but is, of course, fixed for each individual LBFC-system.

Definition 12 (Home Marking, Cyclic System). Let (N, M_0) be a system. A marking $M \in [N, M_0\rangle$ is a *home marking* if it can be reached from any reachable marking, i.e., $\forall M' \in [N, M_0\rangle \colon M \in [N, M'\rangle$. A system (N, M_0) is *cyclic* if its initial marking M_0 is a home marking.

Definition 13 (Cluster). Let $N = (P, T, F)$ be a Petri net and let $v \in P \cup T$. The *cluster* of v, denoted by $[v]_\sim$, is the minimal set of places and transitions that includes

- v, i.e., $v \in [v]_\sim$,
- all output transitions of contained places, i.e., $\forall p \in [v]_\sim \cap P \colon p\bullet \subseteq [v]_\sim$, and
- all input places of contained transitions, i.e., $\forall t \in [v]_\sim \cap T \colon \bullet t \subseteq [v]_\sim$.

The set of all clusters of N is defined by $N/\!\!\sim \; := \{[v]_\sim \mid v \in P \cup T\}$.

It is not hard to show that the set of all clusters $N/\!\!\sim$ forms a partition of the set $P \cup T$ (see [11, Proposition 4.5]). What is more, in free-choice Petri nets, clusters have a very simple structure: they consist of a set of places and a set of transitions such that all places are connected to all transitions as input places and vice versa, i.e., they form a complete bipartite subgraph. Here, the cluster of an element $v \in P \cup T$ can also be determined quite easily: $[v]_\sim = \bullet v \cup (\bullet v)\bullet$ if v is a transition, and $[v]_\sim = v\bullet \cup \bullet(v\bullet)$ if v is a place. As a consequence, in free-choice Petri nets, if one transition of a cluster is enabled in some marking, then all transitions of the cluster are enabled in that marking (see [11, Proposition 4.6]). In free-choice systems, we will therefore also refer to the *set of enabled clusters*.

Definition 14 (Set of Enabled Clusters). Let $N = (P, T, F)$ be a free-choice Petri net and let $M \in \mathbb{N}^P$ be a marking. The *set of clusters enabled in M* is defined by $\mathcal{C}_{en}(N, M) := \{C \in N/\!\!\sim \; \mid \exists t \in C \cap T \colon (N, M)[t\rangle\}$.

3   Shortest Enabling Sequences in T-Systems

As a warmup, we start our analysis of shortest sequences with what we call *enabling sequences*, i.e., sequences that *enable* a given cluster from some reachable marking (and have minimal length with this property). The motivation is as follows: if we want to reach a blocking marking associated to some cluster—that is, a marking which solely enables the transitions of that cluster (cf. Definition 23)—, there are two steps involved. First, we need to *enable* the cluster (if it is not enabled already), and second, we need to *disable* all other clusters different from the target cluster. Thus, understanding how to enable clusters efficiently is an important building block for reaching blocking markings.

Definition 15 (Enabling Sequence). Let (N, M_0) with $N = (P, T, F)$ be a system. For a transition $t \in T$ and some reachable marking $M \in [N, M_0\rangle$, a firing sequence $\sigma \in T^*$ enabled in M is a *t-enabling sequence (from M)* if it results in a marking that enables t, i.e., $(N, M)[\sigma\langle t\rangle$. A t-enabling sequence $\eta \in T^*$ is a *minimal t-enabling sequence* if no strict (permuted) subsequence $\sigma \in T^*$, $\boldsymbol{\sigma} < \boldsymbol{\eta}$, is t-enabling as well.

In other words, a minimal t-enabling sequence enables t and cannot be shortened by removing transitions or permuting them. Of course, minimal t-enabling sequences never contain t itself:

Proposition 1. *Let (N, M_0) with $N = (P, T, F)$ be a system. Let $t \in T$ be some transition and let $\sigma \in T^*$ be a minimal sequence that enables t, i.e., $(N, M_0)[\sigma\langle t \rangle\rangle$. Then, $t \notin \langle \sigma \rangle$.*

We start our investigation with *live T-systems* (instead of considering full LBFC-systems). This first step is justified by the fact that T-systems are, in a precise sense, the basic *building blocks* of LBFC-systems as shown by Desel and Esparza [11, Chapter 7] building on earlier results due to Esparza [15] and Esparza and Silva [13]. A T-net is a Petri net where each place has exactly one input and one output transition. Therefore, T-nets are free-choice.

Definition 16 (T-Net, T-System). A Petri net (P, T, F) is a *T-net* if every place has exactly one input and one output transition, i.e., $\forall p \in P \colon |\bullet p| = 1 = |p \bullet|$. A system (N, M) is a *T-system* if N is a T-net.

T-systems cannot express choices (conflicts) between transitions. As a matter of definition, no pair of transitions $t, t' \in T, t \neq t'$ can compete for tokens from the same input places (formally: $\bullet t \cap \bullet t' = \emptyset$). In general, firing sequences in which all pairs of transitions satisfy this property are known as *biased firing sequences* and have been studied by Desel and Esparza [11, 12]. We recall some of their results and apply them to enabling sequences in live T-systems.

Definition 17 (Biased Firing Sequence). Let (N, M_0) with $N = (P, T, F)$ be a system. A firing sequence $\sigma \in T^*$ is *biased* if $\bullet t \cap \bullet t' = \emptyset$ for every two distinct transitions $t, t' \in \langle \sigma \rangle, t \neq t'$ that occur in σ.

Lemma 1 ([11, Lemma 3.23], [12, Lemma 3.2]). *Let (N, M_0) with $N = (P, T, F)$ be a system. Let $t \in T$ be some transition and let $\sigma_1 \sigma_2 \langle t \rangle$ be a biased firing sequence such that $\sigma_1, \sigma_2 \in (T \setminus \{t\})^*$ be two firing sequences with $\langle \sigma_1 \rangle \supseteq \langle \sigma_2 \rangle$. If $(N, M_0)[\sigma_1 \sigma_2 \langle t \rangle\rangle(N, M)$, then also $(N, M_0)[\sigma_1 \langle t \rangle \sigma_2\rangle(N, M)$.*

Lemma 2 ([11, Lemma 3.24], [12, Lemma 3.3]). *Let (N, M_0) with $N = (P, T, F)$ be a system and let $\sigma \in T^*$ be a biased firing sequence resulting in a marking $M \in [N, M_0\rangle$, i.e., $(N, M_0)[\sigma\rangle(N, M)$. Then, there exists a permutation σ' of σ with $\sigma' = \sigma$ such that $(N, M_0)[\sigma'\rangle(N, M)$ and $\sigma' = \sigma_1 \sigma_2$ with $\sigma_1 \leq \chi_T$ and $\langle \sigma_1 \rangle \supseteq \langle \sigma_2 \rangle$.*

There is another way to rearrange biased firing sequences: every first occurrence of a transition already enabled in the current marking can be moved forward.

Lemma 3. *Let (N, M_0) with $N = (P, T, F)$ be a system. Let $t, t' \in T, t \neq t'$ be two distinct transitions and let $\langle t \rangle \sigma \langle t' \rangle$ be a biased firing sequence such that $\sigma \in (T \setminus \{t, t'\})^*$ and $(N, M_0)[\langle t \rangle \sigma \langle t' \rangle\rangle(N, M)$. If t' is enabled in M_0, i.e., $(N, M_0)[t'\rangle$, then also $(N, M_0)[\langle t', t \rangle \sigma\rangle(N, M)$.*

Proof. Since $(N, M_0)[t'\rangle$, $M_0 \geq \chi_{\bullet t'}$ and because $\langle t \rangle \sigma \langle t' \rangle$ is biased, neither t nor any transition occurring in σ has one of the places $\bullet t'$ as input place, i.e., $\bullet t' \cap (\bullet t \cup \bullet \langle \sigma \rangle) = \emptyset$. Thus, their occurrences do not disable t' and vice versa. Hence, $\langle t', t \rangle \sigma$ is also a firing sequence enabled in M_0, and because it is a permutation, it still results in the same marking M, i.e., $(N, M_0)[\langle t', t \rangle \sigma\rangle(N, M)$. $\qquad\square$

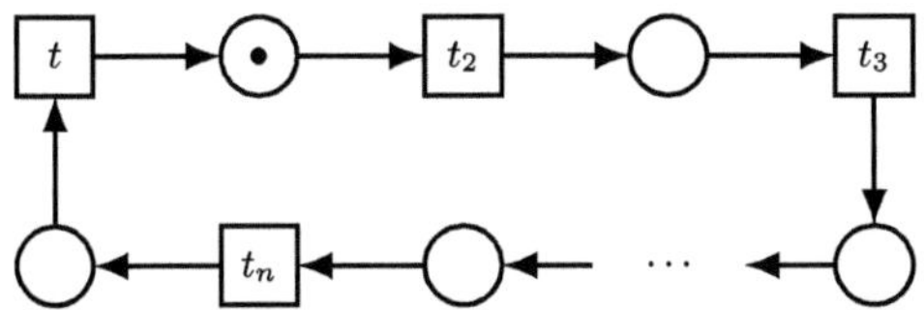

Fig. 2. A live T-system for which the bound in Theorem 1 is tight.

Note that this very much resembles the concept of *expediting* transitions in free-choice systems, cf. [5] and Lemma 6, but the setting here is slightly different.

With Lemmas 1 and 2 we can already show how to shorten biased enabling sequences such that we obtain an enabling sequence which contains each transition at most once.

Lemma 4 (Biased Enabling Sequence Lemma). *Let (N, M_0) with $N = (P, T, F)$ be a system. Let $t \in T$ be some transition and let $\sigma \in T^*$ be a biased sequence that enables t, i.e., $(N, M_0)[\sigma\langle t\rangle\rangle$. Then, there exists a (permuted) subsequence $\eta \in (T \setminus \{t\})^*$ with $\boldsymbol{\eta} \leq \boldsymbol{\sigma}$ such that $(N, M_0)[\eta\langle t\rangle\rangle$ and no transition occurs more than once in η, i.e., $\boldsymbol{\eta} \leq \chi_{T\setminus\{t\}}$.*

Proof. Let us assume that $t \notin \langle\sigma\rangle$ (otherwise choose the prefix before the first occurrence of t as σ, cf. Proposition 1). According to Lemma 2, there exists a permutation $\sigma' \in (T \setminus \{t\})^*$ of σ with $\boldsymbol{\sigma'} = \boldsymbol{\sigma}$ such that $(N, M_0)[\sigma'\langle t\rangle\rangle$ and $\sigma' = \sigma_1\sigma_2$ with $\sigma_1 \leq \chi_T$ and $\langle\boldsymbol{\sigma_1}\rangle \supseteq \langle\boldsymbol{\sigma_2}\rangle$. Then, also $(N, M_0)[\sigma_1\langle t\rangle\sigma_2\rangle$ according to Lemma 1. Choose $\eta = \sigma_1$ and the claim follows.   □

Since all firing sequences in T-systems are biased, we can apply the result above to live T-systems:

Theorem 1 (Shortest Enabling Sequence Theorem). *Let (N, M_0) with $N = (P, T, F)$ be a live T-system and let $t \in T$ be some transition. Then, there exists a t-enabling sequence $\eta \in (T \setminus \{t\})^*$ which fires each transition at most once, i.e., $\boldsymbol{\eta} \leq \chi_{T\setminus\{t\}}$.*

Proof Since (N, M_0) is live, there exists a minimal t-enabling sequence $\eta \in T^*$. Because N is a T-net, every firing sequence in (N, M_0) is biased, including η. Hence, the claim follows by Lemma 4 since η was chosen to be minimal.   □

Note that Theorem 1 does not only prove the existence of a shortest enabling sequence in a live T-system with length of at most $|T| - 1$, but establishes the more general result that *every* minimal enabling sequence satisfies this bound. Also, it is easy to see that this bound is tight: in the live T-system in Fig. 2, every other transition in T must fire to enable transition t.

Actually, for the case of T-systems, we can show an even stronger result: *all* minimal t-enabling sequences are permutations of each other, i.e., their Parikh vectors are identical. This means that in a live marking of a T-system each transition t has a *canonical* enabling sequence (up to the order of transitions).

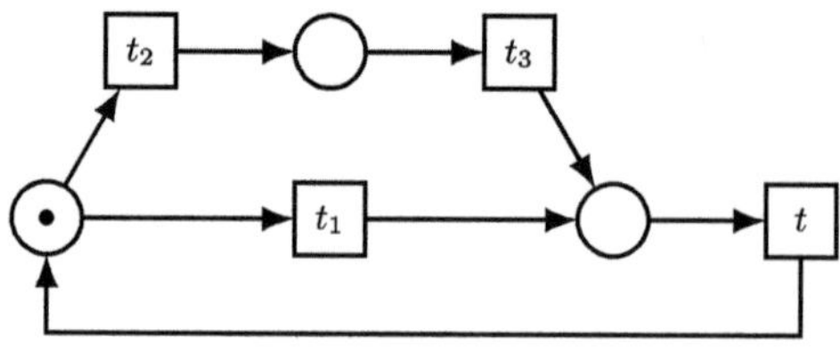

Fig. 3. An LBFC-system with two different minimal t-enabling sequences $\langle t_1 \rangle$ and $\langle t_2, t_3 \rangle$ of different lengths.

Proposition 2. *Let* (N, M_0) *with* $N = (P, T, F)$ *be a live T-system and let* $\eta, \eta' \in T^*$ *be minimal t-enabling sequences for some transition* $t \in T$, *i.e.,* $(N, M)[\eta\langle t \rangle\rangle$ *and* $(N, M)[\eta'\langle t \rangle\rangle$. *Then,* $\boldsymbol{\eta} = \boldsymbol{\eta}'$.

Proof. Let $\eta, \eta' \in T^*$ be two minimal t-enabling sequences. By Lemma 4, both sequences contain each transition at most once. We proceed by induction on $|\eta|$ and show that $\boldsymbol{\eta} = \boldsymbol{\eta}'$. If t is enabled in M_0 (i.e., $|\eta| = 0$), the claim holds trivially since both sequences are empty. Otherwise, there exists an unmarked place $p \in \bullet t$. This place has a *unique* input transition $t_p \in \bullet p$ since N is a T-net. Of course, this transition must occur in both sequences to mark place p. Hence, η and η' have at least one common transition t_p.

Let $\eta = \alpha_1 \langle t_p \rangle \alpha_2$ and $\eta' = \beta_1 \langle t_p \rangle \beta_2$. Based on Lemma 4, we can rearrange α_1 and β_1 into two sequences $\alpha_1^1 \alpha_1^2$ and $\beta_1^1 \beta_1^2$ such that α_1^1 and β_1^1 are both two minimal t_p-enabling sequences enabled in M_0, and by Lemma 3, we obtain $\alpha_1^1 \langle t_p \rangle \alpha_1^2$ and $\beta_1^1 \langle t_p \rangle \beta_1^2$. Since $|\alpha_1^1| < |\eta|$, we can apply the induction hypothesis and obtain $\boldsymbol{\alpha_1^1} = \boldsymbol{\beta_1^1}$. Hence, $\alpha_1^1 \langle t_p \rangle$ and $\beta_1^1 \langle t_p \rangle$ can be fired from M_0 and result in the same marking M_1. From here, the remaining sequences $\alpha_1^2 \alpha_2$ and $\beta_1^2 \beta_2$ are both t-enabling sequences from M_1 and $|\alpha_1^2 \alpha_2| < |\eta|$. Thus, we can again apply the induction hypothesis and obtain $\boldsymbol{\eta} = \boldsymbol{\eta}'$. $\qquad\square$

Of course, we cannot hope for an analogous result for LBFC-systems. The reason is that choices (conflicts) can lead to completely different minimal enabling sequences which are also of different lengths. A simple example is shown in Fig. 3 where two incomparable minimal t-enabling sequences exist.

4   Homing Sequences and Disabling Clusters

Before we turn our attention to enabling sequences in general LBFC-systems, we study how to *disable* clusters different from the target cluster. The key insight is that it is always possible to reach the *home cluster* of an LBFC-system from any reachable marking via an *elementary homing sequence* that fires at most one transition from each cluster. Later, such homing sequences can then be utilized in order to disable all clusters different from the target cluster. Home clusters have first been studied by the third author in a research line on *lucency* [cf. 3,5].

Definition 18 (Home Cluster). Let (N, M_0) with $N = (P, T, F)$ be a system. A cluster $C \in N/\sim$ is a *home cluster* of (N, M_0) if and only if the marking $\chi_{C \cap P} \in \mathbb{N}^P$ is reachable and a home marking, i.e., $\forall M \in [N, M_0\rangle: \chi_{C \cap P} \in [N, M\rangle$. If such a cluster C exists, we say that (N, M_0) has a home cluster.

A home cluster has strong implications on the underlying Petri net structure as we will show next. First, it is known that LBFC-systems can be covered by simple types of subnets, namely by so-called S- and T-components (see Theorems 3 and 4 below). That is, we can find families of basic subnets (we can either choose S-components or T-components for that purpose) which together include all places and transitions of the original net (not necessarily disjointly). An S-component, in turn, is a strongly-connected S-net, where an S-net is the dual concept to a T-net: each transition has exactly one input and one output place. In this way, S-nets cannot express concurrency because transitions can neither synchronize on multiple input places nor distribute tokens to multiple output places. One immediate consequence is that the number of tokens in an S-component is invariant and, since the LBFC-system is live, each S-component must contain at least one token. On the other hand, as a home cluster can be reached from any reachable marking, it must intersect all S-components because in the marking of a home cluster all other places are empty. Let us make these ideas precise.

Definition 19 (S-Net, S-System). A Petri net (P, T, F) is an *S-net* if every transition has exactly one input and one output place, i.e., $\forall t \in T: |{\bullet}t| = 1 = |t{\bullet}|$. A system (N, M) is an *S-system* if N is an S-net.

Definition 20 (S-Component, T-Component). Let $N = (P, T, F)$ be a Petri net. A subnet $N[P' \cup T']$ induced by the sets $P' \subseteq P$ and $T' \subseteq T$ is an *S-component* of N if it is a strongly-connected S-net and contains every input and output transition of a contained place, i.e., $\forall p \in P': {\bullet}p \cup p{\bullet} \subseteq T'$. A subnet $N[P' \cup T']$ induced by the sets $P' \subseteq P$ and $T' \subseteq T$ is a *T-component* of N if it is a strongly-connected T-net and contains every input and output place of a contained transition, i.e., $\forall t \in T': {\bullet}t \cup t{\bullet} \subseteq P'$.

Proposition 3 ([11, Proposition 5.2 (2)]). *Let (N, M_0) be a system and let $N_S = (P', T', F')$ be an S-component of N. Then, for any two markings $M \in [N, M_0\rangle$ and $M' \in [N, M\rangle$, $\sum_{p \in P'} M(p) = \sum_{p \in P'} M'(p)$.*

Proposition 4 ([11, Proposition 5.12 (2)]). *Let (N, M_0) be a system and let $N_T = (P', T', F')$ be a T-component of N. Then, for any sequence of transitions $\sigma \in T'^*$, $(N, M_0)[\sigma\rangle(N, M)$ if and only if $(N_T, M_0|_{P'})[\sigma\rangle(N_T, M|_{P'})$.*

Theorem 2 (Activation of T-Components [11, Theorem 5.20]). *Let $N_T = (P', T', F')$ be a T-component of an LBFC-system (N, M_0) with $N = (P, T, F)$. There exists a firing sequence $\sigma \in (T \setminus T')^*$ with $(N, M_0)[\sigma\rangle(N, M)$ such that $(N_T, M|_{P'})$ is live.*

Theorem 3 (S-Coverability Theorem [11, Theorem 5.6]**).** *Well-formed free-choice Petri nets are covered by S-components, i.e., there exists a family of subnets which are S-components and such that each place and transition of the original net is contained in at least one of these S-components.*

Theorem 4 (T-Coverability Theorem [11, Theorem 5.18]**).** *Well-formed free-choice Petri nets are covered by T-components, i.e., there exists a family of subnets which are T-components and such that each place and transition of the original net is contained in at least one of these T-components.*

Proposition 5. *Let (N, M_0) be an LBFC-system and let $C \in N/{\sim}$ be one of its clusters. If C is a home cluster, then it intersects all S-components of N and this intersection contains precisely one place of C.*

Proof. Let us assume that there is an S-component $N_S = (P', T', F')$ of $N = (P, T, F)$ that does not intersect the home cluster C, i.e., $P' \cap C = \emptyset$. That implies that the marking of the home cluster $\chi_{C \cap P}$ does not mark any places in P'. By Proposition 3, the number of tokens within an S-component is stable. Therefore, there is no possible marking $M \in \mathbb{N}^P$ such that some place in P' is marked and C is a home cluster of the system (N, M). Thus, if C is a home cluster, (N, M_0) cannot be live. This is a contradiction. Hence, C is not a home cluster if it does not intersect all S-components.

Furthermore, an S-component cannot contain two different input places from the home cluster because then, it must contain a common successor transition as well, contradicting the definition of an S-net. Also, the intersection cannot just contain a transition since an S-component must contain *some* input place which is part of the home cluster as well. Hence, each S-component contains precisely one place of the home cluster C. $\qquad\square$

An immediate consequence is that an LBFC-system with a home cluster is safe:

Corollary 1. *An LBFC-system with a home cluster is safe.*

Definition 21 (Homing Sequence). Let (N, M_0) with $N = (P, T, F)$ be a system with a home cluster $C \in N/{\sim}$ and let $M \in [N, M_0\rangle$ be some reachable marking. A firing sequence $\sigma \in T^*$ enabled in M is a *homing sequence* if it results in the marking of the home cluster, i.e., $(N, M)[\sigma\rangle(N, \chi_{C \cap P})$.

According to the definition of a home cluster (Definition 18), its presence alone guarantees the existence of a homing sequence. Next, we show the much stronger result that for every marking a homing sequence can be constructed that contains at most one transition per cluster—in particular, it is of length at most $|N/{\sim}| - 1$. We call such sequences *elementary*:

Definition 22 (Elementary Sequence). Let (N, M_0) with $N = (P, T, F)$ be a system. A firing sequence $\sigma \in T^*$ is *elementary* if it contains at most one transition per cluster and each transition occurs at most once, i.e., $\forall C \in N/{\sim}: \sum_{t \in C \cap T} \boldsymbol{\sigma}(t) \leq 1$.

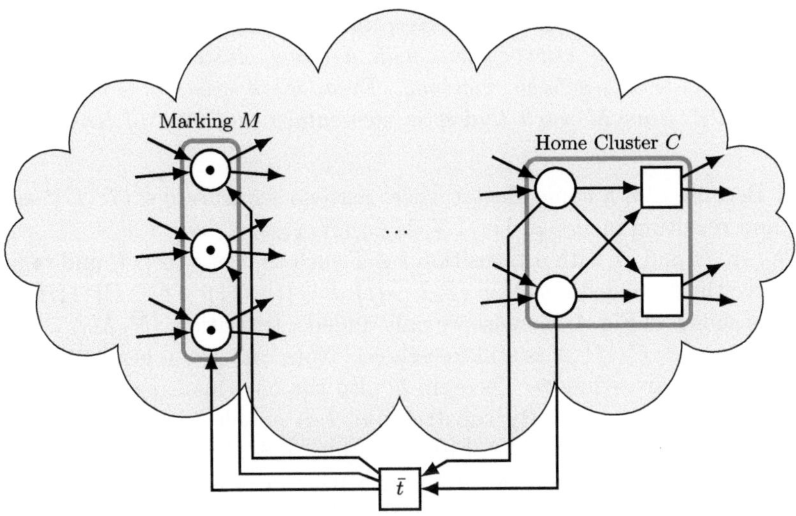

Fig. 4. Extension of the LBFC-system (N, M_0) with home cluster C by a new transition $\bar{t} \notin T$ that reestablishes the marking M from the home cluster marking $\chi_{C\cap P}$.

Proposition 6. *Elementary firing sequences are biased.*

In general, when we want to reach a blocking marking associated to a transition t, i.e., a marking that only enables $[t]_\sim$, it might happen that, on our way, we encounter markings that enable $[t]_\sim$ but also other clusters as well. For the home cluster, however, this is not possible as the following theorem states.

Theorem 5 ([5, Theorem 5.5]). *Let (N, M_0) be a proper (i.e., every transition has at least one input place and one output place) free-choice system with a home cluster $C \in N/\sim$. If all places of the home cluster $C \cap P$ are marked at a reachable marking $M \in [N, M_0\rangle$, then they are the only places marked at M, i.e., $\forall M \in [N, M_0\rangle: M \geq \chi_{C\cap P} \implies M = \chi_{C\cap P}$.*

A direct consequence is that the home cluster is only enabled in $\chi_{C\cap P}$ and then, it is also the only cluster enabled, i.e., $\{C\} = \mathcal{C}_{en}(N, \chi_{C\cap P})$. Also note that well-formed Petri nets are strongly connected and therefore proper. In particular, Theorem 5 applies to LBFC-systems as well.

As a last preparation, we require a result on the realizability of T-vectors in T-systems. It states that in a live T-system it is possible to construct a firing sequence that fires each transition exactly once (in some order).

Lemma 5 ([8, Theorem 5.31], [10, Theorems 6 and 7]). *Let (N, M) with $N = (P, T, F)$ be a T-system. (N, M) is live if and only if χ_T is realizable, i.e., $\exists \sigma \in T^*: (N, M)[\sigma\rangle \wedge \sigma = \chi_T$.*

We are now ready to prove the main result of this section.

Theorem 6 (Shortest Homing Sequence Theorem). *Let (N, M_0) with $N = (P, T, F)$ be an LBFC-system with a home cluster $C \in N/{\sim}$ and let $M \in [N, M_0\rangle$ be a reachable marking. Then, there exists a homing sequence $\eta \in (T \setminus C)^*$ from M such that η is elementary and thus of length at most $|N/{\sim}| - 1$.*

Proof. Because C is a home cluster, there exists a sequence $\eta \in (T \setminus C)^*$ enabled in M and resulting in $\chi_{C \cap P}$, i.e., $(N, M)[\eta\rangle(N, \chi_{C \cap P})$.

We can extend N with a transition $\bar{t} \notin T$ such that $\bullet\bar{t} = C \cap P$ and $\bar{t}\bullet = \langle M \rangle$ to retrieve the extended net $\bar{N} = (P, T \cup \{\bar{t}\}, F \cup \{(p, \bar{t}) \mid p \in C \cap P\} \cup \{(\bar{t}, p) \mid p \in \langle M \rangle\})$ as shown in Fig. 4. Because we only added a transition, $[\bar{N}, M_0\rangle \supseteq [N, M_0\rangle$ and because $\bullet\bar{t} = C \cap P$, $\bar{N}$ is still free-choice. Note that $\chi_{C \cap P}$ is a home marking of (N, M_0) and according to Theorem 5, also the only marking in $[N, M_0\rangle$ that enables the home cluster. By construction, $\bar{t}$ is enabled in $\chi_{C \cap P}$ which is of course also reachable in $(\bar{N}, M_0)$. Firing $\bar{t}$ results in M, i.e., $(\bar{N}, \chi_{C \cap P})[\bar{t}\rangle(\bar{N}, M)$; but, because $M \in [N, M_0\rangle$, the extension with $\bar{t}$ does not introduce additional reachable markings to $[\bar{N}, M_0\rangle$. Hence, $[\bar{N}, M_0\rangle = [N, M_0\rangle$, and $(\bar{N}, M_0)$ is also live and bounded and has a home cluster $C \cup \{\bar{t}\}$.

Due to the well-formedness of $\bar{N}$ and Theorem 4, there exists some T-component $\bar{N}_T = (P', T', F')$ of $\bar{N}$ covering $\bar{t}$, i.e., $\bar{t} \in T'$. By Theorem 2, this T-component can be activated by some firing sequence $\sigma \in ((T \cup \{\bar{t}\}) \setminus T')^*$, i.e., $\exists \sigma \in ((T \cup \{\bar{t}\}) \setminus T')^*$ with $(\bar{N}, M_0)[\sigma\rangle(\bar{N}, M')$ such that $(\bar{N}_T, M'|_{P'})$ is live. Note that σ is also a firing sequence enabled in (N, M_0), i.e., $(N, M_0)[\sigma\rangle$. Since $(\bar{N}_T, M'|_{P'})$ is live, there exists a firing sequence $\sigma' \in T'^*$ enabled in $M'|_{P'}$ that results in some marking which enables $\bar{t}$, and due to Proposition 4, σ' is also a firing sequence in $(\bar{N}, M')$, i.e., $(\bar{N}, M')[\sigma'\langle\bar{t}\rangle\rangle$. But, because of Theorem 5, $\chi_{C \cap P} = \chi_{C \cap P'}$ is the only marking enabling $\bar{t}$. Hence, $(\bar{N}_T, \chi_{C \cap P'})$ is live and safe and firing $\bar{t}$ results in $M|_{P'} = M$, i.e., $(\bar{N}_T, \chi_{C \cap P'})[\bar{t}\rangle(\bar{N}_T, M|_{P'})$.

Using Lemma 5, we can infer that $\chi_{T' \setminus \{\bar{t}\}}$ is realizable from M in $\bar{N}$ (because $\chi_{T'}$ is realizable from $\chi_{C \cap P}$) and results in the marking of the home cluster $\chi_{C \cap P}$. Because our additional transition $\bar{t}$ does not occur in $\chi_{T' \setminus \{\bar{t}\}}$, it is also realizable in N. Therefore, there exists a homing sequence $\eta \in (T \setminus C)^*$ such that $\eta = \chi_{T' \setminus \{\bar{t}\}}$ and because T' contains at most one transition of each cluster, i.e., $|T'| \leq |N/{\sim}|$, η is elementary and the claim follows. $\qquad\square$

With this preparation, we can show how clusters can be disabled by elementary firing sequences in order to reach *blocking markings*.

Definition 23 (Blocking Marking). Let (N, M_0) with $N = (P, T, F)$ be a system. For some transition $t \in T$, a reachable marking $M \in [N, M_0\rangle$ is a *blocking marking* associated to t if t is enabled in M, i.e., $(N, M)[t\rangle$, and every other transition either shares all of its input places with t or is not enabled in M, i.e., $\forall t' \in T : (N, M)[t'\rangle \implies \bullet t' \subseteq \bullet t$.

Of course, in free-choice systems, blocking markings are associated to clusters as either none or all transitions of a cluster are enabled in a marking [11, Proposition 4.6]. Here, a blocking marking M is thus defined by $\forall t' \in T : (N, M)[t'\rangle \implies \bullet t' = \bullet t$. Remarkably, in LBFC-systems, blocking markings are unique:

Theorem 7 (Blocking Theorem [17, Theorem 3.1], [24, Theorem 2.4]**).** *Let* (N, M_0) *with* $N = (P, T, F)$ *be an* LBFC*-system. Then, every cluster* $C \in N/\sim$ *has a unique blocking marking* $M_C \in [N, M_0\rangle$ *associated to it. Furthermore,* M_C *is a home marking which can be obtained from any reachable marking without firing a transition in* C*, i.e.,* $\forall M \in [N, M_0\rangle : \exists \sigma \in (T \setminus C)^* : (N, M)[\sigma\rangle(N, M_C)$.

A much more general property can be shown in presence of a home cluster: the enabled transitions uniquely identify the current marking. This property was termed *lucency* by the third author [cf. 5].

Theorem 8 ([3, Theorem 3]**).** *Let* (N, M_0) *with* $N = (P, T, F)$ *be an* LBFC*-system with a home cluster and let* $M_1, M_2 \in [N, M_0\rangle$ *be two reachable markings. If* M_1 *and* M_2 *enable exactly the same transitions, i.e.,* $\{t \in T \mid (N, M_1)[t\rangle\} = \{t \in T \mid (N, M_2)[t\rangle\}$*, then* $M_1 = M_2$.

Due to the free-choice property, we can translate this result also directly to the set of enabled clusters. Then, we obtain $M_1 = M_2 \iff \mathcal{C}_{en}(N, M_1) = \mathcal{C}_{en}(N, M_2)$.

We are ready to prove the second main result of this section: in every marking of an LBFC-system with a home cluster, it is possible to disable an arbitrary set of clusters via an elementary firing sequence.

Theorem 9 (Disabling Cluster Theorem). *Let* (N, M_0) *with* $N = (P, T, F)$ *be an* LBFC*-system with a home cluster* $C \in N/\sim$ *and let* $M \in [N, M_0\rangle$ *be a reachable marking. For any non-empty subset* $\mathcal{C}'$ *of the clusters ebabled in* M*, i.e.,* $\mathcal{C}' \subseteq \mathcal{C}_{en}(N, M)$*,* $\mathcal{C}' \neq \emptyset$*, there exists an elementary firing sequence* $\sigma \in (T \setminus C)^*$ *leading to a marking* M'*, i.e.,* $(N, M)[\sigma\rangle(N, M')$*, where only the clusters in* $\mathcal{C}'$ *are enabled, i.e.,* $\mathcal{C}_{en}(N, M') = \mathcal{C}'$.

Proof. First observe that for every (non-empty) $\mathcal{C}' \subseteq \mathcal{C}_{en}(N, M)$ there can be at most one marking $M' \in [N, M_0\rangle$ that enables exactly the clusters in $\mathcal{C}'$. This follows directly from Theorem 8. In particular, if $\mathcal{C}' = \mathcal{C}_{en}(N, M)$ we can choose $\sigma = \langle\rangle$ and the claim follows.

Let us assume that $\mathcal{C}' \subset \mathcal{C}_{en}(N, M)$. If $|\mathcal{C}_{en}(N, M)| = 1$, either $\mathcal{C}' = \emptyset$ or $\mathcal{C}' = \mathcal{C}_{en}(N, M)$ must hold. The former case is not possible by assumption and the latter case is already handled above. Therefore, let $|\mathcal{C}_{en}(N, M)| > 1$, and as a consequence, $C \notin \mathcal{C}_{en}(N, M)$ because whenever the home cluster is enabled, it is the only cluster enabled (see Theorem 5). Since N is free-choice, the only way to disable a cluster in $\mathcal{C}_{en}(N, M) \setminus \mathcal{C}'$ is to fire one of its transitions. Let $\eta \in (T \setminus C)^*$ be an elementary homing sequence from M, which exists due to Theorem 6 and results in the marking of the home cluster $\chi_{C \cap P}$. Here, only C is enabled. In particular, no transition of any cluster in $\mathcal{C}_{en}(N, M)$ is enabled in $\chi_{C \cap P}$. Therefore, at least one transition of every cluster in $\mathcal{C}_{en}(N, M)$ occurs in η. In fact, it is exactly one transition of each cluster occurring exactly once in η because η is elementary. Also, there cannot be other enabled clusters in $N/\sim \setminus (\{C\} \cup \mathcal{C}_{en}(N, M))$ of which no transition occurs in η because then, the tokens enabling such an additional cluster would remain in their places after firing η which contradicts Theorem 5 that after firing η only the places of C are marked.

The remainder of the proof is done by induction on the length of η.

Base case: $|\eta| = |\mathcal{C}_{en}(N, M)|$. Then, every transition $t \in \langle \eta \rangle$ is enabled by the premise of the theorem. Because firing η results in $\chi_{C \cap P}$ and every transition occurring in η is enabled, all of their output places are in the home cluster and each transition has a different set of output places, i.e., $\langle \boldsymbol{\eta} \rangle \bullet \subseteq C \cap P$ and $\forall t, t' \in \langle \boldsymbol{\eta} \rangle, t \neq t' : t \bullet \cap t' \bullet = \emptyset$. Now, we can use Lemma 3 to move the transitions of $\mathcal{C}_{en}(N, M) \setminus \mathcal{C}'$ occurring in η forward and by firing them disabling all clusters in $\mathcal{C}_{en}(N, M) \setminus \mathcal{C}'$. Without firing a transition from $\mathcal{C}'$, the home cluster cannot yet be enabled and therefore, we have reached M'.

Induction step: $|\eta| > |\mathcal{C}_{en}(N, M)|$. Now, for every cluster in $\mathcal{C}_{en}(N, M)$, there is exactly one transition occurring in η that is enabled. As above, we can use Lemma 3 to move the transitions of $\mathcal{C}_{en}(N, M) \setminus \mathcal{C}'$ occurring in η forward and by firing them disabling all clusters in $\mathcal{C}_{en}(N, M) \setminus \mathcal{C}'$. This, however, might enable additional clusters not in $\mathcal{C}_{en}(N, M)$ and therefore also not in $\mathcal{C}'$. Let $\mathcal{C}'' \subseteq N/\!\!\sim \setminus (\{C\} \cup \mathcal{C}_{en}(N, M))$ denote this set of additional clusters. In this case (i.e., $\mathcal{C}'' \neq \emptyset$), take the remaining homing sequence $\eta|_{\cup(\mathcal{C}' \cup \mathcal{C}'')}$ with $\left|\eta|_{\cup(\mathcal{C}' \cup \mathcal{C}'')}\right| < |\eta|$ and repeat the induction step until no clusters outside of $\mathcal{C}'$ are enabled anymore. If no additional clusters have been enabled (i.e., $\mathcal{C}'' = \emptyset$), we have already reached M'. $\qquad\square$

Corollary 2. *Let (N, M_0) with $N = (P, T, F)$ be an* LBFC*-system with a home cluster $C \in N/\!\!\sim$ and let $M \in [N, M_0\rangle$ be a reachable marking enabling the transitions of some cluster $C' \in N/\!\!\sim$. Then, the blocking marking M' associated to cluster C' can be obtained by an elementary firing sequence $\eta \in (T \setminus C')^*$, i.e., $(N, M)[\eta\rangle(N, M')$.*

5 Shortest Sequences in LBFC-Systems with a Home Cluster

We next establish the main result of the paper: we show that from every marking in an LBFC-system with a home cluster and for every transition t, we can reach the blocking marking associated to (the cluster of) t via a sequence of linear length, i.e., of length $\mathcal{O}(n)$ where n denotes the number of transitions of the system. Actually, we show something stronger: *every* minimal sequence to the blocking marking is of linear length. We further show that minimal sequences between arbitrary pairs of reachable markings are of quadratic length, i.e., $\mathcal{O}(n^2)$. This improves upon the best known upper bound of cubic length by the Shortest Sequence Theorem of Desel and Esparza [11, Theorem 9.17]. Recall that LBFC-systems with a home cluster are safe (cf. Corollary 1), so the bound of the Shortest Sequence Theorem yields $\mathcal{O}(n^3)$. For improving this bound, we make heavy use of the home cluster (in form of lucency, see Theorem 8).

Let us illustrate our proof strategy for the case of blocking markings. We already saw that we can *disable* clusters via elementary firing sequences. Hence, it suffices to show that we can also *enable* a single cluster via an elementary sequence. By combining the *enabling* and *disabling* sequences, we obtain a sequence which visits each cluster at most twice, i.e., a sequence of linear length. While this works for the case of blocking markings, we require a more sophisticated argument for the case of reaching general markings. To this end, let us introduce the generalized notion of k-*elementary firing sequences*.

Definition 24 (k-Elementary Firing Sequence). Let (N, M_0) with $N = (P, T, F)$ be a system. A firing sequence $\sigma \in T^*$ is k-*elementary* if at most k transitions for each cluster occur in σ, i.e., $\forall C \in N/{\sim} : \sum_{t \in C \cap T} \sigma(t) \leq k$.

Note that a 1-elementary firing sequence is just an elementary firing sequence.

With this definition, our first main result can be stated as follows: *every minimal firing sequence to a blocking marking is 2-elementary*. To prove this, we show that if a minimal sequence would vist some cluster more than twice, then we could rearrange it into a shorter sequence which reaches the blocking marking as well—which, of course, is impossible when we started with a *minimal* sequence. Our main technical tools are the following lemmas which allow us to rearrange firing sequences by *expediting* transitions [5, cf. Definition 5.3 and Lemma 5.4].

Lemma 6. *Let (N, M_0) with $N = (P, T, F)$ be a free-choice system and let $\sigma, \sigma' \in T^*$ be two firing sequences which are both enabled in M_0, i.e., $(N, M_0)[\sigma\rangle$ and $(N, M_0)[\sigma'\rangle$, and which do not share any transition from a common cluster, i.e., for all pairs of transitions $t \in \langle\sigma\rangle$ and $t' \in \langle\sigma'\rangle$, $[t]_\sim \neq [t']_\sim$. Then, the firing sequences $\sigma\sigma'$ and $\sigma'\sigma$ are enabled in M_0, i.e., $(N, M_0)[\sigma\sigma'\rangle$ and $(N, M_0)[\sigma'\sigma\rangle$.*

Proof. By the free-choice property, required tokens for transitions in σ cannot be removed by firing σ' and vice versa. Hence, both $\sigma\sigma'$ and $\sigma'\sigma$ are enabled in M, i.e., $(N, M)[\sigma\sigma'\rangle$ and $(N, M)[\sigma'\sigma\rangle$. $\qquad\square$

Lemma 7. *Let (N, M_0) with $N = (P, T, F)$ be a free-choice system. Let $\sigma \in T^*$ be a firing sequence resulting in a marking $M \in [N, M_0\rangle$, i.e., $(N, M_0)[\sigma\rangle(N, M)$, and let $t \in \langle\sigma\rangle$ be some transition occurring in σ such that $\sigma = \sigma_1\langle t\rangle\sigma_2$ with $\sigma_1, \sigma_2 \in T^*$. Then, there exists a permutation σ' of σ with $\boldsymbol{\sigma'} = \boldsymbol{\sigma}$ such that $(N, M_0)[\upsilon'\rangle(N, M)$ and $\sigma' = \sigma'_1\langle t\rangle\sigma'_2$ with $\boldsymbol{\sigma'_2} \leq \boldsymbol{\sigma_2}$ where for the marking $M' \in [N, M_0\rangle$ reached after firing σ'_1 from M_0, i.e., $(N, M_0)[\sigma'_1\rangle(N, M')$, it holds that $\nexists t' \in \langle\boldsymbol{\sigma'_2}\rangle \setminus [t]_\sim : (N, M')[t\rangle$ and $\mathcal{C}_{en}(N, M') \subseteq \mathcal{C}_{en}(N, M) \cup \{[t]_\sim\}$.*

Proof. Let $M_1 \in [N, M_0\rangle$ be the marking after firing σ_1, i.e., $(N, M_0)[\sigma_1\rangle(N, M_1)$. The proof is conducted by induction on the number of transitions occurring in σ_2 which are enabled after firing σ_1 but outside the cluster $[t]_\sim$. To this end, let $d := |(\langle\boldsymbol{\sigma_2}\rangle \setminus [t]_\sim) \cap (\cup\mathcal{C}_{en}(N, M_1))|$.

Base case: $d = 0$. Then, no transition occurring in σ_2 and outside $[t]_\sim$ is enabled in M_1, i.e., $\nexists t' \in \langle\boldsymbol{\sigma'_2}\rangle \setminus [t]_\sim : (N, M')[t'\rangle$. Set $\sigma'_1 = \sigma_1$ and $\sigma'_2 = \sigma_2$. Then, $M_1 = M'$ and firing $\langle t\rangle\sigma'_2$ might only disable $[t]_\sim$ from $\mathcal{C}_{en}(N, M')$ while possibly enabling other clusters; hence, $\mathcal{C}_{en}(N, M') \subseteq \mathcal{C}_{en}(N, M) \cup \{[t]_\sim\}$.

Induction step: $d > 0$. Let $t_D \in (\langle\boldsymbol{\sigma_2}\rangle \setminus [t]_\sim) \cap (\cup\mathcal{C}_{en}(N, M_1))$ be the first of the d transitions of interest. Then, we have $\sigma_2 = \delta_1\langle t_D\rangle\delta_2$ with $\delta_1, \delta_2 \in T^*$ where δ_1 contains only transitions t' not enabled in M_1, i.e., $\forall t' \in \langle\boldsymbol{\delta_1}\rangle : [t']_\sim \notin \mathcal{C}_{en}(N, M_1)$, and therefore also $[t']_\sim \neq [t_D]_\sim$. By Lemma

6, we can now expedite t_D in front of t. Choose the remaining sequence $\delta_1\delta_2$ with $|(\langle \boldsymbol{\delta_1\delta_2}\rangle \setminus [t]_\sim) \cap (\cup \mathcal{C}_{en}(N, M_1))| < d$ as the new σ_2 and repeat the induction step until the remaining sequence no longer contains an enabled transition outside $[t]_\sim$. $\qquad\square$

Theorem 10 (Shortest Sequences to Blocking Markings). *Let (N, M_0) with $N = (P, T, F)$ be an LBFC-system with a home cluster, let $t \in T$ be some transition and let $M_B \in [N, M_0\rangle$ be the unique blocking marking associated to the cluster $[t]_\sim$. Let $\eta \in (T \setminus [t]_\sim)^*$ be a minimal firing sequence from M_0 to M_B, i.e., $(N, M_0)[\eta\rangle(N, M_B)$ and there is no subsequence $\eta' \in (T \setminus [t]_\sim)^*$, $\boldsymbol{\eta' < \eta}$, satisfying $(N, M_0)[\eta'\rangle(N, M_B)$. Then, η is 2-elementary and thus of length at most $2 \cdot (|N/\!\!\sim| - 1)$.*

Proof. Assume this is false, then we can find an η with the above properties which is not 2-elementary, i.e., there is some cluster $D \in N/\!\!\sim$ such that η can be written as

$$\eta = \sigma_0\langle t_1\rangle\sigma_1\langle t_2\rangle\sigma_2\langle t_3\rangle\sigma_3$$

where $t_1, t_2, t_3 \in D$ and $\sigma_0, \sigma_1, \sigma_2, \sigma_3 \in (T \setminus [t]_\sim)^*$. Now, let $M_1 \in [N, M_0\rangle$ be the marking after firing σ_0, i.e., $(N, M_0)[\sigma_0\rangle(N, M_1)$, let $M_2 \in [N, M_1\rangle$ be the marking after firing $\langle t_1\rangle\sigma_1$, i.e., $(N, M_1)[\langle t_1\rangle\sigma_1\rangle(N, M_2)$, and let $M_3 \in [N, M_2\rangle$ be the marking after firing $\langle t_2\rangle\sigma_2$, i.e., $(N, M_2)[\langle t_2\rangle\sigma_2\rangle(N, M_3)$. Then of course, $(N, M_3)[\langle t_3\rangle\sigma_3\rangle(N, M_B)$. In summary, we have the following situation:

$$M_0 \xrightarrow{\sigma_0} M_1 \xrightarrow{\langle t_1\rangle\sigma_1} M_2 \xrightarrow{\langle t_2\rangle\sigma_2} M_3 \xrightarrow{\langle t_3\rangle\sigma_3} M_B.$$

We now expedite transitions starting at M_3 using Lemma 7. More precisely, we first apply Lemma 7 to the last part $\sigma_2\langle t_3\rangle\sigma_3$ of the sequence η where we fix t_3 (as t in the lemma). This yields a rearranged firing sequence from M_2 to M_B of the form:

$$M_2 \xrightarrow{\langle t_2\rangle\sigma_2'} M_3' \xrightarrow{\langle t_3\rangle\sigma_3'} M_B$$

such that $\mathcal{C}_{en}(N, M_3') \subseteq \mathcal{C}_{en}(N, M_B) \cup \{D\} = \{D, [t]_\sim\}$. In particular, at most two clusters are enabled in M_3'. Of course, $D \in \mathcal{C}_{en}(N, M_3')$ and $D \neq [t]_\sim$ due to the minimality of η (cf. Proposition 1).

Next, we apply Lemma 7 to $\sigma_1\langle t_2\rangle\sigma_2'$ where we fix t_2 (as t in the lemma). This yields a rearranged firing sequence from M_1 to M_3' of the form:

$$M_1 \xrightarrow{\langle t_1\rangle\sigma_1'} M_2' \xrightarrow{\langle t_2\rangle\sigma_2''} M_3'$$

such that $\mathcal{C}_{en}(N, M_2') \subseteq \mathcal{C}_{en}(N, M_3') \cup \{D\} = \{D, [t]_\sim\}$. Again, $D \in \mathcal{C}_{en}(N, M_2')$ and $D \neq [t]_\sim$. Now, if $\mathcal{C}_{en}(N, M_2') = \{D, [t]_\sim\}$, we would also have $\mathcal{C}_{en}(N, M_2') = \mathcal{C}_{en}(N, M_3') = \{D, [t]_\sim\}$. Then, $M_2' = M_3'$ by Theorem 8 such that we could skip firing $\langle t_2\rangle\sigma_2''$ contradicting the minimality of η. Hence, $\mathcal{C}_{en}(N, M_2') = \{D\}$ and $\mathcal{C}_{en}(N, M_3') = \{D, [t]_\sim\}$.

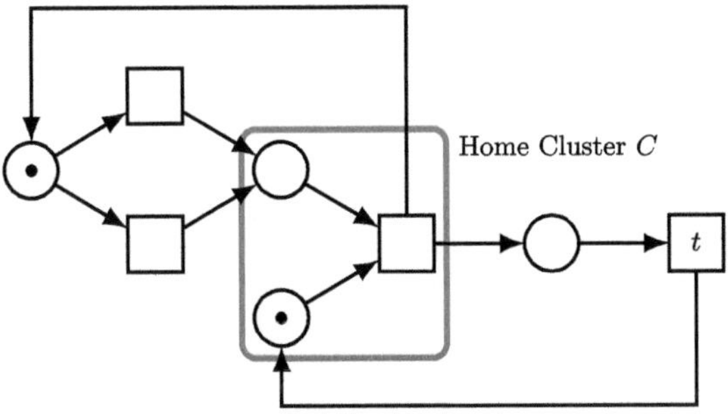

Fig. 5. An LBFC-system with a home cluster C where an elementary firing sequence is not sufficient to reach the blocking marking associated to the transition t.

Finally, we apply Lemma 7 to $\sigma_0\langle t_1\rangle\sigma_1'$ where we fix t_1 (as t in the lemma). This yields a rearranged firing sequence from M_0 to M_2' of the form:

$$M_0 \xrightarrow{\sigma_0'} M_1' \xrightarrow{\langle t_1\rangle\sigma_1''} M_2'$$

such that $\mathcal{C}_{en}(N, M_1') \subseteq \mathcal{C}_{en}(N, M_1') \cup \{D\} = \{D\}$. As we still have $D \in \mathcal{C}_{en}(N, M_1')$, we now obtain $\mathcal{C}_{en}(N, M_1') = \mathcal{C}_{en}(N, M_2')$ and thus $M_1' = M_2'$ by Theorem 8 which eventually results in a contradiction because now we can skip $\langle t_1\rangle\sigma_1''$ so that η could not have been minimal. Hence, η must be 2-elementary. As a last step, we observe that a 2-elementary blocking sequence is of length at most $2 \cdot (|N/\sim| - 1)$ since it does not contain transitions from the cluster to be blocked. $\qquad\square$

The LBFC-system with a home cluster in Fig. 5 demonstrates that there are indeed situations where we need a 2-elementary firing sequence in order to reach a blocking marking associated to some transition t. Here, a 1-elementary firing sequence would not suffice.

It turns out that our proof generalizes from blocking markings to arbitrary reachable markings. In general, we can show that in minimal sequences, we can visit each cluster at most $k + 1$ times where k is the number of clusters enabled in the target marking.

Theorem 11 (Minimal Sequences Are $k + 1$-Elementary). *Let (N, M_0) with $N = (P, T, F)$ be an LBFC-system with a home cluster and let $M \in [N, M_0\rangle$ be a reachable marking with k enabled clusters, i.e., $k := |\mathcal{C}_{en}(N, M)|$. Let $\eta \in T^*$ be a minimal firing sequence from M_0 to M, i.e., $(N, M_0)[\eta\rangle(N, M)$ and there is no subsequence $\eta' \in T^*$, $\boldsymbol{\eta'} < \boldsymbol{\eta}$, satisfying $(N, M_0)[\eta'\rangle(N, M)$. Then, η is $(k + 1)$-elementary.*

Proof. Note that in the case of $k = 1$, this is analogous to the proof of Theorem 10. Although by the premise of this theorem, η could in principle contain a transition of the *single* cluster enabled in M ($k = 1$), this would contradict Proposition 1 in the enabling part of η and thus its minimality.

For the general case where $k > 1$, the proof goes along the same lines as the proof of Theorem 10. Let us assume that for some cluster $D \in N/\sim$ a minimal firing sequence η from M_0 to M fires $k + 2$ transitions from D. Then, η can be written as

$$\eta = \sigma_0 \langle t_1 \rangle \sigma_1 \langle t_2 \rangle \sigma_2 \cdots \langle t_{k+1} \rangle \sigma_{k+1} \langle t_{k+2} \rangle \sigma_{k+2}$$

where $t_1, t_2, \ldots, t_{k+1}, t_{k+2} \in D$ and $\sigma_0, \sigma_1, \sigma_2, \ldots, \sigma_{k+1}, \sigma_{k+2} \in T^*$.

By iteratively applying Lemma 7 from the back to the front and successively fixing the transitions from t_{k+2} to t_1 (as t in the lemma), we obtain a sequence of traversed markings $M_1, M_2, \ldots, M_{k+1}, M_{k+2}$ such that:

$$\mathcal{C}_{en}(N, M_1) \subseteq \mathcal{C}_{en}(N, M_2) \subseteq \cdots \subseteq \mathcal{C}_{en}(N, M_{k+1}) \subseteq \mathcal{C}_{en}(N, M_{k+2}).$$

Also, $|\mathcal{C}_{en}(N, M_1)| \geq 1$ because $D \in \mathcal{C}_{en}(N, M_1)$. However, due to the minimality of η and Thoerem 8, all the inclusions have to be proper. Thus, we obtain $|\mathcal{C}_{en}(N, M_i)| \geq i$ for $1 \leq i \leq k + 2$. Now by the premise of the theorem, we have $|\mathcal{C}_{en}(N, M)| = k$ and with $\mathcal{C}_{en}(N, M_{k+2}) \subseteq \mathcal{C}_{en}(N, M) \cup \{D\}$, we obtain $|\mathcal{C}_{en}(N, M_{k+2})| \leq k + 1$ leading to a contradiction. Hence, η must be $(k + 1)$-elementary. $\qquad\square$

An immediate consequence of the previous theorem is a quadratic upper bound on the length of minimal and thus also shortest firing sequences in LBFC-systems with a home cluster.

Corollary 3 (Shortest Sequences in LBFC-Systems with a Home Cluster). *Let (N, M_0) with $N = (P, T, F)$ be an LBFC-system with a home cluster and let $M \in [N, M_0\rangle$ be a reachable marking. Let $\eta \in T^*$ be a minimal firing sequence from M_0 to M, i.e., $(N, M_0)[\eta\rangle(N, M)$. Then, η is of length at most* $|N/\sim| \cdot (|N/\sim| + 1) \leq |T| \cdot (|T| + 1)$.

Let us remark that our arguments in this section heavily relied on Theorem 8 which, in turn, only holds for LBFC-systems with a *home cluster*.

6   Conclusion

This year not only marks the 100th anniversary of Carl Adam Petri, but also the 31st anniversary of the Shortest Sequence Theorem by Desel and Esparza [11,12] providing a cubic bound as best known upper bound for shortest firing sequences in LBFC-systems.

In this paper, we have proved that the cubic bound from the Shortest Sequence Theorem can be improved to a quadratic one in the presence of a home cluster (Theorem 11). Moreover, in one of our main results (Theorem 10), this cubic bound can be further improved to a linear one for reaching blocking markings. Our results shed new light on the structure of LBFC-systems with a home cluster that can prove useful for algorithmic analysis techniques. For example, in algorithmic contexts where we need to construct firing sequences instead of only deciding reachability, our results provide much better worst-case guarantees. Along the way, we obtained several results of independent interest that might prove useful in future research on free-choice systems. For example, we showed that from every marking, the home cluster can always be reached via an *elementary firing sequence*, i.e., a firing sequence that contains at most one transition from every cluster (Theorem 6). A consequence is that an arbitrary set of enabled clusters can always be disabled via an elementary sequence as well (Theorem 9). To mention just one immediate consequence, this shows that the edit distance between a trace and a sound free-choice workflow net is at most linear (in the length of the trace and the number of transitions of the net). This upper bound can be used in order to improve heuristics for conformance checking in process mining, for example, for alignment computations [9].

Acknowledgments. We are indebted to the anonymous reviewer whose rigorous feedback contributed significantly to improving the quality of this paper.

References

1. van der Aalst, W.M.P.: Verification of workflow nets. In: Azéma, P., Balbo, G. (eds.) ICATPN 1997. LNCS, vol. 1248, pp. 407–426. Springer, Heidelberg (1997). https://doi.org/10.1007/3-540-63139-9_48
2. van der Aalst, W.M.P.: Process mining: data science in action. 2nd edn. Springer, Heidelberg (2016). ISBN: 978-3-662-49850-7. https://doi.org/10.1007/978-3-662-49851-4
3. van der Aalst, W.M.P.: Markings in perpetual free-choice nets are fully characterized by their enabled transitions. In: Khomenko, V., Roux, O.H. (eds.) PETRI NETS 2018. LNCS, vol. 10877, pp. 315–336. Springer, Cham (2018). https://doi.org/10.1007/978-3-319-91268-4_16
4. van der Aalst, W.M.P.: Lucent process models and translucent event logs. Fundamenta Informaticae **169**(1-2), 151–177 (2019). https://doi.org/10.3233/FI-2019-1842
5. van der Aalst, W.M.P.: Free-choice nets with home clusters are lucent. Fundamenta Informaticae **181**(4), 273–302 (2021). https://doi.org/10.3233/FI-2021-2059
6. van der Aalst, W.M.P.: Reduction using induced subnets to systematically prove properties for free-choice nets. In: Buchs, D., Carmona, J. (eds.) PETRI NETS 2021. LNCS, vol. 12734, pp. 208–229. Springer, Cham (2021). https://doi.org/10.1007/978-3-030-76983-3_11
7. Best, E., Desel, J., Esparza, J.: Traps characterize home states in free choice systems. Theor. Comput. Sci. **101**(2), 161–176 (1992). https://doi.org/10.1016/0304-3975(92)90048-K
8. Best, E., Devillers, R.: Petri net primer. In: A Compendium on the Core Model, Analysis, and Synthesis. Springer, Cham (2024). ISBN: 978-3-031-48278-6. https://doi.org/10.1007/978-3-031-48278-6

9. Carmona, J., van Dongen, B.F., Weidlich, M.: Conformance checking: foundations, milestones and challenges. In: van der Aalst, W.M.P., Carmona, J. (eds.) Process Mining Handbook. LNBIP. vol. 448, pp. 155–190. Springer, Cham (2022). ISBN: 978-3-031-08847-6. https://doi.org/10.1007/978-3-031-08848-3_5

10. Commoner, F.G., Holt, A.W., Even, S., Pnueli, A.: Marked directed graphs. J. Comput. Syst. Sci. **5**(5), 511–523 (1971). https://doi.org/10.1016/S0022-0000(71)80013-2

11. Desel, J., Esparza, J.: Free choice petri nets. Cambridge Tracts in Theoretical Computer Science 40. Cambridge University Press, Cambridge (1995). ISBN: 978-0-521-01945-3. https://doi.org/10.1017/CBO9780511526558

12. Desel, J., Esparza, J.: Shortest paths in reachability graphs. J. Comput. Syst. Sci. **51**(2), 314–323 (1995). https://doi.org/10.1006/jcss.1995.1070

13. Esparza, J.: Reduction and synthesis of live and bounded free choice petri nets. Inf. Comput. **114**(1), 50–87 (1994). https://doi.org/10.1006/inco.1994.1080

14. Esparza, J., Nielsen, M.: Decidability issues for petri nets – a survey. J. Inf. Process. Cybern. **30**(3), 143–160 (1994)

15. Esparza, J., Silva, M.: Top-down synthesis of live and bounded free choice nets. In: Rozenberg, G. (ed.) ICATPN 1990. LNCS, vol. 524, pp. 118–139. Springer, Heidelberg (1991). https://doi.org/10.1007/BFb0019972

16. Favre, C., Fahland, D., Völzer, H.: The relationship between workflow graphs and free-choice workflow nets. Inf. Syst. **47**, 197–219 (2015). https://doi.org/10.1016/j.is.2013.12.004

17. Gaujal, B., Haar, S., Mairesse, J.: Blocking a transition in a free choice net and what it tells about its throughput. J. Comput. Syst. Sci. **66**(3), 515–548 (2003). https://doi.org/10.1016/S0022-0000(03)00039-4

18. Meyer, P.J., Esparza, J., Völzer, H.: Computing the concurrency threshold of sound free-choice workflow nets. In: Beyer, D., Huisman, M. (eds.) TACAS 2018. LNCS, vol. 10806, pp. 3–19. Springer, Cham (2018). https://doi.org/10.1007/978-3-319-89963-3_1

19. Schwanen, C.T., Pakusa, W.: On free choice and Wil(l): from petri net theory to process mining. In: Mendling, J., Leemans, S.J.J., van Dongen, B.F., Reijers, H.A. (eds.) Mining a Scientist's Process: Essays Dedicated to Wil van der Aalst on the Occasion of His 60th Birthday, LNCS, vol. 16480, pp. 298–311. Springer, Cham (2026). ISBN: 978-3-032-17618-9. https://doi.org/10.1007/978-3-032-17618-9_22

20. Schwanen, C.T., Pakusa, W., van der Aalst, W.M.P.: A dynamic programming approach for alignments on process trees. In: Delgado, A., Slaats, T. (eds.) Process Mining Workshops. ICPM 2024 International Workshops, LNBIP, vol. 533, pp. 84–97. Springer, Cham (2025). ISBN: 978-3-031-82224-7. https://doi.org/10.1007/978-3-031-5-48222_7

21. Schwanen, C.T., Pakusa, W., van der Aalst, W.M.P.: Alignments meet linear algebra. In: Bergenthum, R., Rivkin, A., van der Werf, J.M.E.M. (eds.) Algorithms & Theories for the Analysis of Event Data (ATAED 2025). CEUR Workshop Proceedings. Aachen: CEUR-WS.org, vol. 3998, pp. 185–200 (2025). https://ceur-ws.org/Vol-3998/paper12.pdf

22. Schwanen, C.T., Pakusa, W., van der Aalst, W.M.P.: Process tree alignments. In: Borbinha, J., Prince Sales, T., Mira Da Silva, M., Proper, H.A., Schnellmann, M. (eds.) Enterprise Design, Operations, and Computing. EDOC 2024. LNCS, vol. 15409, pp. 300–317. Springer, Cham (2025). ISBN: 978-3-031-78337-1. https://doi.org/10.1007/978-3-031-78338-8_16

23. Schwanen, C.T., Pakusa, W., van der Aalst, W.M.P.: Complexity of alignments on sound free-choice workflow nets. In: Amparore, E., Mikulski, Ł. (eds.) Application and Theory of Petri Nets and Concurrency. Petri Nets 2025, LNCS, vol. 15714, pp. 388–410. Springer, Cham (2025). ISBN: 978-3-031-94633-2. https://doi.org/10.1007/978-3-031-94634-9_19
24. Wehler, J.: Simplified proof of the blocking theorem for free-choice petri nets. J. Comput. Syst. Sci. **76**(7), 532–537 (2010). https://doi.org/10.1016/j.jcss.2009.10.001

Coverability Abstraction for the Modular State Space

Sophie Wallner, Julian Gaede, Lukas Zech[iD], and Karsten Wolf[(✉)][iD]

University of Rostock, Schwaansche Str. 2, 18055 Rostock, Germany
`{sophie.wallner,julian.gaede,lukas.zech,karsten.wolf}@uni-rostock.de`

Abstract. Petri net systems modelling distributed systems often provide a component structure, comprising individual components that collaborate occasionally. The *modular state space* [1,6] method for Petri net systems embraces this structure in the exploration of the state space with the aim of completely representing the behavior of a Petri net system while minimizing time and memory resources. The efficient method currently considers bounded Petri net systems with a finite state space.

For unbounded Petri net systems with an infinite state space, the *coverability graph* [7] is a finite representation of the state space that abstracts the infinite behavior on diverging places. As it is related with the state space by a simulation relation [2], the coverability graph is a valid base for verification for unbounded Petri net systems.

This work presents a modular construction algorithm for unbounded Petri net systems, that generates a finite, modular representation of its state space, the *covering modular state space*. The covering modular state space simulates the state space as well and allows efficient modular verification for unbounded Petri net systems. Furthermore, this work examines how verification of standard Petri net system properties such as reachability, existence of deadlocks and liveness, can be performed in the covering modular state space.

Keywords: Petri nets · Coverability Graph · Verification and Model Checking using Nets

1 Introduction

Many large Petri nets systems consist of collaborative components, Petri nets themselves that operate independently and occasionally synchronize for accessing shared exclusive resources. For such Petri net systems, we can build the *modular state space*, an implicit but complete representation of its actual state space. Its construction takes advantage of the component structure. The modular state space method for Petri net systems embraces this structure in the state space exploration with the aim of completely representing the behavior of a Petri net system while minimizing time and memory resources. Therefore, the state spaces of the components are explored independently while a superordinate structure coordinates the synchronization among the components. This

J. Desel and A. Kalenkova (Eds.): PETRI NETS 2026, LNCS 16567, pp. 312–332, 2026.
https://doi.org/10.1007/978-3-032-27879-1_15

approach is based on the work from [1], that we revisited in [6] and made some generalizations.

With respect to verification, the modular state space is a rather promising data structure, as it is distributed and can therefore be processed in parallel. Additionally, it preserves the path structure of the reachability graph and is therefore a valid base for verification of temporal logic properties [14].

The modular structure of a Petri net can be provided as part of the input or automatically extracted as described, for example, in [5].

As the modular state space is only applicable for bounded Petri nets so far, this paper introduces the *covering modular state space* as a finite modular representation of actual state space of an unbounded Petri net system (Sect. 3). Therefore, it combines the modular state space approach with the coverability abstraction technique, which [7] introduced as the *coverability graph* for Petri net system. We present a construction algorithm for the covering state space (Sect. 3.2) that results in a complete and correct modular coverability abstraction of the state space and show how this can be used to verify standard properties and moreover temporal logic $ACTL$-properties (Sect. 3.3) of unbounded Petri net system. This does not require additional constraints on the modular structure. Section 4 provides technical considerations of its implementation.

2   Preliminaries

2.1   Petri Net Systems

Definition 1 (Petri Net (System)). *A Petri net is a tuple $[P, T, F, W]$, where P is a finite set of places, T is a finite set of transitions with $P \cap T = \emptyset$, $F \subseteq (P \times T) \cup (T \times P)$ is the flow relation, and $W : F \to \mathbb{N} \setminus \{0\}$ is the weight function. States of a Petri net are represented by* markings, *which are mappings $m : P \to \mathbb{N}$. A Petri net system is a tuple $N = [P, T, F, W, m_0]$, where $[P, T, F, W]$ is a Petri net and m_0 is the initial marking.*

Definition 2 (Activation of a Transition, Transition Rule). *Transition $t \in T$ of Petri net system N is* activated *in marking m if for all $p \in P$ with $(p, t) \in F : W(p, t) \leq m(p)$, denoted by $m \xrightarrow{t}$. If t is not activated in m, we denote this by $m \xrightarrow{t}\!\!\!\!\!/\,$. Vector $\Delta t \in \mathbb{Z}^{|P|}$ is defined as $\Delta t(p) = W(t, p) - W(p, t)$. If $m \xrightarrow{t}$, the* transition rule *states that firing t leads to marking m' with $m'(p) = m(p) + \Delta t(p)$, denoted by $m \xrightarrow{t} m'$.*

We can extend the transition rule to a sequence of transitions. Marking m' is *reachable* from marking m, if there is a sequence $\sigma \in T^*$ such that $m \xrightarrow{\sigma} m'$. If the sequence is negligible, we denote this by $m \rightsquigarrow m'$. For a set of markings M, we define $RS(M) = \{m' \mid m \rightsquigarrow m', m \in M\}$ as the *reachability set* of M. The *reachability graph* of N gives a structured representation of the reachability set of its initial marking.

Definition 3 (Reachability Graph). *The* reachability graph *of Petri net system N is $R = [V^R, E^R]$, where $V^R = RS(\{m_0\})$ and $(m, t, m') \in E$, iff $m \xrightarrow{t} m'$.*

Paths in the reachability graph represent the system runs, starting from the initial marking. Transition Sequences $\sigma \in T^*$ represent a sequence of actions, according to a path.

2.2   The Modular State Space for Petri Net Systems

The *modular state space* is a complete representation of the reachability graph of a Petri net system [6]. Therefore, we presume a component structure for a Petri net system, where the components themselves are Petri net systems. Their state spaces are explored independently of each other, while a superordinate structure coordinates the synchronization among the components. This approach is based on the work from [1], that we revisited in [6] and made some generalizations. Building the modular state space needs information on the component structure of the Petri net system. This information can be provided in form of a *modular structure*, a blueprint including the components and a synchronization scheme. The components are *instances*, Petri net systems with distinct internal and interface transitions. Internal transitions enact actions inside a single instance, while interface transitions correspond to actions aligned with other instances. A *synchronization vector* lists the according interface transitions of the participating instances.

Definition 4 (Instance, Synchronization Vectors, Modular Structure). *An* instance *is a Petri net system $[N, m_0]$, with $T = T_{internal} \cup T_{interface}$. A modular structure is a tuple $\mathcal{M} = [\mathcal{I}, \mathcal{F}]$, where $\mathcal{I} = \{[N_1, m_{01}], \ldots, [N_\ell, m_{0\ell}]\}$ is a set of disjoint instances and $\mathcal{F} \subseteq (T_{1|interface} \cup \{\bot\}) \times \ldots \times (T_{\ell|interface} \cup \{\bot\})$ is a set of* synchronization vectors. *Instance $[N_j, m_{0j}]$ with $j \in \{1, \ldots, \ell\}$ participates in $f \in \mathcal{F}$ with interface transition $t \in T_{j|interface}$, if $f[j] = t$. If $f[j] = \bot$, $[N_j, m_{0j}]$ does not participate in f.*

Per definition, a synchronization vector contains at most one transition per instance. For modular structure $\mathcal{M} = [\mathcal{I}, \mathcal{F}]$, we can build the according Petri net system $N_{\mathcal{M}} = [P, T, F, W, m_0]$, where $P = \bigcup_{j=1}^{\ell} P_j$ and $m_0 = \bigcup_{j=1}^{\ell} m_{0j}$, as place sets and initial markings of the instances are disjoint. For each synchronization vector $f \in \mathcal{F}$ we introduce a transition, that instance-wise inherits the environment of the corresponding interface transitions. So $T = \bigcup_{j=1}^{\ell} T_{j|internal} \cup \mathcal{F}$ contains those synchronization transitions and the disjoint internal transitions of the instances. Flow relation F and the weight function W are established accordingly.

Given a Petri net system $N_{\mathcal{M}}$ in form of modular structure $\mathcal{M} = [\mathcal{I}, \mathcal{F}]$ with the instances $\mathcal{I} = \{[N_1, m_{01}], \ldots, [N_\ell, m_{0\ell}]\}$, the modular state space contains the local reachability graphs of all instances and a synchronization graph. The *local reachability graph* of an instance $[N_j, m_{0j}]$ with $j \in \{1, \ldots, \ell\}$ represents the behavior of $[N_j, m_{0j}]$ in the context of the modular structure. Therefore, it stores the *projection* of the markings of $N_\mathcal{M}$ to the places of the instance (we call them *local markings*), as well as the internal transition actions and the local effects of synchronizations.

Definition 5 (Local Reachability Graph, Projection of Markings). *Let $R = [V^R, E^R]$ be the reachability graph of $N_\mathcal{M}$. For instance $[N_j, m_{0j}] \in \mathcal{I}$ with $j \in \{1, \dots, \ell\}$, the local reachability graph is defined as $L_j = [V_j^L, E_j^L]$, with $V_j^L = \{\pi_j(m) \mid m \in V^R\}$, where $\pi_j(m) = m \cap (P_j \times \mathbb{N})$ is the projection of marking m to P_j, and $E_j^L = E_{j|internal}^L \cup E_{j|interface}^L$, where $(\pi_j(m), t, \pi_j(m')) \in E_{j|internal}^L$, iff $(m, t, m') \in E^R$ for $t \in T_{j|internal}$, and $(\pi_j(m), t, \pi_j(m')) \in E_{j|interface}^L$, iff $(m, f, m') \in E^R$ for $f \in \mathcal{F}, f[j] = t$.*

A path in the local reachability graph is the local part of a path of the reachability graph. A local sequence $\sigma_j \in T_j^*$ corresponds to a sequence of instance-actions, including both, internal and interface actions. The *synchronization graph* is a coordinating structure for synchronizations among the instances. A synchronization is a simultaneous inter-instance exploration step that is executed if and only if all participating instances of the synchronization vector are synchronization-ready – that is if the according interface transitions are activatable. Here, the specific activating local marking on interface transition t is less relevant than the fact that t can be activated without prior further synchronizations. This allows us to abstract the set of local markings that have the same activatable interface transitions to a *segment*, which is forwardly closed regarding internal transitions, starting from a *generator*, which is also set of local markings (we call *generator elements*).

Definition 6 (Segment, Generator). *Let $L_j = [V_j^L, E_j^L]$ be the local reachability graph of instance $[N_j, m_{0j}]$ for $j \in \{1, \dots, \ell\}$. A segment $O \subseteq V_j^L$ is defined as $O = \mathring{G}$, where $G \subseteq V_j^L$ is the generator of segment O. $\mathring{G}$ with $G \subseteq \mathring{G}$ describes the closure of G regarding internal transitions, i.e. if $m \in \mathring{G}$ and $(m, t, m') \in E_{j|internal}^L$, then $m' \in \mathring{G}$.*

The generator of the initial segment only contains the initial marking of the instance. Consequently, an ℓ-tuple of segments $<O_1, \dots, O_\ell>$, where O_j is a segment for instance $[N_j, m_{0j}]$ for $j \in \{1, \dots, \ell\}$ describes an abstraction of a state of $N_\mathcal{M}$. Based on that, we then can lift the activation of transitions and the transition rule on markings (cf. Definition 2) to *global activation* of synchronization vectors in a tuple of segments and the *synchronization rule* based on *successor segments*. Note, that the denotation is lifted as well.

Definition 7 (Successor Segment, Global Activation, Synchronization Rule). *For instance $[N_j, m_{0j}]$ with $j \in \{1, \dots, \ell\}$, interface transition $t \in T_{j|interface}$ is activated in segment O, if there is a local marking $m \in O$ such that $m \xrightarrow{t}$. For segment O and interface transition t, we define the successor segment as $O^{+t} = \mathring{G}$ where $G = \{m' \mid m \in O, (m, t, m') \in E_{j|interface}^L\}$. If $O \not\xrightarrow{t}$, it holds that $O^{+t} = O$. Synchronization vector $f \in \mathcal{F}$ is globally activated in $<O_1, \dots, O_\ell>$, if $O_j \xrightarrow{t}$ for all instances $[N_j, m_{0j}]$ with $f[j] = t$. If*

316   S. Wallner et al.

$<O_1, \ldots, O_\ell> \xrightarrow{f}$, the synchronization rule *states that the according synchronization leads to* $<O_1^{+t_1}, \ldots, O_\ell^{+t_\ell}>$.

The generators for successor segments may contain multiple elements, as interface transition t of synchronization vector f can be activated in various local markings of O and firing t leads to distinct generator elements. However, the successor segment is unambiguous, as we only permit one interface transition per instance in a synchronization vector. This leads to the definition of the synchronization graph, where a vertex is an ℓ-tuple of segments and an edge corresponds to a synchronization.

Definition 8 (Synchronization Graph). *The* synchronization graph $S = [V^S, E^S]$ *is inductively defined as follows:*
Base: $<\{\mathring{m}_{01}\}, \ldots, \{\mathring{m}_{0\ell}\}> \in V^S$
Step: If $v \in V^S$ *and* $v \xrightarrow{f} v'$, *then* $(v, f, v') \in E^S$ *and* $v' \in V^S$.

A path in the synchronization graph abstracts a set of paths of the reachability graph. A sequence of synchronizations in the synchronization graph abstracts a set of sequences of $N_\mathcal{M}$ that have the same succession of synchronizations. We build the modular state space with an interleaved construction algorithm that explores the segments of the local reachability graphs independently until a synchronization vector is globally activated. A synchronization leads to a change of segments to their successor segments, that are again explored independently afterwards. In this way, we proceed exhaustively. For more detail, we refer to [6] A small example will help to get familiar with the concepts of the modular state space method.

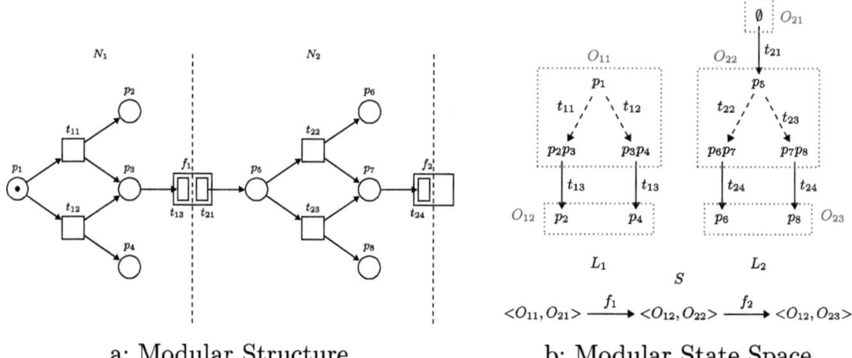

a: Modular Structure   b: Modular State Space

Fig. 1. State Spaces for Example 1.

Example 1. In Fig. 1a, a Petri net based on a modular structure consisting of two instances is depicted. Note that, despite the single connection of the instances through synchronization vector f_1, there is another synchronization vector f_2 where only instance N_2 participates. Figure 1b depicts the modular state space.

2.3   Coverability Abstraction for Unbounded Petri Net Systems

For unbounded Petri net systems, the reachability graph can be infinite, making their verification impractical. To counteract this, the abstraction through the coverability of markings aims to generate a finite representation of the reachability graph, the *coverability graph* [7]. Assume an unbounded Petri net system $N = [P, T, F, W, m_0]$. *Generalized markings* are used to abstract sequences of markings of N on potentially unbounded (diverging) places, where the entry of place $p \in P$ represents the limit of the sequence of the number of tokens on p. The limit for diverging places is represented by ω.

Definition 9 (Generalized Marking). *A generalized marking of N is a mapping $\mu : P \to \mathbb{N} \cup \{\omega\}$. For generalized marking μ, we define $\Omega(\mu) = \{p \mid \mu(p) = \omega\}$.*

For $k \in \mathbb{N}$, we state that $\omega + k = \omega - k = \omega$. The activation of transition $t \in T$ and the transition rule for generalized markings correspond to Definition 2. A sequence of markings of a Petri net system is abstracted by a generalized marking μ, if the sequence *converges* to μ.

Definition 10 (Convergence [11]). *Let μ be a generalized marking of N. A sequence of markings $(m_i)_{i \in \mathbb{N}}$ of N converges to μ, if for any integer number $k \in \mathbb{N}$, there exists an index $i \in \mathbb{N}$ such that, for all following sequence elements m_j with $j > i$ it holds that $m_j(p) > k$, if $p \in \Omega(\mu)$, and $m_j(p) = \mu(p)$, else, for all $p \in P$.*

Note that this definition does not require a strict monotony on diverging places, it just guarantees that at a certain point in the sequence, any value will be exceeded. With this definition of convergence, we now give a definition for the coverability graph.

Definition 11 (Coverability Graph). *The coverability graph $C = [V^C, E^C]$ of N is a directed labeled graph, where $V^C \subseteq (\mathbb{N} \cup \{\omega\})^{|P|}$ is a set of generalized markings with $m_0 \in V^C$. For $\mu \in V^C$, there exists a sequence of reachable markings $(m_i)_{i \in \mathbb{N}}$ of N that converges to μ, and for $\mu \in V^C$ and $\mu \xrightarrow{t}$ for $t \in T$, there exists $\mu' \geq \mu + \Delta t$ with $\mu' \in V^C$ and $(\mu, t, \mu') \in E^C$.*

The reachability graph is a proper coverability graph, as well. This definition is more liberal due to our understanding of convergence than the one from [7], which demands strict monotony in the sequence on the places. Nevertheless, their construction algorithm is also applicable to construct coverability graphs according to our definition, which we describe in the following. First, we define *pumping sequences*, transition sequences leading from a marking to a bigger or equal marking, where it can be executed again, according to the monotony of firing transitions. As this can then be repeated infinitely often, a pumping sequence implies a sequence of increasing markings with potentially diverging places.

Definition 12 (Pumping Sequence). *A transition sequence $\sigma \in T^*$ is a pumping sequence, if $m^* \xrightarrow{\sigma} m$ for $m^* \leq m \in V^R$.*

This holds for generalized markings as well. Diverging places are precisely those where m^* actually becomes larger, what justifies the abstraction of the divergence by an ω-insertion on them. The coverability graph is constructed as follows:

Definition 13 (Coverability Graph Construction). *The construction of* $C = [V^C, E^C]$ *for N is a depth first search state space exploration from* $m_0 \in V^C$, *where for all* $\mu \in V^C$, *we perform the following steps:*

1. Exploration: $\mu \xrightarrow{t} \mu'$, *where* $\mu \xrightarrow{t}$ *for* $t \in T$
2. Pumping Sequence Identification: $\exists \sigma \in T^* : \mu^* \xrightarrow{\sigma} \mu'$ *for* $\mu^* \leq \mu' \in V^C$?
   If σ *exists, update* $\mu'(p) \leftarrow \omega$ *for* $p \in P$ *with* $\mu^*(p) < \mu'(p)$
3. Recording: $E^C \leftarrow E^C \cup \{(\mu, t, \mu')\}$, *if* $\mu' \notin V^C : V^C \leftarrow V^C \cup \{\mu'\}$.

As a Petri net system has a finite set of places, we can insert ω only finitely often, so the construction result is finite.

Lemma 1 (The Coverability Graph is finite [7]). *The coverability graph* $C = [V^C, R^C]$ *of N is finite.*

Note that the construction algorithm does not generate the minimal coverability graph as described in [4,9], and [10]. However, we accept this because our liberal definition allows the application of the coverability graph within the field of verification.

3   Coverability Abstraction for the Modular State Space

3.1   Infinite Behavior in the Modular State Space

So far, the modular state space method from Sect. 2.2 is only applicable for bounded Petri net systems. In the following, we examine how to make the advantages of a component-wise state space generation applicable to unbounded Petri net systems as well, by including the abstraction through coverability from Sect. 2.3. We now assume Petri net system $N_\mathcal{M} = [P, T, F, W, m_0]$ based on the modular structure $\mathcal{M} = [\mathcal{I}, \mathcal{F}]$ to be unbounded. Infinite behavior of $N_\mathcal{M}$ can be identified with pumping sequences, which rely on the comparison of a new marking with previously explored markings. In the modular state space, we store those markings of $N_\mathcal{M}$ only instance-wise in the according local reachability graph. As the instance affiliation of places is unambiguous, one may assume that we can instance-wise compare a new local marking with previously explored local markings to identify local pumping sequences, which then can be composed to a pumping sequence of $N_\mathcal{M}$. However, a pumping sequence of $N_\mathcal{M}$ is not just any combination of local pumping sequences. We emphasize this in Example 2.

Example 2. Figure 2a depicts a modular structure with the coverability graph of the corresponding $N_\mathcal{M}$ in Fig. 2b. The local reachability graph of instance N_2 in Fig. 2c implies two local pumping sequences, one involving t_{21} (participating in f_1) and one involving t_{22} (participating in f_2). The coverability graph however shows that there is only a single pumping sequence in the combination of the instances, falsely flagging the latter local pumping sequence.

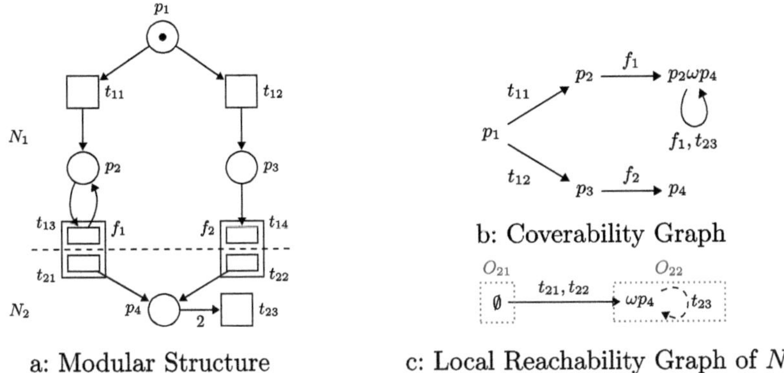

a: Modular Structure   b: Coverability Graph

c: Local Reachability Graph of N_2

Fig. 2. State spaces for Example 2.

In the modular state space, for a sequence σ of $N_{\mathcal{M}}$ the local reachability graph L_j only stores the relevant parts of σ for $[N_j, m_{0j}]$ as local sequences. This is intentional for space-saving reasons. Assume Σ_j as the set of local sequences of L_j. Technically, to reproduce σ in the modular state space we need to identify a selection of local sequences $(\sigma_1, \ldots, \sigma_\ell) \in \Sigma_1 \times \ldots \times \Sigma_\ell$ such that those local sequences align in their succession of synchronizations with the help of the synchronization graph – only then, we can compose σ of them. This combinatorial effort may reverse the performative advantages of the modular state space approach. Nevertheless, for every sequence σ there is a selection of local sequences $(\sigma_1, \ldots, \sigma_\ell)$ for which we can directly state that σ can be composed of them. If the local sequences are *projections* of σ, then their existence directly implies the existence of σ. Therefore, to reproduce σ, it is sufficient to find those projections. This is stated in the following lemma. Before, we apply the concept of projection of markings from Definition 5 to sequences.

Definition 14 (Projection of Sequences). *Let* $\sigma = t_1 t_2 \ldots$ *be a sequence of* $N_{\mathcal{M}}$, *where* $t_i \in \bigcup_{j=1}^{\ell} T_{j|internal} \cup \mathcal{F}$ *for* $i \in \mathbb{N}$. *For instance* $[N_j, m_{0j}]$ *with* $j \in \{1, \ldots, \ell\}$, *the* projection $\pi_j(\sigma) = \pi_j(t_1)\pi_j(t_2) \ldots$ *of* σ *is calculated transition by transition such that*

$$\pi_j(t_i) = \begin{cases} t_i & \text{for } t_i \in T_{j|internal} \\ t & \text{for } t_i = f \in \mathcal{F}, \text{ and } f[j] = t \text{ with } t \in T_{j|interface} \\ \epsilon & \text{for } t_i = f \in \mathcal{F}, f[j] = \bot \text{ or } t_i \in \bigcup_{k=1}^{\ell} T_{k|internal} \setminus T_{j|internal} \end{cases}$$

This definition ensures that the projection of sequence σ of $N_{\mathcal{M}}$ to instance $[N_j, m_{0j}]$ contains the occurring internal transitions as well as the interface transitions according to the synchronizations of sequence σ, where $[N_j, m_{0j}]$ participates. The behavior of the other instances is omitted, as well as synchronizations, where $[N_j, m_{0j}]$ does not participate.

Lemma 2 (Executability of Sequences in the Modular State Space). *Sequence* $\sigma \in (\bigcup_{j=1}^{\ell} T_{j|internal} \cup \mathcal{F})^*$ *such that* $m \xrightarrow{\sigma} m'$ *exists in the reachability* *graph of $N_\mathcal{M}$, if and only if there exists a local sequence σ_j such that $\pi_j(m) \xrightarrow{\sigma_j} \pi_j(m')$ in the local reachability graph for all instances $[N_j, m_{0j}]$.*

Proof. Assume $m \xrightarrow{\sigma} m'$ of $N_\mathcal{M}$. From the construction of the modular state space and Definition 14, it follows directly that $\sigma_j = \pi_j(\sigma)$ such that $\pi_j(m) \xrightarrow{\sigma_j} \pi_j(m')$ exists in the local reachability graphs for all $j \in \{1, \ldots, \ell\}$. Now assume $\pi_j(m) \xrightarrow{\sigma_j} \pi_j(m')$ for all $j \in \{1, \ldots, \ell\}$. There exists a partition $\sigma_j = \sigma_{j1} t_{j1} \sigma_{j2} t_{j2} \ldots t_{jn-1} \sigma_{jn}$, where $\sigma_{ji} \in T^*_{j|internal}$ and $t_{ji} \in T_{j|interface} \cup \{\epsilon\}$ for $i \in \{1, \ldots, n\}$. Note that σ_{ji} is allowed to be empty. From this, we build $\sigma = \sigma_1 f_1 \sigma_2 f_2 \ldots f_{n-1} \sigma_n$, where $\sigma_i = \bigcup_{j=1}^{\ell} \sigma_{ji}$ and $f_i \in \mathcal{F}$ is precisely the synchronization vector, for which it holds that $t_{ji} = f_i[j]$ for participating instances and $t_{ji} = \epsilon$ for non-participating instances. Sequence σ such that $m \xrightarrow{\sigma} m'$ exists in the reachability graph of $N_\mathcal{M}$, as the union of internal transition sequences is always executable, and the contained synchronizations are globally activated. $\square$

So in general, the identification of a pumping sequence in the modular state space can be reduced to the identification of local pumping sequences and ensuring their compatibility by examining that they all are projections of the sequence. Nevertheless, there is an interesting subset of pumping sequences, where the compatibility of the local pumping sequences is directly assured. A pumping sequence $\sigma \in \bigcup_{j=}^{\ell} T^*_{j|internal}$, that only contains internal behavior of the instances, is just a merge of local pumping sequences executable independently of each other. We call them *internal-only pumping sequences* to differentiate them from the ones containing synchronizations, which we call *general pumping sequences* from now on. This distinction becomes relevant in the following subsection as the identifications of the two types of pumping sequences are performed at different steps in the construction algorithm.

3.2   Construction of the Covering Modular State Space

The following construction algorithm adheres to the concept of alternating independent exploration of local behavior and synchronization in order to increase efficiency and abstracts the infinite behavior at the same time. Given a Petri net system $N_\mathcal{M}$ based on modular structure $\mathcal{M} = [\mathcal{I}, \mathcal{F}]$ with the instances $\mathcal{I} = \{[N_1, m_{01}], \ldots, [N_\ell, m_{0\ell}]\}$, the algorithm generates the *covering modular state space* consisting of local coverability graphs of all instances and a synchronization graph. The *local coverability graph* $L_j = [V_j^L, E_j^L]$ for each instance $[N_j, m_{0j}]$ with $j \in \{1, \ldots, \ell\}$ is an abstraction of the instance behavior in the context of the modular structure, where the nodes are projections of generalized markings of $N_\mathcal{M}$ to the places of the instance. We refer to them as local markings (cf. Sect. 2.2) for improved readability. The *synchronization graph* $S = [V^S, E^S]$ is a coordinating structure as in Sect. 2.2. We follow a top-down explanation approach from the main procedure and explain called procedures and used data structures in more detail in the following, as they require further clarifications and correlate with interesting theoretical results. We construct the covering modular state space depth-first from one synchronization graph vertex to another

---

Algorithm 1. Computing the Covering Modular State Space

1: **procedure** cCMSS
2:     **for all** $j \in \{1, \ldots, \ell\}$ **do**
3:         $V_j^L \leftarrow \emptyset, E_j^L \leftarrow \emptyset$ ▷ initialize local coverability graphs
4:         $O_j \leftarrow \text{LOCALFORWARDCLOSURE}(j, m_{0j}, \emptyset)$ ▷ expand initial segment
5:     **end for**
6:     $push(\bigsqcup, [\{[\bot, m_{01}]\}, \ldots, \{[\bot, m_{0\ell}]\}])$
7:     $V^S \leftarrow \{<O_1, \ldots, O_\ell>\}, E^S \leftarrow \emptyset$ ▷ update S and descend recursively
8:     cMCSS_RECUR($<O_1, \ldots, O_\ell>$)
9: **end procedure**

---

along globally activated synchronization vectors. The main procedure implements the synchronization rule (cf. Definition 7) for vertex $v = <O_1, \ldots, O_\ell>$ of S with $v \xrightarrow{f}$ and decays into two significant parts:

1. **Calculating Covering Generators (Line** 4–13**)** New generator elements of the successor segments instance-wise result from synchronization f in v. Beyond that, we prevent infinite behavior induced by general pumping sequences here. Therefore, we use the generator elements of the segments to make sequences of $N_{\mathcal{M}}$ traceable in the local coverability graphs, by relating succeeding generator elements (implemented with procedure SUCCESSOR-GENERATORRELATION (Algorithm 3)) and recording this relation in a stack $\bigsqcup$ with every synchronization. Based on that, procedure INSERTGENERA-TOROMEGAS (Algorithm 4) actually identifies general pumping sequences and abstracts the according infinite behavior.

2. **Expanding Segments (Line** 14–20**)** The successor segments are built instance-wise using procedure LOCALFORWARDCLOSURE (Algorithm 6) for the new generator elements $G_1', \ldots, G_\ell'$. Additionally, we identify and abstract infinite behavior induced by internal-only pumping sequences here.

If the resulting vertex $v' = <O_1^{+t_1}, \ldots, O_\ell^{+t_\ell}>$ is new, we subsequently continue with v'. In this way, we proceed exhaustively. As the succession of vertices is essential for identifying pumping sequences we decided on a recursive approach. Note that for the initial vertex $v_0 \in V^S$, the first step is inappropriate as it does not result from a synchronization and the generators are given by the initial local markings. Therefore, processing v_0 is externalized to Algorithm 1.

Calculating Covering Generators: The Generator Relation. In general, for the identification of general pumping sequences we need to examine that local pumping sequences are projections of the same sequence σ of $N_{\mathcal{M}}$ (Lemma 2) to compose σ of them. In order to perform this examination as efficiently as possible during the construction of the covering modular state space, we would like to make use of the designated generator elements. Given a local sequence σ_j of instance $[N_j, m_{0j}]$, generator elements serve as waypoints of the original sequences σ of $N_{\mathcal{M}}$. By relating succeeding generator elements and storing them in a stack, we make σ traceable in L_j. So, we introduce the *generator relation*

Algorithm 2. Computing the Covering Modular State Space

1: **procedure** cMCSS_RECUR($< O_1, \ldots, O_\ell >$: Current Synchronization Vertex)
2:     $[r_1, \ldots, r_\ell] \leftarrow top(Stack)$
3:     **for all** $f \in \mathcal{F} : t_f$ is globally activated in $<O_1, \ldots, O_\ell>$ **do**
4:         **1. Calculating Covering Generators**
5:         **for all** $j \in \{1, \ldots, \ell\}$ **do**
6:             **if** $f[j] = t$ for $t \in T_{j|interface}$ **then**   ▷ if instance participates in f
7:                 $g'_j \leftarrow$ SUCCESSORGENERATORRELATION(j, t, O_j, r_j)
8:             **else**   ▷ else instance does not participate in f
9:                 $g'_j \leftarrow \{[g, g] \mid [\_, g] \in g_j\}$
10:            **end if**
11:        **end for**
12:        $[g'_1, \ldots, g'_\ell] \leftarrow$ INSERTGENERATOROMEGAS$([g'_1, \ldots, g'_\ell])$ ▷ update relation
13:        $push(\bigsqcup, [g'_1, \ldots, g'_\ell])$
14:        **2. Expanding Segments**
15:        **for all** $j \in \{1, \ldots, \ell\}$ **do**
16:            $O'_j \leftarrow \bigcup_{g:\exists[\_,g]\in g'_j}$ LOCALFORWARDCLOSURE$(j, g, \emptyset)$
17:            **if** $<O'_1, \ldots, O'_\ell> \notin V^S$ **then**
18:                $V^S \leftarrow V^S \cup \{<O'_1, \ldots, O'_\ell>\}$
19:                cMCSS_RECUR$(<O'_1, \ldots, O'_\ell>)$
20:            **end if**
21:            $E^S \leftarrow E^S \cup \{(<O_1, \ldots, O_\ell>, t_f, <O'_1, \ldots, O'_\ell>)\}$
22:        **end for**
23:        $pop(\bigsqcup)$
24:     **end for**
25: **end procedure**

$r_j \subseteq (V_j^L \times V_j^L)$ of instance $[N_j, m_{0j}]$ that captures the succession of generator elements and evolves from a synchronization. Regarding the current synchronization $v \xrightarrow{f} v'$, for participating instances, new generator element $g'_j \in O'_j$ is related to generator element $g_j \in O_j$, if $g_j \xrightarrow{\sigma_j} m_j \xrightarrow{t} g'_j$, where $\sigma_j \in T^*_{j|internal}$ and $f[j] = t$ (denoted as $[g_j, g'_j] \in r'_j$). Procedure SUCCESSORGENERATORRELATION (Algorithm 3) implements the relation. Note that a synchronization may imply multiple firings of t from different local markings of O_j, leading to different generator elements of $O_{j'}$, so r'_j may contain multiple pairs of generator elements. For non-participating instances, every generator $g_j \in O_j$ is related to itself, so $[g_j, g_j] \in r'_j$.   The generator relations of all instances are stored in a global *generator relation stack* $\bigsqcup$ in form of an ℓ-tuple of generator relations. The entries of a stack represent the vertices of S and as two consecutive entries correspond to a synchronization, so the current sequence of S is featured in $\bigsqcup$;

The generator relations of an entry form local markings of L_j. Regarding $v \xrightarrow{f} v'$, the new entry for $\bigsqcup$ is $[r'_1, \ldots, r'_\ell]$, where r'_j is the generator relation of instance $[N_j, m_{0j}]$ for f. With storing the generator relations in $\bigsqcup$ per synchronization, we implicitly store the original sequence σ; the part of $\bigsqcup$ for instance $[N_j, m_{0j}]$ corresponding to the projection of σ. A local pumping sequence may evolve

---

Algorithm 3. Establish Generator Relation Between Succeeding Segments

---

1: **procedure** SUCCESSORGENERATORRELATION(j: instance index, O: segment, t: interface transition, r: generator relation)
2:      $Res \leftarrow \emptyset$
3:      $G \leftarrow \{[g \mid [\_,g] \in r]\}$ $\triangleright$ extract preceding generator elements
4:      **for all** $m \in O$ with $m \xrightarrow{t}$ **do**
5:          $m' \leftarrow m + \Delta t$
6:          $V_j^L \leftarrow V_j^L \cup \{m'\}, E_j^L \leftarrow E_j^L \cup \{(m,t,m')\}$
7:          $Res \leftarrow Res \cup \{[g,m'] \mid g \in G, g \xrightarrow{\sigma_j} m\}$ $\triangleright$ relate m' with appropriate g
8:      **end for**
9:      **return** Res $\triangleright$ fresh generator relation entry r_j' for $\bigsqcup$
10: **end procedure**

---

with every synchronization. So, for every synchronization, we need to consult $\bigsqcup$ to check whether the stored sequence projections of the instances are a local pumping sequence.

Calculating Covering Generators: Coverability of Generators. To check for local pumping sequences, we want to use the information stored in a stack entry. From entry $[r_1, \ldots, r_\ell]$ of $\bigsqcup$, we can reconstruct generator markings of the evolving vertex v, i.e. markings of $N_\mathcal{M}$ which consist of generator elements only.

Definition 15 (Generator Marking). *Let $v = <\mathring{G}_1, \ldots, \mathring{G}_\ell> \in V^S$ be a vertex of S. Marking $m_g \in G_1 \times \ldots \times G_\ell$ is a generator marking of v.*

Generator markings turn out to be useful for the identification of pumping sequences. Any infinite sequence of markings of $N_\mathcal{M}$ arising with synchronizations implies an infinite sequence of generator markings, since the occurring synchronizations can be executed repeatedly, thus creating new generator markings. The following lemma states this implication for general pumping sequences.

Lemma 3 (Identification of General Pumping Sequences on Generator Markings). *Let $\sigma \in (\bigcup_{j=1}^{\ell} T_{j|internal} \cup \mathcal{F})^*$ be a general pumping sequence of $N_\mathcal{M}$, such that $m^* \xrightarrow{\sigma} m'$ for $m^* \leq m' \in RS(\{m_0\})$. Sequence σ implies a general pumping sequence $\sigma' \in (\bigcup_{j=1}^{\ell} T_{j|internal} \cup \mathcal{F})^*$ of $N_\mathcal{M}$, such that $m_g^* \xrightarrow{\sigma'} m_g'$ for $m_g^* \leq m_{g'} \in RS(\{m_0\})$, where m_g^* and m_g' are generator markings such with $m^* \rightsquigarrow m_g^*$ and $m' \rightsquigarrow m_g'$.*

Proof. Assume sequence $\sigma = \sigma_1 f_1 \sigma_2 \ldots \sigma_{n-1} f_n \sigma_n$, where $\sigma_i \in \bigcup_{j=1}^{\ell} T_{j|internal}^*$ and $f_i \in \mathcal{F}$ for all $i \in \{1, \ldots, n\}$. We can split σ such that $m^* \xrightarrow{\sigma_1 f_1 \ldots \sigma_{i-1} f_i} m_g^* \xrightarrow{\sigma_{i+1} f_{i+1} \ldots \sigma_{n-1} f_n} m'$, where m_g^* is a generator marking resulting from synchronization f_i. As $m^* \leq m'$, it holds that $m' \xrightarrow{\sigma_1 f_1 \ldots \sigma_{i-1} f_i} m_g'$, where m_g' is generator marking with $m_g^* \leq m_g'$. So $\sigma' = \sigma_{i+1} f_{i+1} \ldots \sigma_{n-1} f_n \sigma_1 f_1 \ldots \sigma_{i-1} f_i$ is a general pumping sequence such that $m_g^* \xrightarrow{\sigma'} m_g'$. $\qquad\square$

Figure 3 shows this for a sequence with one synchronization. This permits us to identify general pumping sequences and the subsequent insertion of ωs only based on generator markings of $N_{\mathcal{M}}$. Consequently, we use the entries of $\bigsqcup$ to identify local pumping sequences on generator elements. Procedure INSERT-GENERATOROMEGAS (Algorithm 4) implements this proposal. For the current synchronization $v \xrightarrow{f} v'$ and the new entry $[r'_1, \ldots, r'_\ell]$ for $\bigsqcup$, procedure FINDCOVERINGGENERATORELEMENTS (Algorithm 5) recursively scans $\bigsqcup$ for a generator element g_j^* such that $g_j^* \leq g_j'$ for new generator element g_j' that induces a local pumping sequence for $[N_j, m_{0j}]$ We return the stack level of such g_j^*, corresponding to a vertex $v^{n_j} \in V^S$ with generator marking $m_g^* \in v^{n_j}$ that is covered on P_j by g_j' and $v^{n_j} \rightsquigarrow v'$ for the evolving synchronization graph vertex v'. We find those stack levels for each instance. A completely covered generator marking m_g^* occurs if there is a stack level n where each instance $[N_j, m_{0j}]$ covers $m_g^* \in v^n$ on P_j by g_j' and $v^n \rightsquigarrow v'$. By storing the generator relations of all instances together in one stack, we can easily identify such level n of $\bigsqcup$. From the new generator elements of $[r'_1, \ldots, r'_\ell]$, we can build generator marking m_g' for the evolving vertex v'. This justifies the instance-wise insertion of ω for a new generator element g_j' on places $p \in P_j$, where $g_j^*(p) < g_j'(p)$ for each instance. Figure 5 shows different states of $\bigsqcup$ during the exploration of Example 2. After firing f_1 twice (Fig. 5c) the latest entry $[\{(p_2, p_2)\}, \{p4, 2p_4\}]$ implies generator marking $p_2 2p_4$ for that we can find entry $[\{(p_1, p_2)\}, \{\emptyset, p_4\}]$ that contains generator marking $p_2 p_4$, which is covered. As a consequence, we insert ω at p_4. The spurious pumping sequence from Example 2 is exposed by Fig. 5d.

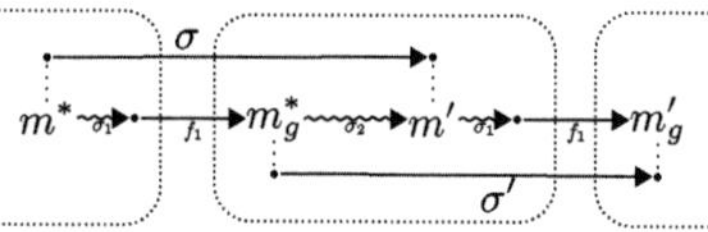

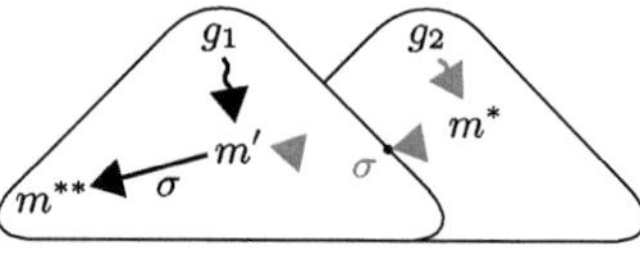

Fig. 3. Shifting Sequence To Generators.   **Fig. 4.** Overlapping Closures.

The application of covering generators ensures a finite number of segments for every local coverability graph. As a consequence, the synchronization graph is finite.

Lemma 4 (The Synchronization Graph of the Covering Modular State Space is finite). *The local coverability graph $L_j = [V_j^L, E_j^L]$ of $[N_j, m_{0j}]$ for $j \in \{1, \ldots, \ell\}$ contains a finite number of segments. As a result, the synchronization graph $S = [V^S, E^S]$ is finite.*

Proof. Lemma 3 states that every general pumping sequence can be detected on generator markings. As P of $N_{\mathcal{M}}$ is finite, there is a finite number of possibilities to insert ω in a generator marking m_g in case of infinite behavior. This limits the set of generator markings and is manifested in a limited set generator elements in the local coverability graph L_j of instance $[N_j, m_{0j}]$. So, the number of segments

Algorithm 4. Insert Omegas In Generator Elements

1: **procedure** INSERTGENERATOROMEGAS($[g_1, \ldots, g_\ell]$: Current Generator Relation)
2:      **for all** $j \in \{1, \ldots, \ell\}$ **and** $[g, g'] \in r_j$ **do**
3:          $Cand_j \leftarrow$ FINDCOVERINGGENERATORELEMENTS($j, g', length(\bigsqcup), g$)
4:      **end for**
5:      **if** $\exists i \in \{n \mid \exists [n, \_] \in Cand_1 \wedge \ldots \wedge \exists [n, \_] \in Cand_\ell\}$ **then**   ▷ covered level
6:          **for all** $j \in \{1, \ldots, \ell\}$ **and** $[i, [g^*, g']] \in Cand_j$ **do**
7:              $g'(p) \leftarrow \omega$ **for all** $p \in P_j$ with $g^*(p) < g'(p)$ ▷ this modifies g_j
8:          **end for**
9:      **end if**
10:     **return** $[g_1, \ldots, g_\ell]$ ▷ modified relation with inserted ω
11: **end procedure**

Algorithm 5. Find Covering Generator Elements

1: **procedure** FINDCOVERINGGENERATORELEMENTS(j: instance index, g': Generator Element, n: Current Index, g^*: Generator Element at Index n)
2:      $Res \leftarrow \{[n, [g^*, g']]\}$ **if** $g' \geq g^*$ **else** $\emptyset$ ▷ g' could cover g^* in level n
3:      **if** $n = 0$ **return** Res **end if**
4:      **for all** $[g, g^*] \in \bigsqcup[n][j]$ **do**
5:          $Res \leftarrow Res \cup$ FINDCOVERINGGENERATORELEMENTS($j, g', n - 1, g$)
6:      **end for**
7:      **return** Res   ▷ set of levels where g' covers
8: **end procedure**

in L_j is finite, as we have a limited number of synchronizations as well. It follows that S is finite, as the set of vertices relies on segments and the edges depict synchronizations.   □

Expanding Segments. Assuming generator relations $r'_1, \ldots, r'_\ell$ where the generator elements have been updated with ωs appropriately for synchronization $<O_1, \ldots, O_\ell> \xrightarrow{f} <O'_1, \ldots, O'_\ell>$. Building on that, we now want to expand segment O'_j from the generator elements for $[N_j, m_{0j}]$ with $j \in \{1, \ldots, \ell\}$. This can be done non-sequentially for all instances, as the local behaviors between two synchronization steps are independent of each other, as described in Sect. 2.2.

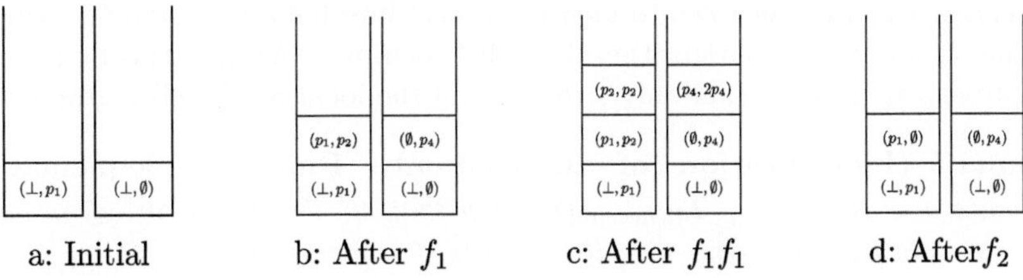

Fig. 5. Generator Relation Stack for Example 2.

Algorithm 6. Building Local Forward Closure for a Given Local Marking

---

1: **procedure** LOCALFORWARDCLOSURE(j: instance index, m: local marking, *Pred*: set of predecessor markings of m)
2:     **if** $m \in Pred$ **then return** $\emptyset$ **end if**   ▷ m already explored
3:     $V_j^L \leftarrow V_j^L \cup \{m\}$ ▷ explored marking is added to L_j
4:     $Res \leftarrow \{m\}$
5:     **for all** $t \in T_{j|internal} : m \xrightarrow{t}$ **do**   ▷ explore internal transitions
6:         $m' \leftarrow m + \Delta t$
7:         **if** $\exists m^* \in Pred : m^* \leq m'$ **then** ▷ identify internal-only pumping sequence
8:             $\forall p \in P_j : m^*(p) < m'(p) : m'(p) \leftarrow \omega$
9:         **end if**
10:        $E_j^L \leftarrow E_j^L \cup \{(m, t, m')\}$
11:        **if** $m' \notin V_j^L$ **then**
12:             $Res \leftarrow Res \cup$ LOCALFORWARDCLOSURE($j, m', Pred \cup \{m\}$)
13:        **else**
14:             $Res \leftarrow Res \cup \{m'' \mid m' \xrightarrow{E^{L\,*}_{j|internal}} m''\}$ ▷ $E^L_{j|internal}$ already known
15:        **end if**
16:     **end for**
17:     **return** Res ▷ local forward closure of m, with ωs inserted
18: **end procedure**

Therefore, procedure LOCALFORWARDCLOSURE (Algorithm 6) builds the forward closure regarding internal transitions for a given local marking. The segment (cf. Definitions 6 and 7) then is the union of those forward closures. Note, that segments in the local coverability graph may overlap. So, if building a forward closure leads us to an already recorded local marking m'_j, we add the already computed forward closure of m'_j regarding internal transitions to the current forward closure (Algorithm 6, l.16). While expanding a segment we also identify internal-only pumping sequences of the instances, for that the cross-instance checking of local pumping sequences is unnecessary (cf. Sect. 3.1). Procedure LOCALFORWARDCLOSURE (Algorithm 6) implements the identification of local pumping sequences for a generator element. Note that for local marking m_j, we record its internal predecessor markings, i.e. local markings from that m_j is reachable only by internal transitions, in a set *Pred*. This is necessary, as V_j^L also contains local markings that are connected to m_j via internal transitions and synchronizations and thus may lead us to spurious local pumping sequences. We traverse *Pred* to identify a local internal-only pumping sequence and insert ωs directly. (Algorithm 6, l.9-10). Lemma 5 justifies this by pointing the preservation of internal-only sequences in the local reachability graph.

Lemma 5 (Identification of Internal-only Pumping Sequences).
Sequence $\sigma \in (\bigcup_{j \in \{1,\dots,\ell\}} T_{j|internal})^*$ such that $m^* \xrightarrow{\sigma} m'$ for $m^* \leq m'$ is an internal-only pumping sequence, if and only if sequence $\pi_j(\sigma) \in T^*_{j|internal}$ is a local pumping sequence such that $\pi_j(m^*) \xrightarrow{\pi_j(\sigma)} \pi_j(m')$ for instance $[N_j, m_{0j}]$.

Proof. For modular state spaces in general, [6] showed that internal-only sequences of $N_{\mathcal{M}}$ appear projected in the local reachability graphs of the instances. The argumentation holds for the covering modular state space as well. $\qquad\square$

Although the identification of internal-only pumping sequences is easy, it is nevertheless of great importance: Internal-only pumping sequences imply infinite behavior within an instance and thus lead to infinitely big segments and infinite local reachability graphs, when undetected. Note, that in the construction algorithm the forward closures of the generator elements are built independently of each other. Naturally, those closures may overlap which means that a local marking of a segment can be reachable from different generator elements, thus actually has different sets of internal predecessor markings. Assume we reach local marking m'_j while building the forward closure from generator element g_j and do not identify a local internal-only pumping sequence, as there is no previously explored marking $m^*_j \in Pred$ such that $m^*_j \leq m'_j$. Besides that, we build the closure from generator element g'_j and explore some local marking m^{**}_j from that marking m'_j is also reachable with internal transitions. Additionally, it holds that $m^{**}_j \leq m'_j$, which implies a local internal-only pumping sequence σ_j such that $m^{**}_j \xrightarrow{\sigma_j} m'_j$. Local marking m'_j is added to the forward closure of g'_j, but as m'_j is already explored, we do not call procedure *localForwardClosure* (Algorithm 6) for m'_j here again and thus do not traverse the alternative set of internal predecessor markings for m'_j, including m^{**}_j, again. Sequence σ_j could be considered to remain undiscovered, we dismiss inserting ωs in m'_j and consequently, segment O'_j gets infinitely big. Fortunately, we can show in the following lemma that the independent exploration of the forward closure of the generator elements does not inhibit finding smaller markings and the according internal-only pumping sequences. Figure 4 provides a graphic representation of the situation.

Lemma 6 (Internal-only Pumping Sequences for Multiple Generator Element Exploration). *For instance $[N_j, m_{0j}]$ with $j \in \{1, \ldots, \ell\}$, w.l.o.g let segment O_j evolve from a set of generator elements $\{g_{j1}, \ldots, g_{jn}\}$ for $n \in \mathbb{N}$, i.e. $O_j = \bigcup_{i=1}^{n} \{\overset{\circ}{g_{ji}}\}$. Local internal-only pumping sequence $\sigma_j \in T^*_{j|internal}$ in O_j, is identified despite building the closures of the generator elements independently.*

Proof. Assume local internal-only pumping sequence $\sigma_j \in T^*_{j|internal}$ such that $m^*_j \xrightarrow{\sigma_j} m'_j$ for the local markings $m^*_j \leq m'_j \in O_j$. We prove the identification of σ_j by induction over the number of generator elements.

Base: For $n = 1$ it holds that $m^*_j, m'_j \in \{\overset{\circ}{g_{j1}}\}$. Sequence σ_j is identified by traversing the set of internal predecessor markings $Pred$ of m'_j, as $m^* \in Pred$ while building $O_j = \{\overset{\circ}{g_{j1}}\}$.

Step: For all generators $g_{j1}, \ldots, g_{jk}$ with $k < n$, all previous local pumping sequences are identified in $\bigcup_{i=1}^{k} \{\overset{\circ}{g_{ji}}\}$. We now build $\{\overset{\circ}{g_{jk+1}}\}$ for generator g_{jk+1}. The following is possible for $m^*_j, m'_j \in O_j$:

- $m^*, m' \in \bigcup_{i=1}^{k} \{\overset{\circ}{g_{ji}}\}$: Both local markings were explored before and sequence σ_j was previously identified.

- $m^*, m' \in \{\overset{\circ}{g_{jk+1}}\}$: Both markings are explored in the current closure of g_{jk+1}. Sequence σ_j is currently identified by traversing $Pred$ of m'_j as $m^*_j \in Pred$.

- $m^* \in \{\overset{\circ}{g_{jk+1}}\}$ and $m' \in \bigcup_{i=1}^{k} \{\overset{\circ}{g_{ji}}\}$: The bigger marking m'_j was already explored from other generators, while the smaller marking m^*_j is explored in the current closure of g_{jk+1}. As $m^*_j \leq m'_j$, sequence σ_j was firable m'_j as well. Firing σ_j in m'_j leads to another marking m''_j such that $m'_j \leq m''_j$. Therefore, σ_j was identified before while traversing $Pred$ of m''_j, as $m'_j \in Pred$.

$\square$

So, the exploration of local behavior from multiple generator element ensures the identification of local internal-only pumping sequences and by that correct and complete ω-insertion for internal-only pumping sequences. Note that, ωs might be inserted based on other local markings, as shown in the proof, but as the main goal is to obtain finiteness and not minimality, we tolerate this. In any case, we always generate finite segments this way. Together with the results on the finiteness of the synchronization graph of the covering modular state space, this has the consequence that the local coverability graphs are finite.

Lemma 7 (The Local Coverability Graphs of the Covering Modular State Space are finite). *Let $[N_j, m_{0j}]$ for $j \in \{1, \ldots, \ell\}$ be an instance and $L_j = [V_j^L, E_j^L]$ its local coverability graph. Then it holds that any segment $O \subseteq V_j^L$ is finite. As a result, $L_j = [V_j^L, E_j^L]$ is finite.*

Proof. From Lemma 5 and Lemma 6 it follows that segment $O \subseteq V_j^L$ is finite. As Lemma 4 states the finite number of segments, L_j is finite. $\square$

The acquired results lead to the conclusion that the covering modular state space actually is a coverability abstraction of the reachability graph of Petri net system $N_{\mathcal{M}}$.

Theorem 1 (Coverability Abstraction of the Covering Modular State Space). *The covering modular state space for a Petri net system $N_{\mathcal{M}}$ based on the modular structure $\mathcal{M} = [\mathcal{I}, \mathcal{F}]$ is a valid coverability abstraction of the reachability graph of $N_{\mathcal{M}}$.*

Proof. We prove this by showing that the covering modular state space is a modular counterpart of a coverability graph (cf. Definition 11). Let $L_j = [V_j^L, E_j^L]$ be the local coverability graph of instance $[N_j, m_{0j}]$ for $j \in \{1, \ldots, \ell\}$. According to [6], we can construct the set of reachable markings $RS(\{m_0\})$ of $N_{\mathcal{M}}$ from the local markings and the synchronization information. For the covering modular state space, we construct the generalized markings M of $N_{\mathcal{M}}$ from the local markings that are generalized markings on the places of their instance the same way. As S and L_j for all instances $[N_j, m_{0j}]$ are finite, M is finite as well and trivially, $m_0 \in M$. The construction of the covering modular state space ensures that every generalized marking $\mu \in M$ actually abstracts a converging sequence of markings, as ωs are inserted correctly and completely and that the transition rule for generalized markings holds. $\square$

3.3   Verification in the Covering Modular State Space

The modular state space method blends in our portfolio approach of our tool LoLA [12] for the verification of Petri net system properties. The covering modular state space method will also be integrated to provide a modular verification attempt for unbounded Petri net systems. Therefore, we formulate standard properties of Petri net systems for the covering modular state space in the following lemma.

Lemma 8 (Standard Properties in a Covering Modular State Space).
Let $L_j = [V_j^L, E_j^L]$ for all $j \in \{1, \dots, \ell\}$ and $S = [V^S, E^S]$ be the covering modular state space of $N_{\mathcal{M}}$. The following propositions hold:

- *If marking $m \in RS(\{m_0\})$ is reachable, then there exists a vertex $v = \,<O_1, \dots, O_\ell> \,\in V^S$ such that $\exists \mu \in \,<O_1 \times \dots \times O_\ell>$ with $\mu \geq m$.*
- *Internal Transition $t \in T_{j|internal}$ is dead if and only if it holds that $m_j \not\xrightarrow{t}$ for all $m_j \in V_j^L$.*
- *Synchronization vector $f \in \mathcal{F}$ is dead, if and only if $v \not\xrightarrow{f}$ all $v \in V^S$.*
- *Place $p \in P_j$ is bounded, if and only if $m_j(p) \neq \omega$ for all $m_j \in V_j^L$.*
- *Vertex $v = \,<O_1, \dots, O_\ell> \,\in V^S$ is a deadlock, if $v \not\xrightarrow{f}$ for all $f \in \mathcal{F}$ and for all segments O_j, there exists a marking $\pi_j(m) \in O_j$ such that $\pi_j(m) \not\xrightarrow{t}$ for all $t \in T_{j|internal}$. If the covering modular state space has a deadlock, $N_{\mathcal{M}}$ has a deadlock as well.*

The presented properties are based on the set of reachable markings. The modular state space method preserves the reachability of markings [6], so the proofs for those properties follow from that and knowledge about the abstraction by the coverability graph. The reachability graph simulates the coverability graph [2]. As Theorem 1 states that the covering modular state space is the modular counterpart of a coverability graph, that preserves paths per construction, it is a necessary and sufficient base for the verification of standard properties expressible in $ACTL$ [3] (i.e. deadness of a transition and boundedness). The covering modular state space can as well be used to verify other properties of the temporal logic $ACTL$ in infinitely large reachability graphs by applying verification methods for temporal logics in the modular state space from [14]. Moreover, there exist properties that rely on the terminal strongly connected components of the reachability graph, such as liveness or reversibility of a Petri net system or the existence of home states. The modular state space does not preserve the strongly connected component structure of the reachability graph per se, their verification requires additional adaptions in the construction. The same holds for the covering modular state space.

4   Implementation

As all other model checking algorithms which perform explicit state space analysis, the proposed method benefits greatly from choosing suitable data structures. Model checking is often not a real-time task but primarily constrained by the available memory. This makes reducing the necessary amount of stored information essential to the real-world application of any such technique. Since a segment consists of local strongly connected components (LSCCs) [6], we can represent a segment as sorted list of LSCC indices. This allows us to efficiently check for segment equality, even if the same segment has different generators. Storing a segment via its LSCCs also enables us to record some information not for each marking, but for the LSCC it is contained in. This improves the repeated information retrieval which occurs during the local exploration (see Algorithm 6 l.14, compare Fig. 4): Instead of iterating over the previously discovered markings, we can just collect the locally reachable LSCCs with their enabled interface transitions. The generator relation as described in Sect. 3.2 is already another optimization. Since one of the central questions during the modular state space exploration is whether a generator element is actually reachable from the generator element it covers, this relation provides us a shortcut to answer this question. Using this relation, we can perform the backwards search on only the generators instead of the full local state space, including the previously explored segments. This also prevents us from having to store all local edges, which would have been a disadvantage compared to traditional state space exploration. Computing the generator relation does not impose additional strain, since it can be easily done during the exploration of a segment as the starting generator element of each search is known. To facilitate the search, each generator relation can be stored as a lookup table of the shape $g' \mapsto \{g^* \mid (g^*, g') \in r_j\}$. The other part of the algorithm that warrants a closer look is the identification of pumping sequences to insert ω in the covering markings. Local pumping sequences can be discovered for a single segment using the same methods as the regular coverability. However, the aforementioned exploitation of LSCCs limits the choice of search technique to depth-first search using Tarjan's algorithm [13] or similar approaches. For each new synchronization graph vertex, previous generators have to be examined as described in Algorithm 5. Since we check each segment separately at first and are only interested in finding a level in $\bigsqcup$, we can optimize the search. After identifying the set of layers where the first instance covers a generator, this candidate set can then be exploited for the following instances: As soon as it is discovered that an instance does not have a covered generator in a given layer, this layer can be removed from the candidate set altogether, since generators in that layer cannot be the start of a generalized pumping sequence. Later instances can also skip the generator comparison for layers not in the candidate set with the same justification and immediately continue with the search. An implementation using these tweaks is feasible, especially if the non-covering modular state space method is available as a starting point. Our model checking tool LoLA [12] can already verify properties of the modular state space, but a quantitative comparison would require a larger number of (preferably real-world) unbounded models, similar to how the Model Checking Contest [8] provides bounded nets. The proposed algorithm and its improvements benefit greatly from properties of the given modular structure, similar to the non-covering approach described in [6]. The more internal-only behavior an instance exhibits, the more reduction is possible by analyzing the modular state space instead of the global. For the covering modular state space, this effect is amplified since the generator relation provides us with even greater benefits by allowing us to skip more local markings in between generators. More internal behavior also increases the chance for internal pumping sequences, which are easier to detect early. Previously described benefits of the modular state space can still can be applied for the coverability analysis. This mainly includes the exploitation of replicated instances to save time and memory since their internal behavior is identical. Established state space reduction techniques such as stubborn sets or symmetric reduction are also still feasible, with further work needed to preserve the desired properties.

5   Conclusion

In this work we introduced the covering modular state space as a finite modular representation of the reachability graph of a Petri net system $N_\mathcal{M}$. The presented construction algorithm extends the existing modular state space approach by abstracting the infinite behavior by identifying pumping sequences and inserting ω accordingly. The pumping sequences are identified based on generator markings in general as we showed, any pumping sequence of $N_\mathcal{M}$ implies a pumping sequence on generator markings. For the subset of pumping sequences only containing instance-internal transitions, the algorithm provides a distinguished method. Although infinite behavior is detected later than in the conventional coverability graph in some cases, the covering modular state space is complete and correct coverability abstraction. Therefore, be used to verify standard properties of unbounded Petri net systems. As the covering modular state space additionally simulates the reachability graph, it can also be used for the verification of temporal logic $ACTL$-properties.

As future work, we want to expand our tool LoLA with an implementation of the covering modular state space to enable verification for unbounded Petri net systems. This can then be used to further study the impact of the modularization on the achievable state space reduction compared to the non-modular coverability graph.

References

1. Christensen, S., Petrucci, L.: Modular analysis of petri nets. Comput. J. **43**(3), 224–242 (2000). https://doi.org/10.1093/comjnl/43.3.224
2. Clarke, E.M., Grumberg, O., Long, D.E.: Model checking. In: International Conference on Foundations of Software Technology and Theoretical Computer Science. Springer, Heidelberg (1999). https://www.cs.cmu.edu/afs/cs/user/emc/www/papers/Books%20and%20Edited%20Volumes/Model%20Checking.pdf

3. Emerson, E.A., Srinivasan, J.: Branching time temporal logic. In: de Bakker, J.W., de Roever, W.-P., Rozenberg, G. (eds.) REX 1988. LNCS, vol. 354, pp. 123–172. Springer, Heidelberg (1989). https://doi.org/10.1007/BFb0013022

4. Finkel, A.: The minimal coverability graph for Petri nets. In: Rozenberg, G. (ed.) ICATPN 1991. LNCS, vol. 674, pp. 210–243. Springer, Heidelberg (1993). https://doi.org/10.1007/3-540-56689-9_45

5. Gaede, J., Overath, J., Wallner, S.: Automatic modularization of place/transition nets. In: Köhler-Bussmeier, M., Moldt, D., Rölke, H. (eds.) Proceedings of the International Workshop on Petri Nets and Software Engineering 2024 co-located with the 45th International Conference on Application and Theory of Petri Nets and Concurrency (PETRI NETS 2024), 24–25 June 2024, Geneva, Switzerland. CEUR Workshop Proceedings, vol. 3730, pp. 53–73. CEUR-WS.org (2024). https://ceur-ws.org/Vol-3730/paper03.pdf

6. Gaede, J., Wallner, S., Wolf, K.: Modular state spaces - a new perspective. In: Kristensen, L.M., van der Werf, J.M. (eds.) Application and Theory of Petri Nets and Concurrency, pp. 312–332. Springer, Cham (2024). https://doi.org/10.1007/978-3-031-61433-0_15

7. Karp, R.M., Miller, R.E.: Parallel program schemata. J. Comput. Syst. Sci. **3**(2), 147–195 (1969). https://doi.org/10.1016/S0022-0000(69)80011-5

8. Kordon, F., et al.: Complete Results for the 2025 Edition of the Model Checking Contest (2025). https://mcc.lip6.fr/2025/results.php

9. Reynier, P.A., Servais, F.: Minimal coverability set for petri nets: karp and miller algorithm with pruning. In: Kristensen, L.M., Petrucci, L. (eds.) Applications and Theory of Petri Nets, pp. 69–88. Springer, Heidelberg (2011). https://doi.org/10.1007/978-3-642-21834-7_5

10. Reynier, P.A., Servais, F.: On the computation of the minimal coverability set of petri nets. In: Filiot, E., Jungers, R., Potapov, I. (eds.) Reachability Problems, pp. 164–177. Springer, Cham (2019). https://doi.org/10.1007/978-3-030-30806-3_13

11. Schmidt, K.: Model-checking with coverability graphs. Formal Methods Syst. Des. **15**(3), 239–254 (1999). https://doi.org/10.1023/A:1008753219837

12. Schmidt, K.: LoLA a low level analyser. In: Nielsen, M., Simpson, D. (eds.) Application and Theory of Petri Nets 2000. Lecture Notes in Computer Science, pp. 465–474. Springer, Heidelberg (2000). https://doi.org/10.1007/3-540-44988-4_27

13. Tarjan, R.: Depth-first search and linear graph algorithms. SIAM J. Comput. **1**(2), 146–160 (1972). https://doi.org/10.1137/0201010

14. Zech, L., Wolf, K.: Verifying temporal logic properties in the modular state space. In: Kristensen, L.M., van der Werf, J.M. (eds.) Application and Theory of Petri Nets and Concurrency, pp. 333–354. Springer, Cham (2024). https://doi.org/10.1007/978-3-031-61433-0_16

Tool Papers

PACO: A Petri Net-Based Tool for Designing, Simulating, and Analyzing Multi-objective Stochastic Processes

Emanuele Chini[1,2]($\boxtimes$), Daniel Amadori[1], Pietro Sala[1],
Sidra Nasir Rajput[1], Matteo Baldi[1], and Mattia Cappelletti[1]

[1] University of Verona, Verona, Italy
`{daniel.amadori,pietro.sala,sidranasir.rajput}@univr.it,`
`{matteo.baldi_02,mattia.cappelletti}@studenti.univr.it`
[2] University of Rome La Sapienza, Rome, Italy
`emanuele.chini@univr.it, emanuele.chini@uniroma1.it`

Abstract. We present PACO, a tool for strategy synthesis and explanation for stochastic processes based on the BPMN 2.0 standard, extended with probabilistic splits and positive impact vectors associated with tasks. The formal semantics is provided by Synchronous Probabilistic Impactful Networks (SPIN), an enriched Petri Net model. At its core, PACO implements an on-the-fly strategy synthesis algorithm for the input SPIN. To improve transparency, PACO includes an explainer module that decomposes synthesized strategies into minimal decision trees attached to individual process choices. PACO is provided as a web-based app that supports design, synthesis, explanation, and interactive simulation of stochastic processes. To further enhance usability at design time, PACO integrates an LLM-assisted design component that helps users create and interpret processes from natural-language descriptions. Finally, tested over a synthetic dataset of processes of increasing complexity, PACO demonstrates that explainable, automated strategy synthesis is practical for realistic BPMN process models despite the inherent computational complexity of the problem.

Keywords: Strategy Synthesis · Multi-Objective Optimization · Explainable Planning · Markov Decision Processes · Probabilistic Planning

1 Introduction

Business Process Model and Notation (BPMN) has emerged as a pivotal formalism in process management, offering a standardized method for detailing business processes across various sectors, including healthcare and industry. Its graphical notation facilitates clear and precise representation of process flows, enabling stakeholders to comprehend, analyze, and improve business operations. In healthcare, BPMN plays a critical role in implementing patient care guidelines

J. Desel and A. Kalenkova (Eds.): PETRI NETS 2026, LNCS 16567, pp. 335–346, 2026.
https://doi.org/10.1007/978-3-032-27879-1_16

[19]. Similarly, in the industrial domain, it aids in efficient management of manufacturing and supply chain processes, ensuring timely delivery of products and services [11]. Over the past decade, increasing attention has focused on Business Process Management (BPM), where selecting appropriate control-flow traces is paramount. These applications demand measurement and employ notions such as cost-awareness [18], energy-awareness [5], and resource-awareness [10], which naturally induce scenarios where multiple measurements must be controlled.

In this work, we introduce PACO, a web-based decision-support instrument for process engineers and management, capable of supporting both day-to-day operational tuning and higher-level strategic planning. Indeed, PACO can perform *strategy synthesis* and explanation for BPMN+CPI processes, an extension of BPMN that makes such questions explicit by annotating models with *Choices*, *Probabilities*, and *Impacts*. The semantic of BPMN+CPI processes is given in terms of Synchronous Probabilistic Impactful Networks (SPIN), a Petri nets-based model that captures concurrency, synchronization, probabilistic branching, and cumulative positive impacts [7]. Given a model and a user-specified multi-dimensional bound vector, PACO finds a strategy (if one exists) that resolves all controllable gateways so that the expected cumulative positive impacts satisfy the bounds. If such a strategy exists, PACO provides an explainer (i.e., a minimal decision tree) for each choice that guide the user in making that decision. In order to complement this static view, PACO provides an interactive simulator that lets users replay executions under the user-defined strategy, observe how stochastic outcomes unfold, and inspect the evolution of impacts along individual runs. Moreover, PACO integrates a large language model (LLM) assistant that translates between natural-language descriptions of processes and the corresponding BPMN+CPI syntax. Process analysts can describe a workflow in domain terms e.g., a healthcare triage-to obtain a candidate BPMN+CPI specification that can be visualized, refined, and analyzed in the same environment. Crucially, any LLM-generated model is subjected to the same grammar-based validation pipeline, ensuring that only syntactically well-formed and semantically meaningful artifacts enter the synthesis and simulation engine.

The paper is structured as follows. Section 2 summarizes the tool goal, workflow, and installation. Section 3 describes architecture and interfaces. Section 4 walks through a representative analysis from unloading shipments in harbors terminal operations. Section 5 compares PACO with existing tools for strategy synthesis and quantitative analysis in stochastic systems and provides the validation.

2 Functionality

This section describes the functionality of PACO from the perspective of a process analyst or process engineer. We focus on what the tool does, abstracting away from algorithmic details that are presented in the companion theory paper. In particular, we describe the supported decision problem, the structure of inputs and outputs, the main analysis capabilities (strategy synthesis, explanation, and simulation), and conclude with installation and usage information.

Supported Decision Problem. PACO operates on BPMN+CPI processes (Choices, Probabilities, and Impacts). The input model consists of: (i) *controllable gateways* representing decisions under the control of the process manager, (ii) *probabilistic gateways* modeling uncertainty due to the environment or external actors, and (iii) *tasks annotated with non-negative, multi-dimensional impact vectors*, optionally including execution durations and gateway delays.

Given a BPMN+CPI model R and a user-defined bound vector $\mathbf{B} \in \mathbb{R}^k_{\geq 0}$, PACO addresses the following decision problem, formalized in [7]:

> *Does there exist a strategy that resolves all controllable gateways such that the expected cumulative impact of the resulting executions is component-wise bounded by* $\mathbf{B}$?

Typical impact dimensions include cost, energy consumption, time, or risk, but the framework is agnostic to their concrete interpretation as long as impacts are non-negative and additive. The current implementation focuses on acyclic BPMN+CPI models; cyclic behavior can be handled through bounded unfoldings, as discussed in [7].

The outputs of PACO are deliberately compact and analyst-oriented:

- a verdict indicating whether a suitable strategy exists;
- if the verdict is positive, a synthesized strategy resolving all reachable controllable gateways;
- the expected cumulative impact vector induced by that strategy;
- a set of local explainers, one per relevant controllable gateway, represented as minimal decision trees;
- optional simulation traces illustrating representative executions under the synthesized strategy.

Strategy Synthesis. The core functionality of PACO is automated strategy synthesis. Internally, the tool employs a deterministic strategy synthesis algorithm that systematically explores the space of all possible strategies while constraining their expected impact to remain below $\mathbf{B}$. Thanks to the positive cumulative impacts, large portions of the strategy space can be pruned early, allowing the tool to conclude either feasibility or infeasibility without constructing the full underlying stochastic model. From the user perspective, strategy synthesis is invoked by providing a bound vector and triggering the analysis. If a strategy exists, PACO reports that the bounds are satisfiable and displays the corresponding expected impacts. If no strategy exists, the tool reports a negative result, allowing the analyst to revise the process model or relax the constraints. This functionality enables analysts to answer quantitative feasibility questions such as whether operational costs, energy usage, or completion times can be kept below prescribed thresholds in expectation, despite probabilistic behavior and concurrency.

Strategy Explanation. To ensure that synthesized strategies are understandable and usable in practice, PACO provides an explainer module that translates strategies into human-readable decision rules. Rather than producing a single monolithic explanation, it associates a *minimal decision tree* with each controllable gateway. This design choice ensures scalability and interpretability even for large processes with many decisions. PACO supports two classes of explainers:

Impact-Based Explainers. Decisions are explained solely in terms of the cumulative impacts accrued so far (e.g., "choose branch A if the accumulated energy is below a given threshold"). These explainers are typically more intuitive and are preferred whenever applicable.

Decision-Based Explainers. Decisions are explained based on the history of past choices and observed outcomes (e.g., "if a specific branch was taken earlier, then choose branch B"). These explainers are more expressive and are guaranteed to exist whenever a strategy exists, but may be less intuitive.

For each controllable gateway, PACO first attempts to construct an impact-based explainer; if this is not possible, it automatically falls back to a decision-based one. The resulting decision trees are minimal with respect to the executions induced by the strategy, ensuring concise and focused explanations.

Interactive Simulation. In addition to synthesis and explanation, PACO provides an interactive simulator. The simulator executes the BPMN+CPI model step by step, sampling probabilistic outcomes and resolving controllable gateways according to a selected strategy or the user input. When a strategy has already been synthesized, the simulator actively supports decision-making by suggesting, at each controllable gateway, the choice prescribed by the strategy together with its associated explainer. This allows analysts to observe how abstract decision rules are applied in concrete executions and to explore both typical and rare scenarios. The simulator can also be used for what-if analysis: analysts may vary probabilities, impacts, or bounds and immediately observe how these changes affect execution behavior and expected outcomes. This makes the simulator a valuable tool for validation, sensitivity analysis, and process redesign.

LLM-Assisted Process Design. To lower the modeling effort and support rapid prototyping, PACO integrates a Large Language Model (LLM) component for process design. The LLM allows analysts to specify processes using structured natural-language descriptions, which are automatically translated into syntactically valid BPMN+CPI models. From the user perspective, the interaction is prompt-based: the analyst provides a textual description of the process logic, including activities, decision points, probabilistic behavior, and quantitative impacts. The LLM then generates a corresponding BPMN+CPI specification that conforms to the formal grammar used by PACO. All generated models are subsequently validated by the parser before being accepted, ensuring that the use of the LLM does not compromise syntactic correctness or semantic consistency. The LLM is intended as a modeling aid rather than as a decision-making component. It does not participate in strategy synthesis, explanation, or simulation, but merely assists in constructing or refining the input process model. This separation ensures that all analytical results produced by PACO including feasibility verdicts, expected impacts, and explainers remain grounded in the formal semantic. In practice, the LLM-assisted workflow is particularly useful during early design phases and for domain experts who may not be familiar with BPMN+CPI syntax. It enables quick iteration over alternative process structures, which can then be rigorously analyzed, explained, and simulated using the core functionalities of PACO.

Installation and Usage. PACO is distributed as a Docker-based application, while the source code is publicly available at https://github.com/danielamadori/PACO; the docker image can be obtained running the command `docker pull danielamadori/paco:latest`.

3   Architecture and Core Capabilities

PACO follows a modular client–server architecture packaged through Docker containers, explicitly separating the interactive graphical interface from the services responsible for parsing, synthesis, explanation, and execution. This separation improves reliability, supports flexible deployment, and ensures that computational components remain independent from the user-facing environment. Figure 1 illustrates the overall architecture of PACO. Section 3.1 describes the frontend workspace, while Sect. 3.2 details the backend services and the simulator component.

3.1   Frontend

The frontend is implemented using Dash [13] and provides the main workspace in which analysts interact with BPMN+CPI models. It offers facilities to import or edit models, configure bound vectors, trigger analyses, and inspect results returned by the backend and simulator. From a structural perspective, the frontend is organized around a routing layer that manages navigation between pages (e.g., the home page, syntax documentation, examples), a set of view components defining page layouts, and controller components registering interactive callbacks. The central interaction surface is the *workbench* on the home page, implemented as a split-pane layout with a sidebar for inputs and configuration and a visualization area rendering BPMN+CPI, Petri nets, and explainer diagrams as SVG graphics.

3.2   Backend and Simulator

The backend is implemented as a FastAPI service [20] running on Python 3.12. It exposes a REST API that supports model validation, semantic construction, strategy synthesis, explanation generation, LLM-assisted modeling, and caches

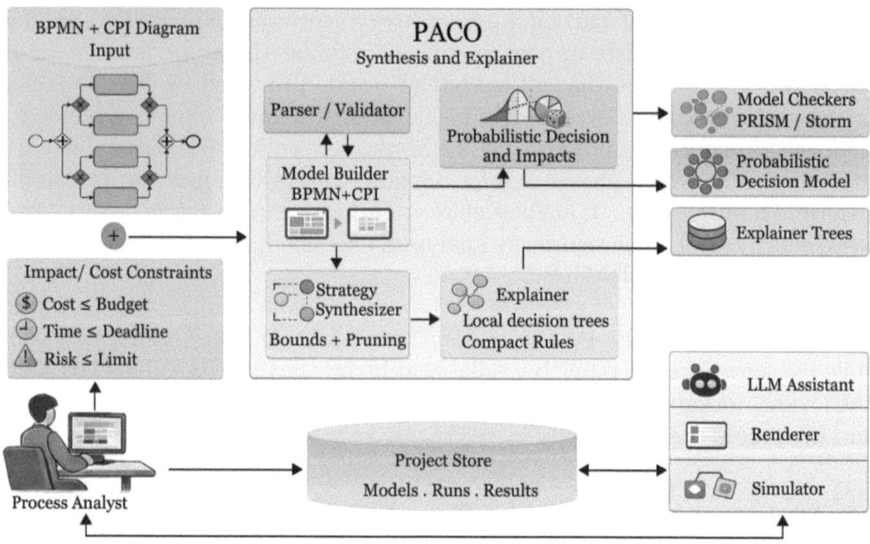

Fig. 1. System architecture of PACO. BPMN+CPI models and bound vectors are processed by the synthesis and explanation engine exposed through a REST API. Strategies, explainers, and execution artifacts are returned to the frontend and optionally cached for reuse.

intermediate artifacts (e.g., BPMN+CPI models, strategies) in a local SQLite database to avoid redundant computation and improve responsiveness. Model validation and parsing rely on an LALR(1) grammar implemented with Lark [1]. The same layer mediates calls to the LLM backend endpoint when LLM-assisted process design is used. During this phase, the backend assigns stable identifiers to controllable and probabilistic gateways, ensuring consistency across parse trees, execution structures, strategies, and explainers. This construction preserves a clear correspondence between BPMN+CPI elements and the underlying SPIN semantics. The graphical artifacts, including diagrams, overlays, and explainer trees, are rendered using Graphviz [2] and PyDot [6] and delivered to the frontend as SVG files. Execution time exploration is delegated to a dedicated simulator service, deployed as a separate component and accessed via a configurable endpoint. Internally, the simulator relies on standard Python libraries and PM4Py [4] to support process execution and analysis. This design keeps the runtime environment simple while leaving room for future extensions.

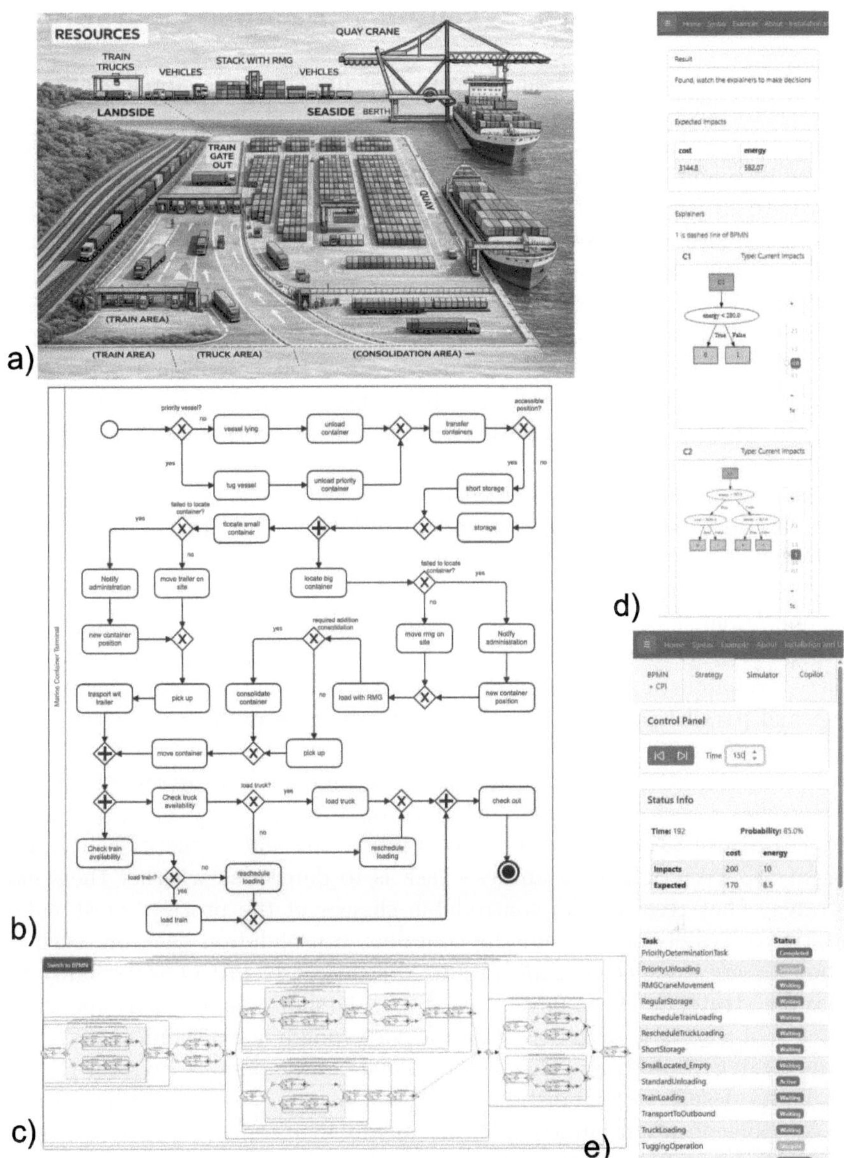

Fig. 2. The figure provides the overview of the marine container terminal analysis workflow. (a) depicts a conceptual layout of the terminal, illustrating the interaction between seaside operations, yard storage areas, consolidation zones, and landside transportation resources for trains and trucks. (b) shows the corresponding BPMN process model. (c) presents the derived SPIN model, where exclusive gateways represent controllable decisions (in orange), e.g. (Color figure online), container consolidation, probabilistic gateways capture uncertainty (in green), e.g., container location failures and transport availability. (d) illustrates the strategy synthesis and explanation view of the web interface, reporting expected impacts and displaying the explainers. (e) shows the simulation panel.

4   Use Case: A Marine Container Terminal Operations

This section presents a realistic use case that illustrates the applicability of the proposed BPMN+CPI modeling and strategy synthesis framework to a complex setting, namely the operation of a marine container terminal. The use case is inspired by quantitative process evaluation approaches for logistics and terminal operations, in particular by the methodology proposed in [9], and is designed to stress the interaction between controllable decisions, probabilistic events, and multiple cumulative impact dimensions.

We consider a container terminal responsible for handling incoming vessels, unloading containers, managing yard storage, and coordinating outbound transportation by rail and truck. The process involves heterogeneous resources (e.g., cranes, tug vessels, trailers, rail-mounted gantry cranes), concurrent activities, and decision points whose resolution directly affects operational cost, processing time, and energy consumption.

The overall workflow is depicted in Fig. 2, which shows the physical layout of the terminal (Fig. 2.a), the corresponding BPMN+CPI process model both in BPMN 2.0 (Fig. 2.b) and the resulting SPIN model (Fig. 2.c). The figure highlights the interaction between seaside operations (vessel handling and unloading), yard operations (storage and retrieval), and landside logistics (rail and truck dispatch). The SPIN model explicitly represents exclusive gateways for controllable choices, probabilistic outcomes for uncertain events (e.g., container location failures), and parallel gateways for concurrent execution paths.

Strategy Synthesis and Explainer. Once the model is constructed, the analyst specifies a vector of bounds reflecting operational constraints: a maximum expected cost of \$3145 per vessel, a total processing and an energy consumption limit of 583 kWh. The core analysis task is to determine whether there exists a strategy that resolves the controllable choices of the process, most notably the priority vessel routing and the container consolidation decision, such that all bounds are satisfied in expectation. For this case, PACO successfully finds a feasible strategy. The analysis result panel (shown in Fig. 2.d) reports the existence of a strategy together with the corresponding expected impacts found by the strategy, confirming that the terminal can be operated within the specified limits despite probabilistic events such as vessel prioritization outcomes and container location failures. Beyond merely synthesizing a strategy, the use case highlights the importance of explainability. For each reachable controllable gateway, the explainer module derives a compact, human-readable decision rule that justifies why a particular branch is chosen. For instance, the explanation attached to the decision of whether put the containers in an accessible location may reveal that accessibility is selected only when the energy consumption is under 280 kWh. These explanations are presented directly on the BPMN+CPI diagram, enabling analysts to understand how local decisions contribute to satisfying global constraints. This is particularly relevant in complex terminal operations, where decisions taken early in the process can have non-obvious effects on downstream parallel activities.

Simulation and What-If Analysis. The simulator (Fig. 2.e) complements synthesis and explanation by enabling an interactive exploration of the process behavior. Through step-by-step execution under the synthesized strategy, analysts can observe concrete realizations of probabilistic outcomes and inspect both typical and rare scenarios, such as repeated container location failures or the simultaneous unavailability of trains and trucks. When a strategy has already been synthesized, the simulator additionally guides the execution by suggesting, at each choice, the decision prescribed by the strategy together with its corresponding explanation, thereby linking simulated behavior directly to the underlying explainer. Moreover, the simulator supports what-if analysis: users can explore alternative strategies, modify probabilities or impact values, and immediately observe how these changes affect accumulated cost, time, and energy. In the context of the terminal use case, this capability is valuable for assessing robustness, identifying bottlenecks, and evaluating the operational impact of infrastructural changes or policy adjustments.

5   Comparison with Other Tools

This section compares PACO with existing tools for strategy synthesis and quantitative analysis in stochastic systems. The focus is on tools that support decision-making under uncertainty with quantitative objectives, rather than on general-purpose simulators or editors. In particular, we consider PRISM [17] and STORM [15], which are widely used in the verification and synthesis of stochastic models. A first distinguishing aspect concerns the handling of costs and impacts. PACO and STORM are designed for non-negative costs or impacts, while PRISM additionally supports negative rewards. Restricting to positive impacts aligns well with business process modeling, where costs and resource consumptions are naturally additive, and enables more efficient synthesis procedures. A second major difference lies in the modeling of parallelism. In PACO, parallelism is expressed directly at the process level through BPMN constructs, making concurrency explicit and intuitive for process designers. In contrast, PRISM and STORM model parallelism through multiple interacting modules that are composed at runtime. While this modular approach is expressive, it often obscures the structure of concurrent business activities and requires additional modeling effort. By preserving BPMN-style parallel gateways, PACO provides a more natural representation for business processes, where synchronization and convergence points play a central role. With respect to computational characteristics, PACO implements a direct, on-the-fly synthesis approach. This contrasts with PRISM and STORM, where BPMN-like models must first be translated into a low-level state-based representation, potentially leading to exponential blow-up. PACO adopts a discrete-time semantics that matches the level of abstraction typically used in BPMN models. The most distinguishing feature of PACO is its support for explainability. None of the considered tools provide native support for explaining synthesized strategies in terms of human-readable decision rules attached to high-level model constructs.

Recent works in *Software and Systems Modeling* contextualize our contribution within BPMN extensions. PE-BPMN [3] and RBPMN [21] follow the "BPMN+X" paradigm, providing formal semantics and executable tooling for privacy- and role-based modeling, respectively. These works confirm the relevance of rigorously defined and tool-supported BPMN extensions. In parallel, simulators such as BIMP/QBP [14] and Petri-net frameworks like GreatSPN [8] and TimeNET [12] focus on simulation and stochastic analysis.

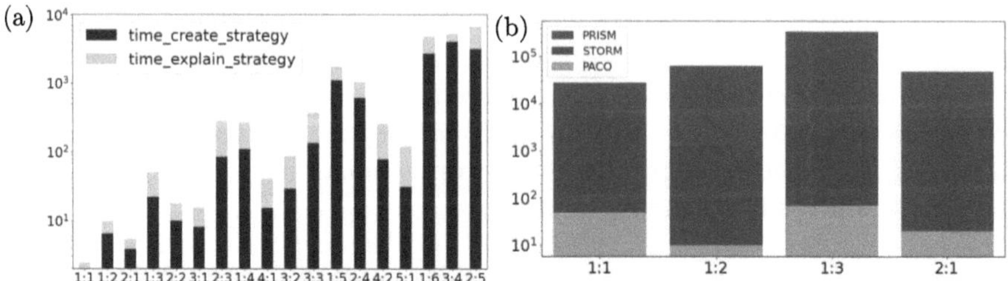

Fig. 3. Log plots for (a) the mean time of finding (purple) and explaining (yellow) strategies for each BPMN+CPI complexity (MNXN:MIX), and (b) for the execution times of strategy finding in PRISM (blue) vs. STORM (green) vs. PACO (Color figure online) (orange).

Validation. To validate PACO and to enable a systematic comparison with PRISM [16] and STORM [15], we adopted the evaluation methodology of [22], where process complexity is measured via two BPMN metrics: *Maximum Nested XORs (MNXN)*, capturing sequential decision depth, and *Maximum Independent XORs (MIX)*, reflecting parallel branching structures. Task impact vectors were generated in $\mathbb{R}_{\geq 0}^{k}$, with $k \in [1, 10]$, across six generation modes differing in cosine similarity, ranging from highly similar to strongly dissimilar ones. A synthetic dataset of 92,896 complete BPMN+CPI processes was obtained by varying MNXN, MIX, the choice-to-nature ratio (10%âĂŞ90%), the impact dimension $k \in [1, 10]$, and the generation mode. Experiments were executed with a timeout of 1.5 min for PACO and 8 min for PRISM and STORM on an Intel Core i9-10980HK CPU (2.40GHz), 32GB RAM, Ubuntu 22.04.3 LTS. Only 2.09% exceeded the timeout threshold, confirming robust scalability within the tested range. Figure 3.(a) reports the mean execution time split between strategy synthesis and explanation across complexity levels. On average, synthesis consumes 52.8% of execution time, while explanation accounts for 47.2%. For simpler processes, explanation dominates the runtime, but as structural complexity increases, its relative cost decreases. Figure 3.(b) compares synthesis time among PACO (mean 460.256 ms, std. 3,231.319 ms), PRISM (mean 46,683.521 ms, std. 58,378.542 ms), and STORM (mean 12,868.587 ms, std. 36,446.926 ms), representing roughly a 100-fold improvement over both frameworks.

Acknowledgments. This work has been carried out while Emanuele Chini was enrolled in the Italian National Doctorate on Artificial Intelligence run by Sapienza University of Rome in collaboration with the University of Verona.

References

1. Lark - parsing library & toolkit (2024). https://github.com/lark-parser/lark. Accessed 20 Apr 2024
2. Bank, S.: Graphviz (2024). https://github.com/xflr6/graphviz. Accessed 20 Apr 2024
3. Belluccini, S., De Nicola, R., Dumas, M., Pullonen-Raudvere, P., Re, B., Tiezzi, F.: Model-based verification of data protection mechanisms in collaborative business processes: S. belluccini et al. Softw. Syst. Model. **24**(2), 489–521 (2025)
4. Berti, A., van Zelst, S., Schuster, D.: PM4PY: a process mining library for python. Software Impacts **17**, 100556 (2023). https://doi.org/10.1016/j.simpa.2023.100556, https://www.sciencedirect.com/science/article/pii/S2665963823000933
5. Cappiello, C., Fugini, M.G., Gangadharan, G., Ferreira, A.M., Pernici, B., Plebani, P.: First-step toward energy-aware adaptive business processes. In: On the Move to Meaningful Internet Systems: OTM 2010 Workshops: Confederated International Workshops and Posters: International Workshops: AVYTAT, ADI, DATAVIEW, EI2N, ISDE, MONET, OnToContent, ORM, P2P-CDVE, SeDeS, SWWS and OTMA. Hersonissos, Crete, Greece, October 25-29, 2010. Proceedings, pp. 6–7. Springer (2010). https://doi.org/10.1007/978-3-642-16961-8_4
6. Carrera, E.: Pydot (2024). https://github.com/pydot/pydot. Accessed 20 Apr 2024
7. Chini, E., Sala, P., Simonetti, A., Zare, O.: Reactive synthesis for expected impacts. Electron. Proc. Theor. Comput. Sci. **409**, 35–52 (2024). https://doi.org/10.4204/eptcs.409.7
8. Chiola, G., Franceschinis, G., Gaeta, R., Ribaudo, M.: Greatspn 1.7: graphical editor and analyzer for timed and stochastic petri nets. Perform. Eval. **24**(1-2), 47–68 (1995)
9. Cimino, M.G., Palumbo, F., Vaglini, G., Ferro, E., Celandroni, N., La Rosa, D.: Evaluating the impact of smart technologies on harbor's logistics via BPMN modeling and simulation. Inf. Technol. Manage. **18**(3), 223–239 (2017)
10. De Leoni, M., Van Der Aalst, W.M., Van Dongen, B.F.: Data-and resource-aware conformance checking of business processes. In: Business Information Systems: 15th International Conference, BIS 2012, Vilnius, Lithuania, May 21-23, 2012. Proceedings 15, pp. 48–59. Springer (2012). https://doi.org/10.1007/978-3-642-30359-3_5
11. Fernandes, J., Reis, J., Melão, N., Teixeira, L., Amorim, M.: The role of industry 4.0 and bpmn in the arise of condition-based and predictive maintenance: a case study in the automotive industry. Appl. Sci. **11**(8), 3438 (2021). https://doi.org/10.3390/app11083438
12. German, R., Kelling, C., Zimmermann, A., Hommel, G.: TimeNet: a toolkit for evaluating non-Markovian stochastic petri nets. Perform. Eval. **24**(1–2), 69–87 (1995)
13. Inc., P.T.: Dash (2024). https://dash.plotly.com/. Accessed 20 Apr 2024
14. Kärgenberg, V.: Online business process model simulator. Ph.D. thesis, Ph. D. Thesis, University of Tartu (2012)

15. Katoen, J.P.: The probabilistic model checking landscape. In: Proceedings of the 31st Annual ACM/IEEE Symposium on Logic in Computer Science, pp. 31–45 (2016)
16. Kwiatkowska, M., Norman, G., Parker, D.: PRISM 4.0: verification of probabilistic real-time systems. In: Gopalakrishnan, G., Qadeer, S. (eds.) Proceedings of the 23rd International Conference on Computer Aided Verification (CAV 2011). LNCS, vol. 6806, pp. 585–591. Springer (2011). https://doi.org/10.1007/978-3-642-22110-1_47
17. Kwiatkowska, M., Norman, G., Parker, D., Santos, G.: PRISM-games 3.0: stochastic game verification with concurrency, equilibria and time. In: Lahiri, S.K., Wang, C. (eds.) CAV 2020. LNCS, vol. 12225, pp. 475–487. Springer, Cham (2020). https://doi.org/10.1007/978-3-030-53291-8_25
18. Magnani, M., Montesi, D.: BPMN: how much does it cost? An incremental approach. In: Alonso, G., Dadam, P., Rosemann, M. (eds.) Business Process Management, pp. 80–87. Springer Berlin Heidelberg, Berlin, Heidelberg (2007). https://doi.org/10.1007/978-3-540-75183-0_6
19. Pufahl, L., Zerbato, F., Weber, B., Weber, I.: BPMN in healthcare: challenges and best practices. Inf. Syst. **107**, 102013 (2022). https://doi.org/10.1016/j.is.2022.102013
20. Ramírez, S.: FastAPI (2020). https://fastapi.tiangolo.com
21. Skouti, T., Seiger, R., Furrer, F.J., Strahringer, S.: RBPMN: the value of roles for business process modeling. Softw. Syst. Model. **23**(6), 1375–1406 (2024)
22. Workneh, T.C., Sala, P., Rizzi, R., Cristani, M.: Business process compliance with impact constraints. Inf. Syst. **129**, 102505 (2025)

IsoNet: Property-Preserving Hierarchical Decomposition of Workflow Nets

Tsung-Hao Huang[1]([✉]) [iD], Lukas M. Jansen[2], Marco Pegoraro[1] [iD],
and Wil M. P. van der Aalst[1] [iD]

[1] Process and Data Science (PADS), RWTH Aachen University, Aachen, Germany
`{tsunghao.huang,pegoraro,wvdaalst}@pads.rwth-aachen.de`
[2] RWTH Aachen University, Aachen, Germany
`lukas.maximilian.jansen@rwth-aachen.de`
`http://www.pads.rwth-aachen.de/`

Abstract. As information systems grow in complexity, standard process analysis frequently encounters computational bottlenecks. While decomposition offers a scalable alternative, its utility depends on the structural and behavioral properties of the fragments. We identify three prerequisites for reliable decomposition: property preservation (e.g., soundness), valid result aggregation, and hierarchical abstraction. This paper presents IsoNet, a tool implementing a decomposition strategy based on subnets that are independent of the rest of the net, apart from designated entry/exit points. IsoNet automatically identifies fragments that can be independently transformed into workflow nets (WF-net) that inherit soundness and free-choiceness properties. By maintaining consistent WF-net semantics both internally and externally, the tool supports a natural zoom-in/zoom-out workflow that simplifies complicated processes while maintaining a connection to the global model.

Keywords: Non-Block-Structured Workflow Nets · Free-Choice Nets · Decomposition · Hierarchical Abstraction · Interactive Visualization · Property Preservation

1 Introduction

Workflow nets (WF-nets) provide a robust mathematical foundation for modeling and verifying business processes. However, as industrial models scale, they often incorporate intricate routing that leads to significant computational challenges. Formal analysis of such models often becomes intractable due to state-space explosion, necessitating the use of decomposition strategies to partition the model into manageable fragments. The effectiveness of such a decomposition depends on how well the fragments support the intended analysis goals. In the context of process mining and formal verification, we argue that three requirements are essential for a reliable and usable decomposition: (1) Property Preservation: Fragments should ideally inherit the behavioral and structural properties

J. Desel and A. Kalenkova (Eds.): PETRI NETS 2026, LNCS 16567, pp. 347–358, 2026.
https://doi.org/10.1007/978-3-032-27879-1_17

of the global model—specifically soundness and free-choiceness—to ensure they can be analyzed as stand-alone WF-nets supported by efficient analysis techniques [6]. (2) Valid Aggregation: The decomposition should satisfy the criteria for a *valid decomposition* [2], allowing local quality metrics to be aggregated on a global level. (3) Hierarchical Abstraction: To manage complexity, the decomposition should support zoom-in/zoom-out functionalities, abstracting complex subnets into single transitions without losing the formal connection to the global process or switching modeling languages.

ISONET addresses these requirements by providing tool support for the decomposition and interactive exploration of WF-nets. Specifically, the tool offers three main contributions: (1) Stand-alone subnets: ISONET automatically identifies subnets that preserve properties (soundness and free-choiceness if it holds for the overall net), allowing fragments to be analyzed independently while maintaining the same WF-net semantics used globally. (2) Interaction & Abstraction: The tool enables users to collapse or expand identified subnets for detailed inspection interactively. This maintains a consistent modeling language both internally and externally, allowing for seamless navigation of "spaghetti" processes. (3) Property-Preserving Analysis: ISONET ensures that soundness and free-choiceness are preserved within subnets. Furthermore, these subnets support valid decomposition [2], ensuring that local analysis results remain globally representative and formally grounded.

The remainder of this paper is structured as follows. Section 2 presents the theoretical foundations of the proposed decomposition approach. Section 3 demonstrates the architecture, functionalities and the user interface of ISONET. Section 4 discusses related work and tools on Petri net decomposition. Section 5 concludes the paper.

2 Foundations

This section introduces the theoretical foundations of ISONET. We focus on the concept of *isolated subnets* and the hierarchical decomposition strategy. For a comprehensive formal treatment of the underlying Petri net theory and ILP constraints, we refer the reader to [9, 10].

Foundations and Isolated Subnets. We consider sound free-choice *Workflow nets (WF-nets)* with a source place i and a sink place o. To enable modular analysis, ISONET identifies subnets that can be treated as independent units. The core construct is the **Isolated Subnet:**

Definition 1 (Isolated Subnet [8–10]). *Let $N = (P, T, F)$ be a WF-net. A subnet $N_s = (P_s, T_s, F_s)$ is an isolated subnet of N if it interfaces with the rest of the net only through a designated set of entry transitions T_{start} (sharing a common preset P_{in}) and exit transitions T_{end} (sharing a common postset P_{out}). Internal nodes in $P_s \cup (T_s \setminus \{T_{start} \cup T_{end}\})$ have no connections to the external nodes.*

As shown in Fig. 1, this structure ensures that once a subnet is activated, its internal progress is independent of the external net. Most importantly, these subnets can be transformed into standalone sound WF-nets[1], preserving structural properties like free-choiceness. This enables IsoNet to perform localized diagnostics that remain valid for the global model.

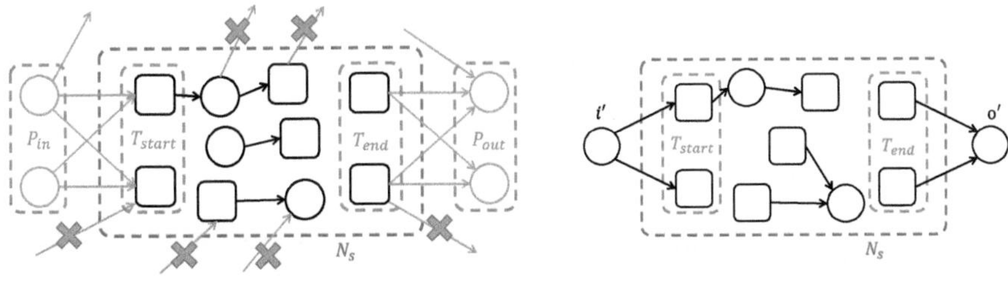

(a) Visualized requirements of an Isolated Subnet   (b) A WF-net transformed from (a)

Fig. 1. An Isolated Subnet and its corresponding sub-WF-net.

Hierarchical Decomposition via ILP. IsoNet partitions a WF-net $N = (P, T, F)$ into a set of pairwise disjoint isolated subnets $D = \{S^1, S^2, ..., S^n\}$ such that $\cup_{1 \leq i \leq n} T^i = T$. To find an optimal decomposition suitable for process mining diagnostics, the tool seeks subnets that are as large as possible without being trivial (i.e., not the entire net).

The tool implements this via an Integer Linear Programming (ILP) formulation. Given a set of initial transitions T_{init}, the ILP maximizes the number of included transitions while enforcing the constraints of Definition 1.

To better illustrate the concept and provide an overview, Fig. 2a shows the decomposition process, specifically visualizing the first two iterations. At each iteration, we begin by defining the sets T_{init} and $T_{excluded}$. The set T_{init} is initialized with the transition that has the highest connectivity, i.e., the largest sum of incoming and outgoing arcs[2]. In the example, the transition labeled b is selected. Transitions that were already assigned in previous iterations are excluded and added to $T_{excluded}$. For the first iteration, $T_{excluded}$ is empty. Next, the ILP solver identifies the maximal non-trivial isolated subnet for transition b, which we denote as S^2; this subnet includes the transitions labeled b and c. These two transitions are then added to $T_{excluded}$ for the subsequent iteration.

Since each isolated subnet can be transformed into a WF-net, the decomposition can be applied recursively to build a hierarchy of subnets until each transition belongs only to its minimal trivial subnet (i.e., the transition itself).

---

[1] Note that every transition is an isolated subnet by definition, so is the whole net (except for the source i and sink o places). We call such nets *trivial minimal or maximal isolated subnets*, respectively.

[2] If multiple transitions share the same connectivity, one is selected at random.

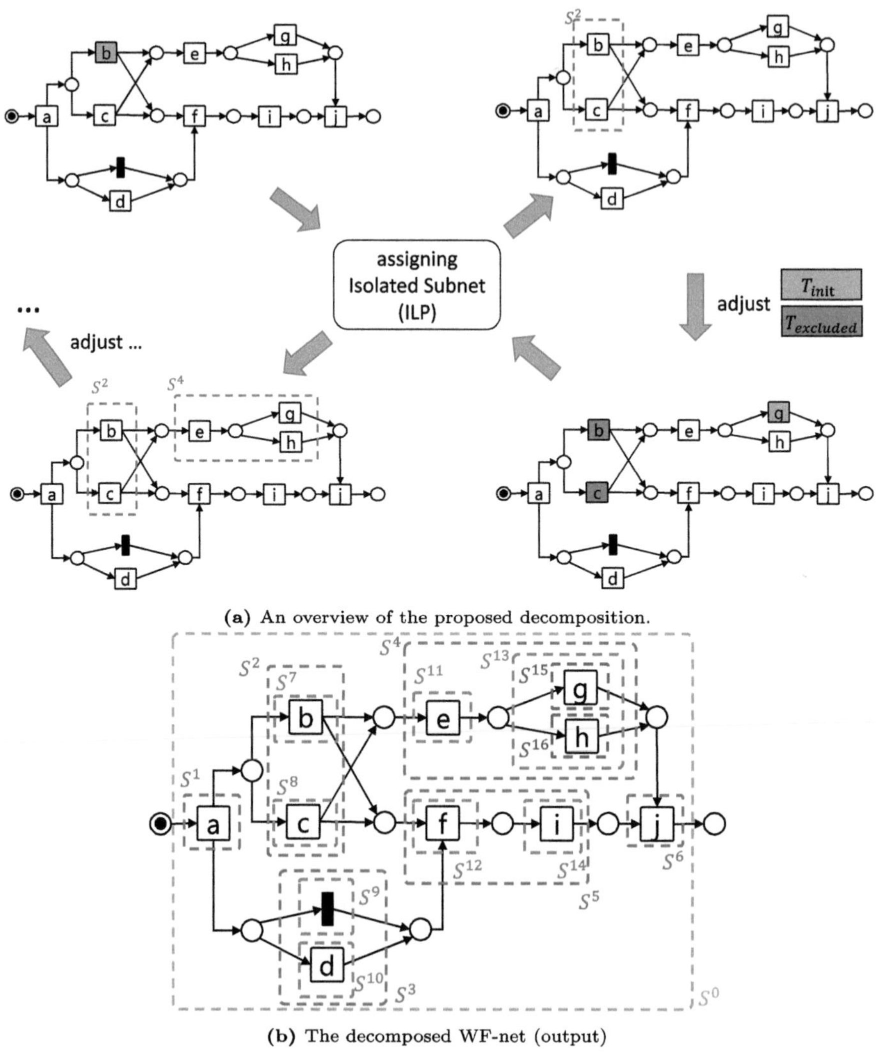

(a) An overview of the proposed decomposition.

(b) The decomposed WF-net (output)

Fig. 2. Overview of the approach behind IsoNet using an example.

This fulfills our goal to provide a hierarchical abstraction. Figure 2b shows the resulting decomposition for the running example. Lastly, the decomposition proposed in this paper can be transformed into a valid decomposition as defined in [2], and can thereby enable distributed conformance checking[3].

---

[3] As the decomposition is transition-bordered and does not include every place, the basic idea is to create connecting components similar to the bridges in [17] so that it fulfills the requirements of a valid decomposition. For more details, we refer readers to [2,17].

3 Architecture and Functionalities

In this section, we present the architecture together with the main functionalities and user interface of IsoNet[4].

3.1 Architecture

The architecture of IsoNet follows a client-server model as illustrated in Fig. 3. The tool is divided into two primary environments: a web-based frontend and backend.

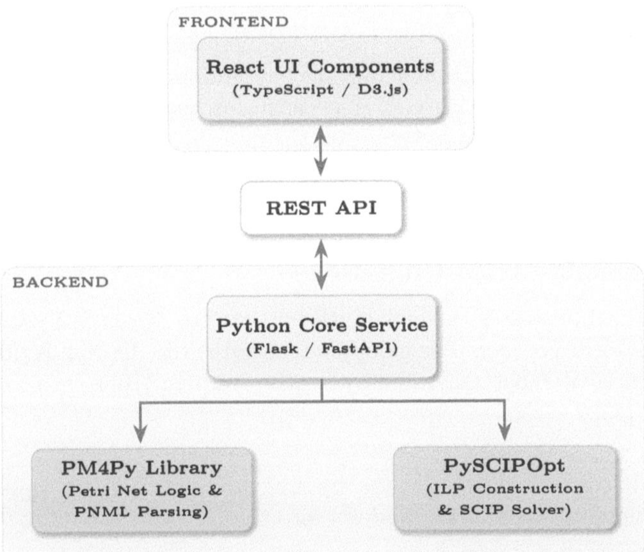

Fig. 3. IsoNet Tool Architecture: The React-based frontend interacts with a Python backend via REST API, utilizing PM4Py for Petri net logic and PySCIPOpt for ILP solving.

Frontend Environment. The frontend is implemented in TypeScript using the React library. To ensure a responsive and lightweight user experience, the frontend is designed to be minimal; its primary responsibility is the orchestration of user interactions and the rendering of visualizations. The layouts of Petri nets and decomposition results are rendered using D3.js, allowing users to interactively explore the subnet hierarchy. Using a browser-based interface, IsoNet remains platform-independent and does not require the user to install local software.

---

[4] https://git.rwth-aachen.de/tsunghao.huang/decomposition_tool.

Backend and Logic Layer. The backend serves as the computational engine of the tool and is implemented in Python. This choice was driven by the integration of specialized libraries essential for Petri net processing and mathematical optimization:

- **PM4Py:** This library handles the core Petri net logic, including the parsing of .pnml files, the generation of net visualizations, and the execution of alignment-based fitness algorithms used for diagnostics.
- **PySCIPOpt** [16]: To compute the maximal isolated subnets, the backend utilizes this interface to the solver. It constructs the Integer Linear Programming (ILP) models described in Sect. 2 and solves them to find optimal decomposition boundaries.

Communications between the two layers are handled via a REST API provided by the backend. This separation allows the backend to be deployed flexibly, either on a server or within a Docker container, ensuring that the heavy ILP computations do not impact the responsiveness of the user interface. Data is interchanged primarily through standard formats like PNML for models and JSON for internal structure definitions, ensuring compatibility with other process mining tools.

3.2　Functionalities and Use Case

This section demonstrates the capabilities of ISONET through a user's workflow using a real-world process. The example illustrates the decomposition of a non-block-structured WF-net generated by Synthesis Miner [10] from a log of a traffic fine process [13].

Decomposition. The user begins by uploading a WF-net (potentially an extremely complex "spaghetti" model[5]) into the Decomposer Interface (Fig. 4).

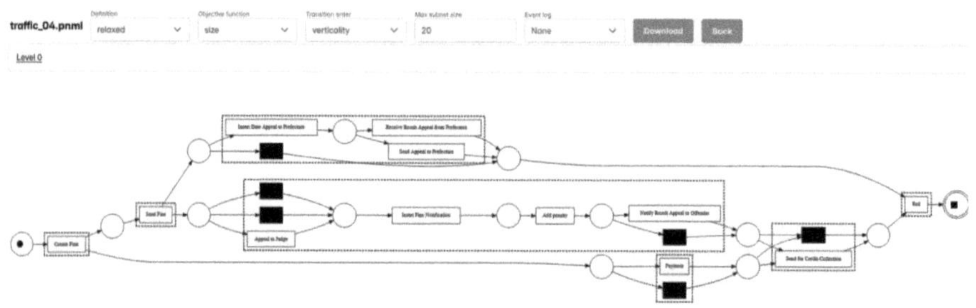

Fig. 4. The decomposition interface of ISONET.

To manage the complexity of the WF-net, the user can configure the decomposition parameters:

---

[5] For the readability, we use a process model of reasonable size. A larger process model with over 300 transitions is shown later in this section.

- **Definition**: Selects either the strict or relaxed definition of an isolated subnet. The strict definition [9, Def. 8] enforces exclusive incoming and outgoing arcs from input places to start transitions and from end transitions to output places, respectively, whereas the relaxed definition (see Defefinition 1) allows additional arcs, potentially affecting the decomposition outcome.
- **Objective Function**: Specifies the ILP objective, such as maximizing the number of transitions in a component or the number of arcs of its boundary transitions. Different objectives emphasize different structural properties.
- **Transition Order**: Determines the order in which transitions are processed, either randomly or based on connectivity. Random ordering may yield different results across runs, while connectivity-based ordering produces consistent results and prioritizes structurally central transitions.
- **Max Subnet Size**: Sets an upper bound on subnet size to limit complexity and improve performance. Larger subnets are further decomposed in subsequent iterations.
- **Event Log**: Allows uploading an XES event log to perform alignment-based conformance checking [4]. Conformance metrics are computed and reported for each decomposed subnet.

Hierarchical Conformance Checking. Once the model is decomposed, the user can perform a zoom-in/zoom-out inspection. As illustrated in Fig. 5, the tool supports navigating the hierarchy:

- **High-Level View (Level 0):** The entire process is abstracted into top-level isolated subnets. If an event log is provided, IsoNet automatically calculates alignment-based fitness for these high-level blocks.
- **Drill-Down (Level 1–4):** If a specific subnet shows low fitness, the user clicks to zoom in. The tool reveals the internal structure of that subnet, which is itself a sound free-choice WF-net. This allows the analyst to pinpoint exactly where deviations occur without losing the global context.

Minimal Subnets Extraction. As WF-nets increase in complexity, manually identifying specific transitions and their smallest independent components becomes a non-trivial task. To assist users, the tool can extract minimal isolated subnets for any chosen transitions. Upon selecting the transitions of interest, the decomposer automatically computes and visualizes the corresponding minimal components.

Figure 6 shows an example using the same WF-net as in Fig. 4. The transitions "Appeal to Judge" and "Notify Result Appeal to Offender" are selected and highlighted. The decomposer extracts the corresponding minimal isolated subnets and presents additional information, including start and end transitions, the subnet structure, and its input and output places (see Definition 1). Since isolated subnets execute independently of the rest of the WF-net, this functionality enables focused analysis of selected transitions without requiring inspection of the entire model.

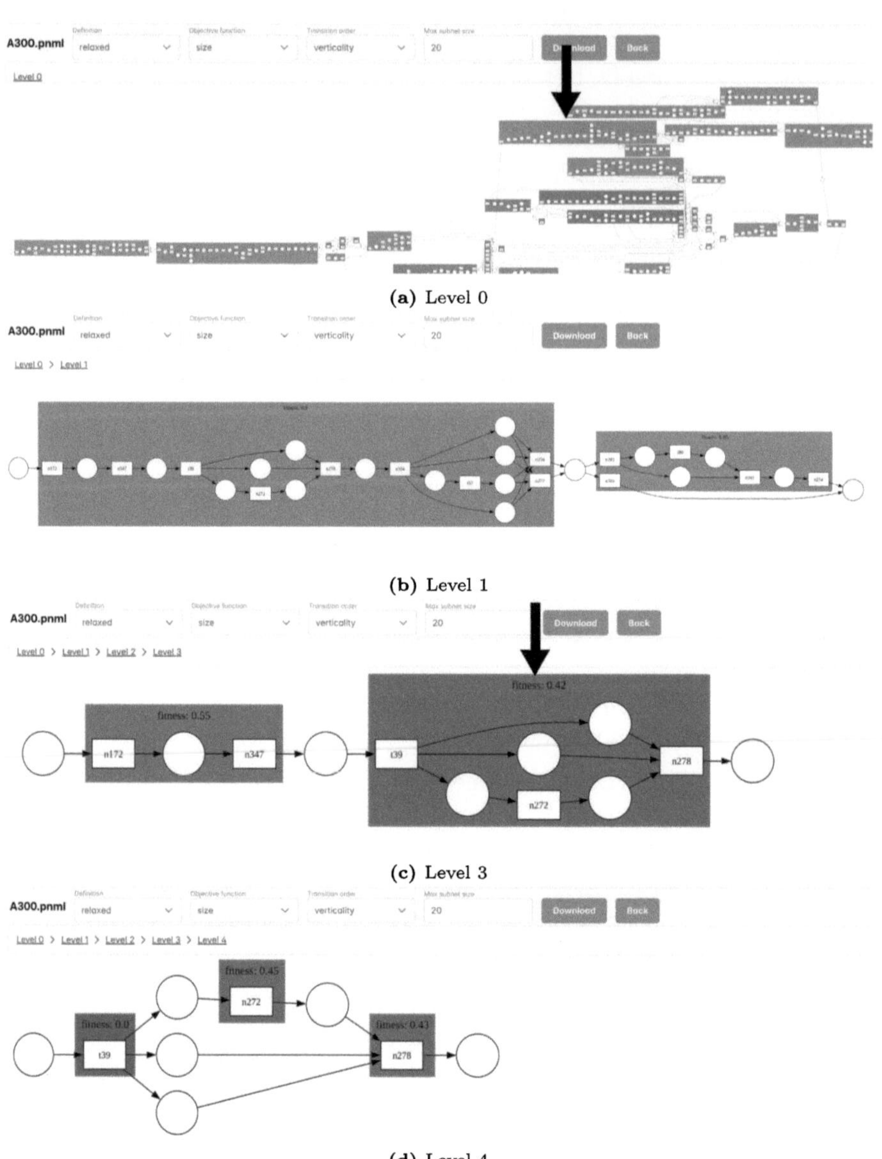

(a) Level 0

(b) Level 1

(c) Level 3

(d) Level 4

Fig. 5. Different decomposition levels with conformance metrics.

Generator. To support the evaluation of process mining algorithms such as [19], ISONET includes a generator for complex, non-block-structured WF-nets and the corresponding logs. Synthetic models with known properties are useful for benchmarking, as discovery techniques are often unavailable for non-block-structured processes. The generator applies randomized transformations based on adapted synthesis rules [6,10], ensuring the resulting models are sound and free-choice by

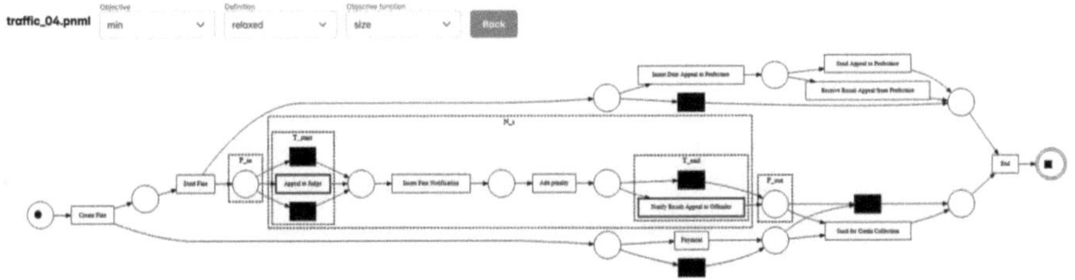

Fig. 6. Examples of minimal isolated subnet extraction in IsoNet.

design. Users can specify the intensity of specific structural patterns by setting transformation counts for each rule, starting from a custom net or a trivial transition. This allows for the controlled generation of"stress-test" models for further decomposition or for simulating event logs to test conformance checking performance. In fact, the model with over 300 transitions shown in 5ais generated by IsoNet.

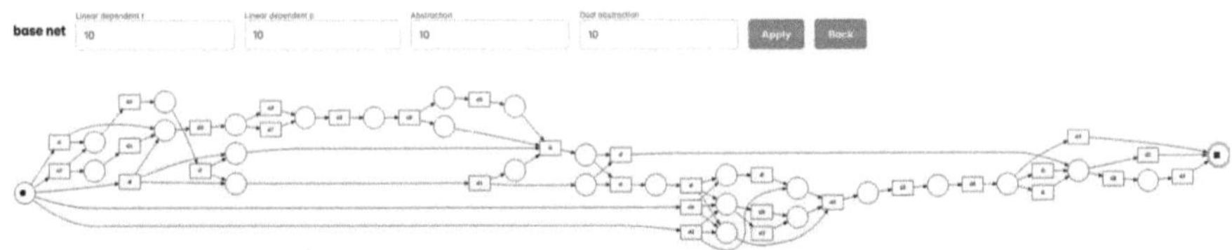

Fig. 7. The Generator Interface in IsoNet.

Figure 7 shows the generator interface, which supports configuration of synthesis rules, generation of the resulting non-block-structured WF-net for further analysis or decomposition, and generation of sample logs from the model.

4   Related Work

Decomposition strategies for Petri nets vary based on their intended analysis goals. Broadly, reduction rules [3,6,11] can be considered a form of decomposition, though they primarily facilitate verification rather than providing identifiable sub-models for diagnostic inspection.

In process mining, early techniques were designed to scale discovery and conformance checking [1,2]. While these improve scalability, they often produce fragmented components that do not behave as stand-alone WF-nets and can yield over-optimistic fitness estimates [17]. Recomposition techniques [12,20] have been proposed to improve the accuracy of these local results, yet they do not focus on providing a hierarchical abstraction for user-led exploration.

To support both diagnostic insights and scalable analysis, the Refined Process Structure Tree (RPST) [5,14,17,18,21,22] has been used to construct hierarchies of single-entry, single-exit (SESE) fragments. While SESE decomposition can be applied to any WF-net, it is optimized for block-structured models and is less informative for nets with multi-entry or multi-exit structures. Conversely, Multiple-entry, multiple-exit (MEME) techniques [7,25] relax these structural assumptions but do not always guarantee that the resulting subnets satisfy the behavioral properties of a sound WF-net.

Comparison with Existing Tools. While various tools have been proposed to fulfill the aforementioned analytical approaches such as process discovery and conformance checking [12,23,24], tools dedicated to interactive decomposition analysis are rare. ISONET complements this by simultaneously supporting hierarchical decomposition, property preservation, and valid conformance result aggregation (through valid decomposition).

The most closely related implementations are the SESE decomposed conformance checking approach in ProM [15,17]. However, ISONET offers two distinct advantages. First, while SESE tools are more suitable for block-structured fragments, ISONET identifies isolated subnets that remain informative for multi-entry/exit structures. Second, unlike existing static implementations, ISONET provides an interactive user interface for hierarchical "zoom-in/zoom-out" exploration, maintaining consistent WF-net semantics across all levels of the process hierarchy.

5   Conclusion

We presented ISONET, a tool for the decomposition and interactive exploration of workflow nets. By identifying isolated subnets that satisfy the requirements for property preservation and valid decomposition, the tool enables independent localized analysis and scalable conformance checking for arbitrary, non-block-structured models. ISONET provides high-level process overviews and hierarchical abstraction, supporting a natural zoom-in/zoom-out workflow that maintains consistent WF-net semantics across all levels of the hierarchy. The tool thus complements established approaches by providing the modularity necessary for the focused diagnostic inspection of complex processes. Future work will explore further integration with automated repair techniques and expanded analysis functionalities for non-free-choice models.

References

1. Aalst, W.M.P.: Decomposing Process Mining Problems Using Passages. In: Haddad, S., Pomello, L. (eds.) PETRI NETS 2012. LNCS, vol. 7347, pp. 72–91. Springer, Heidelberg (2012). https://doi.org/10.1007/978-3-642-31131-4_5
2. van der Aalst, W.M.P.: Decomposing Petri nets for process mining: a generic approach. Distrib. Parallel Databases **31**(4), 471–507 (2013)

3. Aalst, W.M.P.: Reduction Using Induced Subnets to Systematically Prove Properties for Free-Choice Nets. In: Buchs, D., Carmona, J. (eds.) PETRI NETS 2021. LNCS, vol. 12734, pp. 208–229. Springer, Cham (2021). https://doi.org/10.1007/978-3-030-76983-3_11

4. Adriansyah, A.: Aligning Observed and Modeled Behavior. Ph.D. thesis, Technische Universiteit Eindhoven (2014)

5. Choi, Y., Ha, N.L., Kongsuwan, P., Han, K.H.: An alternative method for refined process structure trees (RPST). Bus. Process. Manag. J. **26**(2), 613–629 (2020)

6. Desel, J., Esparza, J.: Free Choice Petri Nets. No. 40, Cambridge University Press (1995)

7. Hauser, R., Friess, M., Küster, J.M., Vanhatalo, J.: An incremental approach to the analysis and transformation of workflows using region trees. IEEE Trans. Syst. Man Cybern. Part C **38**(3), 347–359 (2008)

8. Huang, T.-H., Jansen, L., Pegoraro, M., Park, G., van der Aalst, W.M.P.: Flexible and hierarchical decomposition of workflow nets for process analysis. In: Proceedings of the 38th International Conference on Advanced Information Systems Engineering (CAiSE 2026). Lecture Notes in Computer Science, vol. 16558. Springer (2026)

9. Huang, T.-H., Schneider, E., Pegoraro, M., van der Aalst, W.M.P.: Fast & sound: Accelerating synthesis-rules-based process discovery. In: BPMDS/EMMSAD@CAiSE. Lecture Notes in Business Information Processing, vol. 511, pp. 259–274. Springer, Cham (2024)

10. Huang, T.-H., Schneider, E., Pegoraro, M., van der Aalst, W.M.P.: Flexible and sound: Synthesis Miner. Softw. Syst. Model. (2025)

11. Lee, K.H., Favrel, J.: Hierarchical reduction method for analysis and decomposition of petri nets. IEEE Trans. Syst. Man Cybern. **15**(2), 272–280 (1985)

12. Lee, W.L.J., Verbeek, H.M.W., Munoz-Gama, J., van der Aalst, W.M.P., Sepúlveda, M.: Recomposing conformance: closing the circle on decomposed alignment-based conformance checking in process mining. Inf. Sci. **466**, 55–91 (2018)

13. de Leoni, M.M., Mannhardt, F.: Road Traffic Fine Management Process (2015)

14. de Leoni, M., Munoz-Gama, J., Carmona, J., van der Aalst, W.M.P.: Decomposing Alignment-Based Conformance Checking of Data-Aware Process Models. In: Meersman, R., Panetto, H., Dillon, T., Missikoff, M., Liu, L., Pastor, O., Cuzzocrea, A., Sellis, T. (eds.) OTM 2014. LNCS, vol. 8841, pp. 3–20. Springer, Heidelberg (2014). https://doi.org/10.1007/978-3-662-45563-0_1

15. de Leoni, M., Munoz-Gama, J., Carmona, J., van der Aalst, W.M.P.: Decomposing Alignment-Based Conformance Checking of Data-Aware Process Models. In: Meersman, R., Panetto, H., Dillon, T., Missikoff, M., Liu, L., Pastor, O., Cuzzocrea, A., Sellis, T. (eds.) OTM 2014. LNCS, vol. 8841, pp. 3–20. Springer, Heidelberg (2014). https://doi.org/10.1007/978-3-662-45563-0_1

16. Maher, S., Miltenberger, M., Pedroso, J.P., Rehfeldt, D., Schwarz, R., Serrano, F.: PySCIPOpt: Mathematical Programming in Python with the SCIP Optimization Suite. In: Greuel, G.-M., Koch, T., Paule, P., Sommese, A. (eds.) ICMS 2016. LNCS, vol. 9725, pp. 301–307. Springer, Cham (2016). https://doi.org/10.1007/978-3-319-42432-3_37

17. Munoz-Gama, J., Carmona, J., van der Aalst, W.M.P.: Single-entry single-exit decomposed conformance checking. Inf. Syst. **46**, 102–122 (2014)

18. Polyvyanyy, A., Vanhatalo, J., Völzer, H.: Simplified Computation and Generalization of the Refined Process Structure Tree. In: Bravetti, M., Bultan, T. (eds.) WS-FM 2010. LNCS, vol. 6551, pp. 25–41. Springer, Heidelberg (2011). https://doi.org/10.1007/978-3-642-19589-1_2
19. Schwanen, C.T., Pakusa, W., van der Aalst, W.M.P.: Complexity of alignments on sound free-choice workflow nets. In: Petri Nets. Lecture Notes in Computer Science, vol. 15714. Springer, Cham (2025)
20. Tikhonov, S.E., Mitsyuk, A.A.: A Method to Improve Workflow Net Decomposition for Process Model Repair. In: AIST 2019. LNCS, vol. 11832, pp. 411–423. Springer, Cham (2019). https://doi.org/10.1007/978-3-030-37334-4_37
21. Vanhatalo, J., Völzer, H., Koehler, J.: The refined process structure tree. Data Knowl. Eng. **68**(9), 793–818 (2009)
22. Vanhatalo, J., Völzer, H., Leymann, F.: Faster and More Focused Control-Flow Analysis for Business Process Models Through SESE Decomposition. In: Krämer, B.J., Lin, K.-J., Narasimhan, P. (eds.) ICSOC 2007. LNCS, vol. 4749, pp. 43–55. Springer, Heidelberg (2007). https://doi.org/10.1007/978-3-540-74974-5_4
23. Verbeek, E.: Decomposed process mining with divideandconquer. In: Limonad, L., Weber, B. (eds.) Proceedings of the BPM Demo Sessions 2014 Co-located with the 12th International Conference on Business Process Management (BPM 2014), Eindhoven, The Netherlands, September 10, 2014. CEUR Workshop Proceedings, vol. 1295, p. 86. CEUR-WS.org (2014)
24. Verbeek, H.M.W., van der Aalst, W.M.P., Munoz-Gama, J.: Divide and conquer: a tool framework for supporting decomposed discovery in process mining. Comput. J. **60**(11), 1649–1674 (2017)
25. Zerguini, L.: A novel hierarchical method for decomposition and design of workflow models. Trans. SDPS **8**(2), 65–74 (2004)

A Web-Based Tool for Modeling, Simulation, and Analysis of Petri Nets with Data

Christian Imenkamp[1]([✉]) [ID], Agnes Koschmider[1] [ID], Christoph Matheja[2] [ID], and Andrey Rivkin[3] [ID]

[1] University of Bayreuth, 95447 Bayreuth, Germany
{christian.imenkamp,agnes.koschmider}@uni-bayreuth.de
[2] University of Oldenburg, 26129 Oldenburg, Germany
christoph.matheja@uni-oldenburg.de
[3] Technical University of Denmark, 2800 Kongens Lyngby, Denmark
ariv@dtu.dk

Abstract. This paper introduces an extension of YAPNE (Yet Another Petri Net Editor) – a web-based tool for modeling, simulation, and analysis of data-aware Petri nets. The aim of the tool is to serve as a teaching aid for courses covering formal aspects of process modeling and analysis, as well as courses on process mining. Additionally, the tool's support for PNML, together with its modeling and simulation capabilities, is designed to assist researchers working with Data Petri Nets. The new version of YAPNE provides improved graphical user interface and verification capabilities. Moreover, it comes with (1) improved accessibility in accordance with the Web Content Accessibility Guidelines, and (2) additional simulation and log-generation features. Finally, the paper provides a comprehensive documentation of the tool's architecture, which has not been previously published.

Keywords: Data Petri nets · Modeling · Simulation · Analysis

1   Introduction

Petri nets are oftentimes regarded as a unified framework which, on the one hand, allows to model system behavior and formally analyze its syntactical and behavioral properties, and, on the other hand, ways to scale beyond standard P/T nets by extending the "core" model with such elements and data or time. The increasing use of Petri nets in rapidly expanding fields such as process mining underscores the need for improved modeling, analysis, and simulation tools to address new challenges and requirements from researchers, educators, and industrial stakeholders.

A motivating example illustrating YAPNE's five workflows—Modeling, Import/Export, Simulation, Verification, and Log Generation—is presented in Sect. 2.

J. Desel and A. Kalenkova (Eds.): PETRI NETS 2026, LNCS 16567, pp. 359–370, 2026.
https://doi.org/10.1007/978-3-032-27879-1_18

YAPNE (Yet Another Petri Net Editor) was introduced in [5] as an integrated modeling and analysis environment for Data Petri Nets (DPNs), a data-aware extension of standard P/T nets with data variables and transition guards. While the original tool already provided web-based modeling, simulation, and soundness analysis, it lacked features for broader usability, extensibility, and advanced trace generation.

In this paper, we report on the new version of YAPNE, describing its features, architecture, and how limitations of the previous version have been addressed.

Comparison with the Previous Version. The previously published version of YAPNE [5] provided basic graphical modeling, PNML import/export, interactive simulation (allowing the user to manually fire transitions), and a control-flow-based soundness verification. The new version introduces the following extensions: (1) The tool fully implements the data soundness algorithm from [11]. (2) YAPNE integrates a probabilistic programming-based simulation engine [8,9] for generating event logs with statistical guarantees. The tool also produces simulation reports showing event logs and variable distributions (cf. Fig. 1). (3) The editor has been extended with support for inhibitor, read, and reset arcs, transition priorities and delays, and a sophisticated guard editor. (4) New import/export capabilities include JSON save/load and PNG export. (5) The tool now adheres to WCAG 2.0[1] accessibility standards. Finally, this paper provides a comprehensive report of the tool's functionalities and architecture, which was not covered in [5].

The remainder of this paper is structured as follows. Section 2 introduces a motivating example and presents YAPNE's features organized by workflow. Section 3 presents the software architecture and links it to the feature categories. Section 4 applies the workflows to the running example and evaluates scalability across benchmarks of varying complexity. Section 5 concludes with comparing YAPNE to some related tools and outlining future directions.

2   Tool Functionality

YAPNE is publicly available on https://tinyurl.com/4dd3seep and does not require any additional installation. The tool provides an integrated environment for modeling, simulating, and analyzing DPNs. Table 1 presents a comprehensive overview of the tool's features (beyond those of the standard DPN editor forming the core of YAPNE), organized by functional categories. New functionalities are marked with ★, while improved/revised ones are indicated by ◆. In the following, we elaborate in more detail on the features listed in Table 1.

Motivating Example. Consider an operations analyst in a smart factory who is investigating cases of occasional production interruptions that occurred without apparent mechanical failures. Since DPNs allow capturing both the control-flow and data-flow dimensions while also enabling formal verification of the

---

[1] https://www.w3.org/TR/WCAG20/.

resulting models [10], the analyst can use YAPNE's five workflows to investigate how combinations of sensor values and control logic may lead to such interruptions. The workflows are used as follows: *Modeling* to design the production line as a DPN with sensor variables (*temperature, vibration, load*) and safety-constraint guards; *Import/Export* to store and share the model in PNML; *Simulation* to observe system behavior under different sensor value combinations; *Verification* to check whether all valid configurations allow recovery; and *Log Generation* to produce event logs for process mining analysis. We use this IoT factory scenario as a running example throughout the paper.

In the following, we describe YAPNE's features organized by these five workflows: Modeling, Import/Export, Simulation, Verification, and Log Generation.

Modeling Workflow (F21–F26). YAPNE's core functionality is to provide a modeling environment that supports the creation of standard Petri nets (F21) and their extensions, with the current main focus being on DPNs. The tool includes rapid modeling features (F22) that, when combined with additional hotkeys, allows to efficiently design net graphs. In the "Model" panel, global DPN variables can be specified in terms of their name, type[2], and initial value (F23). When a net element is selected, the "Properties" subpanel provides additional information that is not directly available from the element toolbox on the left-hand side of the canvas. For places, one can define the current marking, a place that is bounded, and the final marking (F21). For transitions, one can specify guards (F26) (the new version of the tool provides comprehensive help on the supported expression syntax), as well as their priority and delay (F25). The latter is relevant for the log-generation functionalities described below (i.e., the system will pause for the given time before completing the firing). For arcs, one can define their weight (F21) and type (F24); the tool currently supports regular, inhibitor, reset, and read arcs. The corresponding behavior is realized in the tool's extended semantics for simulation.

Import/Export Workflow (F9–F12). YAPNE supports multiple formats for model storage and interoperability (F9–F12). For interoperability with other Petri net tools, YAPNE implements PNML import and export following the ISO/IEC 15909-2:2011 standard[3]. The PNML importer includes an integrated layout algorithm based on modified topological sorting [7] that arranges elements in a left-to-right order while minimizing arc crossings. Users can configure spacing, element sizes, and compaction options before applying the layout. YAPNE also comes with its internal JSON format, which fully preserves all model properties, including various extensions. For documentation purposes, YAPNE exports the canvas view as PNG, preserving element positions, labels, token counts, and arc decorations.

---

[2] Currently, YAPNE supports four types: boolean, integer, float, and string.

[3] https://www.iso.org/standard/43538.html.

Table 1. YAPNE feature overview organized by functional category

Category	Feature	Description
Simulation (F1–F5)	F1: Step execution	Fire single transitions manually
	F2: Auto-run	Continuous execution with configurable delay
	F3: Variable tracking★	History of data values across simulation steps
	F4: Conflict resolution★	Priority-based selection among enabled transitions
	F5: WebPPL★	Export the model as a probabilistic program
Verification (F6–F8)	F6: Soundness checking◆	P1 (reachability), P2 (termination), P3 (no dead transitions)
	F7: Counterexamples◆	Visualization of runs leading to violating states
	F8: Guard analysis◆	Detection of unsatisfiable preconditions
Import/Export (F9–F12)	F9: JSON save/load★	Native format preserving all model properties
	F10: PNML import	Parse standard PNML with automatic layout
	F11: PNML export	Serialize to interoperable format
	F12: PNG export★	Rasterized image of current canvas view
Event Logs (F13–F15)	F13: Trace generation★	Simulate multiple cases with arrival distributions
	F14: Format export★	XES, CSV, and JSON output formats
	F15: Data attributes★	Include variable values in generated logs
Accessibility (F16–F20)	F16: Keyboard navigation★	Full keyboard-only operation of canvas
	F17: Screen reader support★	ARIA labels and live announcements
	F18: High contrast mode★	WCAG-compliant color schemes
	F19: Focus management★	Virtual focus for canvas elements
	F20: Skip links★	Bypass navigation to main content
Graphical Editor (F21–F22)	F21: P/T net modeling	Create standard P/T nets
	F22: Rapid modeling◆	Disables drag-n-drop with hotkeys
Extensions (F23–F26)	F23: Data management◆	Global variables and guards
	F24: Arc types★	Support for inhibitor, read, reset arcs
	F25: Other parameters★	Support for priorities and delays
	F26: Editor for conditions★	Sophisticated editor for pre- and post-conditions

Simulation Workflow (F1–F5). YAPNE comes with both step-by-step and automatic simulation capabilities, working for all the types of nets supported by the editor (F1–F5). Based on a given model, its initial marking, and data valuation (when dealing with DPNs), the simulator produces execution traces comprising transition firings, markings, and data variable histories (again, only when DPNs are in place). YAPNE at run-time recognizes the type of a net model it is dealing with and fires transitions based on the execution semantics of the identified Petri net type. Simulation can be performed either by instructing the tool to fire one transition at a time (F1) or by generating an entire run (F2). For both execution modes, YAPNE takes into account defined transition priorities and time delays (F25). YAPNE can resolve conflicts between simultaneously enabled transitions through a priority system (F4). This feature serves as a teaching aid for beginners and can be replaced using mechanisms with more sophisticated approaches (e.g., edge weights, guards). For DPNs, YAPNE leverages the probabilistic programming-based approach from [8,9] for evaluating constraint-based post-conditions to construct next-state variable valuations. For constraints that define valid value ranges, the probabilistic engine samples values according to specified probability distributions within the corresponding bounds. YAPNE also supports exporting the model as a WebPPL probabilistic program [9]. The variable tracking feature (F3) records all value changes throughout the simulation, including probabilistic histories and sampling paths. This provides a comprehensive data evolution history for visual analysis.

Verification Workflow (F6–F8). YAPNE's soundness verification features (F6–F8) are based on the approach from [11]. In a nutshell, this approach relies on symbolic state-space exploration, where the underlying DPN is represented as a labeled transition system whose states symbolically capture reachable configurations, i.e., each state consists of a marking and a formula encoding the possible data valuations. The same algorithm can be seamlessly applied to standard P/T nets.

The soundness verification algorithm operates as follows. The DPN is translated into a labeled transition system (LTS) whose states are pairs (m, φ), where m is a marking and φ is a quantifier-free constraint encoding reachable variable valuations. The LTS is constructed incrementally: for each transition t with precondition $pre(t)$ and post-condition $post(t)$, the algorithm computes $\varphi \wedge pre(t) \wedge post(t)$ and checks satisfiability using Z3, then applies quantifier elimination (Fourier-Motzkin for real-valued domains) to project out intermediate variables. The algorithm terminates for DPNs with real-valued or finite data domains [11]; however, termination is not guaranteed for integer domains. All verification steps are fully exposed in the UI.

YAPNE's soundness checker verifies the following three fundamental properties. Property P1 ensures that every reachable state can eventually reach a designated final (control) state. Property P2 requires that, upon reaching the final marking, no transitions remain enabled. Property P3 checks that every transition can fire in at least one reachable state. These three properties can be verified for both standard P/T nets and DPNs; however, they are not supported for nets with extended arc types (F24) or transition priorities (F25).

The verification procedure generates counterexamples for each violated property (F7). For P1, states that are forward-reachable from the initial configuration but not backward-reachable from final markings constitute deadlock witnesses; the algorithm reconstructs the corresponding trace. For P2, a counterexample is a reachable state (m, φ) where m strictly covers a final marking component-wise, witnessing improper termination. For P3, transitions whose refined variants never label any LTS edge are certified as dead. This counterexample generation augments [11] without altering its behavior.

The tool also provides a detailed verification report, indicating which algorithm from [11] was used, the verification steps performed, and, if a net is deemed unsound, allows the corresponding error to be highlighted directly on the model.

Log Generation Workflow (F13–F15). To support various process mining activities, YAPNE includes a log-generation feature (F13–F15). Given a Petri net model, the generator simulates multiple process instances and records the resulting traces (F13). Users can configure the number of instances[4], arrival distributions (fixed, exponential, or normal), and timestamp-generation parameters. The generator supports three output formats (F14): XES, for compatibility with process mining tools such as ProM and PM4Py, as well as CSV and JSON. For DPNs, the generated trace sets include data attributes derived from variable values at each step (F15), enabling further process mining tasks such as the discovery of decision rules and data-aware process analysis.

The log generation is based on the DPN-to-probabilistic-program translation from [8]. A DPN is translated into a probabilistic program C_{sim} with three components: C_{init} (initialization), $B_{enabled}$ (enabled transition computation), and

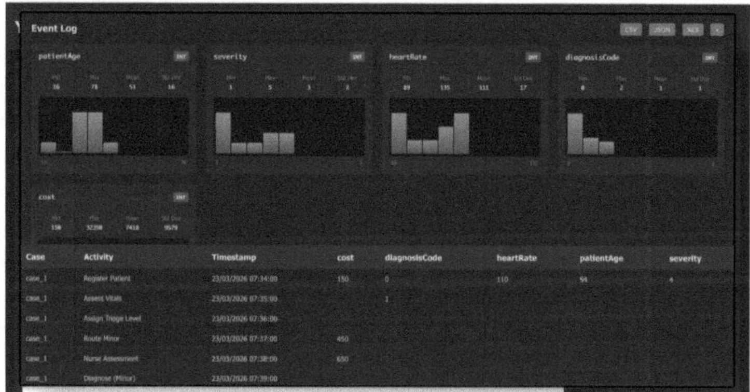

Fig. 1. Simulation report showing the generated event log and variable distributions.

---

[4] YAPNE does not support intra-instance interactions; therefore, each instance is effectively treated as a separate net.

C_{fire} (transition sampling and state update). The correctness of this translation is formally established in [8]: the program produces all legal DPN runs with the probabilities induced by the scheduler. YAPNE uses a uniform scheduler by default and performs the translation automatically using WebPPL [9]; users do not need to write WebPPL programs.

Accessibility (F16–F20). YAPNE adheres to the Web Content Accessibility Guidelines (WCAG) 2.0[5] through a dedicated accessibility layer (F16–F20). The implementation targets all four WCAG principles: perceivable (shadow DOM with ARIA labels), operable (full keyboard navigation and skip links), understandable (high-contrast mode), and robust (screen reader support through ARIA live regions). This ensures that the tool is accessible to users with physical and visual impairments.

3   Software Architecture

YAPNE's architecture is designed according to three principles: complete client-side execution, modularity through extensions, and separation of concerns between model, view, and controller components. The architecture directly supports the feature categories introduced in Table 1.

The system consists of a core Petri net library with layered extensions for DPN support, verification, accessibility, and external integrations. Each layer corresponds to a feature category from Table 1: the `Core Layer` provides basic modeling infrastructure, the DPN Extension adds data-aware simulation capabilities (F1–F5), the `Verification Layer` provides soundness checking (F6–F8), the `Accessibility Layer` enables inclusive access (F16–F20), and the `Integration Layer` handles import/export and event log generation (F9–F15).

Core Layer. The core library provides the essential components for Petri net modeling and visualization. The `PetriNet Model` component manages places, transitions, and arcs as first-class objects with properties such as token counts, arc weights, and labels. It implements traditional Petri net semantics.

`Renderer` provides canvas-based visualization with support for pan, zoom, and element highlighting. It translates the abstract model into visual representations with configurable styling. `Editor` handles user interactions, including element choice, drag-and-drop positioning, and arc creation through click-to-connect gestures. The `PetriNetAPI` serves as an intermediary coordinating high-level operations between model, renderer, and editor.

All core components are designed as "stand-alone". Therefore, it is possible to implement a custom Petri net renderer or editor using the core components for different projects.

---

[5] https://www.w3.org/TR/WCAG20/.

Verification/Accessibility Layer. This layer relies on three components.

- *Z3 Integration for Web Browsers* YAPNE uses the Z3-solver throughout the analysis pipeline (i.e., during the soundness check). Traditional approaches require server-side computation or native bindings, but YAPNE supports client-side constraint solving through WebAssembly technology. The Z3 solver is compiled to WebAssembly and executes within the browser JavaScript engines. However, Z3's WebAssembly build requires SharedArrayBuffer for multi-threaded execution, which browsers restrict for security reasons. YAPNE addresses this through a service worker that adds the required policy HTTP headers. SharedArrayBuffer is supported by all modern browsers, including Chrome, Firefox, and Edge. Therefore, YAPNE's verification features are available in all modern browsers.

- *Expression Evaluation Pipeline* In the probabilistic approach, post-conditions are interpreted as probabilistic assignments rather than hard constraints [8]. The evaluation pipeline first parses the expressions (pre- and post-conditions) to produce an abstract syntax tree (AST). This AST is then translated into probabilistic program code, where current variable values are used as the initial state. Instead of solving for satisfying assignments, the system samples next-state values for primed variables according to the specified probabilistic semantics. The resulting samples are then applied to update the model state.

- *Ally* This component (referred to as `Ally Layer` on Fig. ??) implements WCAG 2.0 compliance for canvas-based interaction (F16–F20). The `CanvasAccessibilityLayer` creates a shadow DOM mirroring canvas content with ARIA attributes, enabling keyboard navigation and screen reader support.

DPN Extension Layer. The Data Petri Net extension follows a non-invasive integration pattern where extension classes inherit from core classes and add data-aware functionality. The `DPN Model` component introduces *DataVariable* objects with typed values and *DataAwareTransition* objects that extend standard transitions with precondition and post-condition expressions.

`DPN Renderer` extends the base renderer to display visual indicators distinguishing data-aware transitions from standard ones and displaying the guard satisfaction status. `DPN API` extends the core API with operations for variable management (i.e., declarations and instantiations), guard evaluation, and constraint-based firing. This layer implements features F1–F5 from Table 1.

Integration Layer. This layer handles alignment with other tools through three components: the `PNML Importer` for model serialization, the `Event Log Generator` for creating trace collections for process mining tools (e.g., ProM, pm4py), and the `PNG Exporter` for canvas rasterization. All components follow the specifications from Sect. 2.

4   Workflow Showcasing and Tool Evaluation

This section showcases YAPNE's workflows on the IoT factory example introduced in Sect. 2. Each workflow is demonstrated through concrete steps in the use case. Additionally, we present a scalability evaluation across benchmarks of varying complexity. The explored use case can be loaded and tested in YAPNE (**"File"** → **"Examples"**).

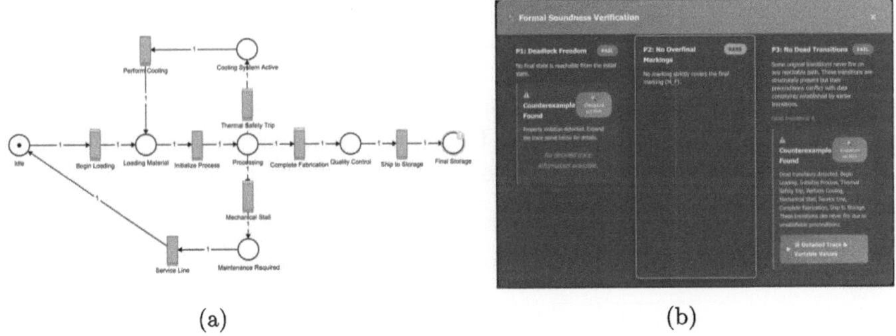

(a)   (b)

Fig. 2. (a) The data Petri net used by the analyst to find the problem in the factory. (b) The result of the soundness verification.

Use Case: *Root-Cause Analysis in an IoT-Enabled Factory.* An operations analyst in a smart factory monitors occasional interruptions in production. Sometimes, the line halts without any apparent mechanical failure. Given that, the goal is to understand how combinations of sensor values and process states lead to these interruptions.

Modeling Workflow. The analyst designs the production line as a DPN (Fig. 2a), where places and transitions represent machine states (idle, processing, cooling, maintenance). Data variables *temperature*, *vibration*, and *load* are defined in the Model panel, with guards enforcing safety constraints (e.g., cooling fires only if *temperature* $> 80 \land vibration > 5$) and post-conditions updating sensor variables.

Import/Export Workflow. The analyst saves the model in JSON, exports it to PNML for interoperability with tools such as ProM, and generates a PNG export for documentation.

Simulation Workflow. The analyst simulates the model with different sensor value combinations using step-by-step mode (F1) to fire individual transitions or auto-run mode (F2) for complete traces. The variable tracking feature (F3) records all value changes for visual analysis.

Verification Workflow. Soundness checking verifies whether all valid sensor configurations allow the line to recover (Fig. 2b). YAPNE constructs the symbolic LTS and checks properties P1–P3 using Z3. If unsound, the tool highlights violating states and provides counterexample traces.

Log Generation Workflow. The analyst generates event logs by translating the DPN into a WebPPL program and sampling traces with statistically representative sensor value distributions. The resulting XES logs enable root-cause analysis by revealing which sensor value combinations correlate with production interruptions.

Scalability Evaluation. Table 2 presents YAPNE's performance across benchmark models of varying complexity. The table reports model complexity (number of places and transitions), verification time, and log generation time.

Table 2. Scalability evaluation across benchmark models of varying complexity.

Benchmark	Complexity	Verif. (s)	Log Gen. (s)
Digital Whiteboard	Simple	0.04	<1
IoT Factory	Simple	0.12	<1
Road Fines	Medium	0.43	2.1
Sepsis	Complex	102	4.8
Hospital Billing	Complex	108	5.3

$|P|$: number of places, $|T|$: number of transitions, $|V|$: number of data variables.
Verification times were reproduced in YAPNE and are consistent with the results reported in [11]; log generation times for 100 traces. All measurements were performed in Chrome.

The verification algorithm scales well for simple and medium-complexity models (Road Fines verified in under one second) but increases significantly for complex models such as Hospital Billing (108 s) and Sepsis (102 s), consistent with [11]. Log generation scales more uniformly: the WebPPL-based simulation generates 100 traces in under 6 s even for complex models, demonstrating that the approach from [8] handles models with 30+ transitions effectively.

Table 3. Comparison with related tools

Feature	YAPNE	ADA	CPN Tools	TAPAAL	LoLA 2	WoPeD	PIPE	LogPPL
Web-based	✓	✓	–	–	–	–	–	–
Data extensions	✓	✓	✓	✓	–	–	✓	–
Soundness verification	✓	✓	✓	✓ ✓	✓	✓	–	–
Data-aware simulation	✓	–	✓	✓	–	–	–	✓
PNML support	✓	✓	✓	✓	–	✓	–	–
Event log generation	✓	–	–	–	–	✓	–	✓
WCAG accessibility	✓	–	–	–	–	–	–	–
Open source	✓	✓	–	✓	✓	✓	✓	✓
No installation	✓	✓	–	–	–	–	–	–

5   Conclusion

The new version of YAPNE provides an extended web-based environment for modeling, simulation, and analysis of standard P/T nets and Data Petri Nets (DPNs). The version extends the previous one [5] with, among other features, an improved data soundness checking algorithm [11], a probabilistic simulation and log generation via WebPPL [8,9], extended arc types, and WCAG 2.0 compliant accessibility. The scalability evaluation in Sect. 4 demonstrates that verification scales well for models with up to 30 transitions and that log generation handles models with 76+ transitions efficiently. The tool adheres to WCAG 2.0 standards to ensure accessibility for users with physical or visual impairments.

YAPNE can be compared to several related tools (see Table 3) that have close set of functionalities and/or used in the contexts similar to those from Sect. 4. CPN Tools (also known as CPN IDE) and TAPAAL are perhaps the tools most closely related to YAPNE. However, while YAPNE does not support as wide a range of net-class extensions or advanced analysis techniques as these tools, it comes with a dedicated algorithm for soundness verification and support for an advanced simulation engine. ADA [3] is the only web-based tool in the comparison that focuses specifically on DPN-related analysis tasks (e.g., first order CTL* model checking), which are currently not supported by YAPNE. At the same time, ADA is primarily aimed at design-time analysis and does not support integrated modeling and simulation. We believe that YAPNE can serve as a versatile resource for both educational purposes and academic research in process analysis and mining. At the same time, a comprehensive usability study is needed to assess and demonstrate the tool's effectiveness in teaching and research contexts. Moreover, YAPNE could be extended with more advanced verification capabilities, such as those currently supported by ADA.

Acknowledgments. This work received funding by the Deutsche Forschungsgemeinschaft (DFG), grant 496119880.

References

1. Dingle, N.J., Knottenbelt, W.J., Suto, T.: Pipe2: a tool for the performance evaluation of generalised stochastic petri nets **36**(4), 34–39 (2009)
2. Dubois, T., Larsen, K.G., Srba, J.: Statistical model checking of stochastic timed-arc petri nets. In: Amparore, E., Mikulski, Ł. (eds.) In Proceedings of PETRI NETS'25. pp. 174–196. Springer, Cham (2025). https://doi.org/10.1007/978-3-031-94634-9_9
3. Felli, P., Montali, M., Winkler, S.: CTL model checking for data-aware dynamic systems with arithmetic. In: Blanchette, J., Kovács, L., Pattinson, D. (eds.) Automated Reasoning, pp. 36–56. Springer, Cham (2022). https://doi.org/10.1007/978-3-031-10769-6_4
4. Freytag, T., Sänger, M.: WoPeD - an educational tool for workflow nets. In: CEUR Workshop Proceedings, vol. 1295, pp. 31–40 (2014). https://ceur-ws.org/Vol-1295/paper6.pdf
5. Imenkamp, C., Kuhn, M., Grüger, J., Matheja, C., Rivkin, A., Koschmider, A.: YAPNE: a tool for modeling and automated verification of data petri nets. In: CEUR Workshop Proceedings, vol. 4088 (2025)
6. Jensen, K., Kristensen, L.M., Wells, L.: Coloured petri nets and CPN tools for modelling and validation of concurrent systems. Int. J. Softw. Tools Technol. Transfer **9**(3), 213–254 (2007)
7. Kitzmann, I., König, C., Lübke, D., Singer, L.: A simple algorithm for automatic layout of BPMN processes. In: 2009 IEEE Conference on Commerce and Enterprise Computing, pp. 391–398 (2009)
8. Kuhn, M., Grüger, J., Matheja, C., Rivkin, A.: Data petri nets meet probabilistic programming. In: Business Process Management, pp. 21–38. Springer, Cham (2024). https://doi.org/10.1007/978-3-031-70396-6_2
9. Kuhn, M., Grüger, J., Matheja, C., Rivkin, A.: LogPPL: a tool for probabilistic process mining. In: CEUR Workshop Proceedings, vol. 3783 (2024)
10. Montali, M.: Automated reasoning for data-aware petri nets. In: Amparore, E.G., Mikulski, L. (eds.) Proceedings of PETRI NETS'25. Lecture Notes in Computer Science, vol. 15714, pp. 1–17. Springer (2025). https://doi.org/10.1007/978-3-031-94634-9_1
11. Suvorov, N.M., Lomazova, I.A.: Verification of data-aware process models: checking soundness of data petri nets. J. Log. Algebraic Methods Program. **138** (2024)
12. Wolf, K.: Petri net model checking with LoLa 2. In: Proceedings of PETRI NETS'18, pp. 351–362 (2018)

Netgrif Platform: A Tool for Executable Models of Object-Centric Processes in Petriflow Language

Gabriel Juhás[1]([✉]) [iD], Juraj Mažári[2] [iD], Tomáš Kováčik[2] [iD],
Milan Mladoniczky[1] [iD], and Matej Chvostek[2] [iD]

[1] Faculty of Informatics, Pan-European University, Bratislava, Slovakia
`gabriel.juhas@paneurouni.com`
[2] NETGRIF, s.r.o., Bratislava, Slovakia
`netgrif@netgrif.com`

Abstract. Traditional process modelling often suffers from separation between data structures, business logic, and presentation layer. In this paper we introduce the Netgrif Platform - an application development platform designed for modelling and execution of object-centric processes using the Petriflow language, that integrates these three layers into a single human-readable and executable model. The Netgrif Platform enables rapid development of business-aligned processes and complex applications by providing this vertical integration of layers.

Keywords: Object-Centric Process · Petri nets · Application development

1 Introduction

Current advancements in process science have established Object-Centric Processes (OCP) [7,11], represented by Object-Centric Petri Nets (OCPN) [1], as a specialized subclass of Colored Petri Nets (CPN) [12] where tokens are restricted to unique object identifiers. While OCPNs excel at capturing the 'forest' of concurrent interactions as a runtime execution model, they often result in complex, monolithic structures that are difficult to re-implement.

To understand the spectrum of object-aware process modeling, we categorize Petri net-based models into three categories based on whether they model the runtime of a system or the definition of object types:

Object-Centric Processes (The Observational Forest): Represented by OCPN, this is a top-down approach where the entire 'forest' of interactions is modeled in a single net. Here, objects are merely IDs (tokens) flowing through a shared environment, primarily used for descriptive analysis.

Funded by the EU NextGenerationEU through the Recovery and Resilience Plan for Slovakia under the project No. 09I03-03-V04-00493.

J. Desel and A. Kalenkova (Eds.): PETRI NETS 2026, LNCS 16567, pp. 371–382, 2026.
https://doi.org/10.1007/978-3-032-27879-1_19

Embedded Object Processes (Trees Embedded in a Forest): Represented by the Nets-within-Nets paradigm [19], where object nets act as tokens within a higher-level system net. While this approach introduces modularity through a hierarchical structure, the object 'trees' remain embedded within and dependent on a predefined global 'forest' (the system net) to orchestrate their lifecycle and cross-object interactions.

Process-Centric Objects (PCO) (The DNA of Trees): Represented by the Petriflow language, this category focuses on modeling object types (classes) by adding a lifecycle directly to the class definition. We recognize Proclets (Lightweight Nets) as an early, seminal attempt at this approach, modeling independent process fragments that interact via messages. Petriflow evolves this concept into autonomous, encapsulated blueprints where classes define the potential relationships and interactions of their objects. PCO approach aligns with the vision given in [18], where the authors extend Petri nets into modular components by formalizing interfaces and composition rules for net modules.

In PCO approach, the classes define the potential relationships and interactions of their objects. The 'forest', i.e. the runtime application is not a static model; it is an emergent running program created by the interaction of objects from independent Process-Centric Classes composed like Lego bricks.

These classes can be utilized in two primary ways. First, as in object-oriented programming, to design an application from scratch. Second, as part of a two-step reverse engineering framework:

1. Step 1: Process Mining from raw object-centric event logs into an OCPN runtime model (The Forest).
2. Step 2: Synthesizing these observations into Process-Centric Classes (The DNA) from the OCPN runtime model.

This transition from a descriptive 'forest' to a prescriptive 'design model' enables the automated generation of scalable, executable software architectures directly from observed real-world behavior. Thus, Process-Centric Object-oriented languages like Petriflow are not competing with **object-centric process runtime models like OCPN**, but rather complementing them by providing a direct path to implementation.

Netgrif Platform was designed for modelling and execution of process-centric objects in language Petriflow [15]. Netgrif Platform is used to run many enterprise applications in different industry fields like healthcare, insurance, telecommunications, electric utility, and others. The platform is also used on multiple universities for teaching and research purposes and management of internal processes, namely on the Faculty of Informatics and Information Technologies of the Slovak University of Technology, the Faculty of Informatics of the Pan-European University in Bratislava, the Faculty of Mathematics and Computer Science of the FernUniversität in Hagen, and the Faculty of Applied Computer Science of the University of Augsburg.

2   Petriflow Language

Inspired by object-oriented programming (OOP), the basic building blocks of the *Petriflow* language are **Petriflow classes**. A Petriflow class serves as a blueprint that encapsulates three integrated layers: *data attributes*, a *lifecycle* in the form of extended Petri nets, and *user interfaces* (UI) as forms associated with tasks.

Data attributes are XML-based objects defined by their type, identifier, title, and various validations. The lifecycle layer is represented by place/transition nets [8–10], enhanced with inhibitor [4], read [21], and reset arcs [2]. Furthermore, Petriflow supports variable arc weights determined by data attribute values or place markings, a concept inspired by self-modifying nets [20]. The UI layer consists of forms, which are subsets of data attributes defining permissions such as required, editable, or read-only for each specific task.

The novelty of this paradigm lies in the vertical integration of these layers within a single model. The Petriflow interpreter automatically handles communication between the client-side device, the application server, and the persistent database. This automation results in compact, all-in-one, human-readable models that ensure business-aligned execution while enabling the realization of complex relationships between different objects.

Relations between **cases** (instances of Petriflow classes) are modeled at the data layer. Data attributes can store references to one or multiple related cases, mirroring concepts like foreign keys in SQL, object references in OOP, or 1:N relationships in class diagrams.

The Petriflow language defines events for classes, cases, lifecycle tasks, and data attributes. These events can be triggered manually by users, automatically by the system, or by *actions*—anonymous Groovy code snippets—as reactions to other events. For example, creating a new case triggers a *create case event*, which generates a unique instance (copy) of the underlying Petri net and data attributes defined in the Petriflow class.

Tasks, represented as transitions, follow semantics based on the *first-consume-then-produce* principle [16]. Unlike standard Petri nets where firing is an atomic operation, here it is decomposed into three discrete logical events:

Assign (Start): Assigns a user to the task and consumes tokens from the preset of the transition.

Finish: Completes the task and produces tokens in the postset of the transition.

Cancel: Aborts the task, unassigns the user, and returns the tokens consumed during the assign event back to the preset.

Data attributes feature *set* events for value modifications and *get* events for data retrieval. Access to these events is governed by role-based access control (RBAC) [5].

When an event is triggered, an action can be executed in either the *pre-phase* or the *post-phase*. Actions in the *pre-phase* operate on the state (data and markings) as it existed before the event, while *post-phase* actions work with the updated state. These actions can trigger a chain of subsequent events across different cases.

By combining data-level references with event synchronization, Petriflow enables sophisticated coordination. For instance, finishing a task in a parent case can programmatically trigger *finish* events for tasks in referenced child cases. Because data attributes can store task references, Petriflow allows the dynamic embedding of referenced task forms as sub-forms within a parent form.

Finally, similar to SQL in relational databases, the Petriflow language provides a powerful query language to create dynamic filters over cases and their tasks [13,14,17].

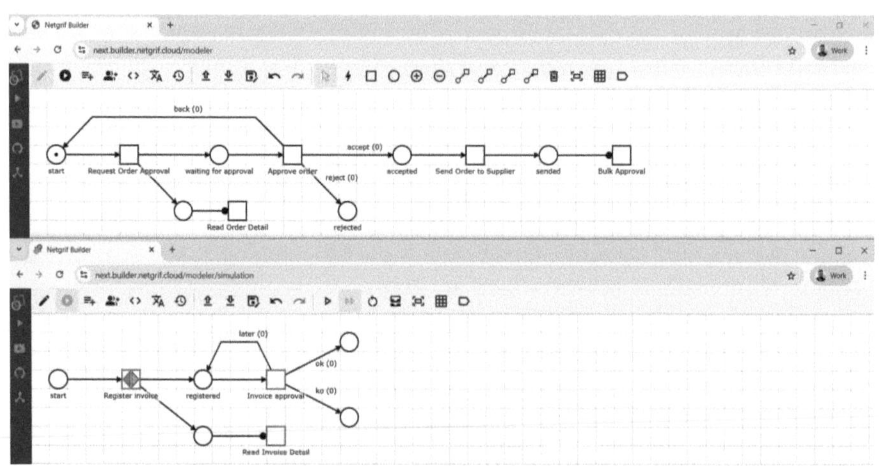

Fig. 1. The Netgrif Application Builder interface showing the Petri net lifecycle for the Order class in editing mode and for Invoice class in simulation mode with transition (task) Register invoice being assigned to a user.

3 Functional Specification of Order and Invoice Management

To illustrate a process-based application in the Petriflow language, created and executed on the Netgrif platform, we use a practical example of managing orders and their associated invoices. The system consists of two main process classes: *Order* and *Invoice*. The life-cycle of the classed is depicted in Fig. 1.

The Order Process Lifecycle. The *Order* class contains data attributes such as *Order ID, Subject,* and *Supplier.* It also includes an attribute for the *Approval Decision,* implemented as an **enumeration map** (with options [accept:Accept], [back:Back], and [reject:Reject]).

When a new *Order* is created, the *Order ID* is automatically set. Following initialization, in form associated with the transition (task) *Request Order Approval* an employee fills out the order details. Subsequently, an approving manager executes the *Order Approval* task. Form of this task contains the enumeration map *Decision*.

The selection in the enumeration map determines the subsequent path of the order through the use of **variable arcs**. Upon finishing the *Order Approval* task an action updates the value of attributes *accept, finish* and *back* determining the weights of the corresponding variable arcs.

Based on the decision, the token is moved accordingly:

- **Accept**: The token moves to the place *accepted* (p_4). Subsequently, the task *Send Order to Supplier* must be executed to move the token to the *sended* state (place p_5).
- **Reject**: The token moves to the place *rejected*.
- **Back**: The token is returned to the *start* place, allowing for modifications.

If accepted, the order moves to the *sended* state (represented by a token in place p_5) by assigning and finishing task *Send Order to Supplier*, making it available for *Bulk Approval* of associated invoices.

Invoice Registration and Linkage. Similar to the *Order* process, the *Invoice ID* is automatically assigned by a script executed in response to the `createCase` event upon the instantiation of a new *Invoice* object. Subsequently, a person with the role *Invoice Processor* can assign themselves the *Register Invoice* task.

In response to the `assign` event of this task, an action is executed to perform a lookup of all active *Order* instances that currently hold a token in the place p_5 (labeled *sended*). This filter ensures that only orders already dispatched to a supplier are available for selection. The *Order ID* values of these retrieved instances are then populated as keys into the *Parent Order ID* enumeration map of the invoice:

```
// Triggered on 'assign' event of Register Invoice
def orders = findCases {
    it.processIdentifier.eq(workspace + "order")
    .and(it.activePlaces.get("p5").eq(1))
}.collectEntries {  [(it.stringId): "Order: " + it.stringId] }
change parent_order_id options { orders }
```

The transmission of the *Invoice ID* to the parent *Order* occurs when the task is finished:

```
// Triggered on 'finish' event of Register Invoice
def parent_order_case=findCase({it._id.eq(parent_order_id.value)});
setData("t1", parent_order_case,
["new_invoice_id": ["value": invoice_id.value, "type": "text"]]);
```

In the *Order* class, a reactive action on the `set` event of the *new_invoice_id* attribute ensures the ID is added to the collection of child cases:

```
// Triggered on 'set' event of new_invoice_id in the Order class
if (new_invoice_id.value !in children_invoice_cases.value) {
    change children_invoice_cases value {
        children_invoice_cases.value + new_invoice_id.value } }
```

Continuation of the Invoice Lifecycle. Following registration, the *Invoice* enters its approval phase. In the form associated with the task *Invoice approval* the approver interacts with an *enumeration map* (with options [OK:Accept], [later:Later], and [KO:Reject]). Similarly to the *Order* approval, a `finish` event action updates the weights of variable arcs (*ok*, *ko*, and *later*) based on the selected decision.

This logic directs the token to the corresponding state: **OK** moves the token to the *accepted* place, **KO** to the *rejected* place, and **later** returns it to the *registered* place for future reconsideration.

Bulk Approval and Sub-form Integration. The final stage is the *Bulk Approval* task in the *Order* process. On the `assign` event, the system identifies and assigns all unassigned *Invoice Approval* (transition t2) tasks:

```
// Triggered on 'assign' event of Bulk Approval
change invoice_approvals value { findTasks{
        (it.caseId.in(children_invoice_cases.value))
        .and(it.transitionId.eq("t2"))
        .and(it.userId.isNull()) }?.collect{it.stringId} }
invoice_approvals.value.each{ id ->
    def task = findTask({it._id.eq(id)}); assignTask(task) }
```

The situation is illustrated in Fig. 2.

The manager interacts with embedded sub-forms. Upon finishing the *Bulk Approval*, the system programmatically completes all child tasks, triggering their individual variable arc logic:

```
// Triggered on 'finish' event of Bulk Approval
def list_of_tasks_ids = invoice_approvals.value;
change invoice_approvals value { []; }
list_of_tasks_ids.each{ id ->
    def task = findTask({it._id.eq(id)}); finishTask(task) }
```

3.1 Discussion and Key Architectural Concepts

The presented example illustrates how complex relationships between independent process instances are managed at the data and event levels within the Petriflow framework. Several key architectural patterns can be identified:

Data-Level Relationships: The system maintains a bidirectional link where the parent *Order* stores a list of child *Invoice* IDs, and each *Invoice* maintains a reference to its parent. This allows for efficient traversal of the object hierarchy.

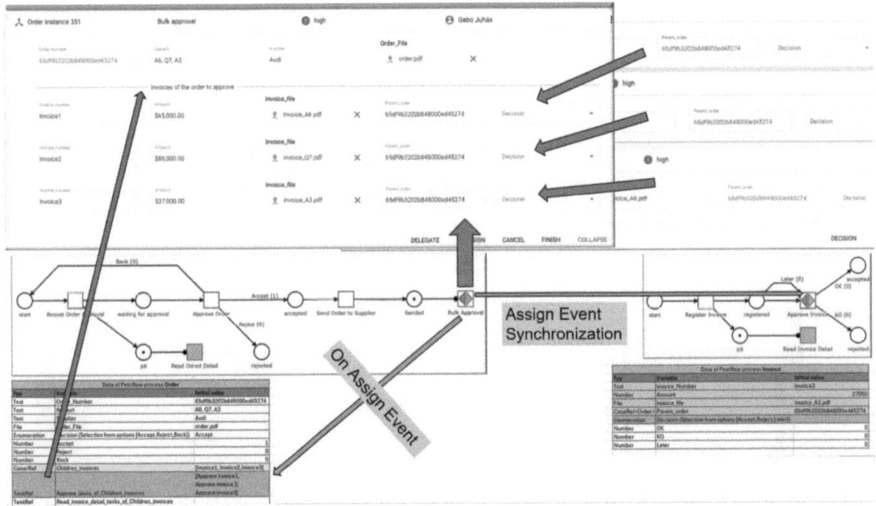

Fig. 2. Detailed synchronization logic between *Order* and *Invoice* objects. The *Assign Event* in the parent process triggers a lookup of child references and dynamically composes the integrated user task.

Dynamic Object and Task Discovery: Using powerful query functions like `findCases` and `findTasks`, the system can dynamically locate objects and their current states (active tasks) based on attributes or place markings.

Event-Based Synchronization: By chaining event triggers (e.g., a `finish` event in one process invoking a `finish` in another), the system achieves seamless synchronization between distinct Petri net instances. Task pointers (Task References) serve as the fundamental mechanism for this inter-process communication.

Hierarchical UI Composition: The ability to store task pointers as data attributes, combined with the platform's rendering engine, allows for the creation of composite forms. Sub-forms of child objects are dynamically embedded into the parent's context, providing a unified user experience for complex operations like bulk approvals.

In conclusion, this approach demonstrates that Petriflow moves beyond traditional monolithic Petri nets by treating process instances as reactive objects that can be dynamically linked, queried, and synchronized through a robust event-driven architecture.

4   Netgrif Platform Architecture

The *Netgrif Platform* provides a model-driven environment for process-based applications. As illustrated in Fig. 3, the architecture is logically divided into the design environment and the execution engine.

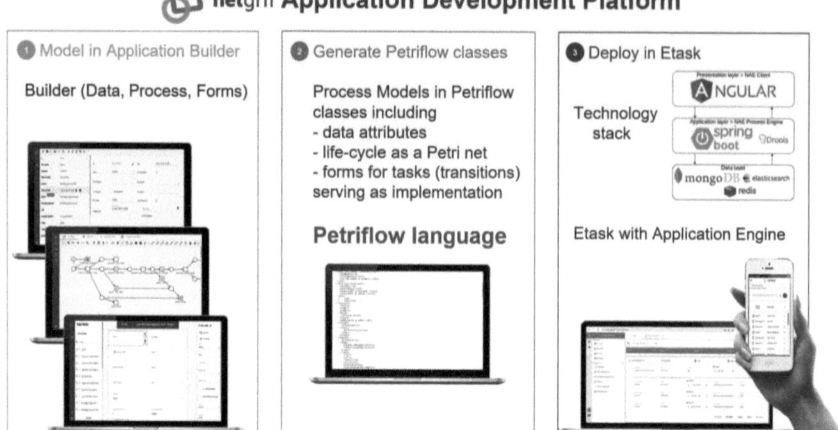

Fig. 3. The Netgrif Application Engine architecture: A cloud-native stack comprising Spring Boot services, MongoDB for persistence, and Elasticsearch for object-centric querying.

4.1 Modeling Component: Netgrif Application Builder

The *Application Builder* is a specialized web-based IDE for visual modeling of *Petriflow classes*. It serves as a standalone environment where designers define the structural aspects of a process: data attributes, Petri net lifecycles, and form layouts. The output is a *Petriflow* XML file—a transportable, technology-agnostic artifact that encapsulates the entire application logic.

The newest version of the *Application Builder* is freely available for public use at https://next.builder.netgrif.cloud.

4.2 Execution Component: Netgrif Application Engine

The core of the execution environment is the *Netgrif Application Engine* (NAE), which acts as a dynamic interpreter for Petriflow models uploaded via the *eTask* interface. The NAE follows a robust three-tier technical stack:

Presentation Layer: An **Angular** client that dynamically renders the UI based on the process marking and form definitions.

Application Layer: A **Java (Spring Boot)** backend that manages the process lifecycle, integrates **Drools** for rule evaluation, and executes Groovy scripts.

Data Layer: A hybrid persistence strategy using **MongoDB** for case states, **Elasticsearch** for advanced querying, and **Redis** for session caching.

The execution environment is accessible after registration at https://etask. netgrif.cloud, where users can create their own workspaces and deploy Petriflow models.

4.3   The Modeling Workflow: Netgrif Application Builder

The *Netgrif Application Builder* (NAB) serves as a specialized IDE for the visual design of *Petriflow classes*. The core functional workflow is centered around three integrated phases:

Data and Lifecycle Definition: In the first phase, the designer defines two independent layers of the class. First, the **data attributes** (e.g., *Order ID, Price*) are specified to form the object's data structure. Second, the **lifecycle** is modeled as an extended Petri net (see Fig. 1), defining abstract states given by markings and tasks given by transitions.

Binding, Roles, and UI Composition: The integration occurs by binding data and permissions to the process. Forms are created as views on data attributes and bound to transitions. NAB allows designers to drag-and-drop attributes into forms, automatically synchronizing them with the data model. Crucially, this phase also includes **Role Definition**, where user roles (e.g., *Manager, Accountant*) are created and assigned to specific transitions. Designers can define fine-grained permissions for each task's events (e.g., who can *assign, cancel*, or *finish* a task), ensuring a robust RBAC model within the application.

Logic Implementation and Interactive Simulation: Complex business rules are scripted via Groovy actions attached to events. To verify the integrated model, NAB provides an **interactive simulation** mode. This allows for manual triggering of assign, finish and cancel events of transitions to play token game to manually check the correctness of the life-cycle.

4.4   Deployment and Runtime Interpretation

Once the XML artifacts are uploaded to *eTask*, the platform immediately instantiates the application without recompilation.

Instance and Task Management. The engine manages concurrent instances (cases) within a unified workspace. It dynamically calculates enabled transitions, i.e. available tasks for each user based on the Petri net marking and RBAC permissions.

Cross-Process Coordination. The runtime's unique capability is demonstrated during the **Bulk Approval** of invoices (see Fig. 4). When a finish event is triggered in the *Order* case, the engine automatically synchronizes this event with all referenced *Invoice* cases, ensuring atomic and consistent state updates across the process hierarchy.

5   Related Work and Comparison

The *Netgrif Platform* occupies a unique niche by integrating formal modeling with rapid application development in *NAB* and execution in *eTask*.

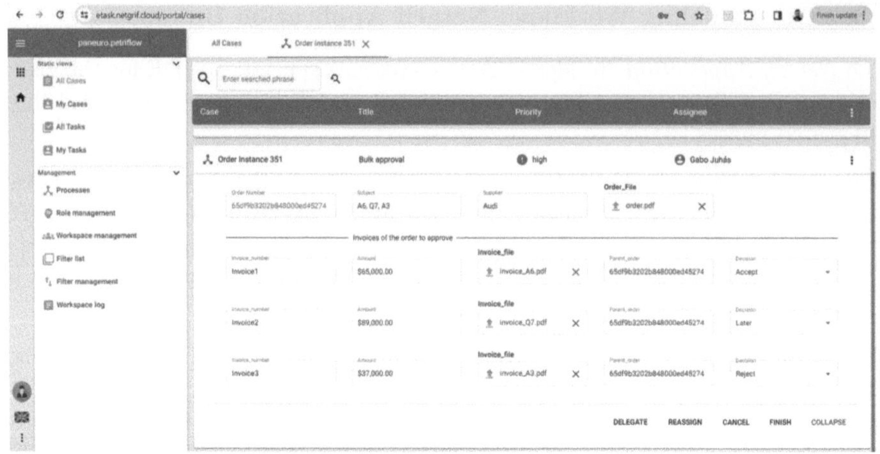

Fig. 4. A dynamic form in eTask demonstrating Bulk Approval: parent process (Order) synchronizing events across multiple child objects (Invoices).

Petri Net Tools: Unlike *CPN Tools* [22], *Renew* [6], or *ProM* [3], which prioritize formal analysis and model checking, the *NAB* is a **generative** environment. Its primary goal is the visual composition of executable *Petriflow classes*, avoiding manual XML coding and bridging the gap between theory and runtime.

BPMN and Process Engines: Compared to business-oriented modelers (*Signavio, Bizagi*) or engines like *Camunda*, Netgrif provides tighter vertical integration. While *Camunda* often requires external services for persistence and frontend, the *eTask* powered by *NAE* natively manages the entire state, including MongoDB persistence and Elasticsearch indexing within a single artifact.

Low-Code Platforms: Unlike *Microsoft Power Apps* or *Mendix*, which often rely on fragmented logic (e.g., combining with Power Automate and external DBs), Netgrif offers a **process-centric object-oriented** paradigm. It treats the Petri net not as documentation, but as the "DNA" of the application, where data, logic, and UI are inseparable. In conclusion, Netgrif is a platform designed for seamless object-lifecycle coordination through a formal foundation.

6   Conclusion

We presented the *Netgrif Platform* as a model-driven environment for object-centric processes using *Petriflow*. Unlike traditional BPM, it provides vertical integration of lifecycles, data, and UIs in executable artifacts. The platform is proven in **academia** for teaching process-driven application development and in **industry** for scalable applications in the healthcare, insurance, utility, telecom, and public sectors.

Evolution and Future Roadmap. The platform's development continues toward a fully integrated ecosystem. While the *Builder* and *eTask* currently require manual XML synchronization, our immediate roadmap includes a unified API-based integration to enable seamless, direct deployment from the modeling to the execution environment. Furthermore, we aim to simplify the query language and introduce graphical modeling for class inheritance, interfaces and cross-class synchronization. By bridging the gap between descriptive object-centric models and prescriptive executable applications, the Netgrif Platform offers a path toward a next-generation paradigm of process-centric software engineering.

References

1. van der Aalst, W.M.P., Berti, A.: Discovering object-centric Petri Nets. In: Fundamenta Informaticae (2020). https://doi.org/10.48550/arXiv.2010.02047
2. van der Aalst, W.M.P., et al.: Soundness of workflow nets with reset arcs. Trans. Petri Nets Other Model. Concurr. **3**, 50–70 (2009). https://doi.org/10.1007/978-3-642-04856-2_3
3. van Dongen, B.F., de Medeiros, A.K.A., Verbeek, H.M.W., Weijters, A.J.M.M., van der Aalst, W.M.P.: The ProM framework: a new era in process mining tool support. In: Ciardo, G., Darondeau, P. (eds.) ICATPN 2005. LNCS, vol. 3536, pp. 444–454. Springer, Heidelberg (2005). https://doi.org/10.1007/11494744_25
4. Alqarni, M., Janicki, R.: On interval process semantics of petri nets with inhibitor arcs. In: Devillers, R., Valmari, A. (eds.) PETRI NETS 2015. LNCS, vol. 9115, pp. 77–97. Springer, Cham (2015). https://doi.org/10.1007/978-3-319-19488-2_4
5. Bergenthum, R., Desel, J., Mauser, S.: Workflow nets with roles. In: EMISA, vol. P-190. LNI. GI, pp. 65–78 (2011). https://www.fernuni-hagen.de/sttp/docs/2011workflownetswithrolesemisasubmitted.pdf
6. Cabac, L., et al.: Renew: the reference net workshop. Trans. Petri Nets Other Models Concurr. II, 1–25 (2009). https://ceur-ws.org/Vol-1372/paper18.pdf
7. De Weerdt, J., Pufahl, L. (eds.) Business Process Management Workshops. LNBIP. Springer, Cham (2024). https://doi.org/10.1007/978-3-031-50974-2
8. Desel, J.: Process modeling using petri nets. In: Process-Aware Information Systems, pp. 147–177. Wiley, New York (2005). https://doi.org/10.1002/0471741442.ch7
9. Desel, J., Juhás, G.: "What Is a Petri net?" Informal answers for the informed reader. In: Ehrig, H., Padberg, J., Juhás, G., Rozenberg, G. (eds.) Unifying Petri Nets. LNCS, vol. 2128, pp. 1–25. Springer, Heidelberg (2001). https://doi.org/10.1007/3-540-45541-8_1 ISBN 978-3-540-45541-7
10. Desel, J., Reisig, W.: The concepts of Petri nets. Softw. Syst. Model. **14**(2), 669–683 (2014). https://doi.org/10.1007/s10270-014-0423-3
11. Gdowska, K., Gómez-López, M.T., Rehse, J.-R. (eds.) Business Process Management Workshops. LNBIP. Springer, Cham (2025). https://doi.org/10.1007/978-3-031-78666-2
12. Jensen, K., Kristensen, L.M.: Coloured Petri Nets. Springer, Heidelberg (2009). https://doi.org/10.1007/b95112
13. Juhás, G., Juhásová, A., Petrovič, L.: Low-code languages in IT education: integrating theory and practice. In: Proceedings of ICETA 2023–21st Year of International Conference on Emerging eLearning Technologies and Applications, pp. 249–257. IEEE (2023). https://doi.org/10.1109/ICETA61311.2023.10343807

14. Juhás, G., et al.: Low-code platforms and languages: the future of software development. In: Proceedings of 20th Anniversary of IEEE International Conference on Emerging eLearning Technologies and Applications, ICETA 2022, pp. 286–293. IEEE (2022). https://doi.org/10.1109/ICETA57911.2022.9974697
15. Juhás, G., et al.: Petriflow language and Netgrif application builder. In: BPM (PhD/Demos), vol. 2973. CEUR Workshop Proceedings, pp. 171–175. CEUR-WS.org (2021). https://ceur-ws.org/Vol-2973/paper_284.pdf
16. Juhás, G.: On semantics of Petri nets over partial algebra. In: Pavelka, J., Tel, G., Bartošek, M. (eds.) SOFSEM 1999. LNCS, vol. 1725, pp. 414–422. Springer, Heidelberg (1999). https://doi.org/10.1007/3-540-47849-3_29
17. Juhás, G., Mladoniczky, M., Petrovic, L.: Practical experience with petriflow: enriched process models serving as implementation. In: Modellierung 2024 - Workshop Proceedings, Potsdam, Germany, 12–15 March 2024. Ed. by Holger Giese and Kristina Rosenthal. Gesellschaft für Informatik e.V., p. 24 (2024). https://doi.org/10.18420/MODELLIERUNG2024-WS-024
18. Kindler, E., Petrucci, L.: Towards a standard for modular petri nets: a formalisation. In: Franceschinis, G., Wolf, K. (eds.) PETRI NETS 2009. LNCS, vol. 5606, pp. 43–62. Springer, Heidelberg (2009). https://doi.org/10.1007/978-3-642-02424-5_5
19. Kummer, O., Wienberg, F., Moldt, D.: Renew: the reference net workshop. In: Tool Presentations, 25th International Conference on Applications and Theory of Petri Nets (2004). https://www2.informatik.uni-hamburg.de/tgi/publikationen/public/data/2003/Kummer+03/Kummer+0
20. Valk, R.: Self-modifying nets, a natural extension of Petri nets. In: Ausiello, G., Böhm, C. (eds.) ICALP 1978. LNCS, vol. 62, pp. 464–476. Springer, Heidelberg (1978). https://doi.org/10.1007/3-540-08860-1_35
21. Vogler, W.: Partial order semantics and read arcs. Theor. Comput. Sci. **286**(1), 33–63 (2002). https://doi.org/10.1016/S0304-3975(01)00234-1
22. Westergaard, M.: The CPN tools 4: multi-formalism and scalability. Appl. Theory Petri Nets Concurr. **7927**, 400–409. LNCS (2013). https://ceur-ws.org/Vol-1021/paper_3.pdf

Petri-Dish: A Petri Net Survey Tool for Education and Research

Marc Kimmel[(✉)], Patrizia Schalk, and Robert Lorenz

University of Augsburg, Universitätsstraße 6a, 86159 Augsburg, Germany
{marc.kimmel,patrizia.schalk,robert.lorenz}@uni-a.de

Abstract. Petri nets are commonly used in various applications, such as software engineering or business process modeling. Naturally, new personnel in these fields need to acquire knowledge about Petri net semantics to use them effectively. Traditional teaching strategies rely on static, non-interactive exercises that require manual preparation and assessment. Although there are tools available to quickly construct Petri nets, none of them focuses specifically on creating and grading exercises. In this paper, we thus present Petri-Dish, an editor for Petri net exercises and case studies. For teaching personnel, this tool offers quick creation of questionnaires and automatic assessment. By providing basic metrics for the constructed Petri nets, it enables its users to prepare exercises in the desired complexity with low effort. Furthermore, Petri-Dish provides an overview of all students' solutions, thus highlighting which topics were not yet well understood. For students, the tool provides an interactive interface that allows free rearrangement of model elements. These features also enable users to easily execute case studies, as demonstrated by two successful projects on Petri net understandability with Petri-Dish.

Keywords: Place Transition Nets · Teaching · Education · Survey Tool

1 Introduction

Petri nets are useful to model, analyze, and improve processes found in fields such as biology [10], systems engineering [6], and process modeling [1]. Consequently, new personnel and students in these fields must acquire a solid understanding of the syntax and semantics of Petri nets to use them effectively. Traditional teaching strategies rely on static and non-interactive exercises. These methods require manual preparation and assessment, making them prone to errors, time-consuming, and delaying feedback for students. There are Petri net tools [19] available that ease the task of constructing nets, but none of them focuses on the creation or automatic grading of exercises. Such a tool would be useful not only in teaching, but also in performing empirical studies on Petri nets in research.

This paper therefore presents *Petri-Dish*, an environment specifically designed to streamline the lifecycle of Petri net exercises and case studies. Petri-

© The Author(s), under exclusive license to Springer Nature Switzerland AG 2026
J. Desel and A. Kalenkova (Eds.): PETRI NETS 2026, LNCS 16567, pp. 383–394, 2026.
https://doi.org/10.1007/978-3-032-27879-1_20

Dish focuses specifically on place/transition nets, since they are the most frequently used Petri net class in practice and teaching. The tool offers the following features and advantages over traditional methods:

- a minimalistic and easy-to-use interface to create Petri nets or import them from PNML files, and to create questions about these nets;
- calculation of metrics based on the Petri net's structure or semantics, like the density of arcs or maximum amount of tokens in a reachable marking;
- automatic evaluation and grading of student submissions, presented in a clear and detailed user interface as shown in Fig. 1; and
- an interactive environment where elements of the net can be moved freely.

Petri-DishâĂŹs interactive design offers a playful experience and integrates established empirical question types, enabling comparisons with existing studies in literature. We evaluated the tool in two case studies investigating structural factors influencing Petri net understandability. The results highlight Petri-Dish's versatility in educational and research settings.

The remainder of this paper details the architecture and functionality of Petri-Dish. Section 2 describes Petri-Dish's functionalities from two points of view: that of instructors creating exercises and that of participants solving them. Afterward, Sect. 3 describes the tool's architecture and its extensibility. Section 4 then highlights two use cases for the tool in the context of education and empirical research, before Sect. 5 compares the tool with other ones in the literature. Finally, Sect. 6 concludes the paper and outlines future work, followed by Sect. 7 on the tool's availability and installation.

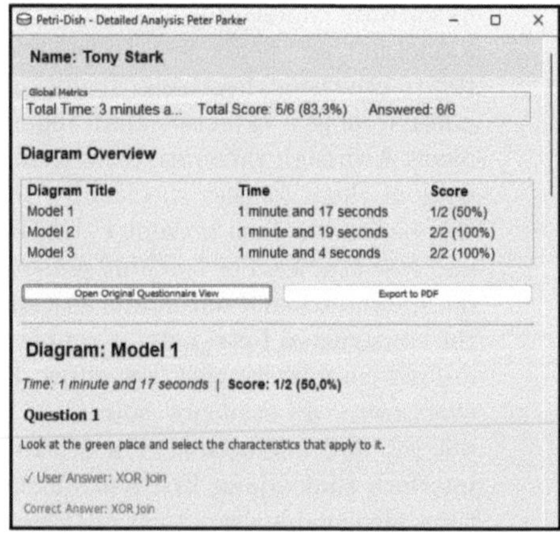

Fig. 1. GUI for evaluating survey responses.

2 Functionality

Petri-Dish simplifies the creation, distribution, and evaluation of exercises and case studies for Petri nets. For exercise creation, Petri-Dish automatically calculates structural and non-structural measures [11,17] for the nets, ranging from simple node counts to complex metrics such as cross-connectivity [20]. To mitigate the effect of the model's layout on students' performance, the tool allows users to freely drag-and-drop model elements. Reflecting these functionalities, Petri-Dish serves two distinct user groups: instructors creating surveys and students completing them. Consequently, Petri-Dish supports two different views:

- the *Survey-Creator* for educators and researchers, which offers extensive functions for creating, distributing, and evaluating surveys; and
- the *Survey-Player* for students or case study participants, a lightweight application dedicated solely to completing surveys.

This section describes the functionalities of both components separately.

Exercise and Survey Creation. Instructors can use the Survey-Creator to design exercises or questions for case studies. Since the exact use case does not matter for Petri-Dish, we call such a collection of questions or exercises a *survey* in the remainder of this section. Surveys contain multiple *pages*. Each page consists of a Petri net and a set of questions on this net. The instructor can create Petri nets by specifying the places, transitions, and arcs of the net, or by importing a net in PNML format. After creating or importing a net, Petri-Dish uses Graphviz algorithms [8] to apply standard layout techniques. After choosing a layout, the instructor can drag and drop model elements to further improve readability. Petri-Dish stores the resulting node positions and shows the net to participants exactly as specified by the instructor. By default, Petri-Dish uses the standard color palette for Petri nets in TikZ format [18]. It is possible to change this color palette or to color each place or transition individually.

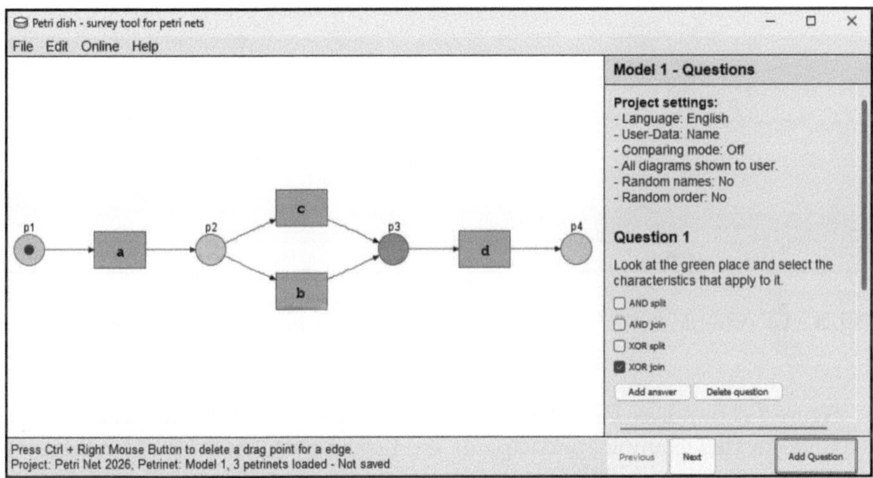

Fig. 2. An example of a survey page in Petri-Dish's Survey-Creator.

Figure 2 shows an example of a survey page within the Survey-Creator with a Petri net on the left side and its associated questions on the right side. This example colors the place p_3 green to easily highlight and refer to the place in a question. Petri-Dish provides question templates for single choice, multiple choice, free text, yes/no questions, and Likert scales. As shown in Fig. 2, instructors can predefine correct answers for single- and multiple-choice questions. This enables automatic grading and facilitates the quantitative evaluation of student submissions. The top right of a page shows the global settings of the survey. These settings include:

- the language of the user interface (currently either English or German);
- whether any personal data will be collected or not;
- whether the participants should compare all nets at the end of the survey;
- whether participants can see all or just a subset of the created pages;
- whether model names are anonymous (e.g. "Model 1") or not; and
- whether Petri-Dish presents the nets to the participants in random order.

To make Petri-Dish more suitable for scientific case studies, instructors can also specify whether participants are allowed to rearrange model elements. As already mentioned, Petri-Dish offers the option to let participants compare all models of the survey after answering all the questions on all pages. This unlocks another type of question, in which participants can compare two models with respect to their language, understandability, or structural properties.

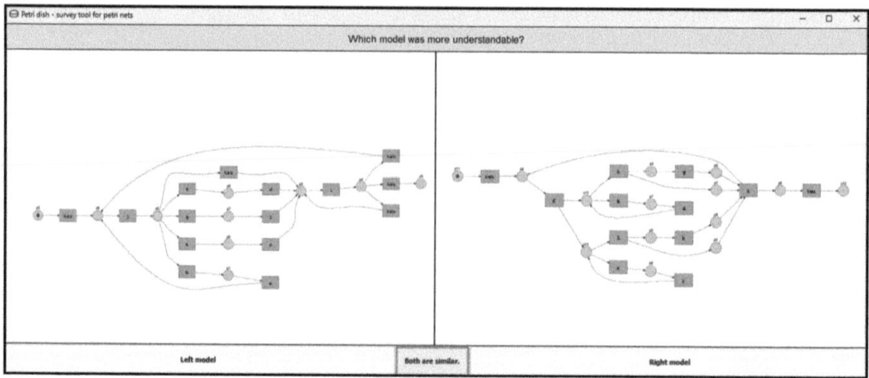

Fig. 3. An example of the pairwise model comparison in Petri-Dish's Survey-Player.

Figure 3 shows how Petri-Dish offers this pairwise model comparison to participants. In this example, participants are tasked with choosing the model that is easier to understand. The selection buttons are located below the models. An additional button between them allows participants to label both nets as equally understandable. During this pairwise comparison, Petri-Dish shows the models by including all changes of the model element positions made by the participant.

Distribution and the Survey-Player. After creating a survey, there are two ways to prepare it for distribution. First, the survey can be stored locally in encrypted `.pdqf` format (**Petri-Dish Question File**) and distributed via email or download. Alternatively, the instructor can upload their survey to an integrated online library. This library is password protected and allows quick access to previously uploaded surveys. Figure 4 presents the online library interface, listing all uploaded

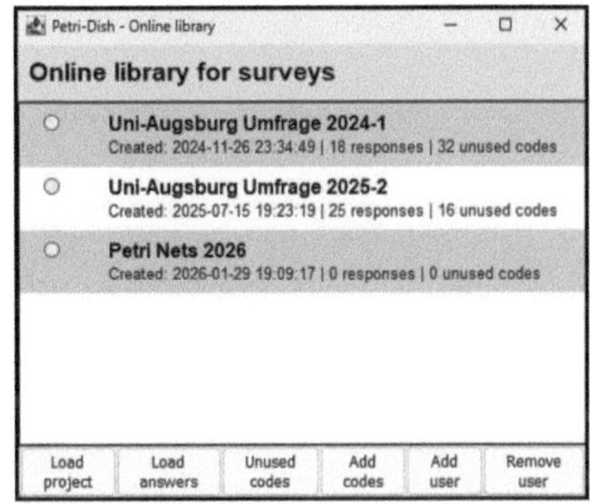

Fig. 4. The online library interface.

surveys. Instructors can access various management functions via the buttons at the bottom of the panel. To facilitate collaboration, the tool allows instructors to invite others to edit selected surveys. Furthermore, it enables the generation of participation codes, which serve as access passwords for a specific survey.

Once distributed in one of the aforementioned ways, participants use the Survey-Player to answer the survey. They can either load a `.pdqf` file or enter their participation code for the online library to access the survey. Participants can then see the Petri nets next to their associated questions and answer them. The interface is identical to the Survey-Creator shown in Fig. 2 without the menu bar or administrative buttons. Participants can zoom in and out of the model to inspect complex parts, and switch between pages at any time. If enabled, they can rearrange nodes and arcs according to their preferences. After completing the survey, participants store their responses in a `.pdaf` file (**Petri-Dish Answer File**) if they are not connected to the online library. Otherwise, Petri-Dish uploads their answers and links them to the corresponding survey. In this way, the instructor has access to all responses and can grade or analyze them.

Data Collection and Evaluation. After all answers have been sent in, the instructor can use the Survey-Creator to evaluate the participants' responses.

Figure 5 shows the results of an exemplary survey with three participants. For each Petri net, it shows the total number of participants who answered the questions, the average time spent with the model, and the average number of correct answers. Furthermore, for each question, it shows how often the participants chose each answer. Petri-Dish highlights correct answers with a star, providing a quick overview of participants' performance. Clicking on a participant's name reveals the detailed panel shown in Fig. 1, which presents page-specific metrics: the time spent on each model, the

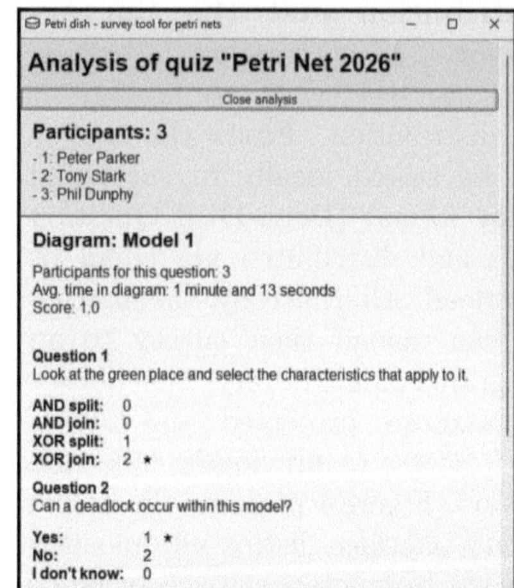

Fig. 5. The evaluation interface.

total score, and the individual question responses. To facilitate participant feedback, Petri-Dish allows instructors to export and distribute performance reports as `.pdf` files. Alternatively, the tool allows exporting all participant data in `JSON` format for advanced analysis in tools such as `R` or `MATLAB`.

3   Architecture

The architecture of Petri-Dish follows four primary objectives:

- platform independence for consistent rendering across operating systems;
- a clear separation between survey creation and participation functionalities;
- data integrity and privacy through a hybrid storage strategy; and
- modularity to support future extensions such as new modeling notations.

The remainder of this section details these design decisions.

Component Structure. Petri-Dish is a `Java 11` desktop application that uses the `Swing` framework for the graphical user interface. We explicitly selected `Java 11` to ensure compatibility with existing university computer labs, which often operate on older runtime environments. This decision removes the dependency on latest software releases, allowing students to run the tool on institutional hardware without administrative updates. We chose a desktop application over web-based solutions to ensure compliance with privacy regulations and to keep sensitive student information on personal devices rather than on remote servers. This approach also prevents inconsistencies between browsers, ensuring that the Petri nets and questions look identical for all participants. We implemented a shared core library as the foundation for both the Survey-Creator and the Survey-Player. This library encapsulates the core logic for graphical rendering,

Petri net interactions, and survey management, ensuring that updates propagate consistently to both components. The Survey-Player extends this foundation with participant-specific features, such as the pairwise comparison interface shown in Fig. 3. In contrast, the Survey-Creator contains exclusive modules for survey design, automated evaluation, and the online library interface. This modular architecture significantly reduces code duplication and facilitates future extensions to Petri-Dish.

Design Patterns and Extensibility. Petri-Dish decouples the internal logic from the user interface by using the *observer pattern* [5]. This ensures automatic updates of the user interface whenever the underlying data changes, and keeps Petri-Dish's components loosely coupled. To easily add more Petri net classes or modeling languages like BPMN or EPC to Petri-Dish, the display of graphical elements follows an *abstract factory pattern*.

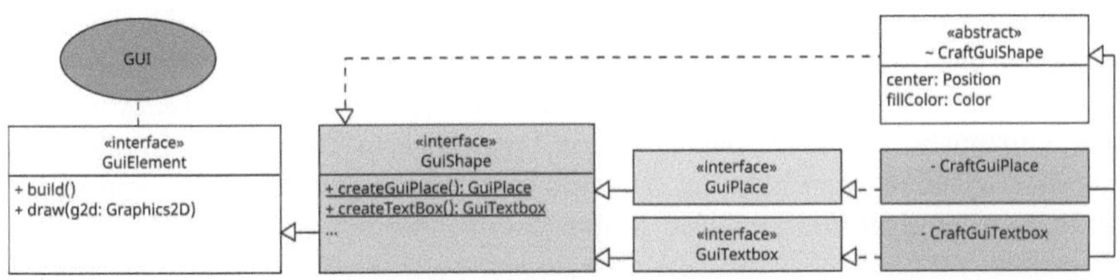

Fig. 6. Class diagram excerpt of the abstract factory pattern used in Petri-Dish.

Figure 6 illustrates the practical application of this pattern. By interacting solely with the generic `GuiElement` interface, the GUI is decoupled from specific notation details. The actual rendering logic is hidden within the `CraftGuiPlace` and `CraftGuiTextbox` classes, highlighted in orange in the figure, which encapsulate the specific drawing instructions needed to render places and transitions on the canvas. These classes implement the contracts defined by the yellow-colored interfaces `GuiPlace` and `GuiTextbox`. Finally, the factory `GuiShape`, shown in red, is responsible for instantiating the correct concrete classes for the chosen notation. The same architectural principle applies to lines: although not explicitly depicted in the figure, line elements also inherit from `GuiElement`. This abstraction allows for various line styles, such as arcs with special arrow tips. This modular design allows developers to add new notations easily without modifying the core logic, preserving the integrity of the existing codebase and ensuring maintainability.

Data and Information Architecture. Petri-Dish uses a hybrid storage strategy that combines local files with optional cloud database support. It offers the proprietary formats `.pdqf` and `.pdaf` for local storage. When using the online library, surveys and results are stored in an SQL database. This database provides centralized access to user accounts, surveys, and responses. This hybrid architecture allows users to choose between local storage and cloud access. Thus, Petri-Dish supports both isolated exercise environments and collaborative research, which define the main use cases for this tool.

4   Use Cases

Petri-Dish was designed as a dual-purpose tool to support both academic research and university teaching. Although these two domains have different primary objectives, both require similar functionalities in tools supporting them. This section outlines how Petri-Dish can support these two use cases by presenting two concrete examples based on our practical experience.

Education: Interactive Learning and Exam Preparation. At our home university, we teach two different courses in *process mining* [1]. Both courses require a basic understanding of place/transition nets. We often observe that students have problems understanding the firing rule or concurrency in Petri nets. Currently, we are preparing to use Petri-Dish in our courses to offer more exercises without overwhelming students or teaching personnel. We aim to task students to analyze fundamental properties of Petri nets, such as:

- **Execution Order:** Determining whether a firing sequence is enabled.
- **Concurrency:** Analyzing whether two transitions can fire simultaneously.
- **Exclusiveness:** Identifying if two transitions compete for a token.
- **Repeatability:** Checking if transitions can reoccur in a firing sequence.
- **Properties:** Assessing whether the model is bounded or contains deadlocks.

These question types can be easily implemented in Petri-Dish in a way that allows automatic grading and feedback. On the students' side, it is easier to click through the exercises in Petri-Dish than to write solutions down and send them in officially. We conducted a small pilot project to test Petri-Dish for this environment. In both of our courses, we recruited students to work with the tool and receive feedback on their ratio of correct answers. Students reported that the tool was intuitive and helped clarify theoretical concepts, making the learning experience more engaging. However, they requested more detailed feedback on their performance. Consequently, a detailed `.pdf` export of individual student responses was added to Petri-Dish, which includes reference solutions and highlighted errors.

Research: Conducting Empirical Studies. The second primary objective of Petri-Dish is to facilitate empirical research by digitizing and automating the data collection process. We conducted two case studies to investigate the impact of structural properties on Petri net understandability, with the corresponding datasets available on GitLab [9]. Research in this direction concerns modeling languages like BPMN and EPCs [3], but we did not find any results on Petri nets. Our study hypothesized that specific complexity metrics [11], redefined for Petri nets [17], would negatively influence a user's ability to understand a net. Petri-Dish ensured result validity through several key features:

- **Controlled Environment:** The automated process ensures validity by enforcing a standardized protocol and minimizing participant deviation.
- **Integrated Complexity Analysis:** The automatic metric calculation avoids errors and directly links model properties to user understanding.
- **Precise Behavioral Insights:** Capturing exact processing times enables the calculation of derived metrics such as task efficiency [3]. This would be infeasible in manual settings, where students put their solutions on paper.
- **Bias Mitigation:** Automatic randomization of the model sequence significantly reduces learning and fatigue effects [2].
- **Robust Subjective Evaluation:** Pairwise comparisons provide preference rankings, a reliable alternative to Likert scales, which are prone to central tendency bias and subjective interpretation inconsistencies.
- **Scalability:** Eliminating manual data entry and grading streamlines the administration of studies with large participant numbers.

Across two case studies, we collected data from 43 participants evaluating 13 distinct Petri nets with 13 questions per model. While the evaluation of the second study is still ongoing, the analysis of the first study has revealed that metrics such as size, cross-connectivity, and concurrency negatively impact objective understandability. Furthermore, we observed that high values for separability and the average connector degree have a similarly negative effect on subjective user acceptance. The results are statistically significant with moderate effect sizes, supporting the validity of these findings. The standardized workflow provided by Petri-Dish facilitates future replication to validate these results.

5   Comparison With Other Tools

To assess the utility of Petri-Dish, we compare it with established methodologies for conducting empirical studies and exercises. This comparison evaluates the tool from two perspectives: the transition from traditional paper-based methods to a digital workflow, and the distinction between Petri-Dish and existing software tools.

Comparison with Paper-Based Approaches. Traditionally, evaluations of model understandability and student exercises rely on pen-and-paper methods. While straightforward, these manual approaches are limited in scope and fail to capture dynamic user interactions. Petri-Dish offers advantages by improving both data quality and administrative efficiency.

A key benefit of Petri-Dish is its ability to capture data inaccessible in paper-based settings. In case studies such as those by Reijers and Mendling [16], the cognitive process leading to an answer is not captured, as only the final result is recorded. Petri-Dish addresses this by recording the exact time spent for each model. While time is a key parameter in case studies, capturing it accurately without a digital tool is difficult. Moreover, Petri-Dish streamlines the workflow through high-level automation. While paper-based exercises require time-consuming manual distribution and grading, Petri-Dish fully automates these steps. Metrics are calculated instantly, and student submissions are graded automatically based on reference solutions. This reduces the administrative overhead, allowing instructors to focus on analysis rather than manual administration.

Despite these advantages, the transition from paper to digital tools introduces a dependency on technical infrastructure. Each participant must have access to a device capable of running `Java` applications. In a modern university context, however, this requirement is rarely a barrier, as students typically possess the necessary hardware and internet connectivity to download and operate the tool.

Comparison with Tool-Supported Approaches. While using software is standard in modern empirical research, the workflow is often split across separate tools for creating, execution, and evaluating surveys. For example, Mendling et al. [13] investigated how personal and model characteristics influence process model understandability. They utilized the ProM framework [14] for metric calculation in the preparation phase, but conducted the actual survey using paper questionnaires, followed by analysis in `SPSS`. This splits the workflow across three separate tools, forcing researchers to manually transfer data and context between preparation, execution, and evaluation. Petri-Dish unifies these steps by combining metric calculation, survey distribution, data collection and evaluation into one platform, simplifying the entire process.

Other studies have moved towards digital data collection, yet they still face an operational disconnect between design and execution. Reijers et al. [15], for example, examined whether syntax highlighting, specifically the coloring of matching split and join nodes, improves the accuracy and speed of process model understandability. They used the `WoPeD` tool [4] to design the models and apply coloring. However, the experiment itself was executed via a separate, custom-built online tool based on `PHP` and `JavaScript`. If a model requires adjustment, it must be edited in the source tool, exported, and re-implemented in the survey interface. Petri-Dish unifies these steps by combining the modeling and questionnaire editor within the Survey-Creator, ensuring changes are applied instantly.

As the discussed case studies demonstrate, survey workflows in this domain typically rely on heterogeneous toolchains. Furthermore, our own research indicates that no specialized tool is currently available to cover the entire workflow of a Petri net-based survey. While generic survey platforms like LimeSurvey [12] or Google Forms [7] cover the general process, they lack domain-specific capabilities. These platforms typically display models as static images, ignoring the underlying logical structure. Petri-Dish, in contrast, stores every net element separately, retaining full awareness of every model element. Because the tool understands the distinction between places, transitions, and arcs, it enables automatic structural validation and interactive features such as the layout rearrangement described in Sect. 2.

Such capabilities are impossible with static image-based tools, which have no information about the net.

6   Conclusion and Future Work

In this paper, we presented Petri-Dish, a tool specifically designed for conducting Petri net exercises and surveys. The tool integrates the design, execution, and analysis of case studies into a single environment. For survey design, the tool calculates complexity metrics automatically, allowing instructors to prepare models with the desired level of complexity. The online library simplifies distribution, and the Survey-Player ensures a standardized workflow for all participants. Finally, Petri-Dish facilitates the analysis of the results by automatically evaluating responses and generating feedback reports for participants. By including features such as automated time tracking, it offers significant advantages over paper-based methods and generic survey tools. Our experience confirms that Petri-Dish significantly reduces administrative workload and prevents measurement errors, making Petri-Dish valuable for teaching and research.

Future work will primarily address user feedback collected during the initial deployment of the tool. The planned enhancements include token-based simulation to further improve user interactivity. We also intend to implement annotation features that allow users to add markers or textual notes directly to the models. For modelers, we plan to integrate a WYSIWYG editor to simplify the creation of survey content. Finally, extending the architecture to support additional modeling notations will broaden the tool's applicability in the field of process modeling. These developments will increase the utility of Petri-Dish for a wider user base and further support research in process modeling.

Availability and Installation. Petri-Dish is publicly available through our GitHub repository [9]. The repository includes installation instructions and datasets collected in the surveys we conducted with the tool. Both the Survey-Creator and the Survey-Player are implemented in **Java**; consequently, a Java Runtime Environment (**JRE**) version 11 or higher is required to execute the software. Additionally, Graphviz [8] must be installed to ensure the availability of all Survey-Creator functionalities. However, the Survey-Player does not require Graphviz, allowing students to run the tool without this dependency. Accessing the online library requires a registered account. To maintain secure access, accounts are currently provided only upon request. Interested users are welcome to request an account via email.

References

1. van der Aalst, W.M.P., Carmona, J.: Process Mining Handbook. LNBIP, vol. 448. Springer, Cham (2022). https://doi.org/10.1007/978-3-031-08848-3
2. Charness, G., Gneezy, U., Kuhn, M.: Experimental methods: between-subject and within-subject design. J. Econ. Behav. Organ. **81** (2012)
3. Dikici, A., Turetken, O., Demirors, O.: Factors influencing the understandability of process models: a systematic literature review. Inf. Softw. Technol. **93**, 112–129 (2018)
4. Freytag, T., Sänger, M.: Woped - a tool for teaching, analyzing and visualizing workflow nets, vol. 1295, p. 31 (2014)
5. Gamma, E., Helm, R., Johnson, R., Vlissides, J.: Design Patterns: Elements of Reusable Object-Oriented Software. Addison-Wesley, Reading (1994)
6. Girault, C., Valk, R.: Petri Nets for Systems Engineering - A Guide to Modeling, Verification, and Applications. Springer, Cham (2003)
7. Google LLC: Google Forms (2026). https://docs.google.com/forms
8. Graphviz Contributors: Graphviz - Graph Visualization Software (2026). https://graphviz.org/. Accessed 30 Jan 2026
9. Kimmel, M.: Git Repository for the Paper: Petri-Dish: A Petri Net Survey Tool for Education and Research (2026). https://gitlab.com/uni-a/petri-dish-a-petri-net-survey-tool-for-education-and-research
10. Koch, I., Reisig, W., Schreiber, F.: Modeling in Systems Biology, The Petri Net Approach, Computational Biology, vol. 16. Springer, Cham (2011)
11. Lieben, J., Jouck, T., Depaire, B., Jans, M.: An improved way for measuring simplicity during process discovery. In: Pergl, R., Babkin, E., Lock, R., Malyzhenkov, P., Merunka, V. (eds.) EOMAS 2018. LNBIP, vol. 332, pp. 49–62. Springer, Cham (2018). https://doi.org/10.1007/978-3-030-00787-4_4
12. LimeSurvey GmbH: LimeSurvey: The Online Survey Tool (2026). https://www.limesurvey.org/
13. Mendling, J., Reijers, H.A., Cardoso, J.: What makes process models understandable? In: Business Process Management, pp. 48–63 (2007)
14. Process Mining Group: ProM Tools: The Process Mining Framework (2026). https://promtools.org/
15. Reijers, H., Freytag, T., Mendling, J., Eckleder, A.: Syntax highlighting in business process models. Decis. Support Syst. **51**(3), 339–349 (2011)
16. Reijers, H., Mendling, J.: A study into the factors that influence the understandability of business process models. IEEE Trans. Syst. Man Cybern. Part A Syst. Hum. **41**, 449–462 (2011)
17. Schalk, P., Burke, A., Lorenz, R.: Navigating complexity: comparing complexity measures with Weyuker's properties. In: 6th International Conference on Process Mining (ICPM), pp. 145–152 (2024)
18. Tantau, T.: The TikZ and PGF Packages. http://sourceforge.net/projects/pgf/
19. University of Hamburg: Petri Nets Tool Database (2026). https://www.informatik.uni-hamburg.de/TGI/PetriNets/tools/db.html. Accessed 30 Jan 2026
20. Vanderfeesten, I., Reijers, H.: On a quest for good process models: the cross-connectivity metric. In: Advanced Information Systems Engineering, pp. 480–494 (2008)

OCPN Studio: Web-Based Modeling, Simulation, and Analysis of Object-Centric Petri Nets

István Koren[1,2]([✉])(iD)

[1] Data Science and Engineering, Eötvös Loránd University, Budapest, Hungary
[2] Process and Data Science, RWTH Aachen University, Aachen, Germany
korenistvan@inf.elte.hu

Abstract. Object-centric process mining (OCPM) addresses the limitations of traditional case-centric process models by capturing complex interactions between multiple object types. However, the advancement of OCPM is often hindered by the difficulty of obtaining high-quality, multi-object data from organizations, heavily necessitating simulation logs. While CPN Tools has long been the primary choice for such tasks, it lacks native support for object-centric constructs and the OCEL 2.0 standard, requiring users to manually implement complex time and data handling logic in Standard ML. This paper presents OCPN Studio, a modern, web-based environment designed specifically for modeling, simulating, and analyzing Object-Centric Petri Nets. By abstracting the underlying CPN formalism and providing native support for object types and date/time constructs, this open-source tool simplifies the modeling process. Our browser-based solution supports hierarchical nets, state-space analysis, and seamless OCEL 2.0 export, lowering the entry barrier for researchers, practitioners, and students alike.

Keywords: Colored Petri Nets · Object-Centric Process Mining · Simulation · OCEL 2.0 · Web Application

1 Introduction

Process mining has emerged as an important discipline bridging data science and process science, enabling organizations to discover, monitor, and improve their operational processes based on event data [1]. Traditional process mining techniques rely on a case-centric paradigm, where each event is associated with a single case identifier. This approach, while effective for straightforward processes, struggles to represent the complex interplay of multiple interacting objects that characterize real-world business processes.

Object-centric process mining (OCPM) addresses these limitations by explicitly modeling the relationships between multiple object types [2]. The Object-Centric Event Log (OCEL) format provides a standardized representation for

J. Desel and A. Kalenkova (Eds.): PETRI NETS 2026, LNCS 16567, pp. 395–405, 2026.
https://doi.org/10.1007/978-3-032-27879-1_21

such data, with OCEL 2.0 introducing support for object attributes, object-to-object relationships, and qualifiers [3]. Object-Centric Petri Nets (OCPNs) serve as the formal foundation for discovering and analyzing object-centric processes [9].

A significant challenge in advancing OCPM research lies in obtaining suitable event data. Organizations are often reluctant to share sensitive process data, and extracting multi-object event logs from existing systems remains technically demanding. Simulation offers a practical alternative: researchers can design process models and generate synthetic logs with known ground truth for algorithm validation and benchmarking.

CPN Tools [7] has long served as the de facto standard for Colored Petri Net modeling and simulation. Its powerful simulation capabilities have made it a valuable tool for generating synthetic event logs. However, CPN Tools presents several challenges for object-centric simulation. First, it requires users to work with Standard ML (SML), a functional programming language with a steep learning curve. Second, generating OCEL-compliant logs requires manual implementation of export logic in SML. Third, CPN Tools must be installed locally, is limited to Windows, and may be less accessible to users requiring assistive technologies such as screen readers.

This paper introduces OCPN Studio[1], an open-source, web-based environment for modeling and simulating OCPNs. The tool lowers the barrier for experimenting with object-centric process models: it offers a modern browser-based interface requiring no installation, where places, transitions, and arcs can be created through drag-and-drop interactions. To facilitate object-centric modeling, it provides native support for object types through record color sets with consistent visual color coding throughout the interface. Rather than requiring users to work directly with low-level CPN inscriptions, the tool provides structured forms for all inscriptions with design-time validation, reducing syntax errors and lowering the learning curve. The simulation engine, implemented in Rust and compiled to WebAssembly for near-native browser performance, employs Rhai as its scripting language. A key distinguishing feature is the built-in OCEL 2.0 export functionality, which automatically transforms simulation traces into standard object-centric event logs without requiring manual export logic. This makes the tool suitable both for researchers generating synthetic event logs and for students learning about (object-centric) Petri nets.

The remainder of this paper is structured as follows. Section 2 introduces the CPN modeling language as used in the tool. Section 3 describes the user interface and construction of Petri net models. Section 4 covers the simulation capabilities. Section 5 presents the analysis features. Section 6 details the technical implementation. Section 7 discusses related tools. Section 8 presents current limitations. Section 9 concludes and outlines future work.

---

[1] Available at https://elte-dsed.github.io/ocpn-studio/ as web-based environment. Source code: https://github.com/ELTE-DSED/ocpn-studio under the MIT license.

2   Modeling Object-Centric Petri Nets

OCPN Studio is based on the formalism of Colored Petri Nets (CPNs) [6], which extend classical Petri nets with typed tokens, arc inscriptions, and guards. While the underlying model is a standard CPN, the tool provides abstractions that make object-centric modeling more accessible.

2.1   Color Sets

Color sets define the types of tokens that can reside in places. The tool supports the standard CPN color set types:

- **Basic types**: UNIT, INT, BOOL, STRING, and integers with ranges (e.g., int with 1..100)
- **Enumerated types**: Finite sets of named values
- **Product types**: Tuples combining multiple types
- **Record types**: Named tuples with field accessors
- **List types**: Sequences of elements of a base type

For object-centric modeling, record types serve as object types. Each record needs to include an id field for unique object identification:

```
colset Aircraft = record id: INT * typeCode: STRING
                       * airline: STRING * carrierType:
                          STRING;
colset Gate = record id: INT;
colset FuelTruck = record id: INT;
```

Listing 1.1. Object type definitions for an airport ground handling model (excerpt)

A distinctive feature is that each color set is assigned a visual color. This color is consistently applied throughout the interface: places display a colored border matching their color set, and arcs inherit colors from connected places. This visual consistency helps users quickly identify which object types flow through different parts of the net. Users can also override the color of individual places, transitions, and arcs to highlight specific model parts independently of their color set.

2.2   Timed Color Sets

OCPN Studio supports timed Petri nets, where tokens carry timestamps indicating when they become available. Any color set can be marked as *timed*, causing its tokens to include availability times. The simulation engine advances a global clock and only enables transitions when all required input tokens have timestamps at or before the current simulation time.

To simplify working with time, the tool introduces a simulation epoch: a real-world datetime that corresponds to simulation time zero. When set, the interface displays times as absolute datetimes rather than abstract millisecond offsets. This is particularly useful when generating event logs for process mining, as it produces realistic timestamps. Additionally, a library of scheduling functions allows calendar-aware time expressions such as `next_workday_at(8, 0)` or `next_weekend_between(at(9), at(17))`, enabling realistic business-hour and shift-based scheduling without manual timestamp arithmetic.

2.3   The Rhai Scripting Language

While CPN Tools uses Standard ML, OCPN Studio employs Rhai [8], an embedded scripting language for Rust. Rhai was chosen primarily because it integrates natively with the Rust/WebAssembly simulation engine, is sandboxed for safe browser execution, and uses C-style syntax familiar to users of JavaScript or Python. Rhai does not support all features of the CPN ML library; in particular, it lacks Standard-ML-style pattern matching. However, the choice is an implementation trade-off for a web-based environment and does not alter the formal modeling power of Colored Petri Nets: all guards, arc inscriptions, and time expressions remain CPN expressions evaluated over typed multisets.

Guards, arc inscriptions, time expressions, and initial markings are all written as Rhai expressions. For example, a guard checking the carrier type of an aircraft (ac) uses intuitive dot notation:

```
ac.carrierType == "low" && ac.airline == "W6"
```

Time expressions use convenience functions for human-readable delays:

```
delay_days(1) + delay_hours(2) + delay_min(30)
```

With these modeling primitives established, we now turn to the practical aspects of constructing CPN models.

3   Construction of CPN Models

OCPN Studio provides a modern, interactive canvas for constructing Petri nets. The interface follows conventions familiar from visual modeling tools while adding features specifically designed for object-centric modeling.

3.1   Visual Editor

The main workspace, shown in Fig. 1 with a model loaded, consists of a zoomable, pannable canvas where users create and arrange net elements. The toolbar provides drag-and-drop creation of places and transitions, while arcs are created by connecting nodes in arc mode. The left sidebar, visible in Fig. 1, offers three switchable views: *Model* for editing element properties and managing declarations, *Simulation* for execution controls and event logs, and *Analysis* for monitors and state-space exploration. The toolbar also provides automatic layout algorithms: Dagre for hierarchical layouts, ELK (Eclipse Layout Kernel), and a specialized OC-Sugiyama algorithm that clusters nodes by object type, making the structure of object-centric nets visually apparent [4].

Each element displays its inscriptions directly on the canvas. Places show their color set name and initial marking; the marking display can be toggled and is visible already at design time, so users can immediately see which places contain tokens before starting a simulation. Transitions display their guard, time inscription, and priority as badges; arcs show their inscriptions. All inscriptions can be repositioned by dragging, allowing users to arrange the visual layout without affecting the model semantics.

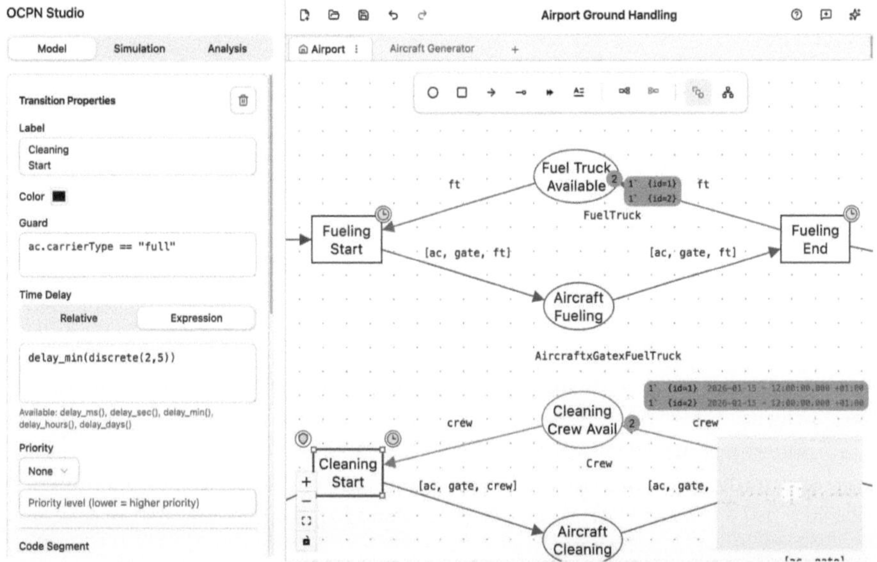

Fig. 1. The OCPN Studio editor interface showing an airport ground handling model (partially visible).

3.2   Property Panels

Selecting an element opens its property panel in the sidebar. Rather than requiring users to write inscriptions in a specific syntax, the tool provides structured forms that guide input and reduce errors.

Place Properties. For places, users select a color set from a dropdown menu that displays available color sets. The initial marking input box adapts to the selected color set type:

- For UNIT types, a simple number input specifies token count
- For basic types (INT, STRING), a text field accepts value lists

- For record types, a dialog allows adding objects field by field (cf. Fig. 2)
- For timed color sets, a dialog supports specifying availability times

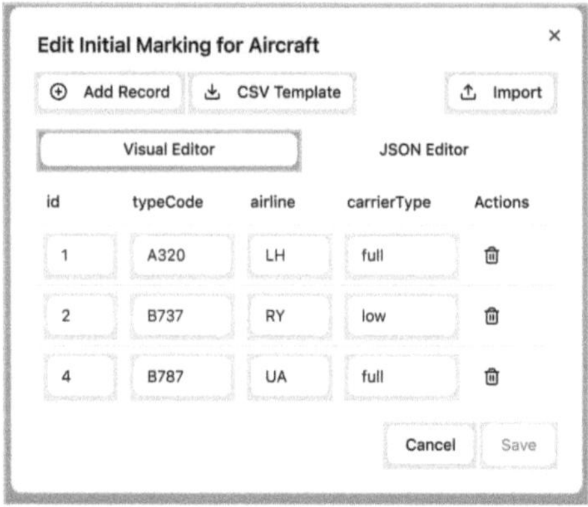

Fig. 2. The initial marking dialog allows users to add and edit object tokens through a structured table interface, eliminating the need to write JSON or Rhai syntax manually.

Transition Properties. Transitions offer fields for guard expressions, time inscriptions, priority selection, and code segments. The time inscription supports two input modes: a structured input with separate fields for days, hours, minutes, seconds, and milliseconds, or a free-form expression field for complex Rhai expressions like `delay_hours(2) + delay_min(rand(0, 30))`.

Arc Properties. Arc inscriptions specify which tokens are consumed or produced. The interface allows both simple variable references and complex multiset expressions. The editor supports the notation `[x,x,x]` for consuming multiple tokens of the same binding, as well as the legacy CPN Tools notation `3`x`.

Validation Feedback. OCPN Studio performs design-time validation and displays visual indicators directly on the canvas. Arcs with type mismatches or undefined variable references show warning icons; hovering over them reveals the specific error. Places referencing undefined color sets and transitions with invalid guards are similarly flagged, helping users catch errors before simulation.

3.3 Declaration Management

The declarations panel provides an organized view of all color sets, variables, priorities, and functions. Users can add, edit, reorder, and delete declarations. The color set editor offers both a visual mode with type-specific forms and a text mode for advanced users who prefer writing definitions directly.

3.4   Hierarchical Modeling

The tool supports hierarchical Petri nets through substitution transitions. A transition can be assigned a *subpage*, a separate Petri net that refines its behavior. Places on the subpage are designated as *port places* (in, out, or in/out) and mapped to *socket places* on the parent net through socket assignments, defining how tokens flow between hierarchy levels. On the canvas, substitution transitions display the subpage name as a clickable tag for direct navigation. Multiple levels of hierarchy are supported; during simulation, the hierarchy is automatically flattened by inlining subpages and merging port/socket place pairs.

3.5   Carl: The AI Assistant

OCPN Studio integrates an AI assistant named Carl that can analyze models and provide suggestions. Carl receives the current Petri net model (as JSON) together with the user's question, and forwards both to the OpenAI API. It can explain CPN concepts and Rhai syntax, describe the current model structure and semantics, suggest improvements for model clarity or correctness, and help debug guard expressions or arc inscriptions. Currently, Carl provides read-only advice; it cannot modify the model directly. Future versions aim to enable direct model modifications based on natural language instructions.

4   Simulation

Once a model has been constructed, the tool provides comprehensive simulation capabilities to execute and observe its behavior.

Users can advance the simulation step by step or run multiple steps automatically. The interface displays a step counter, the current simulation time (absolute datetime if epoch is set), and token markings on all places. A settings dialog allows configuration of steps per run, animation delays, and the simulation epoch. If the simulation engine encounters an error (e.g., a malformed Rhai expression), the interface displays an error notification with diagnostic details.

Event Log. Every transition firing is recorded in the event log with full token details. Each event captures a timestamp (derived from simulation time and epoch), the transition name (as event type), the consumed tokens per input place, and produced tokens per output place. The event log can be expanded to show token details or collapsed for an overview.

OCEL 2.0 Export. After running a simulation, the event log can be exported in the JSON format of the OCEL 2.0 standard. The export logic automatically extracts object types from record color sets, identifies objects by their id field with type prefixes to avoid collisions, maps transitions to event types, generates object-to-event relationships with type-based qualifiers, and tracks object attributes at creation time.

Use Case: Airport Ground Handling. The tool ships with a built-in airport ground handling model that defines object types for aircraft, gates, and fuel trucks, with a dedicated subpage that spawns aircraft tokens during simulation. With a configured simulation epoch (e.g., 2025-01-01), running several hundred simulation steps produces an event log with realistic timestamps and full token details. The resulting log can be exported as OCEL 2.0 JSON and imported into process mining tools such as PM4Py for discovery or conformance checking. This end-to-end cycle (model, simulate, export, analyze) can be completed entirely in the browser without installing any native applications.

5   Analysis

Beyond simulation, the tool provides analysis facilities accessible through the *Analysis* view, which is the third sidebar tab alongside Model and Simulation.

Monitors collect quantitative data during simulation. Five types are available: *Marking Size* monitors track the token count of a place over time, *Transition Count* monitors record how often a transition fires, *Place Breakpoint* and *Transition Breakpoint* monitors pause the simulation when a configurable condition is met, and *Data Collector* monitors evaluate a custom Rhai expression at each step. All observations are summarized in a *Performance Report* showing count, average, minimum, maximum, and standard deviation for each enabled monitor.

State-Space Analysis is supported through full state-space generation. The engine computes the complete reachability graph of a model. Since OCPN models may involve stochastic expressions (e.g., random delays or value distributions), the tool detects sources of non-determinism and presents them to the user, who can choose to apply deterministic overrides so that the state space is well-defined. The interface suggests sensible defaults (e.g., the midpoint for `uniform(a,b)`, or μ for `normal(`μ`,`σ`)`), but users can adjust these values as needed.

The resulting report includes upper and lower token bounds per place (boundedness), live and dead transitions, dead markings, home properties, fairness statistics, and the strongly connected component graph. For models with up to 500 states, the reachability graph can also be visualized in an interactive, layoutable graph view.

6   Implementation

OCPN Studio is implemented as a single-page web application. All modeling and simulation runs entirely on the client side; only the optional AI assistant communicates with an external API (OpenAI).

6.1   Frontend Architecture

The frontend is built with React and TypeScript, using the React Flow library for the interactive graph visualization. State management is centralized in a Zustand store that maintains hierarchically organized Petri nets (with substitution transitions and subpages), global declarations, and simulation configuration. The UI components use Radix primitives styled with Tailwind CSS (shadcn/ui pattern).

6.2   Simulation Engine

The simulation engine, available as a standalone MIT-licensed library, is implemented in Rust and compiled to WebAssembly for browser execution. It uses the Rhai scripting engine for evaluating expressions and supports parsing of color set definitions, compilation of guards and arc expressions to Rhai ASTs, binding enumeration for enabled transition detection, token consumption and production with multiset semantics, and time advancement with timed token handling. Hierarchical nets (Sect. 3.4) are flattened before simulation. The engine also supports full state-space generation, as described in Sect. 5.

In informal testing, the airport ground handling example with hundreds of aircraft objects simulates without perceptible delay in the browser.

6.3   File Formats

The native file format is JSON with the `.ocpn` extension, preserving all model details including visual layout, color set definitions, and simulation configuration. For data interchange, it can import CPN Tools XML files (`.cpn`) and PNML files (ISO/IEC 15909-2), and export to PNML. Both imports are best-effort: CPN Tools models may require manual adjustment of SML expressions to Rhai, and PNML import maps standard sorts to CPN-style color sets but cannot capture OCPN-specific features such as timed tokens or record types. An extensive help dialog provides guidance on translating Standard ML constructs to Rhai.

To position the tool within the ecosystem of existing Petri net and process mining tools, we now review related work.

7   Related Work

Several tools exist for Petri net modeling and simulation; a detailed review is out of scope for this tool paper.

CPN Tools [7] remains the most comprehensive environment for Colored Petri Nets. It provides sophisticated simulation capabilities, state space analysis, and model checking through its integration with the CPN ML library. The developers of CPN Tools now refer users to **CPN IDE** [5], which provides a more modern interface and the JavaScript language but remains a desktop application for Windows only. Both tools lack native support for object-centric event log generation.

Process mining frameworks such as **ProM** and **PM4Py** can handle both Coloured Petri Nets and OCEL 2.0, but they do not focus on modeling and simulation. Instead, they emphasize process discovery, conformance checking, and performance analysis.

OCPN Studio addresses the gap between powerful desktop tools like CPN Tools and modern web-based process mining ecosystems. By providing accessible browser-based modeling with native OCEL 2.0 support, it lowers the barrier for object-centric process simulation.

8   Discussion

While OCPN Studio provides a functional environment for object-centric Petri net modeling and simulation including hierarchical nets, state-space analysis, and design-time validation, it does not yet match the full capability of CPN Tools in all areas.

CPN Tools provides place and transition invariant analysis for conservation and synchronization properties, which are not yet supported. Regarding the expression language, Rhai does not offer Standard-ML-style pattern matching, and while design-time validation catches type mismatches and undefined references, some expression errors are still only detected at simulation time. A formal definition of how OCPN-specific constructs (e.g., object types as record color sets, automatic OCEL mapping) relate to the underlying CPN formalism is planned as future work.

A conceptual extension planned for future versions is object evolution: the ability to model how object attributes change over time during simulation. Currently, object attributes remain fixed as defined in initial markings and can only be changed in code segments in transitions. An object evolution layer would allow simulating external events to update attribute values, with these changes being recorded in the OCEL export for process mining analysis of data perspectives.

9   Conclusion and Future Work

This paper presented OCPN Studio, a web-based environment for modeling, simulating, and analyzing OCPNs. The tool addresses the need for accessible simulation capabilities in object-centric process mining research by providing:

- An intuitive visual editor with structured input forms and design-time validation
- Hierarchical CPNs with substitution transitions and subpages
- Monitors for data collection and a state-space analysis with boundedness, liveness, and deadlock detection
- Color-coded object types with user-overridable colors
- Native OCEL 2.0 export and PNML interoperability
- Performant browser-based simulation via WebAssembly

Several directions for future work emerge from this foundation. The primary planned extension is object evolution, i.e., modeling how object attributes change over time, enabling data-dependent process behavior and richer OCEL exports. A formal definition of the mapping from OCPN-specific constructs to the CPN formalism would strengthen the theoretical grounding. Future versions aim to enable the AI assistant to modify models based on natural language instructions and to extend the help dialog into a dedicated documentation website with examples and templates. Finally, the Rust-based simulation engine could be deployed as a server-side service, bridging the gap from design-time simulation to workflow execution.

By providing an integrated, browser-based environment for object-centric Petri net modeling, simulation, and analysis with native OCEL 2.0 support, the tool removes a practical barrier to advancing object-centric process mining research.

Acknowledgments. The author thanks Wil van der Aalst, Lukas Liß, and Aaron Küsters for their feedback on earlier prototypes of this work, and the anonymous reviewers for their valuable comments, which significantly improved the tool and paper. Funded by the Deutsche Forschungsgemeinschaft (DFG, German Research Foundation) under Germany's Excellence Strategy - EXC-2023 Internet of Production - 390621612.

Disclosure of Interests. The author has no competing interests to declare that are relevant to the content of this article.

References

1. van der Aalst, W.M.P.: Process Mining: Data Science in Action. Springer, 2nd edn. (2016). https://doi.org/10.1007/978-3-662-49851-4
2. van der Aalst, W.M.P.: Object-centric process mining: unraveling the fabric of real processes. Mathematics **11**(12), 2691 (2023). https://doi.org/10.3390/math11122691
3. Berti, A., et al.: OCEL (Object-Centric Event Log) 2.0 Specification (2024). https://doi.org/10.48550/ARXIV.2403.01975
4. Brachmann, T., Koren, I., Liss, L., van der Aalst, W.M.P.: Visualizing object-centric petri nets. In: Business Process Management Workshops (BPM 2025 Workshops). Lecture Notes in Business Information Processing, vol. 535. Springer, Cham (2026), to appear
5. CPN IDE Team: CPN IDE: A Modern Interface for Colored Petri Nets. https://cpnide.org/ (2023). Accessed 22 Mar 2026
6. Coloured Petri Nets. Springer, Heidelberg (2009). https://doi.org/10.1007/b95112
7. Jensen, K., Kristensen, L.M., Wells, L.: Coloured Petri Nets and CPN Tools for modelling and validation of concurrent systems. Int. J. Softw. Tools Technol. Transfer **9**(3–4), 213–254 (2007). https://doi.org/10.1007/s10009-007-0038-x
8. Rhai Contributors: Rhai: An embedded scripting language for Rust. https://rhai.rs/ (2024). Accessed 22 Mar 2026
9. van der Aalst, W.M.P., Berti, A.: Discovering object-centric petri nets. Fund. Inform. **175**(1–4), 1–40 (2020). https://doi.org/10.3233/FI-2020-1946

Author Index